T0345059

ANALYSIS AND SYNTHESIS OF FAULT-TOLERANT CONTROL SYSTEMS

ANALYSIS AND SYNTHESIS OF FAULT-TOLERANT CONTROL SYSTEMS

Magdi S. Mahmoud
King Fahd University of Petroleum and Minerals, Saudi Arabia

Yuanqing Xia
Beijing Institute of Technology, China

This edition first published 2014
© 2014 John Wiley & Sons, Ltd

Registered office
John Wiley & Sons Ltd, The Atrium, Southern Gate, Chichester, West Sussex, PO19 8SQ, United Kingdom

For details of our global editorial offices, for customer services and for information about how to apply for permission to reuse the copyright material in this book please see our website at www.wiley.com.

Library of Congress Cataloging-in-Publication Data

Mahmoud, Magdi S.
  Analysis and synthesis of fault-tolerant control systems / Magdi S. Mahmoud, Yuanqing Xia.
    pages cm
  Includes bibliographical references and index.
  ISBN 978-1-118-54133-3 (cloth)
1. Automatic control.   2. Fault tolerance (Engineering)   3. Control theory.   I. Xia, Yuanqing.   II. Title.
III. Title: Analysis and synthesis of FTCS.
  TJ213.M268428 2013
  629.8–dc23

                                                                                        2013023504

A catalogue record for this book is available from the British Library.

ISBN: 978-1-118-54133-3

Typeset in 10/12pt Times by Aptara Inc., New Delhi, India
Printed and bound in Singapore by Markono Print Media Pte Ltd

1   2014

To my loving wife, Salwa
To the 'M' family:
Medhat, Monda, Mohamed,
Menna, Malak, Mostafa
and Mohamed

MSM

To my honest and diligent
wife, Wang Fangyu
To my lovely daughter,
Xia Jingshu

YX

Contents

Preface

In recent years, we have been witnessing sophisticated control systems designed to meet increased performance and safety requirements for modern technological systems. Technical experience has indicated that conventional feedback control design for a complex system may result in an unsatisfactory performance, or even instability, in the event of malfunctions in actuators, sensors or other system components. In order to circumvent such weaknesses, new approaches to control system design have emerged with the goal of tolerating component malfunctions while maintaining desirable stability and performance properties. These types of control system are often known as "fault-tolerant control systems" (FTCS). The area of fault-tolerant control systems is a complex interdisciplinary research field that covers a diverse range of engineering disciplines, such as modeling and identification, applied mathematics, applied statistics, stochastic system theory, reliability and risk analysis, computer communications, control, signal processing, sensors and actuators, as well as hardware and software implementation techniques.

Modern technological systems rely on sophisticated control systems to meet performance and safety requirements. A conventional feedback control design for a complex system may result in unsatisfactory performance, or even instability, in the event of malfunctions in actuators, sensors or other system components. To overcome such weaknesses, new approaches to control system design have been developed in order to tolerate component malfunctions while maintaining the required levels of stability and performance. This is particularly important for safety-critical systems, such as aircraft, spacecraft, nuclear power plants, and chemical plants processing hazardous materials. In such systems, the consequences of a minor fault in a system component can be catastrophic. Therefore, the demand for reliability, safety and fault tolerance is generally high. It is necessary to design control systems which are capable of tolerating potential faults in these systems in order to improve the reliability and availability while providing desirable performance. More precisely, FTCS are control systems that possess the ability to accommodate component failures automatically. They are capable of maintaining overall system stability and acceptable performance in the event of such failures. In other words, a closed-loop control system which can tolerate component malfunctions, while maintaining desirable performance and stability properties is said to be a fault-tolerant control system [1].

The problem of fault monitoring has always been an area of much importance for research departments in industry. This becomes even more of a priority when we are dealing with nonlinear systems. Monitoring of uncommon behavior of plant and detecting unprecedented changes in systems are essential for maintaining the health of a system, followed by the removal of faulty components, replacement with the better ones, restructuring system architecture, and thus improving overall system reliability. However, with the increasing complexity of modern nonlinear systems, process engineers are facing tough challenges to understand and troubleshoot possible system problems. Highly efficient fault-monitoring methods have become a valuable asset in the life of large systems.

This book is about the analysis and design methods of fault-tolerant control systems. Particular consideration is given to covering wide topics that have been treated in the literature and presenting the results of typical case studies. The key feature is to provide a teaching-oriented volume supported by research.

The terminologies, conventions and notations that have been adopted throughout this book are explicitly presented in place to facilitate smooth readibilty of the different sections. They are quite standard in the scientific media and vary only in form or character.

Magdi S. Mahmoud
Dhahran, Saudi Arabia

Yuanqing Xia
Beijing, China
March 2013

Reference

[1] Zhang, Y., and Jiang, J. (2008) "Bibliographical review on reconfigurable fault-tolerant control systems", *Annual Reviews in Control* **32**, 229–252.

Acknowledgments

The subject matter of fault-tolerant control systems is perhaps one of the most attractive areas of contemporary research and development. It embodies fault diagnosis, fault estimation, fault identification, and fault isolation, to name but a few topics. The topics discussed in this book have constituted an integral part of our academic research investigation over the past few years. The idea of writing the book arose and developed through communication with Dr Nigel Hollingworth. We would like to acknowledge the tireless effort and professional support from Wiley, particularly from Anne Hunt and Tom Carter.

In writing this volume, we have taken the approach of referring within the text to papers or books which we believe have taught us some concepts, ideas and methods. We have further complemented this by adding remarks and notes within and at the end of each chapter to shed light on other related results. We are indebted to the colleagues who introduced us to the subject of fault-tolerant control systems and to the people who made the writing of this book possible.

Magdi Mahmoud owes a measure of gratitude to the management of King Fahd University of Petroleum and Minerals (KFUPM, Saudi Arabia) for continuous encouragement and facilitating all sources of help. Particular appreciation goes to the deanship of scientific research (DSR) for providing a superb competitive environment for research activities through internal funding grants. It is a great pleasure to acknowledge the financial funding afforded by DSR through Project IN121003 and for providing overall support of research activities at KFUPM.

During the past five years, Magdi Mahmoud has had the privilege of teaching various graduate courses at KFUPM. The updated and organized course notes have been instrumental in generating chapters of this book. Valuable comments and suggestions by graduate students have been extremely helpful, particularly from those who attended the courses SE509, SE514, SE517, and SE650, offered by the Systems Engineering Department from 2007 to 2011.

Magdi Mahmoud deeply appreciates the efforts of Muhammad Sabih, Mirza H. Baig, Azhar M. Memon, Haris M. Khalid and Rohmat Widodo as well as Wen Xie from BIT for their unfailing help in preparing portions of the manuscript and performing numerous effective simulations.

The widely-recognized research work of Yuanqing Xia and his students on predictive control and related topics at the School of Automation, Beijing Institute of Technology (BIT) has contributed effectively to several sections of this volume.

Most of all however, we would like to express our deepest gratitude to all the members of our families and especially our wives, Salwa and Wang Fangyu, for their elegant style.

Without their constant love, incredible amount of patience and (mostly) enthusiastic support, this volume would not have been finished.

We would appreciate any comments, questions, criticisms, or corrections that readers may take the trouble of communicating to us at msmahmoud@kfupm.edu.sa, magdim@yahoo.com or yuanqing.xia@gmail.com.

1

Introduction

For more than three decades, the growing demand for safety, reliability, maintainability, and survivability in technical systems has created significant research interest in fault detection and diagnosis (FDD). Such efforts have led to the development of many FDD techniques. For a general exposure to the subject, the reader is directed to [1]–[5].

1.1 Overview

In the literature, fault detection and isolation or fault detection and identification are often used interchangeably and abbreviated as "FDI". To be precise and avoid further confusion, this book adopts the term "FDI" to stand for "fault detection and isolation"; "FDD" is used when the fault identification function is added to FDI. In FTCS designs, fault identification is important; therefore FDD is mainly used throughout this book to highlight the requirement of fault identification.

On a parallel path, research into reconfigurable fault-tolerant control systems has increased progressively since the initial research on restructurable control and self-repairing flight control systems began in the early 1980s (see [6]–[10]). More recently, fault-tolerant control has attracted more and more attention in both industry and academic communities due to increased demands for safety, high system performance, productivity and operating efficiency in wider engineering applications, not limited to traditional safety-critical systems. Several review or survey papers on FTCS have appeared since the 1990s including [11]–[16].

Fault tolerance is no longer limited to high-end systems and consumer products such as automobiles. However it is increasingly dependent on microelectronic and mechatronic systems, on-board communication networks, and software, thus requiring new techniques for achieving fault tolerance. Even though individual research on FTCS has been carried out extensively, systematic concepts, design methods, and even terminology are still not yet standardized. Recently, efforts have been made to unify some terminology [17]. In addition, for historical reasons and because of the complexity of the problem, most of the research on FDD and reconfigurable control (RC) has been treated as two separate fields. More specifically, most of the FDI techniques have been developed as a diagnostic or monitoring tool, rather

Analysis and Synthesis of Fault-Tolerant Control Systems, First Edition. Magdi S. Mahmoud and Yuanqing Xia.
© 2014 John Wiley & Sons, Ltd. Published 2014 by John Wiley & Sons, Ltd.

than as an integral part of FTCS. As a result, some FDD methods may not satisfy the need of controller reconfiguration. On the other hand, most of the research on reconfigurable control is carried out assuming the availability of a perfect FDD. Little attention has been paid to analysis and design with the overall system structure and interaction between FDD and RC.

For example, the following questions are posed:

- From the viewpoint of RC design what are the needs and requirements for FDD?
- What information can be provided by existing FDD techniques for overall FTCS designs?
- How can we analyze systematically the interaction between FDD and RC?
- How can we design FDD and RC in an integrated manner for online and real-time applications?

Many other challenging issues still remain open for further research and development. One of the motivations of this book is to provide an overview of developments in FTCS and to address some challenging problems to attract the attention of future research.

1.2   Basic Concepts of Faults

The terminology used in this book is fairly standard. Below, some basic definitions of faults, failure, disturbances and uncertainties, fault detection, fault isolation, fault identification, and fault diagnosis are given. The interested reader is referred to [18, 19, 20] for more detailed explanation of the above mentioned terminology.

A "fault" is an unpermitted deviation of at least one characteristic property or parameter of a system from the acceptable (standard condition). The closely related term "failure" is regarded as a permanent interruption of a system's ability to perform a required function under specified operating conditions. *Failure* is used for the complete breakdown of a system, while *fault* is used to indicate a deviation from the normal characteristics. As far as detection is concerned, both faults and failures can be treated alike. Moreover, a fault can be treated as an external input or as a parameter deviation which changes the system characteristics. Similar to faults, "disturbances", "uncertainties", and "noises" can also be treated as external inputs. In fault detection and isolation (FDI) terminology, they are termed as "unknown inputs". Unlike faults, these unknown inputs are uncontrolled, unavoidable and are present during normal operation. The effect of the unknown inputs can be incorporated into the controller design and a process can perform well even in the presence of them. Faults, on the other hand, have very severe effects on the process and should be detected.

The process of *fault diagnosis* is referred to as the determination of the size, location, time of detection and type of fault in the process. Based on its performance, a fault diagnosis system (FDS) is regarded as a fault detection (FD), fault detection and isolation (FDI) or fault detection, isolation and analysis (FDIA) system [18]. An FD system is therefore the process of determining the fault in the process and its time of occurrence. An FDI system determines in addition the kind and location of the fault. Similarly, an FDIA, together with detection and isolation, also aims to determine the size and time behavior of the fault. It is worth noting that the existence conditions for fault isolation are more stringent than for fault detection, and even more so in the case of fault identification. Consequently, it is difficult to isolate or identify faults in most situations.

A fault detection system should ideally meet some general requirements. The most important desirable features are:

- early detection of faults (incipient and abrupt)
- successful detection of actuator, component, and sensor faults
- robustness against unknown inputs (external disturbances, measurement noises, and model uncertainties)
- differentiation of faults from unknown inputs so that false alarms are avoided
- less use of online computation so that it can be integrated into large-scale systems easily.

Besides the above important attributes, the design procedure of an FD scheme should be as simple as possible.

1.3   Classification of Fault Detection Methods

There exist a number of techniques used for fault detection (FD) in technical processes or dynamical systems. In this section, we present the widely accepted classification of these techniques.

1.3.1   Hardware redundancy based fault detection

The essence of this scheme is replication of the process component using identical hardware components. Figure 1.1 shows a schematic description of the hardware redundancy. Information about the fault is extracted if there is any deviation of the output of the process component from its redundant pair. Good reliability and the ability to isolate faults are the main advantages of this scheme. The major problems encountered with this scheme are the extra components, increased maintenance cost and additional space required to accommodate the redundant equipment. Thus, its use is limited to a number of key applications, for example, nuclear power plants and flight-control systems [18, 21].

1.3.2   Plausibility test

Figure 1.2 shows a schematic depiction of the plausibility test. The basic idea of this technique is to evaluate the measured process variable with regard to credible, convincing values and

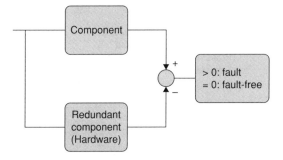

Figure 1.1   Hardware redundancy scheme

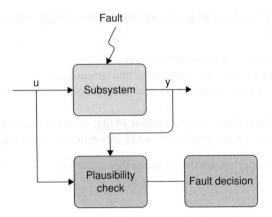

Figure 1.2   Plausibility test scheme

their mutual compatibility. On the assumption that a fault leads to the loss of plausibility, the presence of a fault in a certain variable can be determined using the plausibility check. It can be performed by simple rules with binary logic. The plausibility test is also a kind of limit checking but with a wider tolerance. This test can be viewed as a first step to model-based FD methods. However, it has limited efficacy for detecting faults in a complex process [18, 19].

1.3.3   Signal-based fault diagnosis

Figure 1.3 shows a conceptual depiction of the signal-based FD technique. The central idea of this scheme is to extract the fault information from the process signals. For this purpose, some signal properties (symptoms) are analyzed. These symptoms are generally divided into the time domain characteristics and the frequency domain characteristics of the process signal.

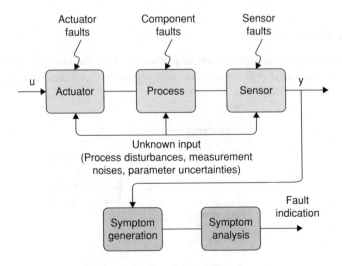

Figure 1.3   Signal-based FD scheme

The time domain characteristics comprise magnitude, mean (arithmetic or quadratic), limit values, trends, statistical moments of the amplitude distributions etc.; the frequency domain characteristics include spectral power densities and frequency spectral lines. Signal-based FD is used under steady-state operation of the process. The efficiency of this scheme is limited when the process is operating in a wide range due to the possible variation of input signals [18].

1.3.4   Model-based fault detection

The intuitive idea of the model-based FD technique is to replace the hardware redundancy by a process model which is implemented in software. The process model runs in parallel with the process itself and is driven by the same process inputs. In this way, the process behavior can be reconstructed online. Analogous to hardware redundancy, this technique is called "software redundancy" or "analytical redundancy" [18]. It is well-known that model-based FD techniques are more powerful than signal-based FD schemes [22, 23] because they use more information about the process.

In a typical model-based FD scheme, there are two stages: residual generation and residual evaluation. In residual generation, the "residual signal" is generated by comparing the process outputs with their estimates. The residual signal carries information about the faults. Since the residual signal, in a real process, is affected by the faults, disturbances, and measurement noises simultaneously, it is required to process the residual signal further to obtain possible information about faults. This is done in the residual evaluation stage.

It is widely accepted that a process model represents the qualitative and quantitative behavior of the process and can be obtained by utilizing well-established techniques from system modeling. The quantitative or analytical model of the process can be represented by a set of differential or difference equations while the qualitative model is expressed in terms of qualitative functions centered around different units in the process. The qualitative models are also known as "knowledge-based models", which include neural networks, petri nets, expert systems, fuzzy logic etc. [22, 23] Based on these arguments, model-based FD schemes can be divided into two classes: knowledge-based and analytical.

Knowledge-based FD techniques are useful where the precise model is not available or is very hard to obtain, for example, large-scale chemical processes and nuclear reactors. An extensive study of knowledge-based FD methods can be found in [22, 24, 25, 26, 27]. Analytical model-based FD techniques, on the other hand, make use of analytical models for the purpose of residual generation. The analytical techniques can be broadly classified as:

- Parity space FD
- Observer-based FD
- Parameter-identification-based FD.

The rest of this section describes these approaches.

1.3.4.1   Parity space approach

Figure 1.4 shows a conceptual diagram of the parity space approach to residual generation. The parity space approach makes use of a parity check on the consistency of the parity equation.

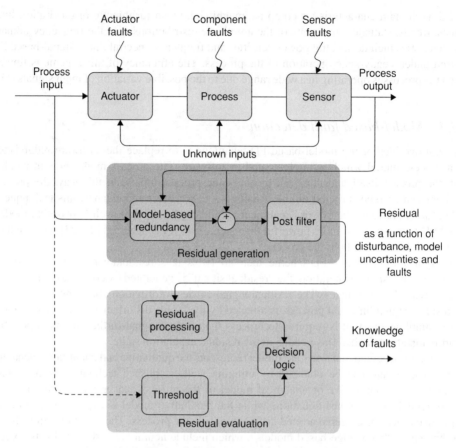

Figure 1.4   Parity space approach

A set of properly modified system equations (also called "parity relations") is derived based on measured signals from the process. These parity relations decouple the residuals from the system states and from each other. This enhances the ability to detect faults. Inconsistency in the parity relations indicates the presence of a fault. In [28], the parity relations were derived based on the state-space model of the system; later, they were derived using the system transfer function [29]–[32].

As mentioned in [18, 23], there exists a close relationship between the parity space approach and the observer-based approach. An extensive study on parity space D is presented in [18], where it is been shown that there exists a one-to-one mapping between the design parameters of observer and parity space based residual generation. Thus, given a set of parity relations, a diagnostic observer can be designed and *vice versa*.

1.3.4.2   Observer-based approach

The observer-based technique, see Figure 1.5, is one of the most commonly applied model-based schemes for detection of faults in a system. In this scheme, the residual signal is obtained

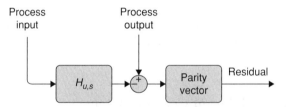

Figure 1.5 Observer-based residual generation

by comparing the process outputs with their estimates. It is worth noting that observers are mainly used by the control community in order to estimate the unmeasured states in the process, while the FDI community use them for diagnostic purposes. The existence conditions for diagnostic observers are more relaxed than for a state observer, however one particular class of diagnostic observer (the fault detection filter (FDF)) can be used for state estimation as well as diagnostic purposes.

1.3.4.3 Parameter identification approach

The parameter identification approach, see Figure 1.6, is also an important FDI technique [22, 33, 34]. In this approach, fault detection is performed based on online parameter estimation. Information about the fault can be extracted by comparing the estimated parameter with the nominal process parameter. Any discrepancy between the two gives an indication of fault. The advantages of this scheme are as follows:

- Several parameters can be estimated with less input and output from the process [19].
- It yields the size of the discrepancy, which is useful for fault analysis [22].

The disadvantage is that an excitation signal is necessary in order to estimate the parameter, which may cause problems in the case of processes running at stationary operating point. Further, the determination of a physical parameter from its mathematical model may not, in

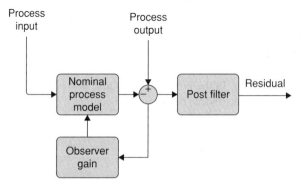

Figure 1.6 Parameter identification scheme

general, give a unique result and is only feasible if the system order is low [22]. There are several parameter estimation techniques available in the literature, among them are the least square (LS) method, the recursive least square(RLS) method, the extended least square (ELS) method etc.

1.4   Types of Fault-Tolerant Control System

Generally speaking, FTCS can be classified into two types: passive fault-tolerant control systems (PFTCS) and active fault-tolerant control systems (AFTCS). In PFTCS, controllers are fixed and are designed to be robust against a class of presumed faults [8]. This approach needs neither FDD schemes nor controller reconfiguration, but it has limited fault-tolerant capabilities. Discussions of PFTCS are beyond the scope of this book and interested readers are referred to [35, 36] and the references therein for recent developments. In the literature, PFTCS are also known as "reliable control systems" or "control systems with integrity".

In contrast to PFTCS, AFTCS react actively to system component failures by reconfiguring control actions so that the stability and acceptable performance of the entire system can be maintained. In certain circumstances, degraded performance may have to be accepted [37]. AFTCS are also referred to as "self-repairing", "reconfigurable", "restructurable", or "self-designing" control systems by some researchers. From the viewpoint of functionality in handling faults, AFTCS were also called fault detection, identification (diagnosis) and accommodation schemes by other researchers. In such control systems, the controller compensates for the impacts of the faults either by selecting a pre-computed control law or by synthesizing a new one online. To achieve a successful control system reconfiguration, both approaches rely heavily on real-time FDD schemes to provide the most up-to-date information about the true status of the system. Therefore, the main goal in a fault-tolerant control system is to design a controller with a suitable structure to achieve stability and satisfactory performance, not only when all control components are functioning normally, but also in cases when there are malfunctions in sensors, actuators, or other system components (for example, in the system itself, in control computer hardware or in software). This book focuses only on aspects pertaining to AFTCS.

1.5   Objectives and Structure of AFTCS

The design objectives for AFTCS include the transient and the steady-state performance for the system, not only under normal operations but also under fault conditions. It is important to point out that the emphasis on system behaviors in these two modes of operation can be significantly different. During normal operation, more emphasis should be placed on the quality of the system behavior. In the presence of a fault, however, how the system survives with an acceptable (probably degraded) performance becomes a predominant issue. Typically, an AFTCS can be divided into four subsystems):

- an FDD scheme;
- a reconfigurable controller;
- a controller reconfiguration mechanism;
- a command/reference governor.

Inclusion of both an FDD scheme and reconfigurable controllers within the overall system structure is the main feature distinguishing an AFTCS from a PFTCS. Key issues in AFTCS are how to design:

- a controller which can easily be reconfigured;
- an FDD scheme with high sensitivity to faults and robustness to model uncertainties, operating condition variations, and external disturbances;
- a reconfiguration mechanism which leads as much as possible to the recovery of the pre-fault system performance in the presence of uncertainties and time-delays in FDD, within the constraints of control inputs and system states.

The critical issue in any AFTCS is the limited amount of time available for the FDD and for the control system reconfiguration. Furthermore, in the case of failure, efficient utilization and management of redundancy (in hardware, software and communication networks), stability, and a transient and a steady-state performance guarantee are some of the important issues to consider in AFTCS.

The overall structure of a typical AFTCS is shown in Figure 1.7. In the FDD module, any fault in the system should be detected and isolated as quickly as possible, and fault parameters, system state/output variables, and post-fault system models need to be estimated online in real-time. Based on the online information about the post-fault system model, the reconfigurable controller should be designed to maintain automatically the stability, desired dynamic performance and steady-state performance. In addition, in order to ensure the closed-loop system tracks a command input trajectory in the event of faults, a reconfigurable feed-forward controller often needs to be synthesized. To avoid potential actuator saturation and to take into consideration the degraded performance after fault occurrence, a command/reference governor may also need to be designed to adjust command input or reference trajectory automatically.

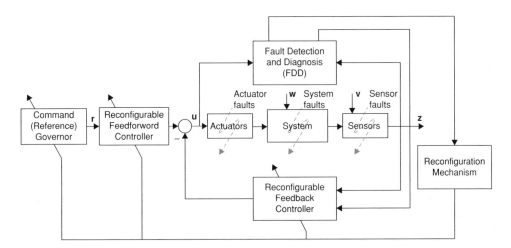

Figure 1.7   General structure of an AFTCS

Based on the described structure, the design objectives of an AFTCS can be stated as:

- to have an FDD scheme that provides, as precisely as possible, information about a fault (time, type and magnitude) and the post-fault model;
- to have a new control scheme (reconfigurable or restructurable) to compensate for the fault-induced changes in the system so that stability and an acceptable closed-loop system performance can be maintained.

It is important to point out that not only do the parameters of the controllers need to be recalculated, but also the structure of the new controllers (in terms of their order, number and type) might be changed. A corresponding AFTCS is often referred to as a "restructurable control system" to emphasize that the controller structure can change. Note that, in the literature, there are generally two ways of classifying AFTCS. One classifies them as reconfigurable versus restructurable; the other differentiates them as accommodation versus reconfiguration. In this book, we adopt the former definition. So long as there is no confusion, we use the term "reconfigurable control" in subsequent sections.

1.6 Classification of Reconfigurable Control Methods

Reconfigurable control methods can be broadly classified into several categories. The two most common categories are those based on control algorithms and those based on fields of application.

1.6.1 Classification based on control algorithms

In the literature, reconfigurable control design methods fall into one of the following approaches: linear quadratic; pseudo-inverse or control mixer; gain scheduling or linear parameter varying; (model reference) adaptive control or model following; eigenstructure assignment; multiple-model; feedback linearization or dynamic inversion; Hoo and other robust controls; model predictive control; variable structure and sliding mode control; generalized internal model control; and intelligent control using expert systems, neural networks, fuzzy logic and learning methodologies. Detailed classification can be carried out according to the following criteria:

- mathematical design tools: These include linear quadratic (LQ), intelligent control (IC), gain scheduling (GS)/linear parameter varying (LPV), adaptive control (AC), feedback linearization (FL)/dynamic inversion (DI), $\mathcal{H}_\infty$ and robust control, qualitative feedback theory (QFT), multiple model (MM), model predictive control (MPC), variable structure control (VSC)/sliding mode control (SMC)and generalized internal model control (GIMC);
- design approaches: These include pre-computed control laws (such as GS/LPV, MM, QFT and GIMC) or online automatic redesign (such as LQ, AC, FL/DI, VSC/SMC and MPC);
- reconfiguration mechanisms: These include optimization, switching, matching, following and compensation;
- types of system to be dealt with, whether linear or nonlinear.

An important criterion for judging the suitability of a control method for AFTCS is its ability to be implemented to maintain an acceptable (nominal or degraded) performance in the impaired system in an on-line real-time setting. In this regard, the following requirements should be satisfied:

- Control reconfiguration must be done under real-time constraints.
- The reconfigurable controller should be designed automatically with little trial-and-error or human interaction.
- The methods selected must provide a solution even if the solution is not optimal.

1.6.2 Classification based on field of application

A large amount of research has been carried out in the framework of aircraft flight control. Several reconfigurable flight control systems have been tested. With rapid advances in microelectronics, mechatronics, smart actuator and sensor techniques, and computing technologies, and motivated by increased demands for high requirements of system performance, product quality, productivity and operating efficiency beyond the conventional safety-critical aerospace and nuclear power systems, FTCS design is becoming an important feature to be considered in commercial product development and system design such as drive-by-wire automobiles. Recently, concepts and methodologies developed in fly-by-wire (FBW) fault-tolerant flight control systems have been extended to a wide range of engineering systems such as automobiles, railway vehicles, surface ships, autonomous underwater vehicles, automated highway systems, petrochemical plants, power systems, robots, medical systems and other industrial systems.

1.7   Outline of the Book

1.7.1   Methodology

Throughout the book, each chapter or section is composed of five parts:

- *mathematical modeling*: in which we discuss the main ingredients of the state-space model under consideration;
- *definitions or assumptions*: in which we state the definitions or constraints on the model variables to pave the way for subsequent analysis;
- *analysis and examples*: which is the core of the section and contains some solved examples for illustration;
- *results*: which are provided most of the time in the form of theorems, lemmas and corollaries.
- *remarks*: which are given to shed some light on the relevance of the developed results *vis-a-vis* published work.

Theorems, lemmas and corollaries are keyed to chapters, for example, *Theorem 3.4* means Theorem 4 in Chapter 3 and so on. Relevant notes and research issues are offered at the end of each chapter for the purpose of stimulating the reader. For convenience, the references are to be found at the end of each chapter. We hope that this way of articulating the information will attract the attention of a wide spectrum of readership.

This book aims to provide a rigorous framework for studying the analysis, stability and control problems of FTC while addressing the dominant sources of difficulties caused by: dimensionality; information structure constraints; parametric uncertainty and time delays. The primary objective is threefold: to review past methods and results from a contemporary perspective; to examine present trends and approaches; and to provide future possibilities, focusing on robust, reliable distributed design methods.

In brief, the main features of the book are:

- It provides an overall assessment of fault-tolerant control algorithms over the past few years.
- It addresses several issues that arise at the interaction of fault detection and fault accommodation.
- It presents key concepts with their proofs, followed by efficient computational methods.
- It considers representative industrial applications and provides the results of simulations tun using MATLAB®.

1.7.2 Chapter organization

Fault-tolerant control systems have been investigated for a long time in the control literature and have attracted increasingly more attention for more than three decades. The literature has grown progressively and quite number of fundamental concepts and powerful tools have been developed from various disciplines. Rapid technological progress raises many fundamental problems that call for further exploration. Among the core issues are those of representation, analysis, design and implementation. In particular, the field still lacks a unified framework that can cope with the core issues in a systematic way. This has motivated us to write the current book, which presents theoretical explorations of several fundamental problems with fault-tolerant control systems.

The book is primarily intended for researchers and engineers in the systems, control and communication community. It can also serve as complementary reading for elective courses for fault-tolerant control systems at the postgraduate level. The material of the book is divided into 10 chapters and an appendix:

- *Chapter 1* is an introductory chapter in which the different concepts and general ideas pertaining to fault-tolerant control (FTC) are presented.
- *Chapter 2* is devoted to a review of fault diagnosis and detection with extra emphasis on the unscented Kalman filter in an integrated design framework for process-fault processing within industrial applications.
- *Chapter 3* introduces robust fault detection methods including distributed fault diagnosis, data-driven detection and robust adaptive estimation.
- *Chapter 4* deals with designing a fault-tolerant control system as an essential component in an industrial process as it enables the system to continue robust operation under some conditions.
- *Chapter 5* outlines fault-tolerant control in nonlinear dynamical systems and presents a survey of various design approaches.
- *Chapter 6* establishes a detailed characterization of the problem of robust fault estimation filter design, encompassing multiconstrained fault estimation and adaptive tracking.

- *Chapter 7* provides results on methods for fault detection in dynamical systems operating over communication networks. This includes modified residual generator schemes, quantized fault-tolerant control, sliding-mode observer and linear-switched systems.
- *Chapter 8* presents some recent architectures and related concepts pertaining to industrial fault-tolerant detection.
- *Chapter 9* gives analysis and design methods of fault estimation for stochastic systems and their applications to unmanned air vehicles, real-time router fault accommodation and fault detection for Markov jump systems.
- *Chapter 10* is devoted to applications of fault-tolerant systems.
- *Appendix A* contains some relevant mathematical lemmas and basic algebraic inequalities that are used throughout the book.

Several illustrative numerical examples are provided within the individual chapters.

1.8 Notes

This book covers a wide spectrum of analysis and design tools for fault-tolerant control (FTC) in dynamical systems. The analytical tools developed in this volume are supplemented with rigorous proofs of closed-loop stability properties and simulation studies. For the purpose of clarity, the book is split into self-contained chapters with each chapter being equipped with illustrative examples, problems and questions. The book is supplemented by appropriate appendices and indexes. At the end of each chapter, a summary of the results with the relevant literature is given and pertinent discussions of future advances and trends are outlined. Some selected problems are left for the reader to solve. The bibliographical notes cannot give a complete review of the literature but indicate the most influential papers or books in which the results presented here have been described; they also provide interested readers with adequate entry points to the still-growing literature in this field.

While organizing the material, we planned that this book would be appropriate for use as graduate-level textbook in applied mathematics as well as other engineering disciplines (electrical, mechanical, civil, chemical, and systems engineering), a good volume for independent study and a suitable reference for graduate students, practicing engineers, interested readers and researchers from a wide spectrum of engineering disciplines, science and mathematics.

References

[1] Basseville, M. (1988) "Detecting changes in signals and systems: A survey", *Automatica*, **24**(3):309–326.

[2] Dash, S., and Venkatasubramanian, V. (2000) "Challenges in the industrial applications of fault diagnostic systems", *Computers and Chemical Engineering*, **24**(2):785–791.

[3] Frank, P. M., Ding, S. X., and Marcu, T. (2000) "Model-based fault diagnosis in technical processes", *Trans. Institute of Measurement and Control*, **22**(1):57–101.

[4] Dochain, D., Marquard, W., Won, S. C., Malik, O. P., and Kinnaert, M. (2006) "Monitoring and control of process and power systems: Towards new paradigms", *Annual Reviews in Control*, **30**(1):69–79.

[5] Witczak, M. (2007) *Modelling and Estimation Strategies for Fault Diagnosis of Nonlinear Systems: From Analytical to Soft Computing Approaches*, Lecture Notes in Control and Information Sciences, **354**, Berlin, Germany: Springer.

[6] Chandler, P. R. (1984) "Self-repairing flight control system reliability and maintainability program: Executive overview", in *Proc. IEEE National Aerospace and Electronics Conference*, 586–590.

[7] Chizeck, H. J., and Willsky, A. S. 1978 "Towards fault-tolerant optimal control", in *Proc. 1978 IEEE Conference on Decision and Control*, **1**, 19–20.

[8] Eterno, J. S., Weiss, J. L., Looze, D. P., and Willsky, A. S. 1985 "Design issues for fault tolerant-restructurable aircraft control", in *Proc. 24th IEEE Conference on Decision and Control*, **1**, 900–905.

[9] Montgomery, R. C., and Caglayan, A. K. (1976) "Failure accommodation in digital flight control systems by Bayesian decision theory", *J. Aircraft*, **13**(2):69–75.

[10] Montgomery, R. C., and Price, D. B. (1976) "Failure accommodation in digital flight control systems for nonlinear aircraft dynamics. *J. Aircraft*, **13**(2):76–82.

[11] Blanke, M., Izadi-Zamanabadi, R., Bøgh, S. A., and Lunau, C. P. (1997) "Fault-tolerant control systems: A holistic view", *Control Engineering Practice*, **5**(5):693–702.

[12] Blanke, M., Frei, C., Kraus, F., Patton, R. J., and Staroswiecki, M. (2000) "What is fault-tolerant control?", in *Proc. 4th IFAC Symposium on Fault Detection, Supervision and Safety for Technical Processes*, 40–51.

[13] Blanke, M., Staroswiecki, M., and Wu, N. E. (2001) "Concepts and methods in fault-tolerant control", in *Proc. American Control Conference*, 2606–2620.

[14] Jiang, J. (2005) "Fault-tolerant control systems: An introductory overview", *Automatica SINCA*, **31**(1): 161–174.

[15] Staroswiecki, M., and Gehin, A. L. (2001) "From control to supervision", *Annual Reviews in Control*, **25**: 1–11.

[16] Steinberg, M. (2005) "Historical overview of research in reconfigurable flight control", in *Proc. IMechE, Part G: J. Aerospace Engineering*, **219**:263–275.

[17] Mahmoud, M., Jiang, J., and Zhang, Y. M. (2003) *Active Fault Tolerant Control Systems: Stochastic Analysis and Synthesis*, Lecture Notes in Control and Information Sciences, **287**, Berlin, Germany: Springer.

[18] Ding, S. X. (2008) *Model-based Fault Diagnosis Techniques: Design Schemes, Algorithms, and Tools*, Berlin, Heidelberg: Springer.

[19] Isermann, R. (2006) *Fault-Diagnosis Systems*, Springer, UK.

[20] Isermann, R., and Balle, P. (1997) "Trends in the application of model-based fault detection and diagnosis of technical processes", *Control Eng. Practice*, **5**, 709–719.

[21] Chen, J., and Patton, R. J. (1999) *Robust Model-Based Fault Diagnosis for Dynamic Systems*, Boston, Mass, USA: Kluwer Academic Publishers.

[22] Frank, P. M. (1996) "Analytical and qualitative model-based fault diagnosis: a survey", *Europ. J. Control*, **2**:6–28.

[23] Frank, P. M. (1994) "On-line fault detection in uncertain nonlinear systems using diagnostic observers: a survey and some new results", *Int. J. Systems Science*, **25**(12):2129–2154.

[24] Sauter, D., Mary, N., Sirou, F., and Thieltgen, A. (1994) "Fault diagnosis in systems using fuzzy logic", in *Proc. 3rd IEEE Conf. Control Application*, Glasgow, UK, 883–888.

[25] Venkatasubramanian, V., Rengaswamy, R., Yin, K., and Kavuri, S. N. (2003) "A review of process fault detection and diagnosis part II: quantitative model based methods", *Computers and Chemical Engineering.*, **27**: 313–326.

[26] Patton, K. (2008) *Artificial Neural Networks for the Modelling and Fault Diagnosis of Technical Processes* Springer, UK.

[27] Chiang, L. H., Russell, E. L., and Braatz, R. D. (2001) *Fault Detection and Diagnosis in Industrial Systems*, Springer, UK.

[28] Chow, E. Y., and Willsky, A. S. (1984) "Analytical redundancy and the design of robust failure detection systems", *IEEE Trans. Automatic Control*, **29**:603–614.

[29] Gertler, J. J., and Singer, D. (1990) "A new structural framework for parity equation based failure detection and isolation", *Automatica*, **26**:381–388.

[30] Gertler, J., and Kunert, M. (1995) "Optimal residual decoupling for robust fault diagnosis", *Int. J. Control.*, **61**:395–421.

[31] Delmaire, G., Cassar, J. P., and Staroswiecki, M. (1994) "Comparison of identification and parity space approach for failure detection in single-input single-output systems", in *Proc. IEEE Conf. Contr. A*, Glasgow, UK, 865–870.

[32] Staroswiecki, M., Cassar, J. P., and Cocquempot, V. (1993) "Generation of optimal structured residuals in the parity space", in *Proc. 12th IFAC World Congress*, Sydney, Australia, 535–542.

[33] Zhang, Y. M., and Jiang, J. (2002) "An active fault-tolerant control system against partial actuator failures", *IEE Proc. Control Theory and Applications*, **149**(1):95–104.

[34] Zhang, Y. M., and Jiang, J. (2002) "Design of restructurable active fault-tolerant control systems", *Preprints of the 15th IFAC World Congress*, Barcelona, Spain.

[35] Hsieh, C. S. (2002) "Performance gain margins of the two-stage LQ reliable control", *Automatica*, **38**:1985–1990.

[36] Jiang, J., and Zhao, Q. (2000) "Design of reliable control systems possessing actuator redundancies", *J. Guidance, Control, and Dynamics*, **23**(4):709–718.

[37] Stengel, R. F. (1991) "Intelligent failure-tolerant control", *IEEE Control Systems Magazine*, **11**(4):14–23.

2

Fault Diagnosis and Detection

2.1 Introduction

Failures can be classified as sudden or incipient. Sudden failures are often the simplest form to diagnose since they usually have a dramatic impact on performance, which can be detected at a number of downstream sensors. When a blast occurs, the impact can be felt everywhere in the vicinity. Incipient or gradual failures are difficult to detect since they manifest themselves as a slow degradation in performance which can only be detected over time. Techniques based on identifying a model and using performance or error metrics to detect failure will find it difficult to identify a failure which occurs at a rate slower (often, much slower) than the rate at which the model drifts under normal conditions.

Another class of failure which has largely been ignored in the fault detection literature is a "pre-existing" failure. This is because a model of correct operation is impossible to identify by looking at a unit after it has already failed, and so most practical fault detection techniques are simply inapplicable to this type of fault. Some recent papers on the same or a similar topic have been published in [1]–[5]. The most recent approaches to fault monitoring and detection can be found in [6]–[9].

The variables shown in Table 2.1 are used throughout this chapter.

2.2 Related Work

This section presents the work which has been conducted in this area of performance monitoring of plant: model-based schemes for fault detection, model-free schemes for fault detection, and probabilistic models with fault detection.

2.2.1 Model-based schemes

The model-based approach is popular for developing FDI techniques [10]. It mainly consists of two stages [11]: residuals are generated by computing the difference between the measured output from the system and the estimated output obtained from the state system estimator (e.g. a Kalman filter). In the second stage, any departure from zero of the residuals indicates a fault

Analysis and Synthesis of Fault-Tolerant Control Systems, First Edition. Magdi S. Mahmoud and Yuanqing Xia.
© 2014 John Wiley & Sons, Ltd. Published 2014 by John Wiley & Sons, Ltd.

Table 2.1   General nomenclature

Symbol	Function
$\bar{x}$	Mean
P_x	Covariance
$2L + 1$	Sigma vectors in unscented Kalman filter (UKF)
QTS	Quadruple tank system
UB	Utility boiler
UT	Unscented transformation
α	Spread of the sigma points around $\bar{x}$
κ	Secondary scaling parameter
β	Incorporate prior knowledge of the distribution of x
λ	Composite scaling parameter
L	Dimension of the augmented state
$\mathfrak{R}^v$	Process-noise covariance
$\mathfrak{R}^n$	Measurement-noise covariance
W_i	Weights
$\mathbf{w}_k$	Stationary process with identity state transition matrix
r_k	Noise
d_k	Desired output
w_k	Nonlinear observation
$\mathfrak{R}^e$	Constant diagonal matrix
λ_{RLS}	Forgetting factor

has likely occurred [12]. However, these methods are developed mainly for linear systems, assuming that a precise mathematical model of the system is available. This assumption, however, may be difficult to satisfy in practice, especially as engineering systems are, in general, nonlinear and are becoming more complex [13].

2.2.2   Model-free schemes

For model-free approaches, only the availability of a large amount of historical process data is assumed. This data can be transformed and presented in different ways as *a priori* knowledge to a diagnostic system. This is known as the "feature extraction process" and it is done to facilitate later diagnosis [14]. The extraction process can proceed mainly as either a quantitative or a qualitative process. Quantitative feature extraction can be statistical or non-statistical. Model-free techniques, such as neural networks, fuzzy logic and genetic algorithms, are used to develop models for FDI techniques. These models can not only represent a wide class of nonlinear systems with arbitrary accuracy, they can also be trained from data. Among these techniques, neural networks are well recognized for their ability to approximate nonlinear functions and for their learning ability [15]. For these reasons, they have been used as models to generate residuals for fault detection [16, 17, 18, 19, 20]. However, it is very difficult to isolate faults with these networks as they are "black boxes" by their nature. Further, it is desirable that fault diagnostic systems should be able to incorporate the experience of the operators [21]. Fuzzy reasoning allows symbolic generalization of numerical data by fuzzy rules and supports the direct integration of the experience of the operators in the FDI decision-making process in order to achieve more reliable fault diagnosis [22, 23]. A rule-based expert system for fault diagnosis in a cracker unit is described in [24]. Optimization algorithms such as genetic algorithms (GA) and particle swarm optimization (PSO), which simulate biological processes to solve search and optimization problems, are also implemented to give a better pictorial view of fault detection and classification.

2.2.3   Probabilistic schemes

Bayesian belief networks (BBN) provide a probabilistic approach to consider cause-and-effect relations between process variables. There have been a few attempts to apply BBNs to fault detection and diagnosis, for example, Mehranbod [25] considers probabilistic sensor fault detection and identification and Kirch and Kroschel [26] propose presenting a BBN model in the form of a set of nonlinear equations and constraints that should be solved for the unknown probabilities. As an inference tool, Rojas-Guzman and Kramer [27] use genetic algorithms for fault diagnosis in a BBN representing a fluid catalytic cracking process. Santoso *et al.* [28] harness the learning capability of BBNs to use process data in an adaptable fault diagnosis strategy. BBNs are also used to perform FDD for discrete events, such as walking [29, 30]. A probabilistic approach with application to bearing-fault detection has also been implemented [31].

2.3   Integrated Approach

Fault is an undesirable factor in any process control industry. It affects the efficiency of the system operation and reduces economic benefits to the industry. The early detection and diagnosis of faults in mission-critical systems is crucial for preventing failure of equipment and loss of productivity and profits, for good management of assets, and for reduction of shutdowns.

For an effective fault diagnosis approach to highly nonlinear systems, we have assumed various faults in the system which have been successfully monitored and estimated through the encapsulation of the unscented Kalman filter (UKF) in various architectures of the multi-sensor data fusion technique. Figure 2.1 shows a proposed implementation plan for fault monitoring using the unscented Kalman filter.

2.3.1   Improved multi-sensor data fusion

The multi-sensor data fusion method has received major attention for various industrial applications. Data fusion techniques combine data, from multiple sensors installed in the plant, and related information, from associated databases, to achieve improved accuracy and more specific inferences that could not be achieved by the use of a single sensor alone. Specifically, in a mission-critical system where timely information is of immense importance, the precision and accuracy achieved through the multi-sensor data fusion technique can be very handy.

A particular industrial process application might have plenty of associated sensor measurements located at different operational levels and with various accuracy and reliability

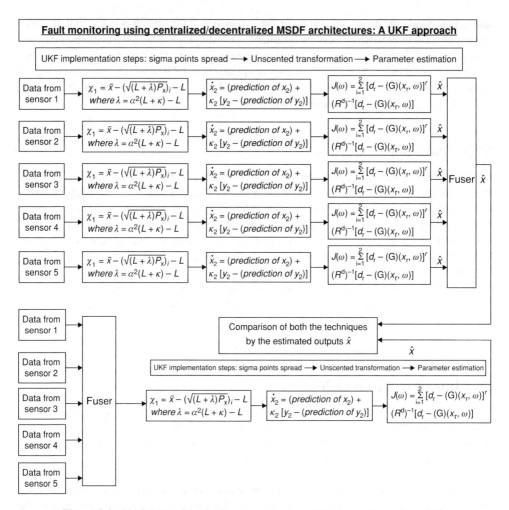

Figure 2.1   Implementation plan for the evaluation of the unscented Kalman filter

specifications. One of the key issues in developing a multi-sensor data fusion system is how the multi-sensor measurements can be fused or combined to overcome the uncertainty associated with individual data sources and obtain an accurate joint estimate of the system state vector. There exist various approaches to resolve this problem, of which the Kalman filter is one of the most significant and applicable solutions.

The unscented Kalman filter (UKF) essentially addresses the approximation issues of the extended Kalman filter (EKF) [32]. The state distribution is represented by Gaussian random variables (GRV) but is specified using a minimal set of carefully chosen sample points that completely capture the true mean and covariance of the GRV. When propagated through a true nonlinear system, it captures the posterior mean and covariance accurately to the second order (Taylor series expansion) for any nonlinearity. The structure of the UKF is elaborated by unscented transformation.

Remark 2.1   *Consider a state-space model given by:*

$$x_t = f(x_t - 1) + \epsilon, \quad x_t \in \mathbf{R}^m \tag{2.1}$$

$$\mathbf{y}_t = g(x_t) + v, \quad \mathbf{y}_t \in \mathbf{R}^D \tag{2.2}$$

Here, the system noise $\epsilon \sim N(0, \Sigma_\epsilon)$ and the measurement noise $v \sim N(0, \Sigma_v)$ are both Gaussian. The EKF linearizes f and g at the current estimate of x_t and treats the system as a non-stationary linear system even though it is not. The UKF propagates several estimates of x_t through f and g, and reconstructs a Gaussian distribution assuming the propagated values came from a linear system. Moreover, in nonlinear processes, when we are using EKF, the probability density function (PDF) is propagated through a linear approximation of the system around the operating point at each time instant. In doing so, the EKF needs the Jacobian matrices which may be difficult to obtain for higher-order systems, especially in the case of time-critical applications. Further, the linear approximation of the system at a given time instant may introduce errors in the state which may lead the state to diverge over time. In other words, the linear approximation may not be appropriate for some systems. Also in the EKF algorithm, during the time-update (prediction) step, the mean is propagated through the nonlinear function, in other words, this introduces an error since, in general, $\bar{y} \neq g(\bar{x})$. In the case of the UKF, during the time-update step, all the sigma points are propagated through the nonlinear function which makes the UKF a better and more effective nonlinear approximator. The UKF principle is simple and easy to implement as it does not require the calculation of the Jacobian matrix at each time step. The UKF is accurate up to second-order moments in the PDF propagation whereas the EKF is accurate up to first-order moments [33].

2.3.2   Unscented transformation

The unscented transformation (UT) is a method for calculating the statistics of a random variable which undergoes a nonlinear transformation [32]. Consider propagating a random variable x (dimension L) through a nonlinear function, $y = f(\mathbf{x})$. Assume x has mean $\bar{\mathbf{x}}$ and covariance $\mathbf{P_X}$. To calculate the statistics of y, we form a matrix χ of $2L + 1$ sigma vectors χ_i according to the following:

$$\chi_0 = \bar{\mathbf{x}} \tag{2.3}$$

$$\chi_i = \bar{\mathbf{x}} + (\sqrt{(L + \lambda)\mathbf{P_X}})_i, i = 1, \ldots, L \tag{2.4}$$

$$\chi_i = \bar{\mathbf{x}} - (\sqrt{(L + \lambda)\mathbf{P_X}})_i - L, i = L + 1, \ldots, 2L \tag{2.5}$$

where $\lambda = \alpha^2(L + \kappa) - L$ is a scaling parameter. The constant α determines the spread of the sigma points around $\bar{\mathbf{x}}$ and is usually set to a small positive value (e.g., $1 \leq \alpha \leq 10^{-4}$). The constant κ is a secondary scaling parameter, which is usually set to $3 - L$, and β is used to incorporate prior knowledge. β is a tunable parameter of the distribution of x (for Gaussian distributions, $\beta = 2$ is optimal, as used in Equation (2.9)). $(\sqrt{(L + \lambda)\mathbf{P_X}})_i$ is the ith column of the matrix square root (that is, the lower-triangular Cholesky factorization). These sigma vectors are propagated through the nonlinear function

$$\xi_i = f(\chi_i), i = 0, \ldots, 2L \tag{2.6}$$

and the mean and covariance for y are approximated using a weighted sample mean and covariance of the posterior sigma points,

$$\bar{y} \approx \sum_{i=0}^{2L} W_i^{(m)} \xi_i, \quad \mathbf{P}_y \approx \sum_{i=0}^{2L} W_i^{(c)} (\xi_i - \bar{y})(\xi_i - \bar{y})^T \tag{2.7}$$

with weights W_i given by:

$$W_0^{(m)} = \frac{\lambda}{L + \lambda} \tag{2.8}$$

$$W_0^{(c)} = \frac{\lambda}{L + \lambda} + 1 - \alpha^2 + \beta \tag{2.9}$$

$$W_i^{(m)} = W_i^{(c)} = \frac{1}{2(L + \lambda)}, i = 1, \ldots, 2L \tag{2.10}$$

Note that this method differs substantially from general Monte Carlo sampling methods, which require orders of magnitude more sample points in an attempt to propagate an accurate (possibly non-Gaussian) distribution of the state. The deceptively simple approach taken with the UT results in approximations that are accurate to the third order for Gaussian inputs for all nonlinearities. For non-Gaussian inputs, approximations are accurate to at least the second order, with the accuracy of the third- and higher-order moments being determined by the choice of α and β.

2.3.3   Unscented Kalman filter

In view of the previous section's discussion, the unscented Kalman Filter (UKF) is a straightforward extension of the UT to the recursive estimation in the following equation:

$$\hat{\mathbf{x}}_k = \mathbf{x}_{k \, predicted} + \kappa_k [y_k - y_{k \, predicted}] \tag{2.11}$$

where the state random variable (RV) is redefined as the concentration of the original state and noise variables:

$$\mathbf{x}_k^a = \begin{bmatrix} \mathbf{x}_k^T & v_k^T & \mathbf{n}_k^T \end{bmatrix} \tag{2.12}$$

The unscented transformation sigma points selection scheme, given by Equations (2.3)–(2.5), is applied to this new augmented state RV to calculate the corresponding sigma matrix, χ_k^a. The UKF equations are given as follows. Note that no explicit calculations of Jacobians or Hessians are necessary to implement this algorithm.

Initialize with:

$$\hat{\mathbf{x}}_0 = E[\mathbf{x}_0], \ \mathbf{P}_0 = E[(\mathbf{x}_0 - \hat{\mathbf{x}})(\mathbf{x}_0 - \hat{\mathbf{x}})^T] \tag{2.13}$$

$$\hat{\mathbf{x}}_0^a = E[\mathbf{x}^a] = [\hat{\mathbf{x}}_0^T \quad 0 \quad 0]^T \tag{2.14}$$

For subsequent iterations, $k \epsilon [1, \ldots, \infty]$.

Calculate the sigma points:

$$\chi_{k-1}^a = [\hat{\mathbf{x}}_{k-1}^a \quad \hat{\mathbf{x}}_{k-1}^a + \gamma\sqrt{P_{k-1}^a} \quad \hat{\mathbf{x}}_{k-1}^a - \gamma\sqrt{P_{k-1}^a}] \tag{2.15}$$

The time-update equations are

$$\chi_{k|k-1}^x = \mathbf{F}(\chi_{k-1}^x, \mathbf{u}_{k-1}, \chi_{k-1}^v) \tag{2.16}$$

$$\hat{\mathbf{x}}_k^- = \sum_{i=0}^{2L} W_i^m \chi_{i,k|k-1}^x \tag{2.17}$$

$$\mathbf{P}_k^- = \sum_{i=0}^{2L} W_i^c (\chi_{i,k|k-1}^x - \hat{\mathbf{x}}_k^-)(\chi_{i,k|k-1}^x - \hat{\mathbf{x}}_k^-)^T \tag{2.18}$$

$$\xi_{k|k-1} = \mathbf{H}(\chi_{k|k-1}^x, \chi_{k-1}^n) \tag{2.19}$$

$$\hat{y}_k^- = \sum_{i=0}^{2L} W_i^m \chi_{i,k|k-1}^Y \tag{2.20}$$

and the measurement-update equations are:

$$\mathbf{P}_{\tilde{y}_k \tilde{y}_k} = \sum_{i=0}^{2L} W_i^c (\xi_{i,k|k-1} - \hat{y}_k^-)(\xi_{i,k|k-1} - \hat{y}_k^-)^T \tag{2.21}$$

$$\mathbf{P}_{x_k y_k} = \sum_{i=0}^{2L} W_i^c (\chi_{i,k|k-1} - \hat{\mathbf{x}}_k^-)(Y_{i,k|k-1} - \hat{y}_k^-)^T \tag{2.22}$$

$$\kappa_k = \mathbf{P}_{x_k y_k} \mathbf{P}_{\tilde{y}_k \tilde{y}_k}^{-1} \tag{2.23}$$

$$\hat{\mathbf{x}}_k = \hat{\mathbf{x}}_k^- + \kappa_k(y_k - \hat{y}_k^-) \tag{2.24}$$

$$\mathbf{P}_k = \mathbf{P}_k^- - \kappa_k \mathbf{P}_{\tilde{y}_k \tilde{y}_k} \kappa_k^T \tag{2.25}$$

where

$$\mathbf{x}^a = \begin{bmatrix} \mathbf{x}^T & v^T & \mathbf{n}^T \end{bmatrix}^T$$

$$\chi^a = \begin{bmatrix} (\chi^x)^T & (\chi^v)^T & (\chi^n)^T \end{bmatrix}^T, \quad \gamma = \sqrt{L + \lambda}$$

Note that λ is the composite scaling parameter, L is the dimension of the augmented state, $\mathfrak{R}^v$ is the process-noise covariance, $\mathfrak{R}^n$ is the measurement-noise covariance, and W_i are the weights.

2.3.4 Parameter estimation

Parameter estimation involves learning a nonlinear mapping $\mathbf{y}_k = \mathbf{G}(\mathbf{x}_k, \mathbf{w})$, where $\mathbf{w}$ corresponds to the set of known parameters. $\mathbf{G}(.)$ may be a neural network or another parameterized function. The extended Kalman filter (EKF) may be used to estimate the parameters by writing a new state-space representation

$$\mathbf{w}_{k+1} = \mathbf{w}_k + \mathbf{r}_k$$

$$\mathbf{d}_k = \mathbf{G}(x_k, w_k) + e_k \tag{2.26}$$

where $\mathbf{w}_k$ corresponds to a stationary process with identity state transition matrix, driven by noise $\mathbf{r}_k$. The desired output $\mathbf{d}_k$ corresponds to a nonlinear observation on $\mathbf{w}_k$.

For optimization, the following prediction error cost is minimized:

$$J(\mathbf{w}) = \sum_{i=1}^{k}[\mathbf{d}_t - \mathbf{G}(\mathbf{x}_t, \mathbf{w})]^T (\mathbf{R}^e)^{-1}[\mathbf{d}_t - \mathbf{G}(x_t, w)] \tag{2.27}$$

Thus, if the "noise" covariance $\mathfrak{R}^e$ is a constant diagonal matrix, then, in fact, it cancels out of the algorithm, and hence can be set arbitrarily (e.g., $\mathfrak{R}^e = 0.5I$). Alternatively, $\mathfrak{R}^e$ can be set to specify a weighted mean square error (MSE) cost. The innovations covariance $\mathbf{E}[\mathbf{r}_k \quad \mathbf{r}_k^T] = \mathbf{R}_k^r$, on the other hand, affects the convergence rate and tracking performance. Roughly speaking, the larger the covariance, the more quickly older data is discarded. There are several options on how to choose $\mathbf{R}_k^r$.

- Set $\mathbf{R}_k^r$ to an arbitrary "fixed" diagonal value, which may then be "annealed" towards zero as training continues.
- Set $\mathbf{R}_k^r = (\lambda_{RLS}^{-1} - 1)\mathbf{P}_{w_i}$, where $\lambda_{RLS} \in (0,1]$ is often referred to as the "forgetting factor". This provides for an approximate exponentially decaying weighting on past data. Note that λ_{RLS} should not be confused with λ used for sigma-point calculation.
- Set

$$\mathbf{R}_k^r = (1 - \alpha_{RM})\mathbf{R}_r^{k-1} + \alpha_{RM}\mathbf{K}_k^w[\mathbf{d}_k - \mathbf{G}(\mathbf{x}_k, \hat{\mathbf{w}})]$$
$$\times [\mathbf{d}_k - \mathbf{G}(\mathbf{x}_k, \hat{\mathbf{w}})]^T (\mathbf{K}_k^w)^T$$

which is a Robbins–Monro stochastic approximation scheme for estimating the innovations. The method assumes that the covariance of the Kalman update model is consistent with the actual update model. Typically, $\mathbf{R}_k^r$ is also constrained to be a diagonal matrix, which implies an independence assumption on the parameters. Note that a similar update may also be used for $\mathbf{R}_k^e$.

2.3.5 Multi-sensor integration architectures

Multi-sensor data fusion can be done at a variety of levels from the raw data or observation level to the feature or state vector level and the decision level. This can lead to the utilization of different configurations or architectures to integrate the data from disparate sensors in an industrial plant and extract the desired monitoring information. Using Kalman filtering as the data fusion algorithm, multiple sensors can be integrated in two key architecture scenarios called the "centralized" method and the "decentralized", or distributed, method. These methods have been widely studied since the beginning of this century [34, 35].

2.3.5.1 Centralized

In the centralized integration method, all the raw data from different sensors is sent to a single location to be fused. This architecture is sometimes called the "measurement fusion integration method" [34, 35], in which observations or sensor measurements are directly fused to obtain a global or combined measurement data matrix $\mathbf{H}^*$. Then, it uses a single Kalman filter to estimate the global state vector based upon the fused measurement. Although this conventional method provides high fusion accuracy to the estimation problem, the large number of states may require high processing data rates that cannot be maintained in practical real-time applications. Another disadvantage of this method is the lack of robustness in the case of failure in a sensor or the central filter itself. For these reasons, parallel structures can often provide improved failure detection and correction, enhanced redundancy management, and decreased costs for multi-sensor system integration. Figure 2.2 shows an implementation of the centralized architecture.

2.3.5.2 Decentralized

There has recently been considerable interest shown in the distributed integration method, in which the filtering process is divided between local Kalman filters that work in parallel to obtain individual sensor-based state estimates and a master filter that combines the local

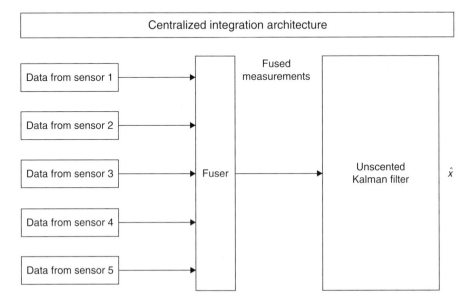

Figure 2.2   Centralized integration method using unscented Kalman filter

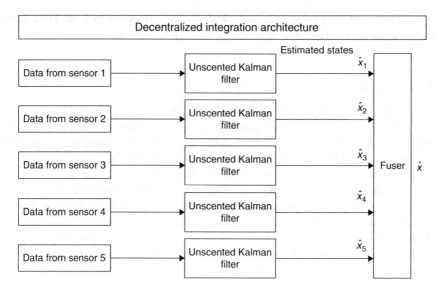

Figure 2.3   Decentralized integration method using unscented Kalman filter

estimates to yield an improved global state estimate. This architecture is sometimes called the "state-vector fusion integration method" [34, 35]. The advantages of this method are greater robustness because of the parallel implementation of fusion nodes and a lower computation load and communication cost at each fusion node. It is also applicable in modular systems where different process sensors can be provided as separate units. On the other hand, distributed fusion is conceptually a lot more complex and is likely to require higher bandwidth than centralized fusion. Figure 2.3 shows an implementation of the decentralized architecture.

2.4   Robust Unscented Kalman Filter

In this section, the problem of robust fault detection filter design is studied using a new approach, which is established by the integration of an optimal residual generator, a residual evaluator and the unscented Kalman filter (UKF). The optimal residual generator is derived with the assumption of no modeling errors and further improved by residual evaluation. We establish a sufficient condition for the existence of a robust fault detection filter and cast the solution in terms of linear matrix inequality (LMI) which helps in the determination of an adaptive threshold for fault detection. We incorporate the UKF for the iterative update of observation noise mean and covariance to provide a robust fault detection filter with tolerance to faults with the help of hypothesis testing computation.

We consider the example of a car-like mobile robot in detailed numerical simulation to demonstrate the effectiveness of this approach.

2.4.1   Introduction

The objective of model-based fault detection is to design a detection mechanism that generates fault-indicating signals. These signals, called "residual signals", are compared with given thresholds to judge whether or not a fault has occurred. For this purpose, many model-based fault detection techniques have been studied over the last two decades [36]. One particularly interesting technique is observer-based fault detection [37]. Observer-based fault detection filters are not only easy to implement in practical systems, but have also been shown by many theoretical studies and applications to be very effective in detecting faults in sensors, actuators, and system components.

Since most continuous dynamical systems are now controlled by digital devices, it is also important to understand the theoretical developments in the digitally sampled data setting. Furthermore, it has been shown that sample-data fault detection problems can be converted to equivalent discrete time detection problems using certain discretization methods [38]. Thus, discrete time fault detection is of great importance and most natural for modern digital implementations. Li and Jaimoukha [39] considered a model-based fault detection and isolation method for linear time-invariant dynamics. There are significant numbers of works addressing discrete time fault detection problems using Kalman filter techniques [40, 41, 42]. Similar to the multi-objective design problem of continuous time fault detection, robustness in discrete time fault detection is also a very important issue but difficult to handle. Many robust filter design techniques, such as H_1 optimization, LMI, parity space, and eigenstructure assignment, have been applied to discrete time fault detection filter design with limited success [43, 44].

Figure 2.4 shows a proposed implementation plan. A robust fault detection filter comprises of a three-step procedure:

1. Introduce a reference system which is an optimal robust fault detection solution, assuming that there is no model uncertainty apart from unknown inputs.

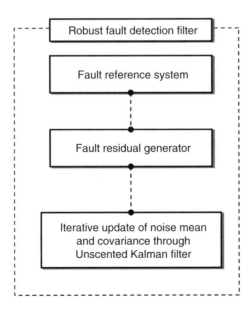

Figure 2.4 Implementation of a robust fault detection filter

2. Synthesize a fault detection filter (FDF) as a residual generator using an LMI formulation of the minimization of the H_∞ norm of the difference between the reference model and the real residual generator.
3. Iteratively update the observation noise and covariance through UKF to estimate that a fault has been detected. Unscented filters are employed in n states of the dynamic system. The residual compares the output of the unscented filters and the output of n states of a healthy model of the plant that contains no faults. This results in n residuals, r_n, summed with hypothesis testing probabilities, which gives the fault (or drift) detection of the system. These drift detections and output of the unscented filters are added to give us the parameter estimation of the system, thus resulting in efficient updating of observation noise and covariance.

2.4.2 Problem formulation

In this section, we study the robust fault detection filter design problem for linear time-invariant (LTI) systems described by:

$$\dot{x}(k) = (\mathbf{A} + \Delta\mathbf{A})x(k) + (\mathbf{B} + \Delta\mathbf{B})\mathbf{u}(k)$$

$$+ \mathbf{B}_f\mathbf{f}(k) + \mathbf{B}_d\mathbf{d}(k) \tag{2.28}$$

$$\mathbf{y}(k) = \mathbf{C}x(k) + \mathbf{D}\mathbf{u}(k) + \mathbf{D}_f\mathbf{f}(k) + \mathbf{D}_d\mathbf{d}(k) \tag{2.29}$$

where $\mathbf{x}(k) \in \mathfrak{R}^n$ is the state, $\mathbf{u}(k) \in \mathfrak{R}^p$ is the control input, $\mathbf{y}(k) \in \mathfrak{R}^q$ are the measurement output vectors, and $\mathbf{d}(k) \in \mathfrak{R}^{n_d}$ and $\mathbf{f}(k) \in \mathfrak{R}^{n_f}$ are the disturbance and fault vectors, respectively. Here, $B_f(k) \in \mathfrak{R}^{n \times n_f}$ and $\mathbf{D}_f(k) \in \mathfrak{R}^{q \times n_f}$ are the component and uninterested fault distribution matrices, respectively, while $\mathbf{B}_d(k) \in \mathfrak{R}^{n \times n_d}$, and $\mathbf{D}_d(k) \in \mathfrak{R}^{q \times n_d}$ are the corresponding disturbance distribution matrices. For simplicity, model uncertainties are assumed to be recast as disturbances. The vector $\mathbf{f}(k)$ represents the faults to be detected and isolated. In Equation (2.28) $A(k)$, $B(k)$, $C(k)$, $D(k)$, $\mathbf{B}_f(k)$, $\mathbf{B}_d(k)$, $\mathbf{D}_f(k)$, $\mathbf{D}_d(k)$ are known matrices with appropriate dimensions. The matrices $\Delta\mathbf{A}$, $\Delta\mathbf{B}$ are modeling errors represented by:

$$\begin{bmatrix} \Delta\mathbf{A} & \Delta\mathbf{B} \end{bmatrix} = \begin{bmatrix} \mathbf{E}_1\Sigma_1\mathbf{F}_1 & \mathbf{E}_2\Sigma_2\mathbf{F}_2 \end{bmatrix} \tag{2.30}$$

where $\mathbf{E}_1$, $\mathbf{E}_2$, $\mathbf{F}_1$, $\mathbf{F}_2$ are known matrices and $\Sigma_1^T\Sigma_1 \leq I$, $\Sigma_2^T\Sigma_2 \leq I$. Denote $\Omega_1 = \{\Delta\mathbf{A}|\Delta\mathbf{A} = \mathbf{E}_1\Sigma_1\mathbf{F}_1, \ \Sigma_1^T\Sigma_1 \leq I\}$ and $\Omega_2 = \{\Delta\mathbf{B}|\Delta\mathbf{B} = \mathbf{E}_2\Sigma_2\mathbf{F}_2, \ \Sigma_2^T\Sigma_2 \leq I\}$. The following assumptions are also used throughout:
  (A1) $\mathbf{A} + \Delta\mathbf{A}$ is asymptotically stable for $\Delta\mathbf{A} \in \Omega_1$.
  (A2) The pairs (A, B) and (C, A) are controllable and detectable, respectively.
  (A3) The matrix in Equation (2.31) has full row rank for all ω.

$$\begin{bmatrix} A - j\omega I & B_d \\ C & D_d \end{bmatrix} \tag{2.31}$$

2.4.3 Residual generation

Generally speaking, a fault detection system consists of two parts: a residual generator and a residual evaluator including a threshold and a decision logic unit. For the purpose of residual generation, we use the fault detection filter in the following form:

$$\dot{\hat{x}}(k) = (A - HC)\hat{x}(k) + (B - HD)u(k) + Hy(k), \tag{2.32}$$

$$\hat{y}(k) = C\hat{x}(k) + Du(k), \tag{2.33}$$

$$r(k) = V(y - \hat{y})(k) \tag{2.34}$$

where $\hat{x}(k) \in \Re^n$ and $\hat{y}(k) \in \Re^p$ represent the state and output estimation vectors, respectively, and $r(k)$ is the residual signal. The design parameters of a robust fault detection filter (RFDF) are the observer gain matrix **H** and the residual weighting matrix **V**. In order to describe the dynamics of RFDF, we first consider the filtering error $e(k) = x(k) - \hat{x}(k)$ and the residual signal $r(k)$. It can be shown that:

$$\dot{e}(k) = (A - HC)e(k) + \Delta Ax(k) + \Delta Bu(k)$$
$$+ (B_f - HD_f)f(k) + (B_d(k) - H_d d(k)) \tag{2.35}$$

$$r(k) = VCe(k) + VD_f f(k) + VD_d d(k) \tag{2.36}$$

Note that the dynamics of the residual signal depends not only on $f(k)$, $d(k)$ and $u(k)$, but also on the state $x(k)$. Thus, the problem of designing an observer-based RFDF, which is one of our main objectives, can be described as one of designing matrices **H**, **V** such that:

- The matrix $\mathbf{A} - \mathbf{HC}$ is asymptotically stable.
- The generated residual $r(k)$ is as sensitive as possible to fault $f(k)$ and as robust as possible to unknown input $d(k)$, control input $u(k)$, and model uncertainties $\Delta\mathbf{A}$, $\Delta\mathbf{B}$.

2.4.4 Residual evaluation

After designing an FDF, the remaining important task for FDI is the evaluation of the generated residual. One of the widely adopted approaches is to choose a "threshold" $J_{th} > 0$ and, based on this, use the following logical relationship for fault detection

$$\|r\|_{2,K} > J_{th} \Rightarrow with\ faults \Rightarrow alarm \tag{2.37}$$

$$\|r\|_{2,K} \leq J_{th} \Rightarrow no\ faults \tag{2.38}$$

where the "residual evaluation function" $\|r\|_{2,K}$ is determined by:

$$\|r\|_{2,K} = [\int_{k_1}^{k_2} \Re^T(k)r(k)dk]^{1/2} \tag{2.39}$$

where $K = k_2 - k_1$, $k \in (k_1, k_2]$ is the finite-time window. Note that the length of the time window is finite (i.e. K instead of ∞). Since an evaluation of a residual signal over the whole time range is impractical, it is desirable that faults are detected as early as possible.

The rest of this chapter provides a detailed implementation and simulation of the proposed scheme.

2.5 Quadruple Tank System

A quadruple tank system (QTS) consists of four interconnected water tanks and two pumps (see Figure 2.5). Its manipulated variables are the voltages to the pumps and its controlled variables are the water levels in the lower two tanks. The quadruple tank system is a multi-input–multi-output (MIMO) system. This system is a prototype of a real-life control problem that we can experiment on and try to solve in the most efficient way, since it deals with multiple variables and thus throws light on large systems in industry.

The system has two control inputs (pump throughputs) that can be manipulated to control the water level in the tanks. The pumps are used to transfer water from a sump into the four overhead tanks. The two tanks at the upper level drain freely into the two tanks at the bottom level and the liquid levels in these bottom two tanks are measured by pressure sensors. The piping system is designed such that each pump affects the liquid levels of both measured tanks. The output of each pump is split into two using a three-way valve. Thus each pump output goes to two tanks: a portion of the flow from Pump 1 is directed into Tank 1 and the rest is directed to Tank 4, which drains into Tank 2; similarly, a portion of the flow from Pump 2 is directed into Tank 2 and the rest is directed to Tank 3, which drains into Tank 1. By adjusting the bypass valves of the system, the proportion of water pumped into different tanks can be changed to adjust the degree of interaction between the pump throughputs and the water levels. The ratio of the split from each pump is controlled by the position of the valve. Because of the large water distribution load, the pumps have each been supplied 12 V.

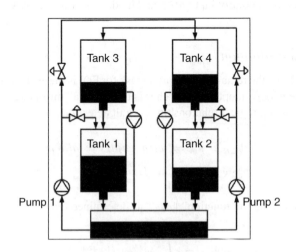

Figure 2.5   Quadruple tank system

2.5.1   Model of the QTS

The mathematical modeling of the quadruple tank process can be obtained by using Bernoulli's law [45]. The constants are denoted in Table 2.2.

A nonlinear mathematical model of the four-tank system is derived based on mass balances and Bernoulli's law. Mass balance for one of the tanks is:

$$A\frac{dh}{dt} = q_{out} - q_{in} \tag{2.40}$$

where A denotes the cross section of the tank and h, q_{in}, and q_{out} denote the water level, the inflow, and the outflow of the tank, respectively. Bernoulli's law is used to establish a relationship between output and height. It states that

$$q_{out} = a\sqrt{2gh} \tag{2.41}$$

where a is the cross section of the outlet pipe (in cm^2) and g is the acceleration due to gravity. A common multiplying factor for a hole of the type being used in this system is the coefficient of discharge, k. We can therefore rewrite Bernoulli's equation as:

$$q_{out} = ak\sqrt{2gh} \tag{2.42}$$

The flow through each pump is split so that a proportion of the total flow travels to each of two tanks. This can be adjusted via one of the two valves shown in Figure 2.5. Assuming that the flow generated is proportional to the voltage applied to each pump, the change in v, and that q_T and q_B are the flows going to the top and bottom tanks, respectively, we are able to come up

Table 2.2   Nomenclature for QTS

Symbol	Function
h_i	Level of water in tank i
a_i	Area of water flowing out from tank i
A_i	Area of tank i
γ_1	Ratio of water diversion between Tanks 1 and 4
γ_2	Ratio of water diversion between Tanks 2 and 3
k_1	Gain of Pump 1
k_2	Gain of Pump 2
v_1	Manipulated Input 1 (Pump 1)
v_2	Manipulated Input 2 (Pump 2)
g	Gravitational constant
a_{leak1}	Leak in pipe of Tank 1
a_{leak2}	Leak in pipe of Tank 2
a_{leak3}	Leak in pipe of Tank 3
a_{leak4}	Leak in pipe of Tank 4
q_{in}	Inflow
q_{out}	Outflow

with the following relationships: $q_B = \gamma k v$ and $q_T = (1 - \gamma)k v$ where $\gamma \epsilon [0, 1]$. Combining all the equations for the interconnected four-tank system we obtain the physical system model.

A fault model can then be constructed by adding extra holes to each tank. The mathematical model of the faulty quadruple tank system can be given as:

$$\frac{dh_1}{dt} = -\frac{a_1}{A_1}\sqrt{2gh_1} + \frac{a_3}{A_1}\sqrt{2gh_3} + \frac{\gamma_1 k_1}{A_1}v_1 + \frac{d}{A_1}$$
$$- \frac{a_{leak1}}{A_1}\sqrt{2gh_1}$$

$$\frac{dh_2}{dt} = -\frac{a_2}{A_2}\sqrt{2gh_2} + \frac{a_4}{A_2}\sqrt{2gh_4} + \frac{\gamma_2 k_2}{A_2}v_2 - \frac{d}{A_2}$$
$$- \frac{a_{leak2}}{A_2}\sqrt{2gh_2}$$

$$\frac{dh_3}{dt} = -\frac{a_3}{A_3}\sqrt{2gh_3} + \frac{(1 - \gamma_2)k_2}{A_3}v_2$$
$$- \frac{a_{leak3}}{A_3}\sqrt{2gh_3}$$

$$\frac{dh_4}{dt} = -\frac{a_4}{A_4}\sqrt{2gh_4} + \frac{(1 - \gamma_1)k_1}{A_4}v_1$$
$$- \frac{a_{leak4}}{A_4}\sqrt{2gh_4}$$

$$\frac{dv_1}{dt} = -\frac{v_1}{\tau_1} + \frac{1}{\tau_1}u_1$$

$$\frac{dv_2}{dt} = -\frac{v_2}{\tau_2} + \frac{2}{\tau_2}u_2 \tag{2.43}$$

2.5.2   Fault scenarios in QTS

Two fault scenarios are created when using the quadruple tank system in the simulation program. In these scenarios, incipient single and multiple tank faults (i.e., leakages) are created by changing system parameters manually at certain times in the simulation. The system inputs, outputs and some states are corrupted by Gaussian noise with zero mean and standard deviation of 0.1.

Scenario I: Leakage in Tank 1
While the system is working in real time, a single incipient fault (i.e., Tank 1 leakage) is created by changing parameter a_{leak1} to 0.81 cm^2 (the value 0.81 is 30 % of the cross-section of the outlet hole of Tank 1) at 350 seconds.

Scenario II: Leakage in Tanks 2 and 3
While the system is working in real time, multiple incipient faults (Tank 2 and 3 leakages) are created by changing parameters a_{leak2} to 1.62 cm^2 and a_{leak3} to 0.54 cm^2 (1.62 is 60 % of the

cross-section of the outlet of Tank 2 and 0.54 is 20 % of the cross-section of the outlet of the Tank 3) at 350 seconds.

2.5.3   Implementation structure of UKF

The elementary implementation structure of an unscented Kalman filter on one of the states of the quadruple tank system can be described as follows:

Step 1: Initialize the parameters and define the conditions according to the system:
  $\mathbf{n} = 6$ (where $\mathbf{n}$ is the number of states)
  $\mathbf{q} = 0$ (where $\mathbf{q}$ shows the standard deviation of process)
  $\mathbf{r} = 1.9$ (where $\mathbf{r}$ shows the standard deviation of measurement)
  $\mathbf{Q} = \mathbf{q}^2 \times \mathbb{I}(\mathbf{n})$ (where $\mathbf{Q}$ represents the covariance of process and $\mathbb{I}(\mathbf{n})$ is the unit matrix of order $\mathbf{n}$)
  $\mathbf{R} = \mathbf{r}^2$ (where $\mathbf{R}$ represents the covariance of measurement)

Step 2: Define the states and measurement equation of the system:
  $\mathbf{f} = (\mathbf{x})[\mathbf{x}(1); \mathbf{x}(2); \mathbf{x}(3); \mathbf{x}(4); \mathbf{x}(5); \mathbf{x}(6)]$ (where $\mathbf{f}$ represents the nonlinear state equations)
  $\mathbf{h} = \mathbf{x}(1)$; (where $\mathbf{h}$ represents the measurement equation)
  $\mathbf{s} = [0; 0; 1; 0; 1; 1]$ (where $\mathbf{s}$ defines the initial state)
  $\mathbf{x} = \mathbf{s} + \mathbf{q} \times randn(6, 1)$ (where $\mathbf{x}$ defines the initial state with noise)
  $\mathbf{P} = \mathbb{I}(\mathbf{n})$ (where $\mathbf{P}$ defines the initial state covariance)
  N (where N presents the total dynamic steps)

Step 3: Upgrade the estimated parameter under observation:
  for $k = 1 : N$
  $\mathbf{x}_1(1) = initializing, \ \mathbf{x}_2(1) = initializing$
  $\mathbf{x}_3(1) = initializing, \ \mathbf{x}_4(1) = initializing$
  $\mathbf{x}_5(1) = initializing, \ \mathbf{x}_6(1) = initializing;$
  $\mathbf{H} = [\mathbf{x}_1(k); \mathbf{x}_2(k); \mathbf{x}_3(k); \mathbf{x}_4(k); \mathbf{x}_5(k); \mathbf{x}_6(k)]$
  $\mathbf{z} = \mathbf{h}(s) + \mathbf{r} \times randn;$ (measurements)
  $sV(:, k) = \mathbf{s};$ (save actual state)
  $zV(:, k) = \mathbf{z};$ (save measurement)
Then inject into the following equation (function):
  $[\mathbf{x}, \mathbf{P}] = ukf(\mathbf{x}, \mathbf{P}, hmeas, \mathbf{z}, \mathbf{Q}, \mathbf{R}, \mathbf{h})$
There will be three functions to complete the process.

Function 1: The unscented Kalman filter
  $[\mathbf{x}, \mathbf{P}] = ukf(\mathbf{x}, \mathbf{P}, hmeas, \mathbf{z}, \mathbf{Q}, \mathbf{R}, \mathbf{h})$
where $[\mathbf{x}, \mathbf{P}] = ukf(\mathbf{x}, \mathbf{P}, h, \mathbf{z}, \mathbf{Q}, \mathbf{R})$ returns the state estimate, $\mathbf{x}$, and state covariance, $\mathbf{P}$).
Note: For a nonlinear dynamic system (for simplicity, noises are assumed as additive):
  $\mathbf{x}_{k+1} = \mathbf{f}(\mathbf{x}_k) + \mathbf{w}_k \mathbf{z}_k = \mathbf{h}(\mathbf{x}_k) + v_k$ where $\mathbf{w} \sim \mathrm{N}(0, \mathbf{Q})$ means $\mathbf{w}$ is Gaussian noise with covariance $\mathbf{Q}$,
$v \sim \mathrm{N}(0, \mathbf{R})$ means v is Gaussian noise with covariance $\mathbf{R}$
Inputs:
  $\mathbf{f}$: function handle for $f(\mathbf{x})$
  $\mathbf{x}$: *a priori* state estimate

P: *a priori* estimated state covariance
h: function handle for $h(\mathbf{x})$
z: current measurement
Q: process noise covariance
R: measurement noise covariance
Outputs:
   x: *a posteriori* state estimate
   P: *a posteriori* state covariance
   $\mathbf{L} = \text{numel}(\mathbf{x})$; (where **L** is the number of states)
   $\mathbf{m} = \text{numel}(\mathbf{z})$; (where **m** is the number of measurements)
   $\alpha = 1e^{-3}$; (default, tunable)
   $k_i = 3 - \mathbf{L}$; (default, tunable)
   $\beta = 2$; (default, tunable)
   $\lambda = \alpha^2 \times (\mathbf{L} + k_i - \mathbf{L})$; (scaling factor)
   $c = \mathbf{L} + \lambda$; (scaling factor)
   $\mathbf{W}_m = (\frac{\lambda}{c} + \frac{0.5}{c} + \text{zeros}(1, 2 \times \mathbf{L}))$; (weights for means)
   $\mathbf{W}_c = \mathbf{W}_m$;
   $\mathbf{W}_c(1) = \mathbf{W}_c(1) + (1 - alpha^2 + \beta)$; weights for covariance
   $c = \sqrt{c}$;
   $\mathbf{P}_1 = \mathbf{P} + \mathbf{Q}$;
   $X = \text{sigmas } (\mathbf{x}, \mathbf{P}, c)$; (sigma points around **x**)
   $[\mathbf{x}_1, X1, \mathbf{P}_1, X2] = ut(fstate, X, \mathbf{W}_m, \mathbf{W}_c, \mathbf{L}, \mathbf{Q})$; (unscented transformation of process)
   $X2 = X - \mathbf{x} (:, \text{ones}(1, 2 \times \mathbf{L} + 1))$;
   $[\mathbf{z}_1, Z1, \mathbf{P}_2, Z2] = ut(hmeas, X, \mathbf{W}_m, \mathbf{W}_c, \mathbf{m}, \mathbf{R}, \mathbf{h})$; (unscented transformation of measurements)
   $P12 = X2 \times \text{diag}(\mathbf{W}_c) \times Z2'$; (transformed cross-covariance)
   $\mathbf{R} = \text{chol}(\mathbf{P}2)$; (where chol is the Cholesky factorization)
   $K = (\frac{P12/\mathbf{R}}{\mathbf{R}'})$; (filter gain)
   $K = P12 \times inv(\mathbf{P}2)$;
   $\mathbf{x} = \mathbf{x} + K \times (\mathbf{z} - \mathbf{z}1)$; (state update)
   $\mathbf{P} = \mathbf{P}1 - K \times P12'$; (covariance update)

Function 2: Unscented transformation
   $function[y, Y, \mathbf{P}, Y1] = ut(\mathbf{F}, X, \mathbf{W}_m, \mathbf{W}_c, n, \mathbf{R}, \mathbf{h})$
Inputs:
   f: nonlinear map
   X: sigma points
   $\mathbf{W}_m$: weights for mean
   $\mathbf{W}_c$: weights for covariance
   n: number of outputs of f
   R: additive covariance
Outputs:
   y: transformed mean
   Y: transformed sampling points
   P: transformed covariance
   $Y1$: transformed deviations

$\mathbf{L} = size(X, 2);$
$y = zeros(n, 1);$
$Y = zeros(n, L);$
$x = [\mathbf{h}];$
for k = 1:$\mathbf{L}$
    $Y(k) = \mathbf{x}(\mathbf{h});$
    $y = y + \mathbf{W}_m(k) \times Y(k);$
end
$Y1 = Y - y; \mathbf{P} = Y1 \times diag(\mathbf{W}_c) \times Y1' + \mathbf{R};$

Function 3: Sigma Points

function $X = sigmas(\mathbf{x}, \mathbf{P}, c)$ Sigma points around reference point
Inputs:
x: reference point
$\mathbf{P}$: covariance
c: coefficient
Outputs:
X: Sigma points
$A = c * chol(\mathbf{P})';$
$Y = \mathbf{x}(:, ones(1, numel(\mathbf{x})));$
$X = [\mathbf{x}Y + AY - A];$

A series of simulation runs was conducted on the quadruple tank system to evaluate and compare the effectiveness of the decentralized and centralized multi-sensor integration approaches based on the unscented Kalman filter (UKF) data fusion algorithm. When performing the different runs, the same fault scenarios were used.

2.5.4   UKF with centralized multi-sensor data fusion

The simulation results of the unscented Kalman filter embedded in the centralized structure of the multi-sensor data fusion technique are depicted in Figure 2.6. Although the centralized structure was able to estimate the fault, there was a considerable offset in the estimation.

2.5.5   UKF with decentralized multi-sensor data fusion

The simulation results of the unscented Kalman filter embedded in the decentralized structure of the multi-sensor data fusion technique are depicted in Figure 2.7 and Figure 2.8. It is clear that the decentralized structure, with increasing precision and a more detailed fault picture, was able to estimate the fault in a much better way than the centralized architecture.

2.5.6   Drift detection

A fault may occur in any phase and in any part of the plant. Critical faults not detected on time can lead to adverse effects. It is clear from Figure 2.9 that the fault is so incipient that the level of water is achieving the same height, except at the beginning. Thus, drift detection can

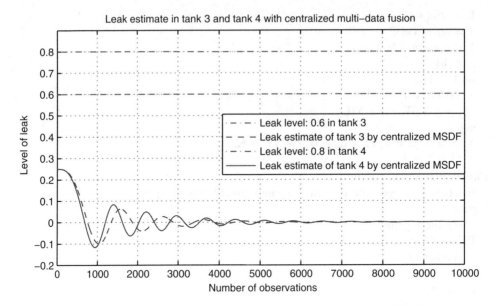

Figure 2.6 Quadruple tank system: Leak estimate for Tanks 3 and 4 with a centralized UKF MSDF approach

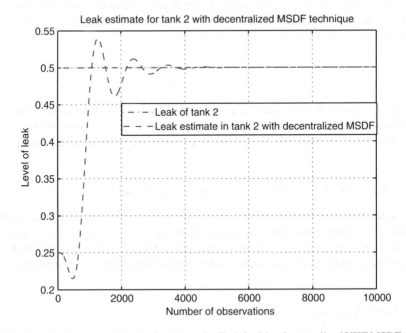

Figure 2.7 Quadruple tank system: Leak estimate for Tank 2 with a decentralized UKF MSDF approach

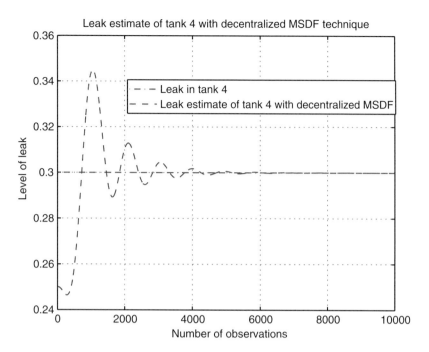

Figure 2.8   Quadruple tank system: Leak estimate for Tank 4 with a decentralized UKF MSDF approach

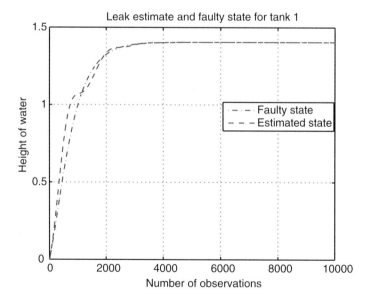

Figure 2.9   Quadruple tank system: Leak and fault estimate for Tank 1

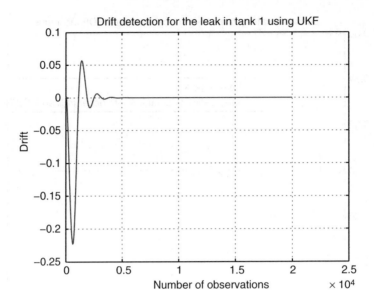

Figure 2.10   Quadruple tank system: drift detection for the leak in Tank 1

give us a better picture for the fault scenario, as shown in Figure 2.10. The kinks showing the middle of the height achievement can alert the engineer to unusual practice in the process.

2.6   Industrial Utility Boiler

The utility boilers in Syncrude Canada are water-tube drum boilers. Since steam is used for generating electricity and process applications, the demand for steam varies. The control objective of the co-generation system is to track the demand for steam while maintaining the pressure and temperature of the header at their respective set points. Table 2.3 lists the variables used in the system.

The principal input variables are u_1, u_2, and u_3. The states are x_1, x_2, x_3, x_4, and x_5. The principal output variables are y_1, y_2, and y_3 [46]. The schematic diagram of the utility boiler can be seen in Figure 2.11.

2.6.1   Steam flow dynamics

Steam flow plays an important role in drum-boiler dynamics. Steam flow from the drum to the header, through the super heaters, is assumed to be a function of the pressure drop from the drum to the header. We use a modified form of Bernoulli's law to represent flow in terms of pressure, with friction [47]. This expression is written as:

$$q_s = K \sqrt{x_2^2 - P_{header}^2} \tag{2.44}$$

Table 2.3   Nomenclature for UB

Symbol	Function
u_1	Feed water flow rate (kg/s)
u_2	Fuel flow rate (kg/s)
u_3	Attemperator spray flow rate (kg/s)
y_1	Drum level (m)
y_2	Drum pressure (kPa)
y_3	Steam temperature (°C)
x_1	Fluid density of system
x_2	Upstream pressure
x_3	Water flow input
x_4	Fuel flow input
x_5	Spray flow input
P_{header}	Downstream pressure
q_s	Steam mass flow rate
V_T	Total volume of system

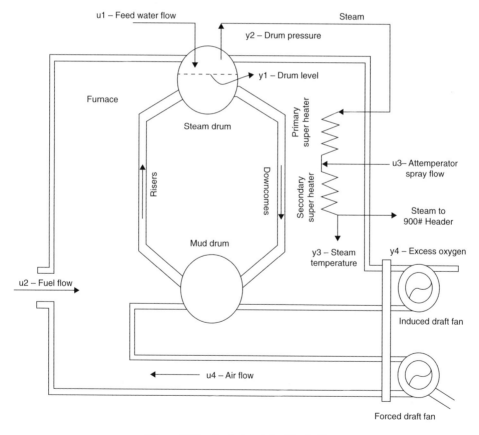

Figure 2.11   An industrial utility boiler

where q_s is the steam mass flow rate, K is a constant, and x_2 and P_{header} are the upstream and downstream pressures, respectively. The constant K is chosen to produce agreement between the measured flow and the pressure drop at a reference condition. For the real system $P_{header} = 6306$ kPa; by measuring the steam flow and drum pressure in the real system, the value of K is identified and the steam flow in the system can be modeled as:

$$q_s = 0.03\sqrt{x_2^2 - 6306^2} \tag{2.45}$$

2.6.2 Drum pressure dynamics

To model the pressure dynamics, we first have to observe the behavior of the system. We apply step inputs to the three inputs at different operating points and look out for a step increase in the feedwater and fuel flow: the system behaves like a first-order system with the same time constant. If we apply a step input only to the spray flow, the system behaves like a first-order system with different time constants. The dynamics for the drum pressure are chosen as follows:

$$\dot{x}_2 = (c_1 x_2 + c_2)q_s + c_3 u_1 + c_4 u_2 \tag{2.46}$$

$$y_2 = x_2 \tag{2.47}$$

Finally, the dynamics of the drum pressure can be modeled as:

$$\dot{x}_2 = (-1.8506 \times 10^{-7} x_2 - 0.0024)\sqrt{x_2^2 - (6306)^2}$$
$$-0.0404 u_1 + 3.025 u_2$$
$$y_2(t) = x_2(t) + p_0 \tag{2.48}$$

where $p_0 = 8.0715$, $p_0 = -0.6449$ and $p_0 = -6.8555$ for low, normal and high loads, respectively. At the three operating points, the initial conditions are $x_{2_0} = 6523.6$, $x_{2_0} = 6711.5$ and $x_{2_0} = 6887.9$ for low, normal and high loads, respectively.

2.6.3 Drum level dynamics

Identification of the water level dynamics is a difficult task. Applying a step input to each input separately shows that the level dynamics is unstable. As the water flow rate increases, the water level increases; as the fuel flow increases, the water level decreases. Three inputs (water flow, fuel flow and steam flow) affect the drum water level. Let x_1 and V_T denote the fluid density and total volume of the system. Then we have

$$\dot{x}_1 = \frac{u_1 - q_s}{V_T} \tag{2.49}$$

where $V_T = 155.1411$. In several experiments, it was observed that the dynamics of the drum level can be given by:

$$y_1 = c_5 x_1 + c_6 q_5 + c_7 u_5 + c_8 u_2 + c_9 \tag{2.50}$$

The constants $c_i, i = 5, \ldots, 9$ should be identified from the plant data. The initial values of x_1 at the three operating points are given by $x_{1_0} = 678.15$, $x_{1_0} = 667.1$, and $x_{1_0} = 654.628$ for low, normal and high loads, respectively.

2.6.4   Steam temperature

In the utility boiler, the steam temperature must be kept at a certain level to avoid overheating the super-heaters. To identify a model for the steam temperature, first-step identification is used:

- When we apply a step to the water flow input, the steam temperature increases, and the steam temperature dynamics behaves like a first-order system.
- When we apply a step to the fuel flow input, the steam temperature increases and the system behaves like a second-order system.
- When we apply a step to the spray flow input, the steam temperature decreases and the system behaves like a first-order system.

Thus, a third-order system is selected for the steam temperature model. This step identification gives an initial estimate for local time constants and gains. By considering the steam flow as an input and applying an input pseudo-random binary sequence (PRBS) at the three operating points, local linear models for the steam temperature dynamics are defined. Combining the local linear models, the following nonlinear model is identified with good fitness for all three operating points:

$$\dot{x}_3(t) = (-0.0211\sqrt{x_2^2 - (6306)^2} + x_4 - 0.0010967 u_1$$
$$+ 0.0475 u_2 + 3.1846 u_3 \tag{2.51}$$

$$\dot{x}_4(t) = 0.0015\sqrt{x_2^2 - (6306)^2} + x_5 + 0.001 u_1$$
$$+ 0.32 u_2 - 2.9461 u_3 \tag{2.52}$$

$$\dot{x}_5(t) = -1.278 \times 10^{-3}\sqrt{x_2^2 - (6306)^2} - 0.00025831$$
$$x_3 - 0.29747 x_4 - 0.8787621548 x_5$$
$$0.00082 u_1 - 0.2652778 u_2 + 2.491 u_3 \tag{2.53}$$

$$y_3 = x_3 + T_0 \tag{2.54}$$

where $T_0 = 443.3579$, $T_0 = 446.4321$, and $T_0 = 441.9055$ for low load, normal load and high load, respectively. At the three operating points, we have $x_{3_o} = 42.2529$, $x_{4_o} = 3.454$, and $x_{5_o} = 3.45082$ for low load; $x_{3_o} = 49.0917$, $x_{4_o} = 2.9012$, and $x_{5_o} = 2.9862$ for normal load; and $x_{3_o} = 43.3588$, $x_{4_o} = -0.1347$, and $x_{5_o} = -0.2509$ for high load. Combining these results gives the model for the utility boiler. In addition, the following limit constraints exist for the three control variables:

$$0 \leq u_1 \leq 120, \quad 0 \leq u_2 \leq 7 \tag{2.55}$$

$$0 \leq u_3 \leq 10 \tag{2.56}$$

2.6.5   Fault model for the utility boiler

To construct a fault model for the utility boiler, extra holes are added to each tank. The mathematical model of the faulty utility boiler can be given as follows where steam pressure faults are present in states 4 and 5:

$$\dot{x}_1(t) = \frac{u_1 - 0.03\sqrt{x_2^2 - (6306)^2}}{155.1411} \tag{2.57}$$

$$\dot{x}_2(t) = (-1.8506 \times 10^{-7} x_2 - 0.0024)\sqrt{x_2^2 - (6306)^2}$$
$$- 0.0404 u_1 + 3.025 u_2 \tag{2.58}$$

$$\dot{x}_3(t) = -0.0211\sqrt{x_2^2 - (6306)^2} + x_4 - 0.0010967 u_1$$
$$+ 0.0475 u_2 + 3.1846 u_3 \tag{2.59}$$

$$\dot{x}_4(t) = 0.0015\sqrt{x_2^2 - (6306)^2} + x_5 - 0.001 u_1$$
$$+ 0.32 u_2 - 2.9461 u_3$$
$$+ (a_{st\ pr})\sqrt{x_2^2 - (6306)^2} \tag{2.60}$$

$$\dot{x}_5(t) = -1.278 \times 10^{-3}\sqrt{x_2^2 - (6306)^2}$$
$$- 0.00025831\ x_3 - 0.29747\ x_4$$
$$- 0.8787621548\ x_5 - 0.00082\ u_1 - 0.2652778$$
$$u_2 + 2.491\ u_3$$
$$+ (a_{st\ pr})\sqrt{x_2^2 - (6306)^2} \tag{2.61}$$

2.6.6   Fault scenarios in the utility boiler

Two fault scenarios are created when using the utility boiler in the simulation program. In these scenarios, steam pressure faults are added in states 4 and 5, resulting in a more uncontrolled nonlinear system.

To ensure the flexibility of UKF-based fault estimation and monitoring, a series of experiments was also performed on the industrial utility boiler system to evaluate and compare the effectiveness of the multi-sensor centralized and decentralized integration approaches based on the unscented Kalman filter data fusion algorithm. A series of simulation runs was performed with a fault in state 4, in the form of increased steam temperature.

2.6.7   UKF with centralized multi-sensor data fusion

The simulation results of the unscented Kalman filter embedded in the centralized structure of the multi-sensor data fusion technique are depicted in Figure 2.12. It can be seen that there is a considerable offset in the estimation.

2.6.8   UKF with decentralized multi-sensor data fusion

The results of the unscented Kalman filter embedded in the decentralized structure of the multi-sensor data fusion technique can be seen in Figure 2.13.

A comparison of the fault estimation of the centralized and decentralized schemes is depicted in Figure 2.14. It is clear that the centralized estimate is better that the decentralized estimate.

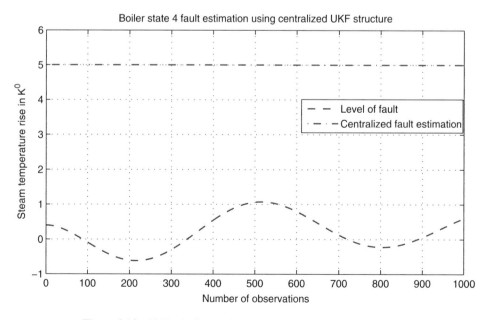

Figure 2.12   Utility boiler: Estimate of state 4 using centralized UKF

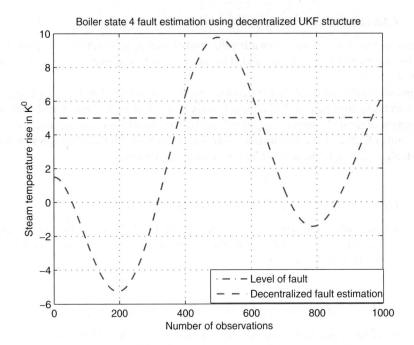

Figure 2.13   Utility boiler: Estimate of state 4 with decentralized UKF

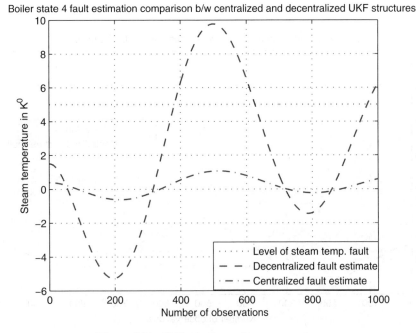

Figure 2.14   Utility boiler: estimate comparisons

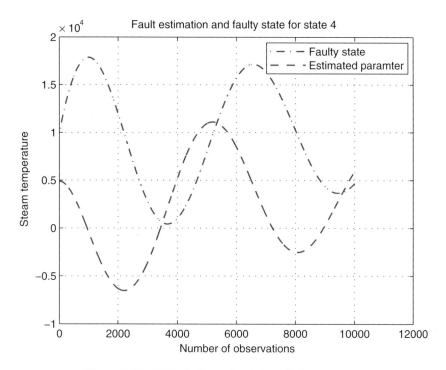

Figure 2.15   Utility boiler: estimated and fault parameters

2.6.9   Drift detection

Several faults may occur in any part of a utility boiler. Critical faults not detected on time can lead to adverse effects. This section shows the drift detection of the faults using unscented Kalman filter.

Figure 2.15 shows the estimated parameter and the fault parameter. It can be seen that there is a difference between them although they follow the same pattern. By drift detection (see Figure 2.16), we can see the prominent kinks in the profile of the faulty parameter estimation, thus giving sufficient sign of action being necessary.

2.6.10   Remarks

In the simulations, a multi-sensor data fusion technique is implemented with both centralized and decentralized structures of UKF. They operate on the quadruple tank system in two leakage fault scenarios and on the utility boiler in an uncontrolled steam pressure fault scenario. In the case of both physical systems, it has been shown that the decentralized structure gives better results than the centralized structure. Drift detection is also effective, showing prominent kinks for the faults. The effectiveness of the decentralized structure for the utility boiler was less than for the quadruple tank system, because of the large steam pressure fault introduced in states 4 and 5.

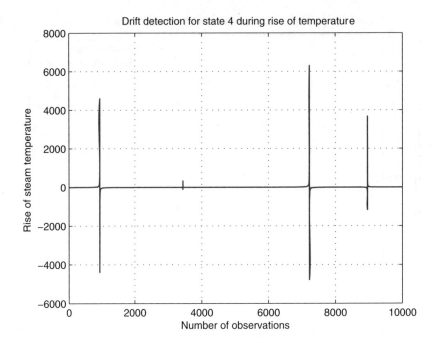

Figure 2.16   Utility boiler: drift detection

2.7   Notes

In this chapter, the problem of designing a fault detection filter has been approached by the integration of three major steps: an optimal residual generator, a residual evaluator, and the implementation of UKF. The optimal residual generator has been derived with the assumption of no modeling errors and is further improved by residual evaluation. A sufficient condition for the existence of a robust fault detection filter has been established and we cast the solution in terms of linear matrix inequality (LMI), which helped in the determination of an adaptive threshold for fault detection. The UKF has been incorporated for the iterative update of observation noise mean and covariance in order to provide for a robust fault detection filter with tolerance to faults, with the help of hypothesis testing computation.

Detailed numerical simulation and experimentation were performed on the design of a robust fault detection filter for an industrial utility boiler. The ensuing results demonstrated the effectiveness of this approach.

References

[1] Hameed, I. A. (2010) "Using the extended Kalman filter to improve the efficiency of greenhouse climate control", *Int. J. Innovative Computing, Information and Control*, **6**(6):2671–2680.
[2] Palangi, H., and Refan, M. H. (2010) "Error reduction of a low cost GPS receiver for kinematic applications based on a new Kalman filtering algorithm", *Int. J. Innovative Computing, Information and Control*, **6**(8):3775–3786.

[3] Ohsumi, A., Kimura Tand Kono, M. (2009) "Kalman filter-based identification of unknown exogenous input of stochastic linear systems via pseudo measurement approach", *Int J. Innovative Computing, Information and Control*, **5**(1):1–16.

[4] Jun, S., Yang, M., Yao, X., and Zhong, R. (2010) "A new two-stage Kalman filter method for chaos system", *ICIC Express Letters*, **4**(2):539–546.

[5] Lu, C. L., Chung, Y. N., Chih-Min Lin, C. M., Yu, C. C., and Chen, T. R. (2010) "Applying Kalman filter-based fusion algorithm to estimation problems", *ICIC Express Letters*, **4**(6A):2109–2114.

[6] Pineda-Sanchez, M., Riera-Guasp, M., Roger-Folch, J., Antonino-Daviu, J. A. et al. (2011) "Diagnosis of induction motor faults in time-varying conditions using the polynomial-phase transform of the current", *IEEE Trans. Ind. Electronics*, **58**(4):1428–1439.

[7] Wolbank, T. M., Nussbaumer, P., Chen, H., and Macheiner, P. E. (2011) "Monitoring of rotor-bar defects in inverter-fed induction machines at zero load and speed", *IEEE Trans. Ind. Electronics*, **58**(5):1468–1478.

[8] Bianchini, C., Immovilli, F., Cocconcelli, M., Rubini, R., and Bellini, A. (2011) "Fault detection of linear bearings in brushless AC linear motors by vibration analysis", *IEEE Trans. Ind. Electronics*, **58**(5):1684–1694.

[9] Henao, H., Fatemi S. M. J. R., Capolino, G. A., and Sieg-Zieba, S. (2011) "Wire rope fault detection in a hoisting winch system by motor torque and current signature analysis", *IEEE Trans. Ind. Electronics*, **58**(5):1707–1717.

[10] Isermann, R. (2006) *Fault-diagnosis Systems: An Introduction from Fault Detection to Fault Tolerance*, Berlin: Springer.

[11] Chow, E. Y., and Willsky, A. S. (1984) "Analytical redundancy and the design of robust failure detection systems", *IEEE Trans. Automatic Control*, **29**(7):603–614.

[12] Gertler, J. J. (1998) *Fault Detection and Diagnosis in Engineering Systems*, New York: CRC Press.

[13] Frank, P. M., Ding, S. X., and Seliger, B. K. (2000) "Current developments in the theory of FDI", in *Proc. IFAC Symposium on Fault Detection, Supervision and Safety of Technical Processes*, Budapest, Hungary, **1**:16–27.

[14] Simani, S., Fantuzzi, C., and Patton, R. (2003) "Model-based fault diagnosis in dynamic systems using identification techniques", *Advances in Industrial Control*, Springer, UK.

[15] Zhang, G. Q. P. (2000) "Neural networks for classification: a survey", *IEEE Trans. Systems, Man and Cybernetics: Part C–Applications and Reviews*, **30**(4):451–462.

[16] Wang, Y., Chan, C. W., and Cheung, K. C. (2001) "Intelligent fault diagnosis based on neuro-fuzzy networks for nonlinear dynamic systems", in *Proc. IFAC Conference on New Technologies for Computer Control*, Hong Kong, China, 101–104.

[17] Watanabe, K., Matsura, I., Abe, M., Kubota, M., and Himmelblau, D. M. (1989) "Incipient fault diagnosis of chemical processes via artificial neural networks", *AICHE J.*, **35**(11):1803–1812.

[18] Venkatasubramanian, V., and Chan, K. (1989) "A neural network methodology for process fault diagnosis", *AICHE J.*, **35**(12):1993–2002.

[19] Ungar, L. H., Powell, B. A., and Kamens, S. N. (1990) "Adaptive networks for fault diagnosis and process control", *Computers and Chem. Eng.*, **14**(4–5):561–572.

[20] Venkatasubramanian, V., Vaidyanathan, R., and Yamamoto, Y. (1990) "Process fault detection and diagnosis using neural networks: Steady state processes", *Computers and Chem. Eng.*, **14**(7):699–712.

[21] de Miguel, L. J., and Blazquez, L. F. (2005) "Fuzzy logic-based decision-making for fault diagnosis in a DC motor", *Engineering Applications of Artificial Intelligence*, **18**(4):423–450.

[22] Chow, M.-Y. (2000) "Special Section on Motor Fault Detection and Diagnosis", *IEEE Trans. Ind. Electron.*, **47**(5):982–1107.

[23] Ramesh, T. S., Davis, J. F., and Schwenzer, G. M. (1992) "Knowledge-based diagnostic systems for continuous process operations based upon the task framework", *Computers and Chem. Eng.*, **16**(2):109–127.

[24] Venkatasubramanian, V., Rengaswamy, R., Yin, K., and Kavuri, S. N. (2003) "A review of process fault detection and diagnosis part II: quantitative model based methods", *Computers and Chemical Engineering.*, **27**:313–326.

[25] Mehranbod, N. (2002) *A Probabilistic Approach for Sensor Fault Detection and Identification*, PhD thesis, Faculty of Drexel University, USA.

[26] Kirch, H., and Kroschel, K. (1994) "Applying Bayesian networks to fault diagnosis", in *Proc. IEEE Conference on Control Applications*, 895–901.

[27] Rojas-Guzman, C., and Kramer, M. (1993) "GALGO: A genetic algorithm decision support tool for complex uncertain systems modeled with Bayesian belief networks", in *Proc. 9th Conf. on Uncertainty in Artif. Intell.*, 368–374.

[28] Santoso, N., Darken, C., and Povh, G. (1999) "Nuclear plant fault diagnosis using probabilistic reasoning", *IEEE Power Engineering Society Summer Meeting*, **2**:714–719.

[29] Nicholson, A., and Brady, J. (1994) "Dynamic belief networks for discrete monitoring", *IEEE Trans. Systems, Man, and Cybernetics*, **24**(11):1593–1599.

[30] Nicholson, A. (1996) *Fall Diagnosis Using Dynamic Belief Networks*, Lecture Notes in Computer Science, Springer, UK.

[31] Zhang, B., Sconyers, C., Byington, C., Patrick, R., et al. (2011) "A probabilistic fault detection approach: application to bearing fault detection ", *IEEE Trans. Ind. Electronics.*, **58**(5):2011–2018.

[32] Wan, E. A., der Merwe, R. V., and Nelson, A. T. (2000) "Dual estimation and the unscented transformation', in *Advances in Neural Information Processing Systems*, S. A. Solla, T. K. Leen, and K. R. Miller (Eds), Cambridge, MA: MIT Press, 666–672.

[33] Van der Merwe, R. (2004) *Sigma-Point Kalman Filters for Probability Inference in Dynamic State-Space Models*, PhD Thesis, Oregon Health and Science University.

[34] Gan, Q., and Harris, C. J. (2001) "Comparison of two measurement fusion methods for Kalman-filter-based multi-sensor data fusion", *IEEE Trans. Aerospace Electr. Systems*, **37**(1):273–280.

[35] Harris, C., Hong, X., and Gan, Q. (2002) *Adaptive Modeling, Estimation and Fusion from Data: A Neurofuzzy Approach*, Springer.

[36] Chen, J., and Patton, R. J. (1999) *Robust Model-Based Fault Diagnosis for Dynamic Systems*, Boston, Mass, USA: Kluwer Academic Publishers.

[37] Frank, P. M., and Ding, X. (1997) "Survey of robust residual generation and evaluation methods in observer-based fault detection systems", *J. Process Control*, **7**(6):403–424.

[38] Izadi, I., Chen, T. W., and Zhao, Q. (2005) "Norm-invariant discretization for sampled-data fault detection", *Automatica*, **41**:1633–1637.

[39] Li, Z., and Jaimoukha, I. M. (2009) "Observer-based fault detection and isolation filter design for linear time-invariant systems", *Int. J. Control*, **82**(1):171–182.

[40] Caliskan, F., and Hajiyev, C. M. (2000) "EKF based surface fault detection and reconfiguration in aircraft control systems", in *Proc. 2000 American Control Conference*, **2**:1220–1224.

[41] Jiang, B., and Chowdhury, F. N. (2005) "Fault estimation and accommodation for linear MIMO discrete-time systems", *IEEE Trans. Control Systems Technology*, **13**(3):493–499.

[42] Zhang, Y. M., and Jiang, J. (1999) "Design of integrated fault detection, diagnosis and reconfigurable control systems", in *Proc. 38th IEEE Conference on Decision and Control*, Phoenix, USA, **4**:3587–3592.

[43] Liu, G. P., Yang, J. B., and Whidborne, J. F. (2001) *Multiobjective Optimization and Control*, Research Studies Press Ltd., Baldock.

[44] Patton, R. J., Chen, J., and Miller J. H. P. (1991) "A robust disturbance decoupling approach to fault detection in process system", in *Proc. 30th Conference on Decision and Control*, Brighton, England, 1543–1548.

[45] Dai, L., and Astrom, K. (1999) Dynamic matrix control of a quadruple tank process, in *Proc. 14th IFAC*, 295–300.

[46] Tan, W., Marquez, H. J., and Chen, T. (2002) "Multivariable robust controller design for a boiler system", *IEEE Trans. Control Systems Technology*, **10**(5):735–742.

[47] Perry, R. H., and Green, D. W. (1997) *Perry's Chemical Engineers' Handbook*, 7th Edition, McGraw-Hill, USA.

3

Robust Fault Detection

3.1 Distributed Fault Diagnosis

A consequence of modern technological advances is the creation of distributed interconnected control systems. Examples of such systems include advanced automotive control systems (which consist of over 50 interconnected microcontrollers coordinating various vehicle functionalities), intelligent vehicle highway systems [1], formation control systems for unmanned aerial vehicles [2], and interconnected critical infrastructure systems (e.g. power generation and distribution systems, telecommunication networks, water distribution networks). Since such interconnected systems need to operate reliably at all times, despite the possible occurrence of faulty behaviours in some subsystems, the design of fault diagnosis and accommodation schemes is a crucial step in achieving reliable and safe operation.

3.1.1 Introduction

In the last 25 years, there have been significant research activities in the design and analysis of fault diagnosis and accommodation schemes (see, for instance, [3]–[6] and the references cited therein). The idea of using adaptive and learning techniques for fault diagnosis and accommodation has been proposed by several researchers [7, 8, 9, 10, 11]. Most of these methods are based on a centralized fault diagnosis architecture. In practice, it is very difficult to address the problem of diagnosing faults in interconnected systems using a centralized architecture because of the constraints on computational capabilities and communication bandwidth. Consequently, in recent years, the area of distributed or decentralized fault diagnosis has attracted increasing attention [2, 12, 13, 14, 15].

In previous works [16, 17], a centralized fault detection and isolation (FDI) methodology for nonlinear uncertain systems has been developed. This chapter significantly extends the previous results by developing a distributed FDI scheme for a class of interconnected nonlinear uncertain systems. Because of the interactions among subsystems and the limitation of information that is available for each subsystem, the problem of distributed FDI is much more challenging. In the distributed FDI architecture we present, a fault diagnostic component is designed for each subsystem in the interconnected system by utilizing local measurements and certain

Analysis and Synthesis of Fault-Tolerant Control Systems, First Edition. Magdi S. Mahmoud and Yuanqing Xia.
© 2014 John Wiley & Sons, Ltd. Published 2014 by John Wiley & Sons, Ltd.

communicated information from neighbouring FDI components associated with its directly interconnected subsystems. The distributed FDI method is presented with an analytical framework aiming at characterizing its important properties. Specifically, the analysis focuses on:

- derivation of adaptive thresholds for distributed fault detection and fault isolation, ensuring the robustness property with respect to interactions among interconnected subsystems and modeling uncertainty;
- investigation of fault detectability and isolability conditions, characterizing the class of faults in each subsystem that are detectable and isolable by the proposed method;
- investigation of stability and learning capability of local adaptive fault isolation estimators (FIEs) designed for each subsystem.

3.1.2   System model

Consider a nonlinear dynamic system composed of M interconnected subsystems with the dynamics of the ith subsystem, $i = 1, \ldots, M$, being described by the following differential equation

$$\dot{z}_i = A_i z_i + \zeta_i(z_i, u_i) + \varphi_i(z_i, u_i, t) + \beta_i(t - T_0) E_i \bar{f}_i(z_i, u_i)$$

$$+ \sum_{j=1}^{M} [h_{ij}(z_i, z_j, u_i, u_j) + d_{ij}(z_i, z_j, u_i, u_j)]$$

$$y_i = \bar{C}_i z_i \tag{3.1}$$

where $z_i \in \mathfrak{R}^{n_i}$, $u_i \in \mathfrak{R}^{m_i}$ and $y_i \in \mathfrak{R}^{l_i}$ are the state vector, input vector and output vector, respectively, of the ith subsystem ($n_i \geq l_i$); $\zeta_i : \mathfrak{R}^{n_i} \times \mathfrak{R}^{m_i} \rightarrow \mathfrak{R}^{n_i}$; $\varphi_i : \mathfrak{R}^{n_i} \times \mathfrak{R}^{m_i} \times \mathfrak{R}^+ \rightarrow \mathfrak{R}^{n_i}$; $\bar{f}_i : \mathfrak{R}^{n_i} \times \mathfrak{R}^{m_i} \rightarrow \mathfrak{R}^{q_i}$; and d_{ij} and $h_{ij} : \mathfrak{R}^{n_i} \times \mathfrak{R}^{n_j} \times \mathfrak{R}^{m_i} \times \mathfrak{R}^{m_j} \rightarrow \mathfrak{R}^{n_i}$ are smooth vector fields. Specifically, the model given by

$$\dot{z}_{N_i} = A_i z_{N_i} + \zeta_i(z_{N_i}, u_i)$$

$$y_{N_i} = \bar{C}_i z_{N_i}$$

is the known nominal model of the ith subsystem with ζ_i being the known nonlinearity and the subscript N_i stands for the 'nominal' variables. The vector field φ_i in Equation (3.1) represents the modeling uncertainty of the ith subsystem, and $\beta_i(t - T_0) E_i \bar{f}_i(z_i, u_i)$ denotes the changes in the dynamics of the ith subsystem due to the occurrence of a fault in the local subsystem.

Specifically, $\beta_i(t - T_0)$ is a step function representing the time profile of a fault which occurs at some unknown time T_0 (i.e. $\beta_i(t - T_0) = 0$ if $t < T_0$ and $\beta_i(t - T_0) = 1$ if $t \geq T_0$); $\bar{f}_i(z_i, u_i)$ is a nonlinear fault function; and E_i is a fault distribution matrix. Additionally, the vector fields h_{ij} and d_{ij} represent the direct interconnection between the ith and the jth subsystems. h_{ij} is the known part of the direct interconnection, and d_{ij} is the unknown part. It is noted that likely many functions h_{ij} and d_{ij} are identically zero in an interconnected system (i.e. many subsystems do not directly influence subsystem i). Moreover, $h_{ii} = 0$ and $d_{ii} = 0$ because the interconnection terms are only defined between two subsystems.

Assumption 3.1   *The constant matrices* $E_i \in \Re^{n_i \times q_i}$ *and* $\bar{C}_i \in \Re^{l_i \times n_i}$ *with* $q_i \leq l_i$ *are of full column rank and satisfy the condition rank* $(\bar{C}_i E_i) = q_i$. *Additionally, all the invariant zeros of the system* $(A_i, E_i, \bar{C}_i)$ *are in the left half plane.*
   Then for $i = 1, \ldots, M$, *there exists a change of coordinates* $x_i = [x_{i1}^T \quad x_{i2}^T]^T = T_i z_i$ *with* $x_{i1} \in \Re^{n_i - l_i}$ *and* $x_{i2} \in \Re^{l_i}$, *such that [14]:*

- $T_i E_i = \begin{bmatrix} 0 \\ E_{i2} \end{bmatrix}$, *where* $E_{i2} \in \Re^{l_i \times q_i}$
- $\bar{C}_i T_i^{-1} = [0 \quad C_i]$, *where* $C_i \in \Re^{l_i \times l_i}$ *is orthogonal.*

Therefore, in the new coordinate system, the system model Equation (3.1) is in the form of

$$\dot{x}_{i1} = A_{i1} x_{i1} + A_{i2} x_{i2} + \rho_{i1}(x_i, u_i) + \phi_{i1}(x_i, u_i, t)$$

$$+ \sum_{j=1}^{M} [H_{ij}^1(x_i, x_j, u_i, u_j) + D_{ij}^1(x_i, x_j, u_i, u_j)]$$

$$\dot{x}_{i2} = A_{i3} x_{i1} + A_{i4} x_{i2} + \rho_{i2}(x_i, u_i) + \phi_{i2}(x_i, u_i, t)$$

$$+ \beta_i(t - T_0) E_{i2} \bar{f}_i(x_i, u_i) \tag{3.2}$$

$$+ \sum_{j=1}^{M} [H_{ij}^2(x_i, x_j, u_i, u_j) + D_{ij}^2(x_i, x_j, u_i, u_j)]$$

$$y_i = C_i x_{i2},$$

where

$$\begin{bmatrix} A_{i1} & A_{i2} \\ A_{i3} & A_{i4} \end{bmatrix} = T_i A_i T_i^{-1}, \begin{bmatrix} \rho_{i1} \\ \rho_{i2} \end{bmatrix} = T_i \zeta_i, \begin{bmatrix} \phi_{i1} \\ \phi_{i2} \end{bmatrix} = T_i \varphi_i,$$

$$\begin{bmatrix} H_{ij}^1 \\ H_{ij}^2 \end{bmatrix} = T_i h_{ij}, \begin{bmatrix} D_{ij}^1 \\ D_{ij}^2 \end{bmatrix} = T_i d_{ij},$$

and the matrix A_{i1} is stable. Let us define $\bar{x}_j$, $\bar{u}_j$ and $\bar{y}_j$ as the vectors comprising of the state variables, input signals, and output variables of those subsystems that have nonzero unknown interconnection terms D_{ij}^1 and D_{ij}^2 with respect to subsystem i. Then, by allowing a more general structure of the fault function, we have

$$\dot{x}_{i1} = A_{i1} x_{i1} + A_{i2} x_{i2} + \rho_{i1}(x_i, u_i) + \eta_{i1}(x_i, \bar{x}_j, u_i, \bar{u}_j, t)$$

$$+ \sum_{j=1}^{M} H_{ij}^1(x_i, x_j, u_i, u_j)$$

$$\dot{x}_{i2} = A_{i3} x_{i1} + A_{i4} x_{i2} + \rho_{i2}(x_i, u_i) + \eta_{i2}(x_i, \bar{x}_j, u_i, \bar{u}_j, t)$$

$$+ \beta_i(t - T_0) f_i(x_i, u_i) + \sum_{j=1}^{M} H_{ij}^2(x_i, x_j, u_i, u_j)$$

$$y_i = C_i x_{i2}, \tag{3.3}$$

where

$$\eta_{i1} \triangleq \phi_{i1} + \sum_{j=1}^{M} D_{ij}^2, \eta_{i2} \triangleq \phi_{i2} + \sum_{j=1}^{M} D_{ij}^2 \, f_i : \Re^{n_i} \times \Re^{m_i} \to \Re^{l_i}$$

is a smooth vector field representing the unstructured nonlinear fault function in each subsystem under consideration, Clearly, Equation (3.1) is a special case of Equation (3.3) with $_i(x_i, u_i) = E_{i2} \bar{f}_i$.

To formulate the fault isolation problem, it is assumed that there are N_i types of fault in the fault set associated with the ith subsystem, $i = 1, \ldots, M$. Specifically, the unknown fault function $f_i(x_i, u_i)$ in Equation (3.3) is assumed to belong to a finite set of fault types given by

$$F_i \triangleq \{f_i^1(x_i, u_i), \ldots, f_i^{N_i}(x_i, u_i)\}. \tag{3.4}$$

Each fault type $f_i^s, s = 1, \ldots, N_i$, is in the form of

$$f_i^s(x_i, u_i) \triangleq [(\theta_{i1}^s(t))^{\top} g_{i1}^s(x_i, u_i), \ldots, (\theta_{il_i}^s(t))^{\top} g_{il_i}^s(x_i, u_i)]^{\top}, \tag{3.5}$$

where $\theta_{ip}^s(t), p = 1, \ldots, l_i$, characterizing the unknown fault magnitude, is a parameter vector assumed to belong to a known compact and convex set Θ_{ip}^s (that is, $\theta_{ip}^s(t) \in \Theta_{ip}^s, \forall t \geq 0$), and g_{ip}^s is a known smooth vector field representing the functional structure of the sth fault affecting the pth component of state vector x_{i2} of the ith subsystem.

For instance, in the case of a leakage fault [16], $\theta_{ip}^s(t)$ characterizes the size of the leakage in a tank and g_{ip}^s represents the functional structure of the fault affecting each state equation.

Remark 3.1 *The fault isolation problem formulated above is motivated by practical consid-erations. In many engineering applications, based on historical data and past experience, the system engineers often have a reasonably good idea of the types of fault that may occur in a particular system. Although different faults may have different nonlinear effects on the system dynamics, for a given type of fault, the uncertainty is often the magnitude of the fault. A well-known fault diagnosis benchmark example, the three-tank system [16], has been considered to motivate the definition of the fault set described by Equation (3.2) and Equation (3.3).*

The objective now is to develop a robust distributed FDI scheme for the class of intercon-nected nonlinear uncertain systems that can be transformed into Equation (3.3). It is worth noting that a new fault, which does not belong to the fault set Equation (3.4), can also be determined based on the presented FDI method if its fault functional structure is sufficiently different. In the discussion, the following assumptions are made:

Assumption 3.2 *The functions η_{i1} and η_{i2} in Equation (3.3), representing the unstructured modeling uncertainty, are unknown nonlinear functions of $x_i, \bar{x}_j, u_i, \bar{u}_j$ and t, but bounded,*

$$|\eta_{i1}(x_i, \bar{x}_j, u_i, \bar{u}_j, t)| \leq \bar{\eta}_{i1}(y_i, \bar{y}_j, u_i, \bar{u}_j, t),$$

$$|\eta_{i2}(x_i, \bar{x}_j, u_i, \bar{u}_j, t)| \leq \bar{\eta}_{i2}(y_i, \bar{y}_j, u_i, \bar{u}_j, t), \tag{3.6}$$

where the bounding functions $\bar{\eta}_{i1}$ and $\bar{\eta}_{i2}$ are known and uniformly bounded in the corresponding compact sets of admissible state variables, inputs and outputs with appropriate dimensions.

Assumption 3.3 *The system state vector x_i of each subsystem remains bounded before and after the occurrence of a fault, i.e. $x_i(t) \in L_\infty, \forall t \geq 0$.*

Assumption 3.4 *The nonlinear terms $\rho_{i1}(x_i, u_i)$ and $\rho_{i2}(x_i, u_i)$ in Equation (3.3) satisfy the following inequalities: $\forall u_i \in U_i$ and $\forall x_i, \hat{x}_i \in X_i$,*

$$|\rho_{i1}(x_i, u_i) - \rho_{i1}(\hat{x}_i, u_i) \leq \sigma_{i1}|x_i - \hat{x}_i| \tag{3.7}$$

$$|\rho_{i2}(x_i, u_i) - \rho_{i2}(\hat{x}_i, u_i)| \leq \sigma_{i2}(y_i, u_i, \hat{x}_i)|x_i - \hat{x}_i| \tag{3.8}$$

where σ_{i1} is a known Lipschitz constant, $\sigma_{i2}(\cdot)$ is a known function that is uniformly bounded, $X_i \subset \Re^{n_i}$ and $U_i \subset \Re^{m_i}$ are compact sets of admissible state variables and inputs, respectively.

Assumption 3.5 *The interconnection terms H_{ij}^1 and H_{ij}^2 satisfy the following condition, i.e. $\forall x_j, \hat{x}_j \in X_j$,*

$$|H_{ij}^1(x_i, x_j, u_i, u_j) - H_{ij}^1(\hat{x}_i, \hat{x}_j, u_i, u_j)| \leq \gamma_{ij}^1|x_j - \hat{x}_j| \tag{3.9}$$

$$|H_{ij}^2(x_i, x_j, u_i, u_j) - H_{ij}^2(\hat{x}_i, \hat{x}_j, u_i, u_j)|$$
$$\leq \gamma_{ij}^2(y_i, y_j, u_i, u_j)|x_j - \hat{x}_j| \tag{3.10}$$

where γ_{ij}^1 is a known Lipschitz constant and γ_{ij}^2 is a known and uniformly bounded function.

Assumption 3.6 *The rate of change of each fault parameter vector $\theta_{ip}^s(t)$ in Equation (3.5) $(s = 1, \ldots, N_i, p = 1, \ldots, l_i)$ is uniformly bounded, i.e. $|\dot{\theta}_i^s(t)| \leq \alpha_i^s$ for all $t \geq 0$, where $\theta_i^s(t) \triangleq [(\theta_{i1}^s(t))^\top, \ldots, (\theta_{il_i}^s(t))^\top]^\top$ and α_i^s is a known constant.*

Assumption 3.7 *The fault function $f_i^s(x_i, u_i)$ satisfies the following condition, i.e. $\forall x_j, \hat{x}_j \in X_j$,*

$$|f_i^s(x_i, u_i) - f_i^s(\hat{x}_i, u_i)| \leq \bar{\omega}_i^s(y_i, u_i)|x_i - \hat{x}_i| \tag{3.11}$$

where $\bar{\omega}_i^s$ is a known and uniformly bounded function.

It is important to observe that Assumption 3.2 characterizes the class of modeling uncertainty under consideration, including various modeling errors in the system's local dynamics (i.e. ϕ_{i1} and ϕ_{i2}) and the unknown part of the interconnection between subsystems (i.e. D_{ij}^1 and

D_{ij}^2). The bounds on the unstructured modeling uncertainty are needed in order to be able to distinguish between the effects of faults and the effects of modeling uncertainty [16, 17]. For instance, in an aircraft engine fault diagnosis application, the modeling uncertainty is the deviation of the actual engine dynamics from a nominal engine model representing the dynamics of a new engine; these deviations result from normal engine component degradation during its service life. Such normal component degradation can be modeled by small changes in certain engine component health parameters (such as, efficiency and flow capacity of the fan, compressor and turbine). Therefore, the bounding function on the modeling uncertainty (that is, $\bar{\eta}_{i1}$ and $\bar{\eta}_{i2}$) can be obtained using knowledge of the normal degradation of these health parameters during a number of flights under the worst case scenario. Additionally, it is worth noting that the modeling uncertainty considered in this chapter is unstructured; distributed fault diagnosis methods in the literature often assume the absence of modeling uncertainty [12] or structured modeling uncertainty [14]. With structured models of the modeling uncertainty, in order to achieve robustness, it is often assumed that certain rank conditions are satisfied by the uncertainty distribution matrix. On the other hand, the utilization of structured uncertainty with additional assumptions on the distribution matrix may allow the design of FDI schemes that completely decouple the fault from the modeling uncertainty.

Assumption 3.3 requires the boundedness of the state variables before and after the occurrence of a fault in each subsystem. Hence, it is assumed that the distributed feedback control system is capable of retaining the boundedness of the state variables of each subsystem even in the presence of a fault. This is a technical assumption required for well-posedness since the distributed FDI design under consideration does not influence the closed-loop dynamics and stability. Also Assumption 3.4 characterizes the type of known nonlinearities of the nominal system dynamics under consideration. Specifically, it is assumed that $\rho_{i1}(x_i, u_i)$ is Lipschitz in u_i and $\rho_{i2}(x_i, u_i)$ satisfies the inequality in Equation (3.8). Note that condition Equation (3.8) is more general than the Lipschitz condition where σ_{i2} is a constant.

On the other hand, Assumption 3.5 requires the interconnection term H_{ij} between subsystems to satisfy a Lipschitz type of condition. Several examples of distributed nonlinear systems with Lipschitz interconnection terms have been considered [1, 12, 18, 19, 20]. Note that H_{ij} is a function of unknown state vectors x_j and x_i.

In Assumption 3.6, known bounds on the rate of change of the fault magnitude $\theta_i^s(t)$ are assumed. In practice, the rate bounds α_i^s can be set by exploiting some *a priori* knowledge on the fault-developing dynamics. Note that the cases of abrupt faults and incipient faults are both covered by the fault model $\beta_i(t - T_0)f_i$ under consideration. Specifically, the fault time profile function $\beta_i(t - T_0)$ is a step function modeling abrupt characteristics of the fault, and the fault magnitude $\theta_i^s(t)$ represents the (possibly time-varying) fault magnitude. For instance, in the case of foreign object damage to the fan of an aircraft engine, the function $\beta_i(t - T_0)$ models the sudden and immediate effect of the damage, and $\theta_i^s(t)$ captures the possibly time-varying development of the fault magnitude following the initial sudden damage. In the specifical case of abrupt faults, we can simply set $\alpha_i^s = 0$ (i.e. θ_i^s is a vector of constants), and the function $\beta_i(t - T_0)f_i$ models the abrupt behaviour of the fault.

Finally, Assumption 3.7 assumes the fault function f_i satisfies the condition given by Equation (3.11). This is needed for the design and analysis of the distributed adaptive FDI algorithm, since the fault function f_i is also a function of the unknown state variables x_i. In the special case that the fault is a function of measurable output y_i and known input u_i, we simply have $\bar{\omega}_i^s = 0$.

Remark 3.2   *An interesting distributed fault estimation method was developed by Yan and Edwards [14] based on sliding-mode observer techniques. Their approach assumes a known bound on the fault function and utilizes a structured model of modeling uncertainty with additional assumptions on the distribution matrices of the modeling uncertainty, which allows the complete decoupling of faults from modeling uncertainty. In this section, we consider a different problem of distributed fault isolation for nonlinear systems with different fault models and unstructured modeling uncertainty based on adaptive estimation techniques. The objective is to detect the occurrence of any faults and to determine if a fault (either in the fault set Equation (3.4) or a new fault that does not belong to that set) has occurred. In addition, in the related work of Zhang, Polycarpou, and Parisini [16, 17], fault diagnosis schemes for nonlinear systems utilizing a centralized architecture were developed. Therefore, in the following discussion, the problem of distributed fault diagnosis for interconnected nonlinear systems is investigated. With the interconnection terms H_{ij}^1 and H_{ij}^2 among subsystems, Equation (3.3), and the presence of unstructured modeling uncertainty, the design and analysis of distributed fault diagnosis methods is much more challenging than that of centralized fault diagnosis methods.*

3.1.3   Distributed FDI architecture

The distributed FDI architecture is comprised of M local FDI components, with one FDI component designed for each of the M subsystems. The objective of each FDI component is to detect and isolate faults in the corresponding local subsystem. Specifically, each local FDI component consists of a fault detection estimator (FDE) and a bank of N_i nonlinear adaptive FIEs, where N_i is the number of nonlinear fault types in the fault set F_i associated with the corresponding ith subsystem of Equation (3.4), $i = 1, \ldots, M$. Under normal conditions, each local FDE monitors the corresponding local subsystem to detect the occurrence of any fault. If a fault is detected in a particular subsystem i, then the corresponding N_i local FIEs are activated to determine the particular type of fault that has occurred in the subsystem.

The FDI architecture for each subsystem follows the generalized observer scheme architectural framework that is well-documented in the fault diagnosis literature [3]. The distributed nature of the FDI method presented here can be better understood if compared with conventional centralized FDI architecture. For M interconnected subsystems and a centralized FDI architecture, $N_1 + N_2 + \ldots, +N_M$ FIEs are needed at the server node. With the distributed FDI architecture, N_i FIEs are needed at the ith subsystem. Hence, the computation is distributed across the subsystems in the network. In addition, it is important to note that, in the proposed distributed FDI architecture, communication is only needed among the FDI components associated with subsystems that are directly interconnected.

3.1.4   Distributed fault detection method

In this section, we investigate the distributed fault detection method, including the designs of each local FDE for residual generation and the corresponding adaptive thresholds for residual evaluation.

Based on the subsystem model described by Equation (3.3), the FDE for each local subsystem is chosen as:

$$\dot{\hat{x}}_{i1} = A_{i1}\hat{x}_{i1} + A_{i2}C_i^{-1}y_i + \rho_{i1}(\hat{x}_i, u_i)$$

$$+ \sum_{j=1}^{M} H_{ij}^1(\hat{x}_i, \hat{x}_j, u_i, u_j)$$

$$\dot{\hat{x}}_{i2} = A_{i3}\hat{x}_{i1} + A_{i4}\hat{x}_{i2} + \rho_{i2}(\hat{x}_i, u_i) + L_i(y_i - \hat{y}_i) \qquad (3.12)$$

$$+ \sum_{j=1}^{M} H_{ij}^2(\hat{x}_i, \hat{x}_j, u_i, u_j)$$

$$\hat{y}_i = C_i \hat{x}_{i2}$$

where $\hat{x}_{i1}$, $\hat{x}_{i2}$ and $\hat{y}_i$ denote the estimated local state and output variables of the ith subsystem, $i = 1, \ldots, M$, $L_i \in \Re^{l_i \times l_i}$ is a design gain matrix, $\hat{x}_i \triangleq [(\hat{x}_{i1})^\top \quad (C_i^{-1}y_i)^\top]^\top$, and $\hat{x}_j \triangleq [(\hat{x}_{j1})^\top \quad (C_j^{-1}y_j)^\top]^\top$ (here $\hat{x}_{j1}$ is the estimate of state vector x_{j1} of the jth interconnected subsystem). The initial conditions are $\hat{x}_{i1} = (0)$ and $\hat{x}_{i2}(0) = C_i^{-1}y_i(0)$. It is worth noting that the distributed FDE in Equation (3.12) for the ith subsystem is constructed based on local input and output variables (i.e. u_i and y_i) and the communicated information $\hat{x}_j$ and u_j from the FDE associated with the jth interconnected subsystem. Note that many distributed estimation and diagnostic methods in literature allow some communication among interconnected subsystems (see, e.g. [2, 12, 14, 15, 21]).

For each local FDE, let $\tilde{x}_{i1} \triangleq x_{i1} - \hat{x}_{i1}$ and $\tilde{x}_{i2} \triangleq x_{i2} - \hat{x}_{i2}$ denote the state estimation errors, and $\tilde{y}_i \triangleq y_i - \hat{y}_i$ denote the output estimation error. Then, before fault occurrence (i.e. for $t < T_0$), using Equation (3.3) and Equation (3.12), the estimation error dynamics are given by

$$\dot{\tilde{x}}_{i1} = A_{i1}\tilde{x}_{i1} + \rho_{i1}(x_i, u_i) - \rho_{i1}(\hat{x}_i, u_i) + \eta_{i1}$$

$$+ \sum_{j=1}^{M} \left[H_{ij}^1(x_i, x_j, u_i, u_j) - H_{ij}^1(\hat{x}_i, \hat{x}_j, u_i, u_j) \right] \qquad (3.13)$$

$$\dot{\tilde{x}}_{i2} = \bar{A}_{i4}\tilde{x}_{i2} + A_{i3}\tilde{x}_{i1} + \rho_{i2}(x_i, u_i) - \rho_{i2}(\hat{x}_i, u_i) + \eta_{i2}$$

$$+ \sum_{j=1}^{M} \left[H_{ij}^2(x_i, x_j, u_i, u_j) - H_{ij}^2(\hat{x}_i, \hat{x}_j, u_i, u_j) \right] \qquad (3.14)$$

$$\tilde{y}_i = C_i(x_{i2} - \hat{x}_{i2}) = C_i\tilde{x}_{i2}, \qquad (3.15)$$

where $\bar{A}_{i4} \triangleq A_{i4} - L_iC_i$. Note that, since C_i is non-singular, we can always choose L_i to make $\bar{A}_{i4}$ stable.

3.1.5   Adaptive thresholds

Next, we investigate the design of adaptive thresholds for distributed fault detection in each subsystem. The following lemma is needed in the analysis:

Lemma 3.1   *[22]: Let $\bar{p}(t), \bar{q}(t) : [0, \infty) \mapsto R$. Then*

$$\dot{\bar{p}}(t) \leq -a\bar{p}(t) + \bar{q}(t), \quad \forall t \geq t_0 \geq 0$$

implies that

$$\bar{p}(t) \leq e^{-a(t-t_0)}\bar{p}(t_0) + \int_{t_0}^{t} e^{-a(t-\tau)}d\tau, \quad \forall t \geq t_0 \geq 0$$

for any finite constant a. A bounding function on the state estimation error vector before fault occurrence (i.e. for $0 \leq t < T_0$) can be obtained:

$$\tilde{x}_1(t) \triangleq [(\tilde{x}_{11})^{\top}, \ldots, (\tilde{x}_{i1})^{\top}, \ldots, (\tilde{x}_{M1})^{\top}]^{\top} \tag{3.16}$$

Specifically, we have the following results.

Lemma 3.2   *Consider the system described by Equation (3.3) and the FDE described by Equation (3.12). Assume that there exists a symmetric positive definite matrix $P_i \in \Re^{(n_i-l_i)\times(n_i \times l_i)}$, for $i = 1, \ldots, M$, such that,*

- *The symmetric matrix*

$$R_i \triangleq A_{i1}^{\top}P_i - P_i A_{i1} - 2P_i P_i - 2\sigma_{i1}\|P_i\|I > 0, \tag{3.17}$$

  where I is the identity matrix.
- *the matrix $Q \in \Re^{M \times M}$, whose entries are given by*

$$Q_{ij} = \begin{cases} \lambda_{\min}(R_i), & i = j \\ -\|P_i\|\gamma_{ij}^1 - \|P_j\|\gamma_{ji}^1, & i \neq j, j = 1, \ldots, M \end{cases} \tag{3.18}$$

is positive definite, where γ_{ij}^1 and γ_{ji}^1 are the Lipschitz constants introduced in Equation (3.9).

Then, for $0 \leq t < T_0$, the state estimation error vector $\tilde{x}_1$ defined by Equation (3.16) satisfies the following inequality:

$$|\tilde{x}_1|^2 \leq \frac{\bar{V}_0 e^{-et}}{\lambda_{\min}(P)} + \frac{1}{2\lambda_{\min}(P)} \int_0^t e^{-e(t-\tau)} \sum_{i=1}^{M} |\bar{\eta}_{i1}|^2 d\tau, \tag{3.19}$$

where the matrix $P \triangleq diag\{P_1, \ldots, P_M\}$, *the constant* $c \triangleq \lambda_{min}(Q)/\lambda_{max}(P)$, *and* $\bar{V}_0$ *is a positive constant to be defined later on.*

Proof. For the ith subsystem, let us consider a Lyapunov function candidate $V_i = \tilde{x}_{i1}^\top P_i \tilde{x}_{i1}$. The time derivative of V_i along the solution of Equation (3.13) is given by

$$\dot{V}_i = \tilde{x}_{i1}^\top (A_{i1}^\top P_i A_{i1}) \tilde{x}_{i1} + 2\tilde{x}_{i1}^\top P_i \eta_{i1}$$

$$+ 2\tilde{x}_{i1}^\top P_i \sum_{j=1}^{M} \left[H_{ij}^1(x_i, x_j, u_i, u_j) - H_{ij}^1(\hat{x}_i, \hat{x}_j, u_i, u_j) \right]$$

$$+ 2\tilde{x}_{i1}^\top P_i [\rho_{i1}(x_i, u_i) - \rho_{i1}(\hat{x}_i, u_i)]. \tag{3.20}$$

Note that

$$x_j - \hat{x}_j = \begin{bmatrix} x_{j1} - \hat{x}_{j1} \\ x_{j2} - C_j^{-1} y_j \end{bmatrix} = \begin{bmatrix} \tilde{x}_{j1} \\ 0 \end{bmatrix}. \tag{3.21}$$

Therefore, based on Equation (3.9) and Equation (3.21), we have

$$2\tilde{x}_{i1}^\top \ P_i \sum_{j=1}^{M} \left[H_{ij}^1(x_i, x_j, u_i, u_j) - H_{ij}^1(\hat{x}_i, \hat{x}_j, u_i, u_j) \right]$$

$$\leq 2|\tilde{x}_{i1}| \cdot \|P_i\| \sum_{j=1}^{M} \gamma_{ij}^1 |x_j - \hat{x}_j|$$

$$= 2\|P_i\| \sum_{j=1}^{M} \gamma_{ij}^1 |\tilde{x}_{i1}| |\tilde{x}_{j1}|. \tag{3.22}$$

Moreover, based on Equation (3.7) and Equation (3.21), we obtain

$$2\tilde{x}_{i1}^\top \ P_i [\rho_{i1}(x_i, u_i) - \rho_{i1}(\hat{x}_i, u_i)]$$

$$\leq 2|\tilde{x}_{i1}| \cdot \|P_i\| |x_i - \hat{x}_i|$$

$$= 2|\tilde{x}_{i1}| \cdot \|P_i\| |\sigma_{i1}| |\tilde{x}_{i1}|$$

$$= \tilde{x}_{i1}^\top [2\sigma_{i1} \|P_i\| I] \tilde{x}_{i1}. \tag{3.23}$$

Additionally, we have

$$2\tilde{x}_{i1}^\top P_i \eta_{i1} \leq |2P_i \tilde{x}_{i1}| |\eta_{i1}| \leq 2\tilde{x}_{i1}^\top P_i P_i \tilde{x}_{i1} + \frac{1}{2} |\eta_{i1}|^2. \tag{3.24}$$

By using Equation (3.20) and Equations (3.22)–(3.24), we obtain

$$\dot{V}_i \leq \left[A_{i1}^\top P_i + P_i A_{i1} + 2 P_i P_i + 2\sigma_{i1} \| P_i \| I \right] \tilde{x}_{i1}$$

$$+ 2\| P_i \| \sum_{j=1}^{M} \gamma_{ij}^! |\tilde{x}_{i1}| |\tilde{x}_{j1}| + \frac{1}{2} |\eta_{i1}|^2. \tag{3.25}$$

Based on Equation (3.17) and the inequality $\tilde{x}_{i1}^\top R_i \tilde{x}_{i1} \geq \lambda_{\min}(R_i) |\tilde{x}_{i1}|^2$, where $\lambda_{\min}(R_i)$ is the minimum eigenvalue of R_i, we obtain:

$$\dot{V}_i \geq -\lambda_{\min}(R_i) |\tilde{x}_{i1}|^2 + 2\| P_i \| \sum_{j=1}^{M} \gamma_{ij}^! |\tilde{x}_{i1}| |\tilde{x}_{j1}| + \frac{1}{2} |\eta_{i1}|^2. \tag{3.26}$$

Consider now the overall Lyapunov function candidate for the interconnected system:

$$V = \sum_{i=1}^{M} V_i = \sum_{i=1}^{M} \tilde{x}_{i1}^\top P_i \tilde{x}_{i1} = \tilde{x}_1^\top P \tilde{x}_1,$$

$$P = diag\{P_1, \ldots, P_M\}.$$

From Equation (3.26) and Equation (3.18), it follows that

$$\dot{V}_i \geq -\sum_{i=1}^{M} \lambda_{\min}(R_i) |\tilde{x}_{i1}|^2 + \sum_{i=1}^{M} \sum_{j=1}^{M} 2\| P_i \| \| \gamma_{ij}^! \| |\tilde{x}_{i1}| |\tilde{x}_{j1}| + \sum_{i=1}^{M} \frac{1}{2} |\eta_{i1}|^2$$

$$= \left[|\tilde{x}_{11}| \quad |\tilde{x}_{21}| \quad \cdots, \quad |\tilde{x}_{M1}| \right] Q \begin{bmatrix} |\tilde{x}_{11}| \\ |\tilde{x}_{21}| \\ \vdots \\ |\tilde{x}_{M1}| \end{bmatrix} + \sum_{i=1}^{M} \frac{1}{2} |\eta_{i1}|^2$$

where the matrix Q is defined by Equation (3.18). Using the Rayleigh principle

$$\lambda_{\min}(P) |\tilde{x}_1|^2 \leq V(t) \leq \lambda_{\max}(P) |\tilde{x}_1|^2$$

and the definition of $V(t)$, we have

$$\dot{V} \leq -\lambda_{\min}(Q) |\tilde{x}_1|^2 + \sum_{i=1}^{M} \frac{1}{2} |\eta_{i1}|^2$$

$$\leq -\frac{\lambda_{\min}(Q)}{\lambda_{\max}(P)} V + \sum_{i=1}^{M} \frac{1}{2} |\eta_{i1}|^2$$

$$= -cV + \sum_{i=1}^{M} \frac{1}{2} |\eta_{i1}|^2.$$

where $\tilde{x}_1$ and the constant c are defined in Equation (3.16) and Lemma 3.2, respectively. Now, based on Lemma 3.1, it can easily be shown that

$$V(t) \leq V(0)e^{-ct} + \frac{1}{2}\int_0^t e^{-c(t-\tau)}\sum_{i=1}^M |\bar{\eta}_{i1}|^2 d\tau.$$

Note that we can always choose a positive constant $\bar{V}_0$ such that $V(0) < \bar{V}(0)$. Thus, based on the definition of $V(t)$ and the Rayleigh principle, Equation (3.19) is proved. ∎

Remark 3.3 *It is also worth noting that a necessary condition for Equation (3.17) is that A_{i1} is Hurwitz. In addition, note that the Matlab linear matrix inequality (LMI) toolbox can be used to find a feasible solution to the matrix inequalities of Equation (3.17) and Equation (3.18). Specifically, the following procedure can be adopted:*

1. *By using the Schur complements, the nonlinear inequalities $-A_{i1}^\top P_i - P_i A_{i1} - 2P_i P_i - 2\sigma_{i1}\|P_i\|I > 0$ can be converted to an LMI form as*

$$\begin{bmatrix} -A_{i1}^\top P_i - P_i A_{i1} - 2\sigma_{i1}\varsigma_i I & \sqrt{2}P_i \\ \sqrt{2}P_i & I \end{bmatrix} > 0 \tag{3.27}$$

and

$$\begin{bmatrix} \varsigma_i I & P_i \\ P_i & \varsigma_i I \end{bmatrix} > 0 \tag{3.28}$$

where ς_i is a positive constant. Then, a suitable solution of P_i can be obtained by solving Equation (3.27) and Equation (3.28) using the Matlab LMI toolbox.
2. *For the matrix P_i found in Step 1, the matrix Q defined in Equation (3.18) is verified. If Q is positive definite, the solution of P_i is valid.*

Now, we analyze the output estimation error $\tilde{y}_i(t)$ of the ith subsystem. For $0 \leq t < T_0$, the solution of Equation (3.14) is given by

$$\tilde{x}_{i2}(t) = \int_0^t e^{\bar{A}_{i4}(t-\tau)}[A_{i3}\tilde{x}_{i1}(\tau) + \eta_{i2}(x_i, \bar{x}_j, u_i, \bar{u}_j, t)]d\tau$$

$$+ \int_0^t e^{\bar{A}_{i4}(t-\tau)}[\rho_{i2}(x_i, u_i) - \rho_{i2}(\hat{x}_i, u_i)d\tau$$

$$+ \int_0^t e^{\bar{A}_{i4}(t-\tau)}\sum_{j=1}^M \left[H_{ij}^2(x_i, x_j, u_i, u_j) - H_{ij}^2(\hat{x}_i, \hat{x}_j, u_i, u_j)\right]d\tau.$$

Therefore, for each component of the output estimation error, i.e. $\tilde{y}_{ip}(t) \triangleq C_{ip}\tilde{x}_{i2}(t)$, $p = 1, \ldots, l_i$, where C_{ip} is the pth row vector of matrix C_i, by applying the triangle inequality, we have

$$\tilde{y}_{ip}(t) \leq \left| \int_0^t C_{ip} e^{\bar{A}_{i4}(t-\tau)} \sum_{j=1}^M [H_{ij}^2(x_i, x_j, u_i, u_j) - H_{ij}^2(\hat{x}_i, \hat{x}_j, u_i, u_j)]d\tau \right|$$

$$+ \left| \int_0^t C_{ip} e^{\bar{A}_{i4}(t-\tau)} [A_{i3}\tilde{x}_{i1} + \eta_{i2}]d\tau \right|$$

$$+ \left| \int_0^t C_{ip} e^{\bar{A}_{i4}(t-\tau)} [\rho_{i2}(x_i, u_i) - \rho_{i2}(\hat{x}_i, u_i)d\tau \right|. \tag{3.29}$$

Based on Equation (3.8), Equation (3.10) and Equation (3.21), we have

$$|H_{ij}^2(x_i, x_j, u_i, u_j) - H_{ij}^2(\hat{x}_i, \hat{x}_j, u_i, u_j)| \leq \gamma_{ij}^2 |\tilde{x}_{j1}|$$

$$|\rho_{i2}(x_i, u_i) - \rho_{i2}(\hat{x}_i, u_i)|. \tag{3.30}$$

Using Equation (3.29) and Equation (3.30), we obtain

$$|\tilde{y}_{ip}(t)| \leq k_{ip} \int_0^t e^{-\lambda_{ip}(t-\tau)} \Big[[\|A_{i3}\| + \sigma_{i2}(y_i, u_i, \hat{x}_{i1})]|\tilde{x}_{i1}||\eta_{i1}|$$

$$+ \sum_{i=1}^M \gamma_{ij}^2 |\tilde{x}_{j1}| \Big] d\tau. \tag{3.31}$$

Since $\bar{A}_{i4}$ is stable, positive constants k_{ip} and λ_{ip} satisfying the above inequality always exist and can be chosen such that $|C_{ip} e^{\bar{A}_{i4}t}| \leq k_{ip} e^{-\lambda_{ip}t}$.

By letting

$$\varrho_i \triangleq [\gamma_{i1}^2, \ldots, \gamma_{i(i-1)}^2, \|A_{i3}\| + \sigma_{i2}, \gamma_{i(i+1)}^2, \ldots, \gamma_{iM}^2]^\top, \tag{3.32}$$

(i.e., the components of ϱ_i are given by $\varrho_{ii} = \|A_{i3}\| + \sigma_{i2}$, and $\varrho_{ij} = \gamma_{ij}^2$ for $j \neq i$), the inequality (3.31) can be rewritten as

$$|\tilde{y}_{ip}(t)| \leq k_{ip} \int_0^t e^{-\lambda_{ip}(t-\tau)} \Big[\sum_{j=1}^M \varrho_{ij} |\tilde{x}_{j1}| + |\eta_{i2}| \Big] d\tau$$

$$\leq k_{ip} \int_0^t e^{-\lambda_{ip}(t-\tau)} [|\varrho_i||\tilde{x}_1| + \bar{\eta}_{i2}]d\tau. \tag{3.33}$$

Now, based on Equation (3.33) and Equation (3.19), we obtain

$$|\tilde{y}_{ip}(t)| \leq k_{ip} \int_0^t e^{-\lambda_{ip}(t-\tau)} [|\varrho_i|\chi(\tau) + \bar{\eta}_{i2}]d\tau, \tag{3.34}$$

where

$$\chi(t) \triangleq \left\{ \frac{\bar{V}_0 e^{-et}}{\lambda_{\min}(P)} + \frac{1}{2\lambda_{\min}(P)} \int_0^t e^{-c(t-\tau)} \sum_{i=1}^M |\bar{\eta}_{i1}|^2 \right\}^{1/2}.$$ (3.35)

Therefore, the decision on the occurrence of a fault (detection) in the ith subsystem is made when the modulus of at least one component of the output estimation error (i.e. $\tilde{y}_{ip}(t)$) generated by the local FDE exceeds its corresponding threshold $v_{ip}(t)$ given by

$$v_{ip}(t) \triangleq \int_0^t e^{-\lambda_{ip}(t-\tau)}[|\varrho_i|\chi(\tau) + \bar{\eta}_{i2}]d\tau.$$ (3.36)

The fault detection time T_d is defined as the first time instant such that $|\tilde{y}_{ip}(T_d)| > v_{ip}(T_d)$, for some $T_d \geq T_0$ and some $p \in \{1, \ldots, l_i\}$, that is,

$$T_d \triangleq \inf \bigcup_{p=1}^{l_i} \{t \geq 0 : |\tilde{y}_{ip}(t)| > v_{ip}(t)\}.$$

Remark 3.4   *It is worth noting that $v_{ip}(t)$ given by Equation (3.36) is an adaptive threshold for fault detection, which has obvious advantage over a constant one. Moreover, the threshold $v_{ip}(t)$ can be easily implemented using linear filtering techniques [16]. Additionally, the constant $\bar{V}_0$ in Equation (3.35) is a (possibly conservative) bound for the unknown initial conditions $V(0)$. However, note that, since the effect of this bound decreases exponentially (i.e. it is multiplied by e^{-ct}), the practical use of such a conservative bound will not affect significantly the performance of the distributed fault detection algorithm.*

The above design and analysis is summarized by the following technical result.

Theorem 3.1   *(Robustness): For the interconnected nonlinear uncertain system described by Equation (3.3), the distributed fault detection method (characterized by FDE Equation (3.12) and adaptive thresholds Equation (3.36) designed for each local subsystem) ensures that each residual component $y_{ip}(t)$ generated by the local FDEs remains below its corresponding adaptive threshold $v_{ip}(t)$ prior to the occurrence of a fault (i.e. for $t < T_0$).*

3.1.6   Distributed fault isolation method

Each local FDI component consists of an FDE and a bank of FIEs. Now, assume that a fault is detected in the ith subsystem at some time T_d; accordingly, at $t = T_d$, the FIEs in the local FDI component designed for the ith subsystem are activated. Each FIE is designed based on the functional structure of one potential fault type in the local subsystem.

Specifically, the following N_i nonlinear adaptive estimators are used as isolation estimators: for $s = 1, \ldots, N_i$,

$$
\begin{aligned}
\dot{\hat{x}}_{i1}^s &= A_{i1}\hat{x}_{i1}^s + A_{i2}C_i^{-1}y_i + \rho_{i1}(\hat{x}_i^s, u_i) \\
&\quad + \sum_{j=1}^{M} H_{ij}^1(\hat{x}_i^s, \hat{x}_j, u_i, u_j) \\
\dot{\hat{x}}_{i2}^s &= A_{i3}\hat{x}_{i1}^s + A_{i4}\hat{x}_{i2}^s + \rho_{i2}(\hat{x}_i^s, u_i) + L_i^s(y_i - \hat{y}_i^s) \\
&\quad + \hat{f}_i^s(\hat{x}_i^s, u_i, \hat{\theta}_i^s) + \Omega_i^s\dot{\hat{\theta}}_i^s + \sum_{j=1}^{M} H_{ij}^2(\hat{x}_i^s, \hat{x}_j, u_i, u_j) \\
\dot{\Omega}_i^s &= \bar{A}_{i4}\Omega_i^s + G_i^s(\hat{x}_i^s, u_i) \\
\hat{y}_i^s &= C_i\hat{x}_{i2}^s,
\end{aligned}
\tag{3.37}
$$

where $\hat{x}_{i1}^s$, $\hat{x}_{i2}^s$ and $\hat{y}_i^s$ denote the estimated state and output variables provided by the sth local FIE, respectively; $L_i^s \in \Re^{l_i \times l_i}$ is a design gain matrix (for simplicity of presentation and without loss of generality, we let $L_i^s = L_i$); $\hat{x}_i^s \triangleq [(\hat{x}_{i1}^s)^{\top} \quad (C_i^{-1}y_i)^{\top}]^{\top}$; and $\hat{x}_j \triangleq [(\hat{x}_{j1})^{\top} \quad (C_j^{-1}y_j)^{\top}]^{\top}$. The function $\hat{f}_i^s(\hat{x}_i^s, u_i, \hat{\theta}_i^s) \triangleq [(\hat{\theta}_{i1}^s)^{\top} \quad g_{i1}^s(\hat{x}_i^s, u_i), \ldots, (\hat{\theta}_{il_i}^s)^{\top} \quad g_{il_i}^s(\hat{x}_i^s, u_i)]^{\top}$ provides the adaptive structure for approximating the unknown fault function $f_i^s(x_i, u_i)$ described by Equation (3.5), and $\hat{\theta}_{ip}^s$ ($i = 1, \ldots, M$ and $p = 1, \ldots, l_i$) is the adjustable parameter vector. The initial conditions are $\hat{x}_{i1}^s(T_d) = 0$, $\hat{x}_{i2}^s(T_d) = 0$, and $\Omega_i^s(T_d) = 0$. It is noted that, according to Equation (3.5), the fault approximation model $\hat{x}_i^s$ is linear in the adjustable weights $\hat{\theta}_i^s$. Consequently, the gradient matrix $G_i^s(\hat{x}_i^s, u_i) \triangleq \partial \hat{f}_i^s(\hat{x}_i^s, u_i, \hat{\theta}_i^s)/\partial \hat{\theta}_i^s = diag[(g_{i1}^s(\hat{x}_i^s, u_i))^{\top}, \ldots, (g_{il_i}^s(\hat{x}_i^s, u_i))^{\top}]$ does not depend on $\hat{\theta}_i^s$. Note that the distributed FIEs (Equation (3.37)) for each local subsystem are constructed based on local measurements (i.e. u_i and y_i) and the communicated information $\hat{x}_j$ and u_j from the FDI component associated with the jth directly interconnected subsystem.

The adaptation in the isolation estimators arises due to the unknown fault magnitude $\theta_i^s \triangleq [(\theta_{i1}^s)^{\top}, \ldots, (\theta_{il_i}^s)^{\top}]^{\top}$. The adaptive law for adjusting $\hat{\theta}_i^s$ is derived using the Lyapunov synthesis approach [22]. Specifically, the learning algorithm is chosen as follows

$$
\dot{\hat{\theta}}_i^s = P_{\Theta_i^s}\{\Gamma\Omega_i^{s\top}C_i^{\top}\tilde{y}_i^s\},
\tag{3.38}
$$

where $\tilde{y}_i^s(t) \triangleq y_i(t) - \hat{y}_i^s(t)$ denotes the output estimation error generated by the sth FIE for the local subsystem, $\Gamma > 0$ is a symmetric, positive-definite learning rate matrix and $P_{\Theta_i^s}$ is the projection operator restricting $\hat{\theta}_i^s$ to the corresponding known set Θ_i^s (in order to guarantee stability of the learning algorithm in the presence of modeling uncertainty [22, 23].

The distributed fault isolation decision scheme stems from the following intuitive principle: If fault s occurs in the ith subsystem, $i = 1, \ldots, M$, at time T_0 and is detected at time T_d, then a set of adaptive threshold functions $\{\mu_{ip}^s(t), p = 1, \ldots, l_i, s = 1, \ldots, N_i\}$ can be designed for the matched sth isolation estimator of the ith subsystem, such that each component of its output estimation error satisfies $|\tilde{y}_{ip}^s(t)| \leq \mu_{ip}^s(t)$, for all $t \geq T_d$. Consequently, such a set of adaptive thresholds $\mu_{ip}^s(t)$ with $s = 1, \ldots, N_i$ can be associated with the output estimation error of each local isolation estimator. In the fault isolation procedure, if, for a particular local isolation estimator $r \in \{1, \ldots, N_i\} \{s\}$, there exists some $p \in \{1, \ldots, l_i\}$, such that the pth component of its output estimation error satisfies $|\tilde{y}_{ip}^r(t)| > \mu_{ip}^r(t)$ for some finite time $t > T_d$, then the possibility of the occurrence of fault r can be excluded.

Based thereon, the following fault isolation decision scheme is developed: if, for each $r \in \{1, \ldots, N_i\} \{s\}$, there exists some finite time $t^r > T_d$ and some $p \in \{1, \ldots, l_i\}$, such that $|\tilde{y}_{ip}^r(t^r)| > \mu_{ip}^r(t^r)$, then the occurrence of fault s in the ith subsystem is concluded.

Remark 3.5   *It is worth noting that the presented FDI method is capable of identifying not only faults defined in the partially unknown fault class F (see Equation (3.4)) but also new faults that do not belong to F (at least one component of the residuals generated by each FIE would exceed its threshold). In addition to the output estimation error, the parameter estimation $\hat{\theta}_i^s$ might also provide some information for fault isolation. However, note that a necessary condition to ensure that the parameter estimation $\hat{\theta}_i^s$ converges to its actual value θ_i^s is the persistency of excitation of signals [22, 23], which is too restrictive in many practical applications. Here we do not assume persistency of excitation.*

3.1.7   Adaptive thresholds for DFDI

The threshold functions $\mu_{ip}^s(t)$ clearly play a key role in the proposed distributed fault isolation decision scheme. The following lemma provides a bounding function for each component of the output estimation error of the matched sth local isolation estimator in the case that fault s occurs in the ith subsystem.

Lemma 3.3   *If fault s in the ith subsystem is detected at time T_d, where $s \in \{1, \ldots, N_i\}$ and $i \in \{1, \ldots, M\}$, then for all $t > T_d$, each component of the output estimation error $\tilde{y}_{ip}^s(t)$ associated with the matched sth local isolation estimator satisfies*

$$|\tilde{y}_{ip}^s(t)| \leq k_{ip} \int_{T_d}^t e^{-\lambda_{ip}(t-\tau)} [|\xi_i| \chi(\tau) + \bar{\eta}_{i2} + \alpha_i^s \|\Omega_i^s\|] d\tau$$

$$+ k_{ip} \omega_{i2} e^{-\lambda_{ip}(t-T_d)} + |(C_{ip}\Omega_i^s)^\top \| \tilde{\theta}_i^s \|, \qquad (3.39)$$

where $\tilde{y}_{ip}^s(t) \triangleq y_{ip}(t) - \hat{y}_{ip}^s(t)$, $p = 1, \ldots, l_i$, $\chi(t)$ is given in Equation (3.35); $\tilde{\theta}_i^s(t) \triangleq \hat{\theta}_i^s(t) - \theta_i^s(t)$ represents the fault parameter vector estimation error; ω_{i2} is a positive constant satisfying $|x_{i2}^s(T_d)| \leq \omega_{i2}$; and $\bar{\varrho}_i$ is defined later on.

Proof.   Denote the state estimation error of the sth local isolation estimator for the ith subsystem by $\tilde{x}_{i1}^s(t) \triangleq x_{i1}(t) - \hat{x}_{i1}^s(t)$ and $\tilde{x}_{i2}^s(t) \triangleq x_{i2}(t) - \hat{x}_{i2}^s(t)$. Using Equation (3.37) and Equation (3.3), in the presence of fault s in the ith subsystem, the state estimation error of the matched sth local FIE satisfies, for $t > T_d$,

$$\dot{\tilde{x}}_{i1}^s = A_{i1} \tilde{x}_{i1}^s + \eta_{i1} + \rho_{i1}(x_i, u_i) - \rho_{i1}(\hat{x}_i^s, u_i)$$

$$+ \sum_{j=1}^M [H_{ij}^1(x_i, x_j, u_i, u_j) - H_{ij}^1(\hat{x}_i^s, \hat{x}_j, u_i, u_j)] \qquad (3.40)$$

$$\dot{\tilde{x}}_{i2}^s = \bar{A}_{i4}\tilde{x}_{i2}^s + A_{i3}\tilde{x}_{i1}^s + \eta_{i2} + \rho_{i2}(x_i, u_i) - \rho_{i2}(\hat{x}_i^s, u_i)$$

$$+ f_i^s(x_i, u_i) - \hat{f}_i^s(\hat{x}_i^s, u_i, \hat{\theta}_i^s)$$

$$- \Omega_i^s\dot{\hat{\theta}}_i^s + \sum_{j=1}^{M}[H_{ij}^2(x_i, x_j, u_i, u_j) - H_{ij}^2(\hat{x}_i^s, \hat{x}_j, u_i, u_j)], \qquad (3.41)$$

where $\bar{A}_{i4}$ is defined in Equation (3.14). Note that

$$f_i^s(x_i, u_i) - \hat{f}_i^s(\hat{x}_i^s, u_i, \hat{\theta}_i^s)$$

$$= G_i^s(x_i, u_i)\theta_i - G_i^s(\hat{x}_i^s, u_i)\theta_i + G_i^s(\hat{x}_i, u_i)\theta_i - G_i^s(\hat{x}_i^s, u_i)\hat{\theta}_i$$

$$= f_i^s(x_i, u_i) - f_i^s(\hat{x}_i^s, u_i) - G_i^s(\hat{x}_i^s, u_i)\tilde{\theta}_i. \qquad (3.42)$$

Using $G_i^s(\hat{x}_i^s, u_i) = \dot{\Omega}_i^s - \bar{A}_{i4}\Omega_i^s$ given by Equation (3.37), we have

$$\dot{\tilde{x}}_{i2}^s = \bar{A}_{i4}(\tilde{x}_{i2}^s + \Omega_i^s\tilde{\theta}_i^s) + A_{i3}\tilde{x}_{i1}^s + \eta_{i2} - \frac{d}{dt}(\Omega_i^s\tilde{\theta}_i^s)$$

$$+ \rho_{i2}(x_i, u_i) - \rho_{i2}(\hat{x}_i^s, u_i) - \Omega_i^s\dot{\hat{\theta}}_i^s$$

$$+ \sum_{j=1}^{M}[H_{ij}^2(x_i, x_j, u_i, u_j) - H_{ij}^2(\hat{x}_i^s, \hat{x}_j, u_i, u_j)]$$

$$+ f_i^s(x_i, u_i) - f_i^s(\hat{x}_i^s, u_i).$$

By letting $\bar{x}_{i2}^s \triangleq \tilde{x}_{i2}^s + \Omega_i^s\tilde{\theta}_i^s$, the above equation can be rewritten as

$$\dot{\bar{x}}_{i2}^s = \bar{A}_{i4}(\bar{x}_{i2}^s + A_{i3}\tilde{x}_{i1}^s + f_i^s(x_i, u_i) - f_i^s(\hat{x}_i^s, u_i)$$

$$+ \sum_{j=1}^{M}[H_{ij}^2(x_i, x_j, u_i, u_j) - H_{ij}^2(\hat{x}_i^s, \hat{x}_j, u_i, u_j)]$$

$$+ \rho_{i2}(x_i, u_i) - \rho_{i2}(\hat{x}_i^s, u_i) + \eta_{i2} - \Omega_i^s\dot{\hat{\theta}}_i^s(t). \qquad (3.43)$$

The solution of Equation (3.43), for $t > T_d$, is given by

$$\bar{x}_{i2}^s(t) = \int_{T_d}^{t} e^{\bar{A}_{i4}(t-\tau)}[A_{i3}\tilde{x}_{i1}^s + \eta_{i2} - \Omega_i^s\dot{\hat{\theta}}_i^s]d\tau$$

$$+ \int_{T_d}^{t} e^{\bar{A}_{i4}(t-\tau)} + \sum_{j=1}^{M}[H_{ij}^2(x_i, x_j, u_i, u_j)$$

$$- H_{ij}^2(\hat{x}_i^s, \hat{x}_j, u_i, u_j)]d\tau$$

$$+ \int_{T_d}^{t} e^{\bar{A}_{i4}(t-\tau)}[\rho_{i2}(x_i, u_i) - \rho_{i2}(\hat{x}_i^s, u_i)]d\tau$$

$$+ \int_{T_d}^{t} e^{\bar{A}_{i4}(t-\tau)}[f_i^s(x_i, u_i) - f_i^s(\hat{x}_i^s, u_i)]d\tau$$

$$+ e^{\bar{A}_{i4}(t-T_d)}\bar{x}_{i2}^s(T_d). \tag{3.44}$$

Using Equation (3.37), Equation (3.3), and the definition of $\bar{x}_{i2}^s(t)$, each component of the output estimation error is given by:

$$\tilde{y}_{ip}^s(t) = C_{ip}\tilde{x}_{i2}^s(t) = C_{ip}(\bar{x}_{i2}^s(t) - \Omega_i^s\tilde{\theta}_i^s). \tag{3.45}$$

Now, based on Equation (3.44) and Equation (3.45), as well as Assumptions 3.2, 3.4 and 3.5, after following similar logic to that reported in the derivation of the adaptive thresholds for fault detection (see Equations (3.21), (3.30) and (3.31)), it can be shown that

$$|\tilde{y}_{ip}^s(t)| \le k_{ip}(t) \int_{T_d}^{t} e^{-\lambda_{ip}(t-\tau)}\Big[\|A_{i3}\||\tilde{x}_{i1}^s| + \bar{\eta}_{i2}$$

$$+ \alpha_i^s\|\Omega_i^s\| + \sum_{j=1}^{M}\gamma_{ij}^2|\tilde{x}_{j1}|\Big]d\tau + |(C_{ip}\Omega_i^s)^\top||\tilde{\theta}_i^s|$$

$$+ k_{ip}\int_{T_d}^{t} e^{-\lambda_{ip}(t-\tau)}\Big[\sigma_{i2}|\tilde{x}_{i1}^s| + \bar{\omega}_i^s|\tilde{x}_{i1}^s|\Big]d\tau$$

$$+ k_{ip}e^{-\lambda_{ip}(t-T_d)}|\bar{x}_{i2}^s(T_d)|,$$

where the constants k_{ip} and λ_{ip} are defined in Equation (3.31). By using $\tilde{x}_1$ given in Equation (3.16) and defining

$$\bar{\varrho}_i \triangleq [\gamma_{i1}^2, \dots, \gamma_{i(i-1)}^2, \|A_{i3}\| + \sigma_{i2} + \bar{\omega}_i^s, \gamma_{i(i+1)}^2, \dots, \gamma_{iM}^2]^\top, \tag{3.46}$$

we have

$$|\tilde{y}_{ip}^s(t)| \le k_{ip}(t) \int_{T_d}^{t} e^{-\lambda_{ip}(t-\tau)}\Big[|\bar{\varrho}_i||\tilde{x}_1| + \bar{\eta}_{i2} + \alpha_i^s\|\Omega_i^s\|\Big]d\tau$$

$$+ |(C_{ip}\Omega_i^s)^\top||\tilde{\theta}_i^s| + k_{ip}e^{-\lambda_{ip}(t-T_d)}|\bar{x}_{i2}^s(T_d)|.$$

Note that Equation (3.40) is in the same form as Equation (3.13). Thus, using the results of Lemma 3.2 (i.e. Equation (3.19)), the above inequality becomes

$$|\tilde{y}_{ip}^s(t)| \le k_{ip}(t) \int_{T_d}^{t} e^{-\lambda_{ip}(t-\tau)}[|\bar{\varrho}_i|\chi + \bar{\eta}_{i2} + \alpha_i^s\|\Omega_i^s\|]d\tau$$

$$+ k_{ip}|\bar{x}_{i2}^s(T_d)|e^{-\lambda_{ip}(t-T_d)} + |(C_{ip}\Omega_i^s)^\top||\tilde{\theta}_i^s|.$$

Where χ is defined by Equation (3.35). Now, the inequality in Equation (3.39) follows directly from the initial condition $\hat{x}_{i2}^s(T_d) = 0$, $\Omega_i^s(T_d) = 0$, and $|x_{i2}^s(T_d)| \le \omega_{i2}$. ■

Although Lemma 3.3 provides an upper bound on the output estimation error of the sth local estimator for subsystem i, the right-hand side of Equation (3.39) cannot be directly used as a threshold function for fault isolation, because $\tilde{\theta}_i^s(t)$ is not available. (As described in Remark 3.5, we do not assume the condition of persistency of excitation.). However, since the estimate $\hat{\theta}_i^s$ belongs to the known compact set $\Theta_i^s(t)$, we have $|\theta_i^s - \hat{\theta}_i^s(t)| \leq \kappa_i^s(t)$ for a suitable $\kappa_i^s(t)$ depending on the geometric properties of set Θ_i^s [8, 16]. Hence, based on the above discussions, the following threshold function is chosen:

$$\mu_{ip}^s(t) = k_{ip} \int_{T_d}^{t} e^{-\lambda_{ip}(t-\tau)}[|\bar{\varrho}_i|\chi + \bar{\eta}_{i2} + \alpha_i^s\|\Omega_i^s\|]d\tau$$

$$+ k_{ip}\omega_{i2}e^{-\lambda_{ip}(t-T_d)} + |(C_{ip}\Omega_i^s)^\top|\kappa_i^s(t). \tag{3.47}$$

Remark 3.6   *Note that the adaptive threshold $\mu_{ip}^s(t)$ can be easily implemented online using linear filtering techniques [8, 16]. The constant bound ω_{i2} is a (possibly conservative) bound for the unknown initial conditions $x_{i2}^s(T_d)$. However, note that, since the effect of this bound decreases exponentially (i.e. it is multiplied by $e^{-\lambda_{ip}(t-T_d)}$), the practical use of such a conservative bound will not significantly affect the performance of the distributed fault isolation algorithm.*

Remark 3.7   *As we can see, the adaptive threshold function described by Equation (3.47) is influenced by several sources of uncertainty entering the fault isolability problem, such as modeling uncertainty (i.e. η_{i1}, η_{i2}), fault parametric uncertainty κ_i^s, unknown fault development rate α_i^s, and unknown initial conditions (i.e. $\bar{V}_0$ and ω_{i2}). Intuitively, the smaller the uncertainty (resulting in a smaller threshold $\mu_{ip}^s(t)$), the easier the task of isolating the faults. On the other hand, as clarified in Section 3.1.9, the ability to isolate a fault depends not only on the threshold $\mu_{ip}^s(t)$, but also on the degree to which the types of fault in each subsystem are different from each other.*

As is well known in the fault diagnosis literature, there is an inherent tradeoff between robustness and fault sensitivity. In Sections 3.1.8–3.1.10, we analyze the fault sensitivity of the distributed fault diagnosis method, including fault detectability and isolability. In addition, we investigate the stability and learning capability of adaptive FIEs.

3.1.8   Fault detectability condition

Theorem 3.2 characterizes (in a non-closed form) the class of faults that are detectable by the proposed distributed fault detection method.

Theorem 3.2   *For the distributed fault detection method described by Equation (3.12) and Equation (3.36), suppose that fault s occurs in the ith subsystem at time T_0, where $s \in \{1, \ldots, N_i\}$ and $i \in \{1, \ldots, M\}$. Then, if there exists some time instant $T_d > T_0$ and some $p \in \{1, \ldots, l_i\}$ such that the fault function $f_i(x_i, u_i)$*

$$|\int_{T_0}^{T_d} C_{ip}e^{\bar{A}_{i4}(T_d-\tau)}f_i(x_i(\tau), u_i(\tau))d\tau| \geq 2v_{ip}(T_d), \tag{3.48}$$

the fault will be detected at time $t = T_d$, that is, $|\tilde{y}_{ip}(T_d)| > v_{ip}(T_d)$.

Proof. In the presence of a fault (that is, for $t \geq T_0$), based on Equation (3.3) and Equation (3.12), the dynamics of the state estimation error $\tilde{x}_{i1} \triangleq x_{i1} - \hat{x}_{i1}$, and $\tilde{x}_{i2} - \hat{x}_{i2}$ satisfies

$$
\begin{aligned}
\dot{\tilde{x}}_{i1} = A_{i1}\tilde{x}_{i1} + \eta_{i1} + \rho_{i1}(x_i, u_i) - \rho_{i1}(\hat{x}_i, u_i) \\
+ \sum_{j=1}^{M}[H_{ij}^1(x_i, x_j, u_i, u_j) - H_{ij}^1(\hat{x}_i, \hat{x}_j, u_i, u_j)]
\end{aligned}
\tag{3.49}
$$

$$
\begin{aligned}
\dot{\tilde{x}}_{i2} = \bar{A}_{i4}\tilde{x}_{i2} + A_{i3}\tilde{x}_{i1} + \eta_{i2} + \beta_i f_i \\
+ \sum_{j=1}^{M}[H_{ij}^2(x_i, x_j, u_i, u_j) - H_{ij}^2(\hat{x}_i, \hat{x}_j, u_i, u_j)] \\
+ \rho_{i2}(x_i, u_i) - \rho_{i2}(\hat{x}_i, u_i).
\end{aligned}
\tag{3.50}
$$

Therefore, for each component of the output estimation error, that is, $\tilde{y}_{ip}(t) \triangleq C_{ip}\tilde{x}_{i2}(t)$, $p = 1, \ldots, l_i$, we have

$$
\begin{aligned}
\tilde{y}_{ip}(t) = \int_0^t C_{ip}e^{\bar{A}_{i4}(t-\tau)}[A_{i3}\tilde{x}_{i1} + \eta_{i2} + \beta_i f_i(x_i, u_i)]d\tau \\
+ \int_0^t C_{ip}e^{\bar{A}_{i4}(t-\tau)}[\rho_{i2}(x_i, u_i) - \rho_{i2}(\hat{x}_i, u_i)]d\tau \\
+ \int_0^t C_{ip}e^{\bar{A}_{i4}(t-\tau)}\sum_{j=1}^{M}[H_{ij}^2(x_i, x_j, u_i, u_j) - H_{ij}^2(\hat{x}_i, \hat{x}_j, u_i, u_j)]d\tau
\end{aligned}
$$

Note that Equation (3.49) is in the same form as Equation (3.13). Therefore, from Lemma 3.2, we have $|\tilde{x}_1(t)| \leq \chi(t)$, where $\chi(t)$ is defined in Equation (3.35). Then, by applying the triangle inequality and Equation (3.30), we obtain:

$$
\begin{aligned}
|\tilde{y}_{ip}(t)| \geq |\int_0^t [C_{ip}e^{\bar{A}_{i4}(t-\tau)}\beta_i f_i(x_i, u_i)]d\tau| \\
- k_{ip}\int_0^t e^{-\lambda_{ip}(t-\tau)}\Big[\|A_{i3}\||\tilde{x}_{i1}| + |\bar{\eta}_{i2}| \\
+ \sum_{j=1}^{M}\gamma_{ij}^2|\tilde{x}_j| + \sigma_{i2}|\tilde{x}_{i1}|\Big]d\tau \\
\geq |\int_0^t C_{ip}e^{\bar{A}_{i4}(t-\tau)}\beta_i f_i(x_i, u_i)d\tau| \\
- k_{ip}\int_0^t e^{-\lambda_{ip}(t-\tau)}[|\varrho_i|\chi + \bar{\eta}_{i2}]d\tau,
\end{aligned}
\tag{3.51}
$$

where ϱ_i is defined in Equation (3.32). By substituting Equation (3.36) into Equation (3.51), we have

$$|\tilde{y}_{ip}(t)| \geq \left| \int_0^t C_{ip} e^{\bar{A}_{i4}(t-\tau)} \beta_i f_i(x_i(\tau), u_i(\tau)) d\tau \right| - v_{ip}(t). \tag{3.52}$$

Based on the property of the step function β_i, if there exists $T_d > T_0$, such that Equation (3.48) is satisfied, then it is concluded that $|\tilde{y}_{ip}(T_d)| > v_{ip}(T_d)$, implying that the fault is detected at time $t = T_d$.   ∎

Remark 3.8   *It is crucial to observe from Equation (3.48) that the integral on the left-hand side stands for the filtered fault function. According to the fault detectability theorem (Theorem 3.2), if the magnitude of the filtered fault function on the time interval $[T_0, T_d]$ becomes sufficiently large, then the fault in the ith subsystem can be detected. The result also shows that if a fault function $f_i(x_i, u_i)$ changes sign over time then it may be difficult (or impossible) to detect.*

3.1.9   Fault isolability analysis

For our purpose, a fault in each subsystem is considered to be isolable if the distributed fault isolation scheme is able to reach a correct decision in finite time. To quantify this, we introduce the fault mismatch function between the sth fault and the rth fault in the ith subsystem, for $i = 1, \ldots, M$:

$$h_{ip}^{sr}(t) \triangleq C_{ip} \left(\Omega_i^s \theta_i^s - \Omega_i^r \hat{\theta}_i^r \right), \tag{3.53}$$

where $r, s = 1, \ldots, N_i, r \neq s$ and $p = 1, \ldots, l_i$.

The term $h_{ip}^{sr}(t)$ can be interpreted as a filtered version of the difference between the actual fault function $G_i^s \theta_i^s$ and its estimate $G_i^r \hat{\theta}_i^r$ associated with the rth local isolation estimator whose structure does not match the actual fault s in the local subsystem. Recall that each local FIE is designed based on the functional structure of one of the nonlinear faults in the fault class associated with the local subsystem. Consequently, if fault s occurs, its estimate $G_i^r \hat{\theta}_i^r$ generated by FIE r is determined by the structure of FIE r, which in turn is determined by fault r. Therefore, the fault mismatch function $h_{ip}^{sr}(t)$, defined as the ability of the rth local FIE to learn fault s in the local subsystem, offers a measure of the difference between fault s and fault r associated with the local subsystem.

Theorem 3.3 characterizes the class of isolable faults in each subsystem.

Theorem 3.3   *Consider the distributed fault isolation scheme described by Equation (3.37) and Equation (3.47). Suppose that fault $s(s = 1, \ldots, N_i)$, occurring in the ith subsystem at time T_0, is detected at time T_d. Then, fault s is isolable if, for each $r \in \{1, \ldots, N_i\} \{s\}$, there exists some time $t^r > T_d$ and some $p \in \{1, \ldots, l_i\}$, such that the fault mismatch function*

$h_{ip}^{sr}(t^r)$ *satisfies*

$$|h_{ip}^{sr}(t^r)| \geq 2k_{ip} \int_{T_d}^{t^r} e^{-\lambda_{ip}(t^r-\tau)}[|\bar{\varrho}_i|\chi + \bar{\eta}_{i2}]d\tau$$

$$+ k_{ip} \int_{T_d}^{t^r} e^{-\lambda_{ip}(t^r-\tau)}[\alpha_i^s \|\Omega_i^s\| + \alpha_i^r \|\Omega_i^r\|]d\tau$$

$$+ (C_{ip}\Omega_i^r)^\top |\kappa_i^r + 2\omega_{i2}k_{ip}e^{-\lambda_{ip}(t^r-T_d)}. \tag{3.54}$$

Proof. Denote the state estimation errors of the rth local isolation estimator for subsystem i by $\tilde{x}_{i1}^r(t) \triangleq x_{i1}(t) - \hat{x}_{i1}^r(t)$ and $\tilde{x}_{i2}^r \triangleq x_{i2}(t) - \hat{x}_{i2}^r(t)$. Using Equation (3.37) and Equation (3.3), in the presence of fault s in the ith subsystem for $t > T_d$, we have

$$\dot{\tilde{x}}_{i1}^r = A_{i1}\tilde{x}_{i1}^r + \rho_{i1}(x_i, u_i) - \rho_{i1}(\hat{x}_i^r, u_i) + \eta_{i1}$$

$$+ \sum_{j=1}^{M} [H_{ij}^1(x_i, x_j, u_i, u_j) - H_{ij}^1(\hat{x}_i^r, \hat{x}_j, u_i, u_j)] \tag{3.55}$$

$$\dot{\tilde{x}}_{i2}^r = \bar{A}_{i4}\tilde{x}_{i2}^r + A_{i3}\tilde{x}_{i1}^r + \eta_{i2} + \rho_{i2}(x_i, u_i) - \rho_{i2}(\hat{x}_i^r, u_i)$$

$$+ \sum_{j=1}^{M} [H_{ij}^2(x_i, x_j, u_i, u_j) - H_{ij}^2(\hat{x}_i^r, \hat{x}_j, u_i, u_j)]$$

$$+ G_i^s(x_i, u_i)\theta_i^s - G_i^s(\hat{x}_i^s, u_i)\theta_i^s + G_i^s(\hat{x}_i^s, u_i)\theta_i^s$$

$$- G_i^r(\hat{x}_i^r, u_i)\hat{\theta}_i^r - \Omega_i^r\dot{\hat{\theta}}_i^r. \tag{3.56}$$

By substituting $G_i^s(\hat{x}_i^s, u_i) = \dot{\Omega}_i^s - \bar{A}_{i4}\Omega_i^s$ and $G_i^r(\hat{x}_i^r, u_i) = \dot{\Omega}_i^r - \bar{A}_{i4}\Omega_i^r$ into Equation (3.56), we obtain

$$\dot{\tilde{x}}_{i2}^r = \bar{A}_{i4}\left(\tilde{x}_{i2}^r + \Omega_i^r\hat{\theta}_i^r - \Omega_i^s\theta_i^s\right) + A_{i3}\tilde{x}_{i1}^r + \eta_{i2} - \Omega_i^s\dot{\theta}_i^s$$

$$- \frac{d}{dt}(\Omega_i^r\hat{\theta}_i^r - \Omega_i^s\theta_i^s) + \rho_{i2}(x_i, u_i) - \rho_{i2}(\hat{x}_i^r, u_i)$$

$$+ \sum_{j=1}^{M} [H_{ij}^2(x_i, x_j, u_i, u_j) - H_{ij}^2(\hat{x}_i^r, \hat{x}_j, u_i, u_j)]$$

$$+ f_i^s(x_i, u_i) - f_i^s(\hat{x}_i^s, u_i).$$

By defining $\bar{x}_{i2}^r(t) \triangleq \tilde{x}_{i2}^r(t) + \Omega_i^r\hat{\theta}_i^r - \Omega_i^s\theta_i^s$, the above equation can be rewritten as follows

$$\dot{\bar{x}}_{i2}^r = \bar{A}_{i4}\bar{x}_{i2}^r + A_{i3}\tilde{x}_{i1}^r + f_i^s(x_i, u_i) - f_i^s(\hat{x}_i^s, u_i)$$

$$+ \sum_{j=1}^{M} [H_{ij}^2(x_i, x_j, u_i, u_j) - H_{ij}^2(\hat{x}_i^r, \hat{x}_j, u_i, u_j)]$$

$$+ \eta_{i2} - \Omega_i^s\dot{\theta}_i^s + \rho_{i2}(x_i, u_i) - \rho_{i2}(\hat{x}_i^r, u_i). \tag{3.57}$$

The pth component of the output estimation error generated by the rth local FIE for subsystem i (i.e. $\tilde{y}_{ip}^r(t) \triangleq y_{ip}(t) - \hat{y}_{ip}^r(t)$, $p = 1, \ldots, l_i$) is given by

$$\tilde{y}_{ip}^r(t) = C_{ip}\tilde{x}_{i2}^r(t) = C_{ip}(\bar{x}_{i2}^r(t) - \Omega_i^r\hat{\theta}_i^r + \Omega_i^s\theta_i^s)$$
$$= C_{ip}\bar{x}_{i2}^r(t) + h_{ip}^{sr}(t).$$

Invoking the triangle inequality, we have

$$|\tilde{y}_{ip}^r(t)| \geq |h_{ip}^{sr}(t)| - |C_{ip}\bar{x}_{i2}^r(t)|. \tag{3.58}$$

Note that Equation (3.57) is in a similar form to Equation (3.43). Therefore, by using Equation (3.57) and Equation (3.58) and by following similar logic to that reported in the proof of Lemma 3.3, we have

$$|\tilde{y}_{ip}^r(t)| \geq |h_{ip}^{sr}(t)| - \int_{T_d}^t |C_{ip}e^{\bar{A}_{i4}(t-\tau)}|[|\bar{\varrho}_i|\chi(\tau) + |\eta_{i2}| + |\Omega_i^s\dot{\theta}_i^s|]d\tau$$

$$-|C_{ip}e^{\bar{A}_{i4}(t-T_d)}||x_{i2}^r(T_d)|$$

$$\geq |h_{ip}^{sr}(t)| - k_{ip}\int_{T_d}^t e^{-\lambda_{ip}(t-\tau)}[|\bar{\varrho}_i|\chi(\tau) + |\eta_{i2}| + |\Omega_i^s\dot{\theta}_i^s|]d\tau$$

$$-k_{ip}e^{-\lambda_{ip}(t-T_d)}|x_{i2}^r(T_d)|.$$

Taking into account the corresponding adaptive threshold $\mu_{ip}^r(t)$ given by Equation (3.47), we can conclude that, if Equation (3.54) is satisfied at time $t = t^r$, we obtain $|\tilde{y}_{ip}^r(t^r)| > \mu_{ip}^r(t^r)$, which implies that the possibility of the occurrence of fault r in subsystem i can be excluded at time $t = t^r$. ∎

Remark 3.9 *According to Theorem 3.3, if, for each $r \in \{1, \ldots, N_i\} \{s\}$, the fault mismatch function $h_{ip}^{sr}(t^r)$ satisfies Equation (3.54) for some time $t^r > 0$, then the pth component of the output estimation error generated by the rth FIE of subsystem i would exceed its corresponding adaptive threshold at time $t = t^r$, i.e. $|\tilde{y}_{ip}^r(t^r)| > \mu_{ip}^r(t^r)$, hence excluding the occurrence of fault r in subsystem i. Therefore, Theorem 3.3 characterizes (in a non-closed form) the class of nonlinear faults that are isolable in each subsystem by the proposed robust distributed FDI scheme.*

3.1.10 Stability and learning capability

We now investigate the stability and learning properties of adaptive FIEs.

Theorem 3.4 *Suppose that fault s, occurring in the ith subsystem, is detected at time T_d, where $s \in \{1, \ldots, N_i\}$ and $i \in \{1, \ldots, M\}$. Then, the distributed fault isolation scheme described by Equations (3.37), (3.38) and (3.47) guarantees that:*

- *for each local fault isolation estimator q, $q = 1, \ldots, N_i$, the estimate variables $\hat{x}_{i1}^q(t)$, $\hat{x}_{i2}^q(t)$, and $\hat{\theta}_i^q(t)$ are uniformly bounded;*

- *there exists a positive constant $\bar{\kappa}_i$ and a bounded function $\bar{\xi}_i^s(t)$, such that, for all finite time $t_f > T_d$, the output estimation error of the matched sth local isolation estimator satisfies*

$$\int_{T_d}^{t_f} |\tilde{y}_i^s(t)|^2 dt \leq \bar{\kappa}_i + 2 \int_{T_d}^{t_f} |\bar{\xi}_i^s(t)|^2 dt. \tag{3.59}$$

Proof. Let us first address the signal boundedness property. The state estimation error and output estimation error of the qth FIE for the ith subsystem are defined as $\tilde{x}_{i1}^q(t) \triangleq x_{i1}(t) - \hat{x}_i^q(t)$, $\tilde{x}_{i2}^q(t) \triangleq x_{i2}(t) - \hat{x}_{i2}^q(t)$, and $\tilde{y}_i^q \triangleq y_i(t) - \hat{y}_i^q(t)$, respectively. Using Equation (3.37) and Equation (3.3), for $t > T_d$, the output estimation error is $\tilde{y}_i^q = C_i \tilde{x}_{i2}^q$, and the state estimation error satisfies

$$\dot{\tilde{x}}_{i1}^q = A_{i1}\tilde{x}_{i1}^q + \eta_{i1} + \rho_{i1}(x_i, u_i) - \rho_{i1}(\hat{x}_i^q, u_i)$$

$$+ \sum_{j=1}^{M} [H_{ij}^1(x_i, x_j, u_i, u_j) - H_{ij}^1(\hat{x}_i^q, \hat{x}_j, u_i, u_j)] \tag{3.60}$$

$$\dot{\tilde{x}}_{i2}^q = \bar{A}_{i4}\tilde{x}_{i2}^q + A_{i3}\tilde{x}_{i1}^q + \eta_{i2} + \rho_{i2}(x_i, u_i) - \rho_{i2}(\hat{x}_i^q, u_i)$$

$$+ G_i^s(x_i, u_i)\theta_i^s - G_i^s(\hat{x}_i^s, u_i)\theta_i^s + G_i^s(\hat{x}_i^q, u_i)\theta_i^s - G_i^q(\hat{x}_i^q, u_i)\theta_i^q - \Omega_i^q \dot{\hat{\theta}}_i^q$$

$$+ \sum_{j=1}^{M} [H_{ij}^2(x_i, x_j, u_i, u_j) - H_{ij}^2(\hat{x}_i^q, \hat{x}_j, u_i, u_j)]. \tag{3.61}$$

By substituting $G_i^s(\hat{x}_i^s, u_i) = \dot{\Omega}_i^s - \bar{A}_{i4}\Omega_i^s$ and $G_i^q(\hat{x}_i^q, u_i) = \dot{\Omega}_i^q - A_{i4}\Omega_i^q$, given by Equation (3.37), into Equation (3.61) and by defining $\bar{x}_{i2}^q \triangleq \tilde{x}_{i2}^q - \Omega_i^s \theta_i^s + \Omega_i^q \hat{\theta}_i^q$, we obtain

$$\dot{\bar{x}}_{i2}^q = \bar{A}_{i4}\bar{x}_{i2}^q + A_{i3}\tilde{x}_{i1}^q + \rho_{i2}(x_i, u_i) - \rho_{i2}(\hat{x}_i^q, u_i)$$

$$+ \sum_{j=1}^{M} [H_{ij}^2(x_i, x_j, u_i, u_j) - H_{ij}^2(\hat{x}_i^q, \hat{x}_j, u_i, u_j)] + \eta_{i2}$$

$$+ f_i^s(x_i, u_i) - f_i^s(\hat{x}_i^s, u_i) - \Omega_i^s \dot{\theta}_i^s. \tag{3.62}$$

Note that Equation (3.60) is in the same form as Equation (3.13). Therefore, based on the results of Lemma 3.2 (that is, Equation (3.19)) and Assumptions 3.2 and 3.3, we have $\tilde{x}_{i1}^q \in L_\infty$, $\tilde{x}_{j1} \in L_\infty$, and $\hat{x}_{i1}^q \in L_\infty$. Additionally, based on logic similar to that in the proof of Lemma 3.2, we know that $\rho_{i2}(x_i, u_i) - \rho_{i2}(\hat{x}_i^q, u_i)$, $H_{ij}^2(x_i, x_j, u_i, u_j) - H_{ij}^2(\hat{x}_i^q, \hat{x}_j, u_i, u_j)$ and $f_i^s(x_i, u_i) - f_i^s(\hat{x}_i^s, u_i)$ are bounded. Moreover, with parameter projection Equation (3.38), we have $\hat{\theta}_i^q \in L_\infty$. Furthermore, because η_{i2}, Ω_i^s, and $\dot{\theta}_i^s$ are bounded (Assumptions 3.2 and 3.6) and $\bar{A}_{i4}$ is a stable matrix, using Equation (3.62) we can obtain $\bar{x}_{i2}^q \in L_\infty$. The definition of $\bar{x}_2^q$ allows us to conclude that $\tilde{x}_2^q \in L_\infty$ and $\hat{x}_2^q \in L_\infty$. This concludes the first part of the theorem.

Now, let us prove the second part of the theorem, concerning the learning capability of the local FIE when it matches the sth fault in the local subsystem, i.e. $q = s$. In this case, the solution of Equation (3.62) can be written as $\bar{x}_{i2}^s(t) = \xi_{i1}^s(t) + \xi_{i2}^s(t)$, $\forall t \geq T_d$, where ξ_{i1}^s and ξ_{i2}^s are the solutions of the following differential equations, respectively,

$$\dot{\xi}_{i1}^s = \bar{A}_{i4}\xi_{i1}^s + A_{i3}\tilde{x}_{i1}^s + \rho_{i2}(x_i, u_i) - \rho_{i2}(\hat{x}_i^s, u_i)$$

$$+ \sum_{j=1}^{M}[H_{ij}^2(x_i, x_j, u_i, u_j) - H_{ij}^2(\hat{x}_i^s, \hat{x}_j, u_i, u_j)] + \eta_{i2}$$

$$+ f_i^s(x_i, u_i) - f_i^s(\hat{x}_i^s, u_i) - \Omega_i^s\dot{\theta}_i^s, \quad \xi_{i1}^s(T_d) = 0$$

$$\dot{\xi}_{i2}^s = \bar{A}_{i4}\xi_{i2}^s, \quad \bar{A}_{i4}\xi_{i2}^s(T_d) = \tilde{x}_{i2}^s(T_d) = x_{i2}^s(T_d).$$

Using the definition of $\bar{x}_{i2}^s$, we have $\tilde{x}_{i2}^s = \xi_{i1}^s(t) + \xi_{i2}^s(t) - \Omega_i^s\tilde{\theta}_i^s$. Therefore,

$$\tilde{y}_i^s(t) = C_i\tilde{x}_{i2}^s = C_i[\xi_{i1}^s(t) + \xi_{i2}^s(t)] - C_i\Omega_i^s\tilde{\theta}_i^s. \tag{3.63}$$

Now, consider a Lyapunov function candidate $V_i = \frac{1}{2\Gamma^s}(\tilde{\theta}_i^s)^2 + \int_t^\infty |C_i\xi_{i2}^s(\tau)|^2 d\tau$. The time derivative of V_i along the solution of Equation (3.38) is given by $\dot{V}_i = \frac{1}{\Gamma^s}\tilde{\theta}_i^s P_{\Theta^s}\{\Gamma^s\Omega_i^{s\top}C_i^\top\tilde{y}_i^s\} - |C_i\xi_{i2}^s|^2 - \frac{1}{\Gamma^s}\tilde{\theta}_i^s\dot{\theta}_i^s$. Clearly, since $\theta_i^s \in \Theta^s$, when the projection operator P is in effect, it always results in smaller parameter errors that decrease $\dot{V}_i$ [22, 8]. Therefore, using Equation (3.63) and completing the squares, we obtain

$$\dot{V}_i \leq (\tilde{y}_i^s)^\top C_i\Omega_i^s\tilde{\theta}_i^s - |C_i\xi_{i2}^s|^2 - \frac{1}{\Gamma^s}\tilde{\theta}_i^s\dot{\theta}_i^s$$

$$= (\tilde{y}_i^s)^\top(-\tilde{y}_i^s + C_i\xi_{i1}^s + C_i\xi_{i2}^s) - |C_i\xi_{i2}^s|^2 - \frac{1}{\Gamma^s}\tilde{\theta}_i^s\dot{\theta}_i^s$$

$$\leq -\frac{|\tilde{y}_i^s|^2}{2} + |C_i\xi_{i1}^s|^2 + \frac{1}{\Gamma^s}|\tilde{\theta}_i^s||\dot{\theta}_i^s|. \tag{3.64}$$

Let $\bar{\xi}_i^s \triangleq \left(|C_i\xi_{i1}^s|^2 + \frac{1}{\Gamma^s}|\tilde{\theta}_i^s||\dot{\theta}_i^s|\right)^{\frac{1}{2}}$. Integrating Equation (3.64) from $t = T_d$ to $t = t_f$, we obtain $\int_{T_d}^{t_f}|\tilde{y}_i^s(t)|^2 dt \leq \bar{\kappa}_i + 2\int_{T_d}^{t_f}|\bar{\xi}_i^s(t)|^2 dt$, where $\bar{\kappa}_i \triangleq \sup_{t_f \geq T_d}\{2[V_i(T_d) - V_i(t_f)]\}$. ∎

Theorem 3.4 guarantees the boundedness of all the variables involved in the local adaptive FIEs when a fault is detected in the corresponding subsystem. Moreover, the performance measure given by Equation (3.59) shows that the ability of the matched local isolation estimator to learn the post-fault system dynamics is limited by the extended L_2 norm of $\bar{\xi}_i^s(t)$, which, in turn, is related to the modeling uncertainties η_{i1} and η_{i2}, the parameter estimation error $\tilde{\theta}_i^s$, and the rate of change of the time-varying bias θ_i^s.

The stability and learning capability of local adaptive FIEs designed for each subsystem are also established. A simulation example of interconnected pendulums mounted on carts shows the effectiveness of the distributed FDI method. One direction for future research is to consider interconnected nonlinear systems with more general nonlinearities and interconnection terms. The interconnection term considered in this chapter is assumed to be partially known. The case of completely unknown interconnection effects should be investigated to extend the applicability of the distributed FDI method. Another interesting research topic is the integration of fault diagnosis with fault-tolerant control techniques to compensate for the effect of faults using online diagnostic information (e.g., see [24]).

3.2 Robust Fault Detection Filters

This section addresses three problems:

- determination of a reference residual model;
- design of an adaptive threshold;
- unscented Kalman filter-based iterative update of noise mean and covariance.

3.2.1 Reference model

Consider LTI Equations (2.28)–(2.29) with observer-based FDF Equations (2.32)–(2.34). On the basis of the work of Ding [25], we formulate the nominal case optimal FDF design problem as the following optimization problem:

$$\min_{H,V} \ J,$$

$$J = \frac{\| V(D_d + (sI - A + HC)^{-1}(B_d - HD_d)) \|_\infty}{\inf_{\omega \in \Omega} \sigma_i V(D_f + (j\omega I - A + HC)^{-1}(B_f - HD_f))}$$

$$\Omega \subseteq [0, \infty) \tag{3.65}$$

Assuming that Assumptions 3.1–3.3 are satisfied, a state space analytical solution of the optimal FDF problem can be obtained by direct use of Ding's Theorem 2 [25].

Theorem 3.5 *Given Equations (2.28)–(2.29) with* $\Delta A = 0$ *and* $\Delta B = 0$, *suppose Assumptions 3.1–3.3 hold true and*

$$H^* = (B_d D_d^T + YC^T)Q^{-1} \tag{3.66}$$

$$V^* = Q^{-1/2}. \tag{3.67}$$

Solve the optimization problem Equation (3.65), where $Q = D_d D_d^T$, $Y \geq 0$ *is a solution of the algebraic Riccati equation.*

$$Y(A - B_d D_d^T Q^{-1}C)^T + (A - B_d D_d^T Q^{-1}C)Y$$

$$- YC^T Q^{-1}CY + B_d(I - D_d^T Q^{-1}D_d)^2 B_d^T = 0 \tag{3.68}$$

Using Theorem 3.5, we obtain a solution of optimal FDF design for LTI systems without modeling errors, which can be used as the reference residual model of RFDF design.

Remark 3.10 *H^* and V^*, given in Equation (3.66) and Equation (3.67), solve the optimal fault detection filter design problem for the LTI system in Equation (2.28) and Equation (2.29)* *with $\Delta A = 0$ and $\Delta B = 0$. The corresponding reference residual model is given by:*

$$\dot{x}_f(k) = (A - H^*C)x_f(k) + (B_f - H^*D_f)f(k)$$
$$+ (B_d - H^*D_d)d(k), \tag{3.69}$$

$$r_f = V^*Cx_f(k) + V^*D_f f(k) + V^*D_d d(k) \tag{3.70}$$

Lemma 3.4   *Consider the uncertain LTI system*

$$\dot{x}(k) = (A + \Delta A)x(k) + (B + \Delta B)w(k), \tag{3.71}$$

$$z(k) = Cx(k) + Dw(k) \tag{3.72}$$

with $\Delta A = \Omega_1$ and $\Delta B = \Omega_2$. Given $\gamma > 0$, if there exist scalars, $\varepsilon_1 > 0$, $\varepsilon_2 > 0$, and a matrix $P > 0$ such that the following equation holds:

$$\begin{bmatrix} PA + A^T P + \varepsilon_1 F_1^T F_1 & PB & C^T & PE_1 & PE_2 \\ B^T P & -\gamma^2 I + \varepsilon_2 F_2^T F_2 & D^T & D^T & 0 \\ C & D & -I & 0 & 0 \\ E_1^T P & 0 & 0 & -\varepsilon_1 I & 0 \\ E_2^T P & 0 & 0 & 0 & -\varepsilon_2 I \end{bmatrix} < 0, \tag{3.73}$$

then for any $\Delta A \in \Omega_1$, $\Delta B \in \Omega_2$, the system is asymptotically stable and satisfies $\|z\|_2 < \gamma \|w\|_2$. Now, the RFDF design problem formulated earlier can be solved by repeatedly using Lemma 3.4.

Theorem 3.6   *Given $\gamma > 0$, if there exist scalars $\varepsilon_1 > 0$, $\varepsilon_2 > 0$ and matrices $P_1 > 0$, $P_2 > 0$, $P_3 > 0$, Y_1, V such that the LMI*

$$[N_{ij}]_{9 \times 9} < 0 \tag{3.74}$$

holds, then system is asymptotically stable and H_∞ performance is satisfied.

$$N_{11} = P_1 A + A^T P_1 - Y_1 C - C^T Y_1, N_{15} = P_1 B_f - Y_1 D_f,$$
$$N_{16} = P_1 B_d - Y_1 D_d, N_{17} = C^T V^T, N_{18} = P_1 E_1,$$
$$N_{19} = P_1 E_2, N_{22} = P_2 A_0 + A_0^T P_2, N_{25} = P_2 B_{of},$$
$$N_{26} = P_2 B_{od}, N_{27} = -C_0^T, N_{33} = P_3 A + A^T P_3 +$$
$$\varepsilon_1 F_1^T F_1, N_{34} = P_3 B, N_{35} = P_3 B_f, N_{36} = P_3 B_d,$$
$$N_{38} = P_3 E_1, N_{39} = P_3 E_2, N_{44} = -\gamma^2 I + \varepsilon_2 F_2^T F_2,$$

$$N_{55} = -\gamma^2 I, \, N_{57} = D_f^T V^T - D_{of}^T, \, N_{66} = -\gamma^2 I,$$

$$N_{67} = D_d^T V^T - D_{od}^T, \, N_{77} = -I, \, N_{88} = -\varepsilon_1 I, \, N_{99} = -\varepsilon_2 I,$$

$$N_{ij} = 0, otherwise. \tag{3.75}$$

3.2.2 Design of adaptive threshold

Consider the designed residual generation system:

$$\dot{x}(k) = Ax(k) + Bu(k) + B_f f(k) + B_d d(k) + \Delta Bu(k)$$
$$+ \Delta Ax(k), \tag{3.76}$$

$$\dot{e}(k) = (A - HC)e(k) + (Bd - HD_d)d(k) + \Delta Ax(k)$$
$$+ \Delta Bu(k) + (B_f - HD_f)f(k) \tag{3.77}$$

$$r(k) = VCe(k) + VD_f f(k) + VD_d d(k) \tag{3.78}$$

By selecting Equation (2.39) as the residual evaluation function, we have:

$$\| r \|_{2,K} = \| r_d(k) + r_u(k) + r_f(k) \|_{2,K}, \tag{3.79}$$

where $r_d(k)$, $r_u(k)$ and $r_f(k)$ are defined as follows:
$r_d(k) = r(k) \, |_{u=0, f=0},$
$r_u(k) = r(k) \, |_{d=0, f=0},$
$r_f(k) = r(k) \, |_{d=0, u=0} \, .$
Moreover, the fault-free residual evaluation function is:

$$\| r_d + r_u \|_{2,K} \leq \| r_d \|_{2,K} + \| r_u \|_{2,K} \leq J_{th,d} + J_{th,u} \tag{3.80}$$

where

$$J_{th,d} = sup_{\Delta A \in \Omega_1, \Delta B \in \Omega_2, d \in L_2} \| r_d \|_{2,K}$$
$$J_{th,u} = sup_{\Delta A \in \Omega_1, \Delta B \in \Omega_2} \| r_u \|_{2,K} \tag{3.81}$$

We choose the threshold J_{th} as:

$$J_{th} = J_{th,d} + J_{th,u} \tag{3.82}$$

where $J_{th,d}$ is constant and can be evaluated off-line, while $u(k)$ is assumed to be known online and $J_{th,u}$ can be evaluated online by:

$$J_{th,u} = \gamma_u \| u \|_{2,K} \tag{3.83}$$

where

$$\gamma_u = sup_{\Delta A \in \Omega_1, \Delta B \in \Omega_2} \frac{\| r_u \|_{2,K}}{\| u \|_2} \tag{3.84}$$

and γ_u can be determined using Lemma 3.4. Under the assumption of $d \in L_2$, we can further have

$$sup_{\Delta A \in \Omega_1, \Delta B \in \Omega_2, d \in L_2} \| r_d(k) \|_{2,K} = M_T (M_T > 0). \tag{3.85}$$

Therefore, the threshold can be determined by

$$J_{th} = M_T + \gamma_u \| u \|_{2,K} \tag{3.86}$$

Note that the defined threshold consists of two parts: the constant part M_T and part $J_{th,u}$ which depends on the plant input u and can be calculated online. Therefore, changing the plant input u implies a new determination of the threshold. In this sense, the threshold is adaptive.

Remark 3.11 *It is pointed out that the above design of fault detection system can achieve a false-alarm rate of zero, at the expense of a miss rate above zero, since the worst case modeling errors and unknown inputs are considered in either the residual generator design or the threshold implementation. To overcome this problem, one may select a threshold smaller than J_{th} based on some acceptable false-alarm rate.*

3.2.3   Iterative update of noise mean and covariance

The discrete-time unscented Kalman filter (UKF) is considered here for the noise and covariance upgrade. To achieve it, fault detection and estimation is performed through UKF, which also ensures stability. Consider the following assumption:

Assumption 3.8   The transfer function matrix $C[zI - (A - KC)]^{-1}B$ is strictly positive real, where $H \in \mathfrak{R}^{n \times r}$ is chosen such that $A - KC$ is stable.

Remark 3.12   *For a given positive definite matrix $Q > 0 \in \mathfrak{R}^{n \times n}$, there exists matrices $P = P^T > 0 \in \mathfrak{R}^{n \times n}$ and a scalar R such that:*

$$(A - HC)^T P(A - HC) = -Q \tag{3.87}$$

$$PB = C^T R \tag{3.88}$$

To detect a fault with UKF, the structure of the UKF is followed; it uses a nonlinear transformation to calculate the statistics of a random variable. The following system is constructed:

$$\hat{x}(k) = A\hat{x}(k) + g(u(k), y(k)) + B\xi_H f(u(k), y(k), \hat{x}(k))$$

$$+ K(y(k) - \hat{y}(k)) \tag{3.89}$$

$$\hat{y}(k) = C\hat{x}(k) \tag{3.90}$$

where $\hat{x}(k) \in \Re^n$ is the state estimate, the input is $u \in \Re^m$, and the output is $y \in \Re^r$. The pair (A, C) is observable. The nonlinear term $g(u(k), y(k))$ depends on $u(k)$ and $y(k)$ which are directly available. The function $f(u(k), y(k), x(k)) \in \Re^r$ is a nonlinear vector function of $u(k)$, $y(k)$ and $x(k)$. $\xi(k) \in \Re$ is a parameter which changes unexpectedly when a fault occurs. Since it has been assumed that the pair (A, C) is observable, a gain matrix K can be selected such that $A - KC$ is a stable matrix. We define:

$$e_x(k) = x(t) - \hat{x}(k), \quad e_y(k) = y(k) - \hat{y}(k). \tag{3.91}$$

Then, the error equations can be given by:

$$e_x(k + 1) = (A - KC)e_x(k) + B[\xi(k)f(u(k), y(k), x(k))$$
$$-\xi_H f(u(k), y(k), \hat{x}(k))], \tag{3.92}$$
$$e_y(k) = Ce_x(k). \tag{3.93}$$

The convergence of the above filter is guaranteed by the following theorem.

Theorem 3.7 *Under Assumption 3.8, the filter is asymptotically convergent when no fault occurs ($\xi(k) = \xi_H$), i.e., $\lim_{k \to \infty} e_y(k) = 0$.*

Proof. Consider the Lyapunov function

$$V(e(k)) = e_x^T(k)Pe_x(k) \tag{3.94}$$

where P is the solution of Equation (3.87), Q is chosen such that $\rho_1 = \lambda_{min}(Q) - 2\|C\|.|R|\xi_H L_0 > 0$. Along the trajectory of the fault-free system, Equation (3.92), the corresponding Lyapunov difference $e(k)$ is:

$$\Delta V = E\{V(e(k + 1)|e_k, p_k)\} - V(e(k))$$
$$= E\{e^T(k + 1)P_i e(k + 1)\} - e^T(k)P_i e(k)$$
$$= (A_e e_x + B_L u_e)^T P(A_e e_x + B_L u_e) - e_x^T(k)Pe_x(k)$$
$$= e^T(k)[(P(A - KC) + (A - KC)^T P)$$
$$+ PB\xi_H[f(u(k), y(k), x(k))$$
$$- f(u(k), y(k), \hat{x}(k))]]e(k) \tag{3.95}$$

From Equation (3.87), one can further obtain that

$$\Delta V \leq -e_x^T(k)Qe_x(t) + 2\|e_y(k)\|.|R|\xi_H L_0\|e_x(k)\|$$
$$\leq -\rho_1\|e_x\|^2 < 0 \tag{3.96}$$

Thus, $\lim_{k \to \infty} e_x(k) = 0$ and $\lim_{k \to \infty} e_y(k) = 0$. This completes the proof. ∎

The UKF essentially addresses the approximation issues of the extended Kalman filter (EKF) [26, 27, 28]. The basic difference between the EKF and UKF stems from the manner in which Gaussian random variables (GRV) are presented through system dynamics. In the EKF, the state distribution is approximated by a GRV, which is then propagated analytically through the first-order linearization of the nonlinear system. This can introduce large errors in the true posterior mean covariance of the transformed GRV, which may lead to sub-optimal performance and divergence of the filter. The UKF addresses this problem by using a deterministic sampling approach. The state distribution is again approximated by a GRV, but it is now represented using a minimal set of carefully chosen sample points. These sample points completely capture the true mean and covariance of the GRV, and when propagated through the *true* nonlinear system, capture the posterior mean and covariance accurately to second order (Taylor series expansion) for any nonlinearity. The EKF, in contrast, only achieves first-order accuracy.

3.2.4 Unscented transformation (UT)

The structure of the UKF is elaborated by UT for calculating the statistics of a random variable which undergoes a nonlinear transformation [28]. Consider propagating a random variable x (dimension L) through a nonlinear function, $y = f(x)$. Assume x has mean $\bar{x}$ and covariance P_x. To calculate the statistics of y, we form a matrix $\mathcal{X}$ of $2L + 1$ sigma vectors $\mathcal{X}_i$ according to:

$$
\begin{aligned}
\mathcal{X}_0 &= \bar{x}, \\
\mathcal{X}_i &= \bar{x} + (\sqrt{(L+\lambda)P_x})_i, \quad i = 1, \ldots, L \\
\mathcal{X}_i &= \bar{x} - (\sqrt{(L+\lambda)P_x})_i - L, \quad i = L+1, \ldots, 2L
\end{aligned}
\tag{3.97}
$$

where $\lambda = \alpha^2(L + \kappa) - L$ is a scaling parameter. The constant α determines the spread of the sigma points around $\bar{x}$ and is usually set to a small positive value ($1 \leq \alpha \leq 10^{-4}$). The constant κ is a secondary scaling parameter, which is usually set to $3 - L$, and β is used to incorporate prior knowledge of the distribution of x (for Gaussian distributions, $\beta = 2$ is optimal). $\sqrt{(L+\lambda)P_x})_i$ is the ith column of the matrix square root (that is, the lower-triangular Cholesky factorization). These sigma vectors are propagated through the nonlinear function $\mathcal{Y}_i = f(\mathcal{X}_i), i = 0, \ldots, 2L$. Now the mean and covariance for y are approximated using a weighted sample mean and covariance of the posterior sigma points:

$$
\bar{y} \approx \sum_{i=0}^{2L} W_i^m \mathcal{Y}_i,
$$

$$
\mathbf{P}_y \approx \sum_{i=0}^{2L} W_i^c (\mathcal{Y}_i - \bar{y})(\mathcal{Y}_i - \bar{y})^T,
$$

$$
W_0^{(m)} = \frac{\lambda}{L + \lambda},
$$

$$W_0^{(c)} = \frac{\lambda}{L + \lambda} + 1 - \alpha^2 + \beta,$$

$$W_i^{(m)} = W_i^{(c)} = \frac{1}{2(L + \lambda)}, i = 1, \ldots, 2L.$$

A block diagram illustrating the steps in performing the UT is shown in Figure 2.4.

Remark 3.13   *Note that this method differs substantially from general Monte Carlo sampling methods, which require orders of magnitude more sample points in an attempt to propagate an accurate (possibly non-Gaussian) distribution of the state. The deceptively simple approach taken with the UT results in approximations that are accurate to the third order for Gaussian inputs for all nonlinearities. For non-Gaussian inputs, approximations are accurate to at least the second order, with the accuracy of the third- and higher-order moments being determined by the choice of α and β.*

In view of the foregoing, the UKF is an extension of the UT to the following recursive estimation:

$$\hat{x}_k = x_{k_{prediction}} + \kappa_k \left[y_k - y_{k_{prediction}} \right] \tag{3.98}$$

where the state random variables (RV) are redefined as the concentration of the original state and noise variables: $x_k^a = [x_k^T \ v_k^t \ n_k^t]^T$. The UT sigma point selection scheme is then applied to this new augmented state RV to calculate the corresponding sigma matrix, $\mathcal{X}_k^a$. The UKF equations are given below. Note that no explicit Jacobian or Hessian calculations are necessary to implement this algorithm. Initialize with:

$$\hat{x}_0 = E[x_0],$$

$$P_0 = E[(x_0 - \hat{x}_0)(x_0 - \hat{x}_0)^T],$$

$$\hat{x}_0^a = E[x^a] = [\hat{x}_0^T \ 0 \ 0]^T. \tag{3.99}$$

For $k \in [1, \ldots, \infty]$, calculate the sigma points:

$$\mathcal{X}_{k-1}^a = [\hat{x}_{k-1}^a \ \ \hat{x}_{k-1}^a + \gamma \sqrt{P_{k-1}^a} \ \ \hat{x}_{k-1}^a - \gamma \sqrt{P_{k-1}^a}] \tag{3.100}$$

The UKF time-update equations are:

$$\mathcal{X}_{k|k-1}^x = F(\mathcal{X}_{k-1}^x, \ u_{k-1}, \ \mathcal{X}_{k-1}^v),$$

$$\hat{x}_k^- = \sum_{i=0}^{2L} W_i^m \mathcal{X}_{i,k|k-1}^x,$$

$$P_k^- = \sum_{i=0}^{2L} W_i^c (\mathcal{X}_{i,k|k-1}^x - \hat{x}_k^-)(\mathcal{X}_{i,k|k-1}^x - \hat{x}_k^-)^T,$$

$$\mathcal{Y}_{k|k-1} = \mathbf{H}(\mathcal{X}^x_{k|k-1}, \mathcal{X}^n_{k-1}),$$

$$\hat{y}^-_k = \sum_{i=0}^{2L} W^m_i \mathcal{Y}_{i,k|k-1} \tag{3.101}$$

The UKF measurement-update equations are:

$$P_{\tilde{y}_k \tilde{y}_k} = \sum_{i=0}^{2L} W^c_i (\mathcal{Y}_{i,k|k-1} - \hat{y}^-_k)(\mathcal{Y}_{i,k|k-1} - \hat{y}^-_k)^T,$$

$$P_{x_k y_k} = \sum_{i=0}^{2L} W^c_i (\mathcal{X}_{i,k|k-1} - \hat{x}^-_k)(\mathcal{Y}_{i,k|k-1} - \hat{y}^-_k)^T,$$

$$\kappa_k = P_{x_k y_k} P^{-1}_{\tilde{y}_k \tilde{y}_k},$$

$$\hat{x}_k = \hat{x}^-_k + \kappa_k(y_k - \hat{y}^-_k),$$

$$P_k = P^-_k - \kappa_k P_{\tilde{y}_k \tilde{y}_k} \kappa^T_k \tag{3.102}$$

where

$$x^a = [x^T \quad v^T \quad n^T]^T,$$

$$\mathcal{X}^a = [(\mathcal{X}^x)^T \quad (\mathcal{X}^v)^T \quad (\mathcal{X}^n)^T]^T, and$$

$$\gamma = \sqrt{L + \lambda}. \tag{3.103}$$

In addition, λ is the composite scaling parameter, L is the dimension of the augmented state, $\mathfrak{R}^v$ is the process-noise covariance, $\mathfrak{R}^n$ is the measurement-noise covariance, and W_i are the weights.

The time-update equations are:

$$\chi^x_{k|k-1} = \mathrm{E}(\chi^x_{k-1}, u_{k-1}, \chi^v_{k-1}),$$

$$\hat{x}^-_k = \sum_{i=0}^{2L} W^m_i \chi^x_{i,k|k-1},$$

$$P^-_k = \sum_{i=0}^{2L} W^c_i (\chi^x_{i,k|k-1} - \hat{x}^-_k)(\chi^x_{i,k|k-1} - \hat{x}^-_k)^T,$$

$$Y_{k|k-1} = \mathbf{H}(\chi^x_{k|k-1}, \chi^n_{k-1}), \quad \hat{y}^-_k = \sum_{i=0}^{2L} W^m_i \chi^Y_{i,k|k-1}. \tag{3.104}$$

The measurement-update equations are:

$$P_{\tilde{y}_k \tilde{y}_k} = \sum_{i=0}^{2L} W^c_i (Y_{i,k|k-1} - \hat{y}^-_k)(Y_{i,k|k-1} - \hat{y}^-_k)^T,$$

$$P_{x_k y_k} = \sum_{i=0}^{2L} W_i^c (\chi_{i,k|k-1} - \hat{x}_k^-)(Y_{i,k|k-1} - \hat{y}_k^-)^T,$$

$$\kappa_k = P_{x_k y_k} P_{\bar{y}_k \bar{y}_k}^{-1}, \quad \hat{x}_k = \hat{x}_k^- + \kappa_k(y_k - \hat{y}_k^-),$$

$$P_k = P_k^- - \kappa_k P_{\bar{y}_k \bar{y}_k} \kappa_k^T \tag{3.105}$$

where $x^a = [x^t \quad v^t \quad n^t]^t$, $\chi^a = [(\chi^x)^t \quad (\chi^v)^t \quad (\chi^n)^t]^t$ and $\gamma = \sqrt{L + \lambda}$. In addition, λ is the composite scaling parameter, L is the dimension of the augmented state, $\Re^v$ is the process-noise covariance, $\Re^n$ is the measurement-noise covariance, and W_i are the weights.

The iterative update of observation noise mean and covariance is given by μ^+ and $\Re^+$ respectively:

$$\mu^+ = \frac{\sum_{n=1}^N \int [z_n - S_n(x_n)] \alpha_n(x_n \beta_n(x_n)) dx_n}{N \sum_{n=1}^N \int \alpha_n(x_n) \beta_n(x_n) dx_n} \tag{3.106}$$

$$\bar{z}_n = z_n - S_n(x_n) - \mu, \tag{3.107}$$

$$\Re^+ = \frac{\sum_{n=1}^N \int \bar{z}_n \bar{z}_n^T \alpha_n(x_n \beta_n(x_n)) dx_n}{N \sum_{n=1}^N \int \alpha_n(x_n) \beta_n(x_n) dx_n}, \tag{3.108}$$

where z_n is the state with noise under observation, S_n is the estimated state $\times$ the C matrix, α_n is the estimated state, and N is the number of iterations.

3.2.5   Car-like mobile robot application

This system has been evaluated and tested on a car-like mobile robot [29]. The underside body diagram and a photograph of the robotic vehicle can be seen in Figure 3.1.

Figure 3.1   Car-like mobile robot

The following is a basic model that represents the physical system of a car-like mobile robot.

$$dotx = [cos\theta - \frac{btan\phi}{l}sin\theta]v_u \tag{3.109}$$

$$\dot{y} = [sin\theta + \frac{btan\phi}{l}cos\theta]v_u \tag{3.110}$$

$$\dot{\theta} = \frac{tan\phi}{l}v_u \tag{3.111}$$

$$\dot{v}_u = \frac{v_u(b^2m + J)tan\phi}{(cos\phi)^2[l^2m + (b^2m + J)(tan\phi)^2]}(\frac{1}{\tau_s}\phi$$

$$+ c_s u_2) + \frac{l^2(cos\phi)^2}{(cos\phi)^2[l^2m + (b^2m + J)(tan\phi)^2]}F_D \tag{3.112}$$

$$\dot{F}_D = -\frac{R_a}{L_a}F_D - \frac{K_mK_b + R_ab_m}{L_aN_m^2R_w^2}v_u + \frac{K_mN_w}{L_aN_mR_w}u_1 \tag{3.113}$$

$$\dot{\phi} = \frac{1}{\tau_s}\phi + c_s u_2 \tag{3.114}$$

The car-like mobile robot may experience various faults that reduce its performance and reliability. The faults can occur in components such as sensors, actuators, controllers, communication system elements and the actual physical platform. Since good driving control is essential, the fuel must be provided appropriately – the fuel is the controller of the vehicle. With these factors in mind, the faults considered in this study are those that cause battery voltage drop, i.e. controller fault.

3.2.5.1   Fault scenarios in a car-like mobile robot

Fault scenarios are created using the car-like mobile robot in the simulation program. In these scenarios, sensor fault, actuator fault, and controller fault are considered.

Scenario I: Sensor Fault
While the system is working in real time, sensor faults are introduced into the system. The sensor faults are considered as x^f and y^f in state 1 and state 2 respectively.

Scenario II: Actuator Fault
While the system is working in real time, actuator faults are introduced into the system. $v_{u_{arf}}$ and u^f are considered as the actuator faults of the system in state 3 and state 6 respectively.

A DC motor, which converts DC current into torque, drives the rear wheels of the vehicle. The more current that is sent through the windings in the motor, the more torque it produces and the faster the shaft turns. Since current is proportional to voltage, the shaft speed is proportional to the voltage across the motor. However, motors are typically controlled by digital electronics with that output only 0 V and 5 V. However, if 0 V or 5 V is applied, the motor will run at only two speeds. To control the voltage using a digital signal, a pulse width modulation (PWM) signal is used. PWM signals take advantage of the fact that DC motors cannot respond as quickly as digital electronics can operate. So, if the motor is given a signal that is changing between 0 V and 5 V at a speed of 2 kHz with a 50 % duty cycle (defined as the ratio of high time to the period of the signal), the motor cannot possibly turn on and off so quickly. The result is that the motor acts as if it is receiving a 2.5 V. In general, the voltage seen by the motor is the average over one period of the PWM signal.

Scenario III: Controller Fault

While the system is working in real time and getting the input for driving the dynamics of the system, a fault is introduced in the driving force. $\frac{K_m N_w}{L_a N_m R_w} u_{cl_f}$ is considered in state 5 of the system.

3.2.5.2 Modeling the fault scenarios

Considering the fault scenarios, the fault operating modes are as follows:

- *Mode 1:* Battery voltage rise above 2.5 V
- *Mode 2:* Battery voltage drop of 2.5 V
- *Mode 3:* Decoupling of encoder on x-axis
- *Mode 4:* Decoupling of encoder on y-axis
- *Mode 5:* Effect on armature due to voltage drop
- *Mode 6:* Lack of controller gain due to voltage drop.

In the models M_1 and M_2 for Mode 1 and Mode 2, a battery voltage drop is indicated by a negative value added to the right of Equation (3.111). This study examined two different magnitudes of voltage drop.

$$\frac{d\theta}{dt} = \frac{tan\phi}{l} v_u - v_{u_{a_f}},$$

$$where \begin{cases} v_{u_{a_f}} = 0 \text{ if } no\, fault, v_{u_{a_f}} \in \Re \\ v_{u_{a_f}} \neq 0 \text{ if } faulty, v_{u_{a_f}} \in \Re \end{cases} \tag{3.115}$$

Note that $v_{u_{a_f}}$ is the armature fault of the DC motor reflecting M_1 and M_3.

In the models M_3 and M_4 for Mode 3 and Mode 4, the decoupling of a wheel encoder on the x-axis or the y-axis is indicated by a zero value for the corresponding measurement in $y_k = h(x_k, v_k)$.

$$\frac{dx}{dt} = [cos\theta - \frac{btan\phi}{l} sin\theta]v_u - x^f,$$

$$where \begin{cases} x^f = 0 \text{ if } no\, fault, x^f \in \Re \\ x^f \neq 0 \text{ if } faulty, x^f \in \Re \end{cases} \tag{3.116}$$

x^f represents the sensor fault in the x-axis reflecting M_3.

$$\frac{dy}{dt} = [sin\theta + \frac{btan\phi}{l}cos\theta]v_u - y^f$$

$$where \begin{cases} y^f = 0 \ if \ no \ fault, \ y^f \in \Re \\ y^f \neq 0 \ if \ faulty, \ y^f \in \Re \end{cases} \quad (3.117)$$

y^f represents the sensor fault in the y-axis reflecting M_4.

In the model M_5 for Mode 5, a battery voltage drop affects the armature, resulting in a fault indicated by a negative value added to the right of Equation (3.111).

$$\frac{d\theta}{dt} = \frac{tan\phi}{l}v_u - v_{u_{arf}}$$

$$where \begin{cases} v_{u_{arf}} = 0 \ \ if \ no \ fault, \ v_{u_{arf}} \in \Re \\ v_{u_{arf}} \neq 0 \ \ if \ faulty, \ v_{u_{arf}} \in \Re \end{cases} \quad (3.118)$$

$v_{u_{arf}}$ is the armature fault of the DC motor reflecting M_5.

In the model M_6 for Mode 6, a lack of controller gain due to voltage drop is indicated by a negative value added to the right of Equation (3.113).

$$\frac{dF_D}{dt} = -\frac{R_a}{L_a}F_D - \frac{K_mK_b + R_ab_m}{L_aN_m^2R_w^2}v_u$$

$$+ \frac{K_mN_w}{L_aN_mR_w}u_1 - \frac{K_mN_w}{L_aN_mR_w}u_{cl_f}$$

$$where \begin{cases} \frac{K_mN_w}{L_aN_mR_w}u_{cl_f} = 0 \ if \ no \ fault, \\ \frac{K_mN_w}{L_aN_mR_w}u_{cl_f} \neq 0 \ if \ faulty, \end{cases} \quad (3.119)$$

Similarly,

$$\frac{K_mN_w}{L_aN_mR_w}u_{cl_f}$$

is the controller fault for the DC motor reflecting M_6.

The UKF UKF_i, corresponding to node i, was designed on the basis of model M_i of the nonlinear discrete system which describes the system for fault i.

The estimation vector $\hat{y}_{k,i}$ is compared with the measurements vector y_k, thus yielding a residuals vector $r_{k+1,i} = y_{k+1} - y_{k\hat{+}1,i}$ that is processed by the fault isolation module to determine the mode in which the robot is operating (represented by M_i).

The conditional probability that hypothesis H_i is true at time $k + 1$ is given by the following expression:

$$P_{k_1}(H_i) = \frac{g(y_{k+1}/M_i, [y_0 \ldots, y_k]).Pk(H_i)}{\sum_{j=0}^{n}(g(y_{k+1}/M_j, [y_0 \ldots, y_k]).P_k(H_j))} \tag{3.120}$$

In Equation (3.120), $g(.)$ is the conditional probability density function of the measurement y_{k+1}, conditioned on the model M_i and the previous measurements. This function is determined by the expression:

$$g(y_{k+1,i}/M_i, [y_0^T \ldots, y_k^T]) = \frac{e^{-\frac{1}{2}D_{k+1,i}}}{(2\pi)^{\frac{m}{2}}|S_{k+1,i}|^{\frac{1}{2}}} \tag{3.121}$$

where D_{k+1} is the Mahalanobis distance at time $k + 1$, defined as:

$$D_{k+1,j} = \mathfrak{R}_{k+1,i}^T S_{k+1,i}^{-1} r_{k+1,i} \tag{3.122}$$

and m is the number of elements in the measurement vector.

3.2.5.3 Simulation of the faults

In the first stage, fault estimation was implemented through the residual generation and evaluation function. Figure 3.2 shows the step response of all the nonlinear states of the car-like mobile robot. Figures 3.3–3.8 show, for each of the six nonlinear modes of the robot, simulation results for the fault estimation and comparison with the state profile.

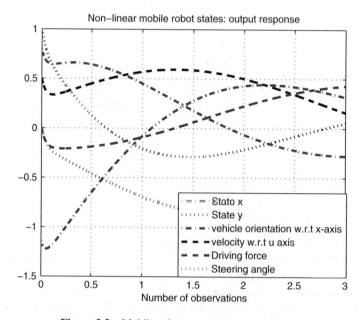

Figure 3.2 Mobile robot: output response of states

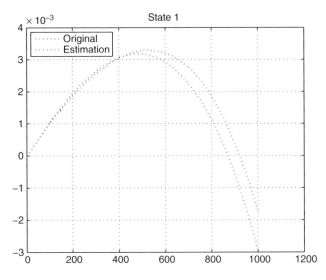

Figure 3.3 Mobile robot: fault estimation for Mode 1

The next stage involved updating the noise mean and covariance using UKF. The backward probability element β was calculated (see Figure 3.9) followed by the covariance (see Figure 3.10).

We consider here the hypothesis-based fault detection of Mode 4 (the actuator fault). Using Equation (3.120), the hypothesis equation for Mode 4 is:

$$P_k(H_4) = \frac{g(y_{k+1}/M_i, [y_0 \ldots, y_k]).Pk(H_i)}{\sum_{j=0}^{n}(g(y_{k+1}/M_j, [y_0 \ldots, y_k]).P_k(H_j))} \tag{3.123}$$

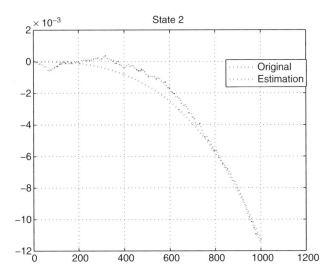

Figure 3.4 Mobile robot: fault estimation for Mode 2

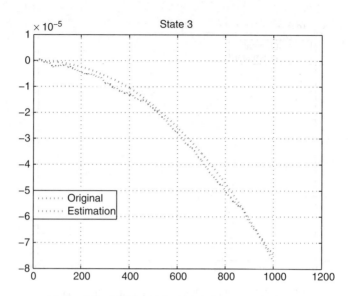

Figure 3.5   Mobile robot: fault estimation for Mode 3

In the implementation of UKF, hypothesis-testing helped to effectively detect a Mode 4 fault (Figure 3.11).

The simulation result for the estimation of the actuator fault using the UKF is shown in Figure 3.12. It can be seen that, with an increasing precision and a more detailed fault picture, UKF was able to estimate the fault in a much better way. It even spotted a very small kink resulting from the actuator fault.

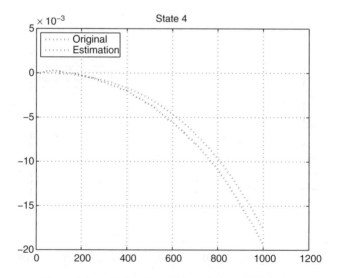

Figure 3.6   Mobile robot: fault Estimation for Mode 4

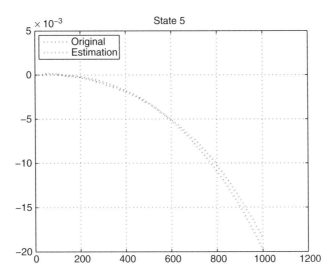

Figure 3.7 Mobile robot: fault Estimation for Mode 5

A fault may occur in any phase and in any part of the plant. Critical faults not detected on time can lead to adverse effects.

In the following discussion, the fault detection of the faults caused by any drift using unscented Kalman filter is clarified.

It can be seen from Figure 3.13 (for Mode 1) that the fault is so incipient that, apart from in the beginning, it is showing the same profile. Thus, drift detection can give us a better picture for the fault scenario. This can also be seen in case of Mode 5, as shown in Figure 3.14. The

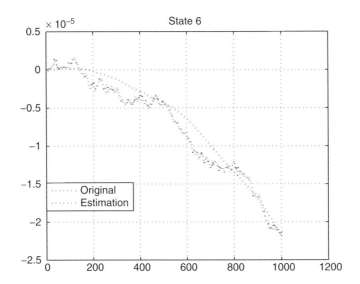

Figure 3.8 Mobile robot: fault Estimation for Mode 6

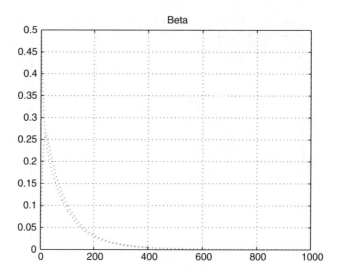

Figure 3.9   Backward probability in noise update

kinks showing the middle of the profile can alert the engineer to some unusual practice going on in the process.

3.3   Simultaneous Fault Detection and Control

We have learned from the preceding sections that fault diagnosis (FD) has received much attention for complex modern automatic systems (such as cars, aircraft, rockets, and so on) since the 1970s (see the work of Wang, Yang, and Liu and the references therein [30]). In the

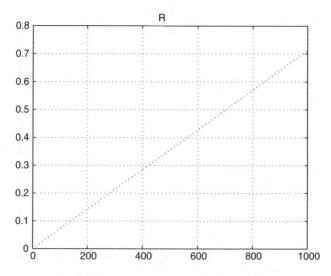

Figure 3.10   Iterative update of noise covariance

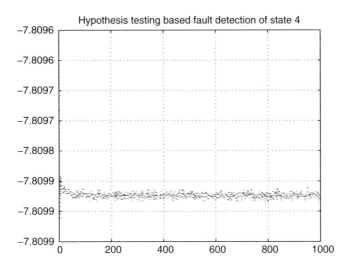

Figure 3.11   Mobile robot: Fault detection of Mode 4 using hypothesis testing

FD research field, diagnostic systems are often designed separately from control algorithms, although it is highly desirable that both the control and the diagnostic modules are integrated into one system module [31]. Hence, the problem of simultaneous fault detection and control (SFDC) has attracted a lot of attention in the last two decades, both in research and in application domains [32]–[35]. The simultaneous design unifies the control and detection units into a single unit which results in less complexity than with separate design. Hwang,

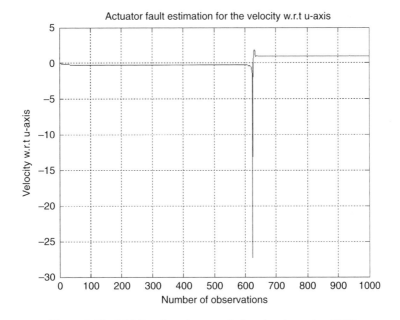

Figure 3.12   Mobile robot: Actuator fault estimation using UKF

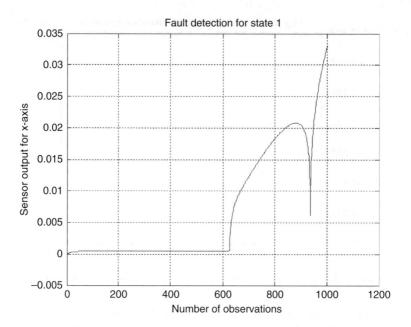

Figure 3.13   Mobile robot: Fault detection for Mode 1

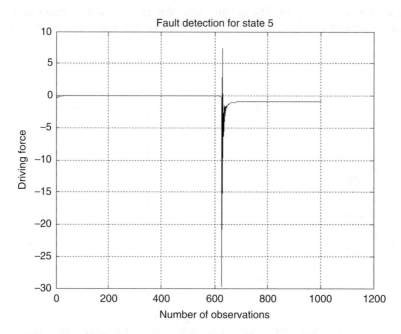

Figure 3.14   Mobile robot: Fault detection for Mode 5

Peng, and Hsu [36] studied the implementation of an integrated control and diagnosis system for an advanced hard disk drive. The robust integrated control and diagnosis approach using H_∞-optimization techniques was applied to Boeing 747-100/200 aircraft by Marcos and Balas [37]. Ding [31] presented a brief survey of the integrated design of feedback controllers and fault detectors.

3.3.1 Introduction

This section presents an H_∞ formulation of the SFDC problem using a joint dynamic-observer detector and state-feedback controller with an average dwell time (ADT) approach for continuous-time switched linear systems. We design a single unit, called a "detector/controller", where the detector is a dynamic observer and the controller is a state feedback mechanism, that produces detection signals (used to detect faults) and control signals (used to satisfy certain control objectives). Our method has major advantages in contrast with the previous results because we obtain strict LMI conditions for designing the dynamic observer parameters and controller gain. Indeed, we use the dynamic observer to design for the SFDC problem based on strict LMI conditions.

3.3.2 System model

Consider the following continuous-time switched linear system:

$$G : \begin{cases} \dot{x}(t) = A_\sigma x(t) + B_{1\sigma} u(t) + B_{2\sigma} d(t) + B_{3\sigma} f(t) \\ y(t) = C_{1\sigma} x(t) + D_{11\sigma} d(t) + D_{12\sigma} f(t) + D_{13\sigma} u(t), \\ z(t) = C_{2\sigma} x(t) + D_{21\sigma} d(t) + D_{22\sigma} f(t) + D_{23\sigma} u(t) \end{cases} \tag{3.124}$$

where, $x(t) \in \mathfrak{R}^n$ is the state, $u(t) \in \mathfrak{R}^m$ is the control input, $y(t) \in \mathfrak{R}^p$ is the measured output, and $z(t) \in \mathfrak{R}^l$ denotes the regulated output on which we want to impose certain robustness or performance objectives. $d(t) \in \mathfrak{R}^r$ are assumed to be finite energy disturbance modeling errors caused by exogenous signals, linearization or parameter uncertainties. For instance, model uncertainties can be presented as

$$d(t) = \Delta A x(t) + \Delta B_1 u(t). \tag{3.125}$$

Moreover, the unknown input $f(t) \in \mathfrak{R}^q$ is a possible fault. With $f(t)$ set to zero, Equation (3.124) describes the normal mode, that is, a fault-free system. σ is a piecewise constant function of time, called a switching signal, which takes its values from a finite set $l = \{1, 2, \ldots, N\}$ where $N > 1$ is the number of subsystems. At an arbitrary time t, σ may be dependent on t or $x(t)$, or both, or other logic rules. For a switching sequence $t_0 < t_1 < t_2 < \ldots, \sigma$ is continuous and may be either autonomous or controlled. When $t \in [t_k, t_{k+1})$, we say that the $\sigma(t_k)$th subsystem is active; therefore, the trajectory $x(t)$ of Equation (3.124) is the trajectory of the $\sigma(t_k)$th subsystem. The matrices $A_i, B_{1i}, B_{2i}, B_{3i}, C_{1i}, C_{2i}, D_{11i}, D_{12i}, D_{13i}, D_{21i}, D_{22i}$, and $D_{23i}(\forall \sigma = i \in l)$ are of the appropriate dimensions.

Remark 3.14 *The fault distribution matrices B_{3i}, D_{12i}, D_{22i} can be determined according to which faults are to be detected. For sensor faults, they are*

$$\begin{cases} B_{3i} = 0 \\ D_{12i} = D_{22i} = I \end{cases}, \tag{3.126}$$

for actuator faults, they are

$$\begin{cases} B_{3i} = B_{1i} \\ D_{12i} = D_{13i} \\ D_{22i} = D_{23i} \end{cases}. \tag{3.127}$$

The following model is proposed for the joint detector (dynamic observer) and controller (state feedback):

$$F : \begin{cases} \dot{\hat{x}}(t) = A_\sigma \hat{x}(t) + B_{1\sigma} u(t) + \eta(t) \\ \hat{y}(t) = C_{1\sigma} \hat{x}(t) + D_{13\sigma} u(t) \\ r(t) = y(t) - \hat{y}(t) \\ u(t) = -k_\sigma \hat{x}(t) \end{cases}, \tag{3.128}$$

where $\eta(t) \in \Re^n$ is the correction signal, the dynamics of which is given by:

$$\begin{cases} \dot{x}_d(t) = A_{d\sigma} x_d(t) + B_{d\sigma} r(t) \\ \eta(t) = C_{d\sigma} x_d(t) + D_{d\sigma} r(t) \end{cases}, \tag{3.129}$$

where $\hat{x}(t) \in \Re^n$ is the estimation of $x(t)$, $\hat{y}(t) \in \Re^p$ is the observer output, $x_d(t) \in \Re^n$ is an auxiliary vector, $K_i \in \Re^{m \times n}$ is the controller gain, $r(t) \in \Re^p$ is the residual signal, and the constant matrices $A_{di}, B_{di}, C_{di}, D_{di}$ are the observer parameters to be designed later.

Remark 3.15 *The switching of the observer should coincide exactly with the switching of the system; therefore, the parameters of the detector/controller, Equation (3.128), depend on the system modes.*

Now, substituting the detector/controller, Equation (3.128), into the Equation (3.124) results in the following closed-loop system equations:

$$\begin{cases} \dot{x}_{cl}(t) = \bar{A}_\sigma x_{cl}(t) + \bar{B}_{d\sigma} d(t) + \bar{B}_{f\sigma} f(t) \\ r(t) = \bar{C}_{1\sigma} x(t) + \bar{D}_{1d\sigma} d(t) + \bar{D}_{1f\sigma} f(t), \\ z(t) = \bar{C}_{2\sigma} x(t) + \bar{D}_{2d\sigma} d(t) + \bar{D}_{2f\sigma} f(t) \end{cases} \tag{3.130}$$

where

$$\bar{A}_i = \begin{bmatrix} A_i - B_{1i} K_i & D_{di} C_{1i} & C_{di} \\ 0 & A_i - D_{di} C_{1i} & -C_{di} \\ 0 & B_{di} C_{1i} & A_{di} \end{bmatrix},$$

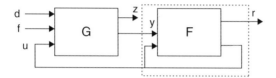

Figure 3.15 Block diagram of the SFDC problem

$$\bar{B}_{di} = \begin{bmatrix} D_{di}D_{11i} \\ B_{2i} - D_{di}D_{11i} \\ B_{di}D_{11i} \end{bmatrix}, \bar{B}_{fi} = \begin{bmatrix} D_{di}D_{12i} \\ B_{3i} - D_{di}D_{12i} \\ B_{di}D_{12i} \end{bmatrix}, \tag{3.131}$$

$$\bar{C}_{2i} = \begin{bmatrix} C_{2i} - D_{23i}K_i & C_{2i} & 0 \end{bmatrix}, \bar{D}_{2di} = D_{21i},$$

$$\bar{C}_{1i} = \begin{bmatrix} 0 & C_{1i} & 0 \end{bmatrix}, \bar{D}_{1di} = D_{11i}, \bar{D}_{1fi} = D_{12i},$$

$$\bar{D}_{2fi} = D_{22i}, \quad x_{cl}(t) = \begin{bmatrix} \hat{x}^T(t) & e^T(t) & x_d^T(t) \end{bmatrix}^T,$$

and $e(t) = x(t) - \hat{x}(t)$.

Given the plant G, the SFDC problem is that of finding a block F which, in addition to stabilizing the closed-loop system of Equation (3.130), simultaneously achieves certain control objectives (robustness of regulated output to faults and disturbances) and some detection objectives (sensitivity of the residual to faults and robustness to disturbances). The block diagram of the SFDC problem is depicted in Figure 3.15.

In the next section, the SFDC design problem is transformed into an H_∞ optimization problem.

3.3.3 Problem formulation

The SFDC problem to be addressed in this paper can be stated as follows: Given Equation (3.124), design a detector/controller Equation (3.128) such that the closed-loop system of Equation (3.130) is stable, the effects of disturbances on regulated output $z(t)$ and residual output $r(t)$ are minimized, the effects of faults on $z(t)$ are minimized, and the effects of faults on $r(t)$ are maximized.

More specifically, we wish to find a detector/controller such that the closed-loop system is stable and the following performance indices are satisfied:

$$\begin{aligned}
&(i) \sup \frac{\|z(t)\|_2}{\|d(t)\|_2} < \gamma_1, \gamma_1 > 0, \\
&(ii) \sup \frac{\|z(t)\|_2}{\|f(t)\|_2} < \gamma_2, \gamma_2 > 0, \\
&(iii) \sup \frac{\|r(t)\|_2}{\|d(t)\|_2} < \gamma_3, \gamma_3 > 0, \\
&(iv) \sup \frac{\|r_e(t)\|_2}{\|f(t)\|_2} = \frac{\|r(t) - r_f(t)\|_2}{\|f(t)\|_2} < \gamma_4, \gamma_4 > 0,
\end{aligned} \tag{3.132}$$

where $r_f = W_f f$, with W_f being an appropriately chosen stable weighting matrix. Condition *(iv)* is a standard H_∞ model matching problem [38] which means that the residual signal $r(t)$ robustly tracks a filtered version of the fault signals, $W_f f$.

Let a minimal state-space realization of the tracking filter W_f be given by

$$
\begin{aligned}
x_F &= A_F x_F + B_F f \\
r_f &= B_F x_F + D_F f
\end{aligned}
\tag{3.133}
$$

where A_F, B_F, C_F, and D_F are assumed to be known real constant matrices with the appropriate dimensions. Then, we can get the following state-space representation for fault signal f to r_e:

$$
\begin{aligned}
\dot{\xi} &= \tilde{A}\xi + \tilde{\xi}f \\
r_e &= \tilde{C}\xi + \tilde{D}f
\end{aligned}
\tag{3.134}
$$

where

$$
\begin{bmatrix} \tilde{A} & \tilde{B} \\ \tilde{C} & \tilde{D} \end{bmatrix} =
\left[
\begin{array}{c|cccc|c}
A_F & & 0 & & & B_F \\
\hline
 & A_i - B_{1i}K_i & D_{di}C_{1i} & C_{di} & & D_d D_{12i} \\
0 & 0 & A_i - D_{di}C_{1i} & -C_{di} & & B_{3i} - D_{di}D_{12i} \\
 & 0 & B_{di}C_{1i} & A_{di} & & B_{di}D_{12i} \\
\hline
 & 0 & -C_{1i} & 0 & & \\
\hline
C_F & & & & & D_F - D_{12i}
\end{array}
\right]
$$

$$
\xi = \begin{bmatrix} x_F \\ x_{cl} \end{bmatrix}.
\tag{3.135}
$$

The following discussion makes use of the projection and matrix inversion lemmas presented in Appendix A.

3.3.4   Simultaneous fault detection and control problem

There are four performance indices (Equation 3.132) that must be satisfied simultaneously to solve the SFDC problem. In this discussion, we first transform the performance indices into the LMI feasibility conditions (Theorems 3.8–3.11). Then, in Corollary 3.1, a feasible solution to the SFDC problem is obtained by considering all of Theorems 3.8–3.11 simultaneously. If the LMI conditions in each theorem involve the product of Lyapunov matrices and system state space matrices, we have to take equal Lyapunov matrices in Corollary 3.1, which leads to a conservatism problem. Hence, we employ Lemma A.3 (the projection lemma) to reduce conservatism in the SFDC problem by introducing additional matrix variables, so as to avoid the coupling of Lyapunov matrices with the system matrices. Note that our results are based on an ADT approach.

3.3.4.1   Minimizing the effect of disturbance on regulated output

Design objective *(i)* is transformed to LMI feasibility constraints in the following theorem:

Theorem 3.8   *Consider the switched linear system in Equation (3.130) and let $\alpha > 0$, $\lambda > 0$, $\gamma_{1i} > 0$, and $\mu \geq 1$ be given constants. The corresponding system is asymptotically stable* *and guarantees the performance index* (i) *with* $\gamma_1 = \max\{\gamma_{1i}\}$ *for any switching signal satisfying ADT Equation (A.13) if there exist symmetric positive definite matrices* T_{1i} *and matrices* X_{11i}, Y_{11i}, Q_i, S_i, A_{ki}, B_{ki}, C_{ki}, D_{ki}, *and* M_i *such that*, $\forall i \in l$, *the following LMIs hold:*

$$
\begin{bmatrix}
\alpha T_{1i} + E_{11} & T_{1i} + E_{12} & E_{13} & E_{14} \\
\bullet & E_{22} & E_{23} & 0 \\
\bullet & \bullet & E_{33} & 0 \\
\bullet & \bullet & \bullet & -I
\end{bmatrix} < 0,
\tag{3.136}
$$

$$
T_{1i} < \mu T_{1j},
\tag{3.137}
$$

where

$$
E_{11} = \begin{bmatrix}
\Xi_1 & \bullet & \bullet \\
C_{ki}^T & \Xi_2 & \bullet \\
C_{1i}^T D_{ki}^T & A_i^T - C_{1i}^T D_{ki}^T + A_{ki}^T & \Xi_3
\end{bmatrix}
$$

$$
\Xi_1 = (Q_i^T A_i^T - M_i^T B_{1i}^T) + (Q_i^T A_i^T - M_i^T B_{1i}^T)^T
$$

$$
\Xi_2 = (Y_{11i}^T A_i^T - C_{ki}^T) + (Y_{11i}^T A_i^T - C_{ki}^T)^T
$$

$$
\Xi_3 = (A_i^T X_{11i} - C_{1i}^T B_{ki}^T) + (A_i^T X_{11i} - C_{1i}^T B_{ki}^T)^T
$$

$$
E_{22} = \begin{bmatrix}
Q_i + Q_i^T & 0 & 0 \\
\bullet & Y_{11i}^T + Y_{11i} & S_i + I \\
\bullet & \bullet & X_{11i} + X_{11i}^T
\end{bmatrix}, \quad
E_{23} = \lambda \begin{bmatrix}
D_{ki} D_{11i} \\
B_{2i} - D_{ki} D_{11i} \\
X_{11i}^T B_{2i} - B_{ki} D_{11i}
\end{bmatrix}
$$

$$
E_{12} = \lambda \begin{bmatrix}
Q_i^T A_i^T - M_i^T B_{1i}^T & 0 & 0 \\
C_{ki}^T & Y_{11i}^T A_i^T - C_{ki}^T & A_{ki}^T \\
C_{1i}^T D_{ki}^T & A_i^T - C_{1i}^T D_{ki}^T & A_i^T X_{11i} - C_{1i}^T B_{ki}^T
\end{bmatrix} - \begin{bmatrix}
Q_i & 0 & 0 \\
0 & Y_{11i} & I \\
0 & S_i^T & X_{11i}^T
\end{bmatrix}
$$

$$
E_{13} = \begin{bmatrix}
D_{ki} D_{11i} + (Q_i C_{2i}^T - M_i^T D_{23i}^T) D_{21i} \\
B_{2i} - D_{ki} D_{11i} + Y_{11i}^T C_{2i}^T D_{21i} \\
X_{11i}^T B_{2i} - B_{ki} D_{11i} + C_{2i}^T D_{21i}
\end{bmatrix}, \quad
E_{14} = \begin{bmatrix}
Q_i C_{2i}^T - M_i^T D_{23i}^T \\
Y_{11i}^T C_{2i}^T \\
C_{2i}^T
\end{bmatrix}
$$

$$
E_{33} = D_{21i}^T D_{21i} - \gamma_{1i}^2 I
\tag{3.138}
$$

The control gain K_i *and dynamic observer parameters* A_{di}, B_{di}, C_{di}, *and* D_{di} *are given by*

$$
D_{di} = D_{ki}
$$

$$
C_{di} = (C_{Ki} - D_{di} C_{1i} Y_{11i})(Y_{21i})^{-1}
$$

$$
B_{di} = X_{21i}^{T\,-1}(X_{11i}^T D_{di} - B_{ki}),
\tag{3.139}
$$

$$
A_{di} = X_{21i}^{T\,-1}(A_{ki} - X_{11i}^T(A_i - D_{di} C_{1i})Y_{11i} - X_{21i}^T B_{di} C_{1i} Y_{11i} + X_{11i}^T C_{di} Y_{21i})Y_{21i}^{-1}
$$

$$
K_i = M_i Q^{-1}
$$

where X_{21i} *and* Y_{21i} *are invertible matrices satisfying the condition*

$$
Y_{21i}^T X_{21i} = S_i - Y_{11i} X_{11i}.
\tag{3.140}
$$

Proof. By Lemma A.4, Equation (3.130) under switching signals with ADT τ_a is asymptotically stable and satisfies performance index *(i)* if Equation (A.11) and Equation (A.12) hold. Choose the Lyapunov function $V_i(t)$ as follows:

$$V_i(t) = x_{cl}^T(t)P_{1i}x_{cl}(t), \forall \sigma(t) = i \in l, \tag{3.141}$$

where P_{1i} is the positive definite symmetric matrix and $x_{cl}(t)$ is defined in Equation (3.131). From Equation (A.12) and Equation (3.141), the following inequality is obtained:

$$
\begin{bmatrix} X_{cl}^T & d \end{bmatrix}^T \left(\begin{bmatrix} I & 0 \\ \bar{A}_i & \bar{B}_{di} \end{bmatrix}^T (\phi_c \otimes P_{1i}) \begin{bmatrix} I & 0 \\ \bar{A}_i & \bar{B}_{di} \end{bmatrix} \right.
$$
$$
\left. + \begin{bmatrix} 0 & I \\ \bar{C}_{2i} & \bar{D}_{2di} \end{bmatrix}^T M \begin{bmatrix} 0 & I \\ \bar{C}_{2i} & \bar{D}_{2di} \end{bmatrix} \right) \begin{bmatrix} X_{cl} \\ d \end{bmatrix} < 0, \tag{3.142}
$$

where

$$M = \begin{bmatrix} -\gamma_1^2 I & 0 \\ 0 & I \end{bmatrix}, \phi_c = \begin{bmatrix} \alpha & 1 \\ 1 & 0 \end{bmatrix}, \tag{3.143}$$

$\bar{A}_i$, $\bar{B}_{di}$, $\bar{C}_{2i}$, and $\bar{D}_{2di}$ are defined in Equation (3.131). The following condition is sufficient to imply Equation (3.142):

$$\begin{bmatrix} I & 0 \\ \bar{A}_i & \bar{B}_{di} \end{bmatrix}^T (\phi_c \otimes P_{1i}) \begin{bmatrix} I & 0 \\ \bar{A}_i & \bar{B}_{di} \end{bmatrix} + \begin{bmatrix} 0 & I \\ \bar{C}_{2i} & \bar{D}_{2di} \end{bmatrix}^T M \begin{bmatrix} 0 & I \\ \bar{C}_{2i} & \bar{D}_{2di} \end{bmatrix} < 0. \tag{3.144}$$

Equation Equation (3.144) is reformulated as

$$N_U^T Z N_U < 0, \tag{3.145}$$

where N_U and Z are defined as:

$$Z = \begin{bmatrix} \alpha P_{1i} + \bar{C}_{2i}^T \bar{C}_{2i} & P_{1i} & \bar{C}_{2i}^T \bar{D}_{2di} \\ * & 0 & 0 \\ * & * & \bar{D}_{2di}^T \bar{D}_{2di} - \gamma_{1i}^2 I \end{bmatrix}, N_U = \begin{bmatrix} I & 0 \\ \bar{A}_i & \bar{B}_{di} \\ 0 & I \end{bmatrix}. \tag{3.146}$$

If we choose N_V in Equation (A.9b) as:

$$N_V = \begin{bmatrix} \lambda I & 0 \\ -I & 0 \\ 0 & I \end{bmatrix} \rightarrow V = \begin{bmatrix} I & \lambda I & 0 \end{bmatrix} \tag{3.147}$$

and use Lemma A.3, then it can be concluded that inequality Equation (3.144) is equivalent to

$$
Z + \begin{bmatrix} \bar{A}_i^T \\ -I \\ \bar{B}_{di}^T \end{bmatrix} \begin{bmatrix} X_i & \lambda X_i & 0 \end{bmatrix} + \begin{bmatrix} X_i \\ \lambda X_i^T \\ 0 \end{bmatrix} \begin{bmatrix} \bar{A}_i & -I & \bar{B}_{di} \end{bmatrix} < 0. \tag{3.148}
$$

By partitioning X_i as follows:

$$
X_i = \begin{bmatrix} X_{1i} & 0 \\ 0 & X_{2i} \end{bmatrix}, X_{2i} = \begin{bmatrix} X_{11i} & X_{12i} \\ X_{21i} & X_{22i} \end{bmatrix}, \tag{3.149}
$$

and using Schure complement [39], the following inequality is obtained:

$$
\begin{bmatrix} \alpha P_{1i} + \Theta_1 & P_{1i} + \Theta_2 & \Theta_3 & \bar{C}_{2i}^T \\ \bullet & \Theta_4 & \Theta_5 & 0 \\ \bullet & \bullet & \Theta_6 & 0 \\ \bullet & \bullet & \bullet & -I \end{bmatrix} < 0
$$

$$
\Theta_1 = \left(\bar{A}_i^T \begin{bmatrix} X_{1i} & 0 \\ 0 & X_{2i} \end{bmatrix} \right) + \left(\bar{A}_i^T \begin{bmatrix} X_{1i} & 0 \\ 0 & X_{2i} \end{bmatrix} \right)^T
$$

$$
\Theta_2 = \lambda \bar{A}_i^T \begin{bmatrix} X_{1i}^T & 0 \\ 0 & X_{2i}^T \end{bmatrix} - \begin{bmatrix} X_{1i}^T & 0 \\ 0 & X_{2i}^T \end{bmatrix}
$$

$$
\Theta_3 = \bar{C}_{2i}^T D_{2di} + \begin{bmatrix} X_{1i}^T & 0 \\ 0 & X_{2i}^T \end{bmatrix} \bar{B}_{2i}
$$

$$
\Theta_4 = \lambda \begin{bmatrix} -X_{1i} - X_{1i}^T & 0 \\ 0 & -X_{2i} - X_{2i}^T \end{bmatrix}
$$

$$
\Theta_5 = \lambda \begin{bmatrix} X_{1i}^T & 0 \\ 0 & X_{2i}^T \end{bmatrix} \bar{B}_{2i}
$$

$$
\Theta_6 = \bar{D}_{2di}^T \bar{D}_{2di} - \gamma_{1i}^2 I \tag{3.150}
$$

Now, define new matrices Y_i, Q_i, Π_{1i}, Π_{2i}, and $\tilde{\Pi}_{1i}$ as follows:

$$
Y_i = X_{2i}^{-2} = \begin{bmatrix} Y_{11i} & Y_{12i} \\ Y_{21i} & Y_{22i} \end{bmatrix}, \Pi_{1i} = \begin{bmatrix} Y_{11i} & I \\ Y_{21i} & 0 \end{bmatrix}, Q_i = X_{1i}^{-1}, \tag{3.151}
$$

$$
\Pi_{2i} = X_{2i} \Pi_{1i} = \begin{bmatrix} I & X_{11i} \\ 0 & X_{21i} \end{bmatrix}, \tilde{\Pi}_{1i} = diag(Q_i, \Pi_{1i}).
$$

Note that from Equation (3.150), we find $X_{1i} + X_{1i}^T > 0$ and $X_{2i} + X_{2i}^T > 0$; therefore, $Y_i + Y_i^T > 0$, $X_{11i} + X_{11i}^T > 0$, $X_{22i} + X_{22i}^T > 0$, and $Y_{11i} + Y_{11i}^T > 0$, which imply nonsingularity of X_{1i}, X_{2i}, X_{11i}, X_{22i}, Y_{11i}. Also, without loss of generality, we assume that X_{21i} and Y_{21i} are nonsingular. Therefore, $\tilde{\Pi}_{1i}$ is nonsingular. It should be mentioned that nonsingularity of X_{21i}

is guaranteed by assuming $X_{21i} = I$ easily. Then, Lemma A.5 yields $Y_{21i} = -X_{22i}^{-1} X_{21i} Y_{11i}$, which is nonsingular because the matrices X_{22i}, X_{21i} and Y_{11i} are invertible.

Now, if we perform a congruence transformation with $diag(\tilde{\Pi}_{1i}^T, \tilde{\Pi}_{1i}^T, I, I)$ on inequalities Equation (3.150), we obtain

$$
\begin{bmatrix}
\alpha T_{1i} + E_{11} & T_{1i} + E_{12} & E_{13} & E_{14} \\
\bullet & E_{22} & E_{23} & 0 \\
\bullet & \bullet & E_{33} & 0 \\
\bullet & \bullet & \bullet & -I
\end{bmatrix} < 0.
\tag{3.152}
$$

where E_{11}, E_{12}, E_{13}, E_{14}, E_{22}, E_{23}, and E_{33} are defined in Equation (3.138) and

$$
\begin{aligned}
A_{ki} &= X_{11i}^T((A_i - D_{di}C_{1i})Y_{11i} - C_{di}Y_{21i}) + X_{21i}^T((B_{di}C_{1i})Y_{11i} + A_i Y_{21i}), \\
B_{ki} &= X_{11i}^T D_{di} - X_{21i}^T B_{di}, \\
C_{ki} &= D_{di}C_{1i}Y_{11i} + C_{di}Y_{21i}, \\
D_{ki} &= D_{di}, \\
M_i &= K_i Q_i, \\
T_{1i} &= \tilde{\Pi}_{1i}^T P_{1i} \tilde{\Pi}_{1i}^T.
\end{aligned}
\tag{3.153}
$$

Therefore, inequality Equation (3.136) is obtained. Note that from $P_{1i} > 0$, we find that T_{1i} must be positive definite. Now, by replacing Equation (3.141) in Equation (A.11) and pre-multiplying it by $\tilde{\Pi}_{1i}^T$ and post-multiplying it by $\tilde{\Pi}_{1i}$, Equation (3.137) is concluded. This completes the proof. ∎

3.3.4.2   Minimizing the effect of fault on regulated output

The LMI constraints for performance index *(ii)* are given in the following theorem:

Theorem 3.9   *Consider the switched linear system of Equation (3.130) and let $\alpha > 0$, $\lambda > 0$, $\gamma_{2i} > 0$, and $\mu \geq 1$ be given constants. The corresponding system is asymptotically stable and guarantees performance index* (ii) *with $\gamma^2 = \max\{\gamma_{2i}\}$ for any switching signal satisfying ADT Equation (A.13) if there exist symmetric positive definite matrices T_{2i} and matrices X_{11i}, Y_{11i}, Q_i, S_i, A_{ki}, B_{ki}, C_{ki}, D_{ki}, and M_i such that, $\forall i \in l$, the following LMIs hold:*

$$
\begin{bmatrix}
\alpha T_{2i} + E_{11} & T_{2i} + E_{12} & E_{13} & E_{14} \\
\bullet & E_{22} & E_{23} & 0 \\
\bullet & \bullet & E_{33} & 0 \\
\bullet & \bullet & \bullet & -I
\end{bmatrix} < 0.
\tag{3.154}
$$

$$
T_{2i} < \mu T_{2j},
\tag{3.155}
$$

where

$$E_{13} = \begin{bmatrix} D_{ki}D_{12i} + (Q_iC_{2i}^T - M_i^T D_{23i}^T)D_{21i} \\ B_{3i} - D_{ki}D_{12i} + Y_{11i}^T C_{2i}^T D_{21i} \\ X_{11i}^T B_{3i} - B_{ki}D_{12i} + C_{2i}^T D_{21i} \end{bmatrix}, E_{14} = \begin{bmatrix} Q_iC_{2i}^T - M_i^T D_{23i}^T \\ Y_{11i}^T C_{2i}^T \\ C_{2i}^T \end{bmatrix}$$

$$E_{23} = \begin{bmatrix} D_{ki}D_{12i} \\ B_{3i} - D_{ki}D_{12i} \\ X_{11i}^T B_{3i} - B_{ki}D_{12i} \end{bmatrix}, E_{33} = D_{22i}^T D_{22i} - \gamma_{2i}^2 I,$$

and E_{11}, E_{12}, E_{22}, are defined in Equation (3.138). The filter gains A_{di}, B_{di}, C_{di}, D_{di}, and the controller gain K_i are obtained from Equation (3.139).

Proof. The proof of this theorem is similar to that of Theorem 3.8, so it is omitted for the sake of brevity. ∎

3.3.4.3 Minimizing the effect of disturbance on residual

The following theorem gives the LMI constraints for performance index *(iii)*:

Theorem 3.10 *Consider the switched linear system of Equation (3.130) and let $\alpha > 0$, $\lambda > 0$, $\gamma_{3i} > 0$, and $\mu \geq 1$ be given constants. The corresponding system is asymptotically stable and guarantees the performance index* (iii) *with $\gamma_3 = \max\{\gamma_{3i}\}$ for any switching signal satisfying ADT Equation (A.13) if there exist symmetric positive definite matrices T_{3i} and matrices X_{11i}, Y_{11i}, Q_i, S_i, A_{ki}, B_{ki}, C_{ki}, D_{ki}, and M_i such that, $\forall i \in l$, the following LMIs hold:*

$$\begin{bmatrix} \alpha T_{3i} + E_{11} & T_{3i} + E_{12} & E_{13} & E_{14} \\ \bullet & E_{22} & E_{23} & 0 \\ \bullet & \bullet & E_{33} & 0 \\ \bullet & \bullet & \bullet & -I \end{bmatrix} < 0, \tag{3.156}$$

$$T_{3i} < \mu T_{3j}, \tag{3.157}$$

where

$$E_{13} = \begin{bmatrix} D_{ki}D_{11i} \\ B_{2i} - D_{ki}D_{11i} + Y_{11i}^T C_{1i}^T D_{11i} \\ X_{11i}^T B_{2i} - B_{ki}D_{11i} + C_{1i}^T D_{11i} \end{bmatrix}, E_{14} = \begin{bmatrix} 0 \\ Y_{11i}^T C_{1i}^T \\ C_{1i}^T \end{bmatrix},$$

$$E_{23} = \begin{bmatrix} D_{ki}D_{11i} \\ B_{2i} - D_{ki}D_{11i} \\ X_{11i}^T B_{2i} - B_{ki}D_{11i} \end{bmatrix}, E_{33} = D_{11i}^T D_{11i} - \gamma_{3i}^2 I,$$

and E_{11}, E_{12}, E_{22} are defined in Equation (3.138). The filter gains A_{di}, B_{di}, C_{di}, D_{di}, and the controller gain K_i are obtained from Equation (3.139).

Proof. The proof of this theorem is similar to that of Theorem 3.8, so it is omitted for the sake of brevity.   ∎

3.3.4.4 Maximizing the effect of fault on residual

The LMI constraints for performance index *(iv)* are given in the following theorem:

Theorem 3.11   *Consider the switched linear system of Equation (3.134) and let $\alpha > 0$, $\lambda > 0$, $\gamma_{4i} > 0$ and $\mu > 1$ be given constants. The corresponding system is asymptotically stable and guarantees performance index* (iv) *with $\gamma_4 = \max\{\gamma_{4i}\}$ for any switching signal satisfying ADT Equation (A.13) if there exist symmetric positive definite matrices T_{4i} and matrices X_{11i}, Y_{11i}, Q_i, Q_F, S_i, A_{ki}, B_{ki}, C_{ki}, D_{ki}, and M_i, such that, $\forall i \in l$, the following LMIs hold:*

$$
\begin{bmatrix}
\alpha T_{4i} + E_{11} & T_{4i} + E_{12} & E_{13} & E_{14} \\
\bullet & E_{22} & E_{23} & 0 \\
\bullet & \bullet & E_{33} & 0 \\
\bullet & \bullet & \bullet & -I
\end{bmatrix} < 0,
\tag{3.158}
$$

$$
T_{4i} < \mu T_{4j},
\tag{3.159}
$$

where

$$
E_{11} =
\begin{bmatrix}
\Gamma_1 & \bullet & & \bullet & & \bullet \\
0 & \Gamma_2 & & \bullet & & \bullet \\
0 & C_{ki}^T & \Gamma_3 & & \bullet \\
0 & C_{1i}^T D_{ki}^T & A_i^T - C_{1i}^T D_{ki}^T + A_{ki} & \Gamma_4
\end{bmatrix},
$$

$$
\Gamma_1 = (Q_F^T A_F^T) + (Q_F^T A_F^T)^T,
$$

$$
\Gamma_2 = (Q_i^T A_i^T - M_i^T B_{1i}^T) + (Q_i^T A_i^T - M_i^T B_{1i}^T)^T,
$$

$$
\Gamma_3 = (Y_{11i}^T A_i^T - C_{ki}^T) + (Y_{11i}^T A_i^T - C_{ki}^T)^T,
$$

$$
\Gamma_4 = (A_i^T X_{11i} - C_{1i}^T B_{ki}^T) + (A_i^T X_{11i} - C_{1i}^T B_{ki}^T)^T,
$$

$$
E_{22} = -\lambda
\begin{bmatrix}
Q_F + Q_F^T & 0 & 0 & 0 \\
\bullet & Q_i + Q_i^T & 0 & 0 \\
\bullet & \bullet & Y_{11i}^T + Y_{11i} & S_i + I \\
\bullet & \bullet & \bullet & X_{11i} X_{11i}^T
\end{bmatrix},
$$

$$
E_{23} = \lambda
\begin{bmatrix}
B_F \\
D_{ki} D_{12i} \\
B_{3i} - D_{ki} D_{12i} \\
X_{11i}^T B_{3i} - B_{ki} D_{12i}
\end{bmatrix},
$$

$$E_{12} = \lambda \begin{bmatrix} Q_F^T A_F^T & 0 & 0 & 0 \\ 0 & Q_i^T A_i^T - M_i^T B_{1i}^T & 0 & 0 \\ 0 & C_{ki}^T & Y_{11i}^T A_i^T - C_{ki}^T & A_{ki}^T \\ 0 & C_{1i}^T D_{ki}^T & A_i^T - C_{1i}^T D_{ki}^T & A_i^T X_{11i} - C_{1i}^T B_{ki}^T \end{bmatrix}$$
$$- \begin{bmatrix} Q_F & 0 & 0 & 0 \\ 0 & Q_i & 0 & 0 \\ 0 & 0 & Y_{11i} & I \\ 0 & 0 & S_i^T & X_{11i}^T \end{bmatrix},$$

$$E_{13} = \lambda \begin{bmatrix} Q_F^T C_F^T (D_F - D_{12i}) + B_F \\ D_{ki} D_{12i} \\ B_{3i} - D_{ki} D_{12i} - Y_{11i}^T C_{1i}^T (D_F - D_{12i}) \\ X_{11i}^T B_{3i} - B_{ki} D_{12i} - C_{1i}^T (D_F - D_{12i}) \end{bmatrix}, \quad E_{14} = \begin{bmatrix} Q_F^T C_F^T \\ 0 \\ -Y_{11i}^T C_{1i}^T \\ -C_{1i}^T \end{bmatrix},$$

$$E_{33} = (D_F - D_{12i})^t (D_F - D_{12i}) - \gamma_{4i}^2 I.$$

The filter gains A_{di}, B_{di}, C_{di}, D_{di}, and the controller gain K_i are obtained from Equation (3.139).

Proof. Assume

$$X_i = diag(X_F, X_{1i}, X_{2i})$$
$$\tilde{\Pi}_{1i} = diag(Q_F, Q_i, \Pi_{1i}),$$
$$Q_F = X_F^{-1}$$

Now, similar to the proof of Theorem , by applying Lemma A.4 to closed-loop Equation (3.134) and using congruence transformation $diag(I, \tilde{\Pi}_{1i}^T, \tilde{\Pi}_{1i}^T, I, I)$, the inequalities of Equation (3.158) and Equation (3.159) are obtained. ∎

At this point, all control and detection objectives given in Equation (3.132) have been transformed into LMI feasibility constraints. The next corollary unifies the aforementioned theorems and provides a procedure for solving the SFDC problem.

Corollary 3.1 *Given positive scalars γ_1, γ_2, γ_3, α, λ, and $\mu \geq 1$, a feasible solution to the SFDC problem is obtained by solving a sequence of convex optimization problems*

$$\min_{T_{1i}, T_{2i}, T_{3i}, T_{4i}, X_{11i}, Y_{11i}, Q_i, Q_F, S_i, A_{ki}, B_{ki}, C_{ki}, D_{ki}, M_i} \gamma_4 , \quad (3.160)$$
s.t. Equations (3.136), (3.137), (3.154)−(3.159) hold

Proof. Collect Theorems 3.8–3.11. ∎

Remark 3.16   *The presented SFDC problem can also be solvable under the arbitrary switching signals. In fact, selection of $\mu = 1$ leads to $T_{1i}, \equiv T_1, T_{2i} \equiv T_2, T_{3i} \equiv T_3, T_{4i} \equiv T_4, X_i \equiv X$, and $\tau^* = 0$; then, the closed-loop system of Equation (3.130) includes a common Lyapunov functional and the switching signals can be arbitrary.*

Remark 3.17   *In some previous works (e.g., [30] and [40]), iterative LMI algorithms, which need more computational time than LMI methods, are applied for conservatism reduction in FD problems. These algorithms depend heavily on the initial conditions of the iterations and are not globally convergent. To avoid these problems, the extended LMIs are applied to present strict LMI conditions for solving our problem. Indeed, as can be seen from Theorems 3.8–3.11, we reduce the conservatism in our SFDC problem by introducing additional matrix variables and eliminating the coupling of Lyapunov matrices with the system state matrices.*

Remark 3.18   *As mentioned before, applying the static observer in designing SFDC has some problems that persuade us to employ the dynamic observer. To increase understanding of the advantages of using the dynamic observer in SFDC design, we compare its structure with the static observer.*

Consider, the joint static-observer detector and state-feedback controller for Equation (3.124) as follows:

$$
F : \begin{cases}
\dot{\hat{x}}(t) = A_\sigma \hat{x}(t) + B_{1\sigma} u(t) + L_\sigma(y(t) - \hat{y}(t)) \\
\hat{y}(t) = C_{1\sigma} \hat{x}(t) \\
r(t) = y(t) - \hat{y}(t) \\
u(t) = -K_\sigma \hat{x}(t)
\end{cases} . \tag{3.161}
$$

By Equation (3.124) and Equation (3.161), the closed-loop system dynamics can be derived as follows:

$$
\begin{cases}
\dot{x}_{cl}(t) = \bar{A}_\sigma x_{cl}(t) + \bar{B}_{d\sigma} d(t) + \bar{B}_{f\sigma} f(t) \\
r(t) = \bar{C}_{1\sigma} x(t) + D_{11\sigma} d(t) + D_{12\sigma} f(t) , \\
z(t) = \bar{C}_{2\sigma} x(t) + D_{21\sigma} d(t) + D_{22\sigma} f(t)
\end{cases} \tag{3.162}
$$

where

$$
\bar{A}_i = \begin{bmatrix} A_i - B_{1i} K_i & B_{1i} K_i \\ 0 & A_i - L_i C_{1i} \end{bmatrix}, \bar{B}_{di} = \begin{bmatrix} B_{2i} \\ B_{2i} - L_i D_{11i} \end{bmatrix},
$$

$$
\bar{B}_{fi} = \begin{bmatrix} B_{3i} \\ B_{3i} - L_i D_{12i} \end{bmatrix}, x_{cl}(t) = \begin{bmatrix} x(t)^T & e(t)^T \end{bmatrix}^T
$$

$$
\bar{C}_{1i} = \begin{bmatrix} 0 & C_{1i} \end{bmatrix}, \bar{C}_{2i} = \begin{bmatrix} C_{2i} - D_{23i} K_i & D_{23i} K_i \end{bmatrix}
$$

For comparison, consider a static observer-based SFDC for performance index *(i)* (the effect of disturbance on regulated output). The closed-loop Equation (3.162) is asymptotically stable and satisfies performance index *(i)*, if there exist positive definite symmetric matrices P_{1i} and matrices X_{1i}, X_{2i} and N_i such that, $\forall i \in l$, the following inequalities are satisfied:

$$
\begin{bmatrix}
\alpha P_{1i} + \Lambda_{1i} & \Lambda_{2i} & \Lambda_{3i} & \Lambda_{4i} \\
\bullet & \Lambda_{5i} & \Lambda_{6i} & 0 \\
\bullet & \bullet & -\gamma_i^2 & D_{21i}^T \\
\bullet & \bullet & \bullet & -I
\end{bmatrix} < 0,
$$

$$
P_{1i} < \mu P_{1j}, \tag{3.163}
$$

where

$$
\psi_i = \begin{bmatrix}
A_i^T X_{1i} - K_i^T B_{1i}^T X_{1i} & 0 \\
K_i^T B_{1i}^T X_{1i} & A_i^T X_{2i} - C_{1i}^T L_i^T X_{2i}
\end{bmatrix}, N_i = L_i^T X_{2i},
$$

$$
\Lambda_{1i} = \psi_i + \psi_i^T,
$$

$$
\Lambda_{2i} = \lambda \psi_i + P_{1i} - \begin{bmatrix} X_{1i}^T & 0 \\ 0 & X_{2i}^T \end{bmatrix},
$$

$$
\Lambda_{3i} = \begin{bmatrix} X_{1i}^T B_{2i} \\ X_{2i}^T B_{2i} - N_i^T D_{11i} \end{bmatrix}, \quad \Lambda_{4i} = \begin{bmatrix} C_{2i}^T - K_i^T D_{23i} \\ K_i^T D_{23i} \end{bmatrix},
$$

$$
\Lambda_{5i} = \lambda \begin{bmatrix} -X_{1i} - X_{1i}^T & 0 \\ 0 & -X_{2i} - X_{2i}^T \end{bmatrix},
$$

$$
\Lambda_{6i} = \begin{bmatrix} \lambda X_{1i}^T B_{2i} \\ \lambda (X_{2i}^T B_{2i} - N_i^T D_{11i}) \end{bmatrix}
$$

In this regard, we note that Equation (3.163) is derived from a procedure similar to the procedure given in the proof of Theorem 3.8. It must be emphasized that Equation (3.163) is not a strict LMI because of the nonlinear terms K_i^T, B_i^T, X_{1i}. In this situation, the SFDC design problem can be solved in one step using an equality constraint. It should be pointed out that the equality constraint is applicable only if $D_{23i} = 0$ in Equation (3.163). However, if $D_{23i} \neq 0$, then a two-step procedure can be considered. Using a dynamic observer, one does not encounter such problems because extra auxiliary dynamics are introduced with the new state variable x_d. In fact, the advantage of using this new auxiliary state variable is that we can apply more degrees of freedom by employing A_{di}, B_{di}, and C_{di} in the closed-loop system dynamics and therefore the conditions for designing the controller and observer parameters can be presented in terms of strict LMI conditions.

After designing the SFDC problem, the remaining important task is to evaluate the generated residual. Similar to the approach proposed by Casavola, Famularo, and Franze [38], the residual evaluation function $J_r(L)$ and the threshold J_{th} can be selected as

$$
J_r(L) = \|r(t)\|_{2,L} = \left(\int_0^L r(v)^T r(v) dv \right)^{1/2},
$$

$$
J_{th} = \sup_{d(t) \in l_2, u(t), f(t) = 0} J_r(L) \tag{3.164}
$$

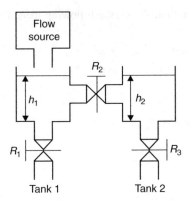

Figure 3.16   A two-tank system

where L is the evaluation time window. Note that the length of the time window is finite. Because an evaluation of the residual signal over the whole time range is impractical, it is desired that faults will be detected as early as possible [41]. Based on Equation (3.164), the occurrence of faults can be detected by the following logic rule.

$$J_r(L) > J_{th} \Rightarrow \text{Faults} \Rightarrow \text{Alarm}$$

$$J_r(L) \le J_{th} \Rightarrow \text{No Faults}, \tag{3.165}$$

3.3.5   Two-tank system simulation

To illustrate the effectiveness of the proposed method, an example is given in this section.

Consider the liquid level control system shown in Figure 3.16. The system consists of two tanks, one flow source, two outlet pipes, and one connecting pipe [42]. The pipes contain valves that can be opened or closed by an external controller. Based on the status of each valve (open or closed), there are eight different system modes. Consider the following three valve configurations.

$$Mode1 \quad R_2 : ON \quad R_1, R_3 : OFF$$

$$Mode2 \quad R_1, R_2 : ON \quad R_3 : OFF$$

$$Mode3 \quad R_2, R_3 : ON \quad R_1 : OFF$$

It is assumed that flow through the valves is laminar, which implies that the relation between the flow rate in the valves and the height of the liquid is linear. Depending on the value of the tank capacity C_T and the pipe resistance R in each mode, the behavior of the system is governed by a specific differential equation. The state-space representation of the system is given by

$$\begin{cases} \dot{x}(t) = A_\sigma x(t) + B_{1\sigma} u(t) \\ y(t) = C_{1\sigma} x(t) \\ z(t) = C_{2\sigma} \end{cases},$$

where the state $x(t) = [h_1(t) \quad h_2(t)]^T$ contains the heights of the liquid in the tanks. Consider the following values for the system parameters:

$$C_{T1} = 5\,\text{m}^2, \quad C_{T2} = 3\,\text{m}^2,$$
$$R_1 = R_2 = 300\,\frac{\text{s}}{\text{m}^2}, \quad R_3 = 100\,\frac{\text{s}}{\text{m}^2}.$$

For the three modes, one can obtain

$$A_1 = \begin{bmatrix} -0.0007 & 0.0007 \\ 0.0011 & -0.0011 \end{bmatrix}, \quad B_{11} = \begin{bmatrix} 0.2 \\ 0 \end{bmatrix},$$

$$A_2 = \begin{bmatrix} -0.0013 & 0.0007 \\ 0.0011 & -0.0011 \end{bmatrix}, \quad B_{12} = \begin{bmatrix} 0.2 \\ 0 \end{bmatrix},$$

$$A_3 = \begin{bmatrix} -0.0007 & 0.0007 \\ 0.0011 & -0.0044 \end{bmatrix}, \quad B_{11} = \begin{bmatrix} 0.2 \\ 0 \end{bmatrix},$$

$$C_{1i} = [0 \quad 1], \, C_{2i} = C_{1i}, \, D_{13i} = D_{23i} = 0,$$

$$(\forall i \in \{1, 2, 3\})$$

The disturbance and fault models are assumed as

$$B_{3i} = B_{2i} = B_{1i}, \, D_{11i} = D_{12i} = D_{21i} = D_{22i} = 0,$$

where the fault matrix $B_{3i} = B_{1i}$ for actuator faults. Note that $B_{2i} = B_{1i}$ too, as is common in industrial applications where disturbances enter the system by corrupting the input signal.

Reference model parameters for the residual are selected as

$$A_F = -0.06, \quad B_F = 0.04, \quad C_F = 0.01, \quad D_F = 0.$$

By solving the optimization problem in Corollary 3.1 for a given $\gamma_1 = \gamma_2 = \gamma_3 = e - 4, \lambda = 0.1, \alpha = 0.06, \mu = 1.1$, one can obtain $\gamma_4 = 2.129e - 4$ and $\tau_a^* = 1.5885$. Furthermore, the dynamic observer parameters and the control gains were obtained using Equation (3.139) as follows:

$$\left[\begin{array}{c|c} A_{d1} & B_{d1} \\ \hline C_{d1} & D_{d1} \end{array}\right] = \left[\begin{array}{cc|c} -3.029 & 1.57e5 & -4.35e3 \\ -0.008 & -10.65 & -1.042 \\ -0.157 & -1.06e4 & -125.05 \\ -0.002 & 98.35 & -2.913 \end{array}\right], \quad K_1^T = \begin{bmatrix} 69.34 \\ 3.562e3 \end{bmatrix},$$

$$\left[\begin{array}{c|c} A_{d2} & B_{d2} \\ \hline C_{d2} & D_{d2} \end{array}\right] = \left[\begin{array}{cc|c} -3.029 & 1.57e5 & -4.32e3 \\ -3.003 & -1.55e5 & -1.041 \\ -0.1554 & -1.051e4 & -125.057 \\ -0.0022 & 96.76 & -2.9134 \end{array}\right], \quad K_2^T = \begin{bmatrix} 69.438 \\ 3.254e3 \end{bmatrix},$$

$$\left[\begin{array}{c|c} A_{d3} & B_{d3} \\ \hline C_{d3} & D_{d3} \end{array}\right] = \left[\begin{array}{cc|c} -2.867 & 1.49e5 & -4.14e3 \\ -0.00084 & -9.74 & -1.047 \\ -0.1601 & -9.54e3 & -125.62 \\ -0.0021 & 94.66 & -2.771 \end{array}\right], \quad K_3^T = \begin{bmatrix} 53.454 \\ 2.735e3 \end{bmatrix}.$$

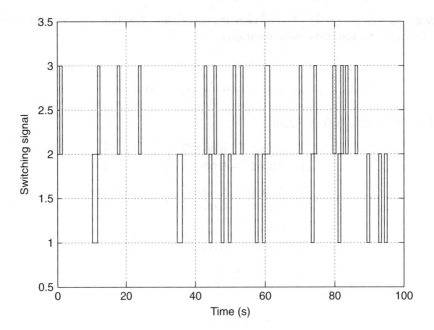

Figure 3.17 Switching signal

The disturbance $d(t)$ is assumed to be $(0.5e^{-0.04t}\cos(0.7\pi t) + 0.1)u(t)$. The fault signal $f(t)$ is simulated as a pulse signal with amplitude 1 that occurred from 40 to 60 s. The switching signal is generated by choosing ADT $\tau_a = 2 > 1.5885$ as shown in Figure 3.17. The residual signal $r(t)$ is shown in Figure 3.18.

The threshold can be determined as $J_{th} = 0.04104$ for $L = 100$. Figure 3.19 shows evaluation of the residual evaluation function $J_r(t)$, where the solid line is the case with a fault f_k and the dashed line is the fault-free case. The simulation results show that $J_r(t) = 0.04168 > 0.04104$ for $t = 44.4$, which means that the fault f_k can be detected 4.4 s after its occurrence.

The regulated output $z(t)$ of the closed-loop system is shown in Figure 3.20, from which it can be concluded that the effects of disturbance $d(t)$ and fault $f(t)$ on the regulated output $z(t)$ have been attenuated.

3.4 Data-Driven Fault Detection Design

In this section, we study data-driven based robust fault detection (FD) problems for time-delay systems with unknown inputs. The robustness and sensitivity of residual signals to the unknown inputs as well as to the faults in terms of L_2 are examined. First, the weighting matrix is selected for an appropriate design of filter, then a fault detection filter is designed with a Lyapunov–Krasovskii function (LKF) with time delay. The main results include the detailed derivation of these steps followed by its implementation on a four-tank system.

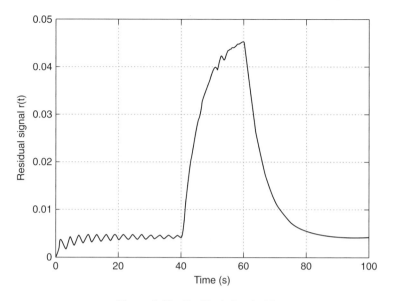

Figure 3.18   Residual signal $r(t)$

3.4.1   Introduction

Data-driven based fault detection has always been an area of interest in mission-critical systems and process control industries. In such system structures, the whole model or dynamics are not usually available and so system reliability is at risk. To overcome this, data is generated from the assumed model of the system and then compared with measured data from the physical

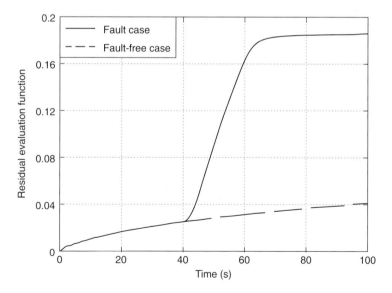

Figure 3.19   Evaluation of residual evaluation function

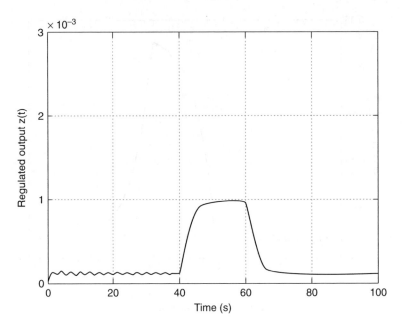

Figure 3.20   Regulated output $z(t)$

system to create residuals that relate to specific faults. For example, in fault prognosis, models based on data-driven structures are derived from statistical learning techniques, most of which originated from theories of pattern recognition. Data-driven models are usually developed by collecting the input–output data; they can process a wide variety of data types and exploit the similarities in the data that cannot be discovered by physical models. Based on this, Orchard and Vachtsevanos presented a non-line particle-filtering-based framework for failure prognosis in nonlinear, non-Gaussian systems [43]. A nonlinear state-space model of the plant and a particle-filtering (PF) algorithm is used to estimate the probability density function (PDF) of the state in real time. The state PDF estimate is then used to predict the evolution in time of the fault indicator, obtaining as a result the PDF of the remaining useful life (RUL) for the faulty subsystem. The hidden Markov model (HMM) and hidden semi-Markov model (HSMM) belong to the set of data-driven models. Baruah and Chinnam [44] first pointed out that a standard HMM could be applied to the area of prognosis in machining processes. An integrated fault diagnostic and prognostic approach for monitoring bearing health and CBM was introduced by Zhang *et al.* [45]. The proposed scheme consists of principle component analysis, HMM, and an adaptive stochastic fault prediction model. Camci [46] proposed an integrated diagnostics and prognostic architecture that employed support vector machines and HMM. However, HMMs have some inherent limitations. One is the assumption that successive system behavior observations are independent. The other is the Markov assumption itself, that the probability of a given state at time t only depends on the state at time t_1, which is sometimes untenable in practical applications. In order to cope with the inaccurate duration modeling of HMMs, some researchers have proposed HSMMs to model the state duration explicitly [47, 48].

The research and application of robust fault detection in automated processes have received considerable attention during recent decades and a great number of results have been achieved [49, 50, 51]. In the past three decades, many significant results concerning fault detection and isolation (FDI) problems have been developed (see, for example, the work of Frank *et al.* [52], Mangoubi *et al.* [53], and Patton [54] and references therein).

Most of the achievements have been for delay-free systems. However, time delays are frequently encountered in industry and are often the source of performance degradation in a system [51]. This section focuses on the fault detection filter design for time-delay linear time-invariant (LTI) systems with unknown inputs. Although, time delay is an inherent characteristic of many physical systems, such as rolling mills, chemical processes, water resources, and biological, economic and traffic control systems, very little FDI research has been carried out for them [25, 55, 56, 57]. Jiang and Chowdhury [56] deal with the nominal case of fault identification (without considering the influence of model uncertainty and unknown inputs). Frank *et al.* [52] formulate the fault detection filter (FDF) design problem as a two-objective nonlinear programming problem where no analytic solution can be constructed in general. Jiang, Staroswiecki, and Cocquempot [57] extends the results to the discrete-time case. The authors of an earlier study [25] also developed an FDF design approach based on H_∞-filtering but the most important and difficult issue, concerning the selection of a "reference residual model", has not been successfully solved.

3.4.2 Problem formulation

With reference to Figure 3.21, we now concentrate our attention on the fault detection problems for time-delay systems.

The system model under consideration is given by:

$$\dot{x}(t) = Ax(t) + \sum_{i=1}^{N} A_i x(t - d_{xi}(t))$$

$$+ Bu + \sum_{i=1}^{L} B_i u(t - d_{ui}) + B_f f + B_d d \tag{3.166}$$

$$y(t) = Cx + Du + D_f f + D_d d \tag{3.167}$$

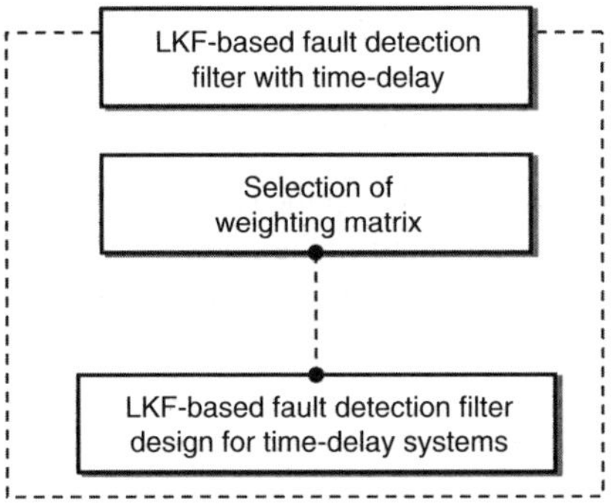

Figure 3.21 Fault detection filter with time-delay

where $x(t) \in \Re^n$, $u(t) \in \Re^m$, and $y(t) \in \Re^p$ are the state, input and output vectors, respectively, and $A_i \in \Re^{n \times n}$ (i = 1,2,3,...,N), $B_i \in \Re^{n \times m}$ (i = 1,2,3,...,L), $C \in \Re^{p \times n}$, $B_f \in \Re^{n \times q}$, $B_d \in \Re^{n \times s}$, $D_f \in \Re^{p \times q}$ and $D_d \in \Re^{p \times s}$ are known constant matrices. In addition, $f \in \Re^q$ and $d \in \Re^s$ are faults and disturbance vectors, respectively. Assume that the time-varying delays satisfy:

$$d_{xi}(t) \le \bar{d}_x < \infty, \, d_{ui}(t) \le \bar{d}_u < \infty$$
$$\dot{d}_{xi}(t) \le \bar{m}_{xi} < 1, \, \dot{d}_{ul}(t) \le \bar{m}_{ul} < 1 \quad (3.168)$$

We propose to use the following fault detection filter for the purpose of residual generation:

$$\dot{\hat{x}} = A\hat{x}(t) + \sum_{i=1}^{N} A_i \hat{x}(t - d_{xi}(t))$$

$$+ Bu + \sum_{i=1}^{L} B_i u(t - d_{ui}) + H(y - \hat{y}) \quad (3.169)$$

$$r = Ce + D_f f + D_d d \quad (3.170)$$

Denoting $e = x - \hat{x}$, then the dynamics of the fault detection filter can be expressed by:

$$\dot{e} = (A - HC)e + \sum_{i=1}^{N} A_i e(t - d_i(t))$$

$$+ (B_f - HD_f)f + (B_d - HD_d)d \quad (3.171)$$

$$r = Ce + D_f f + D_d d \quad (3.172)$$

Proceeding further, we introduce the following system:

$$\dot{e}_f = (A - HC)e_f + \sum_{i=1}^{N} A_i e_f(t - d_i(t))$$

$$+ (B_f - HD_f)f \quad (3.173)$$

$$r_f = Ce_f + D_f f \quad (3.174)$$

which describes the influence of the faults on the residual signals.

We now formulate the issues pertaining to the problem of designing a fault detection filter (FDF) for the time-delay system of Equations (3.166)–(3.167).

3.4.3   Selection of weighting matrix

Given a performance index which describes the sensitivity of the residual signals to the faults, find an observer gain matrix, denoted by H_f, such that r_f, is optimal in the sense of the given performance index. The weighting matrix is then set as:

$$\dot{e}_f = (A - HC)e_f + \sum_{i=1}^{N} A_i e_f(t - d_i(t))$$

$$+ (B_f - HD_f)f \tag{3.175}$$

$$r_f = Ce_f + D_f f \tag{3.176}$$

3.4.4 Design of FDF for time-delay system

The design problem is phrased as follows: Given constants $\beta > 0$ and $\gamma \geq y_m in$ and a weighting matrix $W_f(s)$, find an observer gain matrix H such that the system of Equations (3.171)–(3.172) is stable and Equation (3.168) holds.

The design problem is phrased as follows: *Given a constant $\gamma \geq y_m in$ and a weighting matrix $W_f(s)$, find an observer gain matrix H and matrix V such that β is minimized under conditions that Equation (3.175) and Equation (3.176) is stable as well as Equation (3.168) holds, that is*

$$min_{H,V}\ \beta \tag{3.177}$$

such that the observer gain is:

$$\int_0^\infty r_e^T r_e dt \leq \beta^2 \int_0^\infty f^T f dt + \gamma^2 \int_0^\infty d^T d dt \tag{3.178}$$

the rate of change of the error is:

$$\dot{e} = (A - HC)e + \sum_{i=1}^{N} A_i e(t - d_i(t))$$

$$+ (B_f - HD_f)f + (B_d - HD_d)d \tag{3.179}$$

and the residual is:

$$r = Ce + D_f f + D_d d \tag{3.180}$$

We call the system in Equations (3.171)–(3.172) a robust fault detection filter for the time-delay system of Equations (3.166)–(3.167) if the observer gain matrix solves the optimization problem.

3.4.5 LMI design approach

In the following discussion, an LMI approach is developed to solve the robust fault detection filter design problem. To this end, the following sub-problems will be solved:

- evaluation of the influence of the faults and selection of the weighting matrix;
- solutions of the RFDF problem.

The following theorem establishes the main result:

Theorem 3.12   *Given scalars $\varrho > 0$ and $\mu > 0$, the system of Equations (3.171)–(3.172) with $u(.) = 0$ is delay-dependent asymptotically stable with L_2-performance bound γ if there exist symmetric matrices $0 < P, 0 < W_a, 0 < W_c, 0 < Q, 0 < R$, weighting matrices N_a, N_c, N_s, M_a, M_c, M_s and a scalar $\gamma > 0$ satisfying the following LMI:*

$$
\begin{bmatrix}
\Upsilon_{01} & \Upsilon_{02} & \Upsilon_{03} & \varrho M_a & \varrho N_a & P\Gamma_0 & G_o^T & PA_0^T W \\
\bullet & \Upsilon_{04} & \Upsilon_{05} & \varrho M_c & \varrho N_c & 0 & G_{do}^T & PA_{do}^T W \\
\bullet & \bullet & \Upsilon_{06} & \varrho M_s & \varrho N_s & 0 & 0 & 0 \\
\bullet & \bullet & \bullet & -\varrho W_a & 0 & 0 & 0 & 0 \\
\bullet & \bullet & \bullet & \bullet & -\varrho W_c & 0 & 0 & 0 \\
\bullet & \bullet & \bullet & \bullet & \bullet & -\gamma^2 I & \Phi_0^T & P\Gamma_0^t W \\
\bullet & \bullet & \bullet & \bullet & \bullet & \bullet & -I & 0 \\
\bullet & \bullet & \bullet & \bullet & \bullet & \bullet & \bullet & -\varrho W
\end{bmatrix} < 0
\qquad (3.181)
$$

where

$$
\begin{aligned}
\Upsilon_{o1} &= PA_o + A_o^t P + Q + R + N_a + N_a^t + M_a + M_a^t \\
\Upsilon_{o2} &= PA_{do} - 2N_a + N_c^t + M_c^t, \ W = W_a + W_c \\
\Upsilon_{o3} &= N_a - M_a + N_s^t + M_s^t \\
\Upsilon_{o4} &= -(1 - \mu)Q - 2N_c - 2N_c^t \\
\Upsilon_{o5} &= N_c - M_c - 2N_s^t \\
\Upsilon_{o6} &= -R + N_s + N_s^t - M_s - M_s^t
\end{aligned}
\qquad (3.182)
$$

Proof. In terms of $\xi(t) = [e_f^t(t) \quad e_f^t(t - \tau(t)) \quad e_f^t(t - \varrho)]^t$ and using the classical Leibniz rule $e_f(t - \theta) = e_f(t) - \int_{t-\theta}^t \dot{e}_f(s)ds$ for any matrices N_a, N_c, N_s, M_a, M_c, M_s of appropriate dimensions, the following equations hold:

$$
2\xi^t(t)2N[-\int_{t-\tau(t)}^t \dot{e}_f(s)ds + e_f(t) - e_f(t - \tau)] = 0
\qquad (3.183)
$$

$$
2\xi^t(t)(M - N)[-\int_{t-\varrho}^t \dot{e}_f(s)ds + e_f(t) - e_f(t - \varrho)] = 0
\qquad (3.184)
$$

Expansion of Equation (3.183) and Equation (3.184) gives:

$$e'_f(t)[N_a + N'_a + M_a + M'_a]e_f(t) + 2e'_f(t)[-2N_a + M'_c$$
$$+ N'_c]e_f(t - \tau(t))) + 2e'_f(t)[N_a + M_a + N'_s + M'_s]e_f(t - \varrho)$$
$$+ 2e'_f(t - \tau(t))[-2N_c - 2N'_c]e_f(t - \tau(t))2e'_f(t - \tau(t))[N_c$$
$$- 2N'_s - M_c]e_f(t - \varrho) + 2e'_f(t - \varrho)[N_s + N'_s - M_s - M'_s]$$

$$e_f(t - \varrho) - 2\xi'(t)2N \int_{t-\tau(t)}^{t} \dot{e}_f(s)ds - 2\xi'(t)(M - N)$$

$$\int_{t-\varrho}^{t} \dot{x}(s)ds = 0 \tag{3.185}$$

Consider now the Lyapunov–Krasoviskii functional (LKF):

$$V(t) = V_o(t) + V_a(t) + V_c(t) + V_m(t)$$

$$V_o(t) = e'_f P e_f(t),$$

$$V_a(t) = \int_{-\varrho}^{0} \int_{t+s}^{t} \dot{e}_f{}'(\alpha)(W_a + W_c)\dot{e}_f(\alpha)d\alpha ds,$$

$$V_c(t) = \int_{t-\varrho}^{t} e'_f(s)R e_f(s)ds,$$

$$V_m(t) = \int_{t-\tau(t)}^{t} e'_f(s)Q e_f(s)ds \tag{3.186}$$

where $0 < P = P', 0 < W_a = W'_a, 0 < W_c = W'_c, 0 < Q = Q', 0 < R = \Re'$ are the matrices of appropriate dimensions. The first term in Equation (3.186) is standard for nominal systems without delay. The second and fourth terms correspond to the delay-dependent conditions. The third term is introduced to compensate for the enlarged time interval from $t - \varrho \longrightarrow t$ to $t - \tau \longrightarrow t$. A straightforward computation gives the time-derivative of $V(e_f)$ with $w(t) = 0$ as:

$$\dot{V}_o^t = 2e'_f P \dot{e}_f(t)$$
$$= 2e'_f(t)P[A_o e_f(t) + A_{do}e_f(t - \tau)]$$

$$\dot{V}_a(t) = \varrho \dot{e}'_f(W_a + W_c)\dot{e}_f(t) - \int_{t-\varrho}^{t} \dot{e}'_f(s)(W_a$$
$$+ W_c)\dot{e}_f(s)ds$$

$$\dot{V}_c^t = e'_f(t)R e_f(t) - e'_f(t - \varrho)R e_f(t - \varrho)$$

$$\dot{V}_m(t) = e'_f(t)Q e_f(t) - (1 - \dot{\tau})e'_f(t - \tau(t))$$
$$Q e_f(t - \tau(t))$$
$$\leq e'_f(t)Q e_f(t) - (1 - \mu)e'_f(t - \tau(t))$$
$$Q e_f(t - \tau(t)) \tag{3.187}$$

From Equation (3.186) and Equation (3.187), using Equation (3.184), we obtain:

$$\dot{V}(t) \leq e_f^t(t)[PA_o + A_o^t + Q + R + N_a + N_a^t + M_a$$
$$+ M_a^t]e_f(t) + 2e_f^t[PA_{do} - 2N_a + M_c^t + N_c^t]e_f(t - \tau)$$
$$+ 2e_f^t(t)[N_a - M_a + N_s^t + M_s^t]e_f(t - \varrho) + 2e_f{}^t(t - \tau)$$
$$[N_c - 2N_s^t - M_c]e_f(t - \varrho) - e_f{}^t(t - \tau)[(1 - \mu)Q$$
$$+ 2N_c + 2N_c^t e_f(t - \tau(t)] + e_f{}^t(t - \varrho)[-R + N_s$$
$$+ N_s^t - M_s - M_s^t]e_f(t - \varrho) + \varrho e_f{}^t(t)(W_a$$
$$+ W_c)\dot{e}_f(t) - \int_{t-\varrho}^{t} \dot{e}_f{}^t(s)$$
$$(W_a + W_c)\dot{e}_f(s)ds - 2\xi^t(t)2N \int_{t-\tau(t)}^{t} \dot{e}_f(s)ds$$
$$-2\xi^t(t)(-N) \int_{t-\varrho}^{t} \dot{e}_f(s)ds - 2\xi^t(t)M \int_{t-\varrho}^{t} \dot{e}_f(s)ds \tag{3.188}$$

Proceeding to manipulate the terms in Equation (3.188), we arrive at:

$$-2\xi^t(t)2N \int_{t-\tau(t)}^{t} \dot{e}_f(s)ds + 2\xi^t(t)N \int_{t-\varrho}^{t} \dot{e}_f(s)ds$$
$$= -2\xi^t(t)N \int_{t-\tau(t)}^{t} \dot{e}_f(s)ds + 2\xi^t(t)N \int_{t-\varrho}^{t-\tau(t)} \dot{e}_f(s)ds$$

$$\tag{3.189}$$

Considering

$$\int_{t-\varrho}^{t} \dot{e}_f^t(s)(W_a + W_c)\dot{e}_f(s)ds$$
$$= \int_{t-\varrho}^{t} \dot{e}_f^t(s)(W_a)\dot{e}_f(s)ds + \int_{t-\varrho}^{t} \dot{e}_f^t(s)(W_c)\dot{e}_f(s)ds$$
$$= \int_{t-\varrho}^{t} \dot{e}_f^t(s)(W_a)\dot{e}_f(s)ds + \int_{t-\tau(t)}^{t} \dot{e}_f^t(s)(W_c)\dot{e}_f(s)ds$$
$$+ \int_{t-\varrho}^{t-\tau(t)} \dot{e}_f^t(s)(W_c)\dot{e}_f(s)ds \tag{3.190}$$

then $\dot{V}(t)$ becomes

$$\dot{V}(t) \le \xi^t \Upsilon_o \xi(t) - \int_{t-\varrho}^t \dot{e}_f^t(s) W \dot{e}_f(s) ds$$

$$+ \xi^t(t) \begin{bmatrix} \varrho A_o^t \\ \varrho A_{do}^t \\ 0 \end{bmatrix} W \begin{bmatrix} \varrho A_o^t \\ \varrho A_{do}^t \\ 0 \end{bmatrix}^t \xi(t)$$

$$-2\xi^t(t) N \int_{t-\tau(t)}^t \dot{e}_f(s) ds$$

$$-2\xi^t(t)(-N) \int_{t-\varrho}^{t-\tau(t)} \dot{e}_f(s) ds$$

$$-2\xi^t(t) M \int_{t-\varrho}^t \dot{e}_f(s) ds. \tag{3.191}$$

Add and subtract the terms

$$\xi^t(t)[\varrho M W_a^{-1} M^t + \varrho N W_c^{-1} N^t]\xi \tag{3.192}$$

and consider the following terms:

$$\varrho \xi^t(t) M W_a^{-1} M^t \xi(t) + \varrho \xi^t(t) N W_c^{-1} N^t \xi(t)$$

$$\varrho \xi^t(t) M W_a^{-1} M^t \xi(t) - \tau(t) \xi^t(t) N W_c^{-1} N^t \xi(t)$$

$$-(\varrho - \tau(t)) \xi^t(t) N W_c^{-1} N^t \xi(t)$$

$$-2\xi^t(t) N \int_{t-\tau(t)}^t \dot{e}_f(s) ds + 2\xi^t(t) N \int_{t-\varrho}^{t-\tau(t)} \dot{e}_f(s) ds$$

$$-2\xi^t(t) M \int_{t-\varrho}^t \dot{e}_f(s) ds + \int_{t-\varrho}^{t-\tau(t)} \dot{e}_f^t(s) W_c \dot{e}_f ds$$

$$+ \int_{t-\tau(t)}^t \dot{e}_f^t(s) W_c \dot{e}_f ds + \int_{t-\varrho}^t \dot{e}_f^t(s) W_a \dot{e}_f ds. \tag{3.193}$$

After some manipulation, the terms in Equation (3.193) become:

$$\xi^t(t)[\varrho M W_a^{-1} M^t + \varrho N W_c^{-1} N^t]\xi$$

$$- \int_{t-\tau(t)}^t [\xi^t N + \dot{e}_f^t W_c] W_c^{-1} [\xi^t N + \dot{e}_f^t W_c]^t ds$$

$$- \int_{t-\varrho}^{t-\tau(t)} [-\xi^t N + \dot{e}_f^t W_c] W_c^{-1} [-\xi^t N + \dot{e}_f W_c]^t ds$$

$$- \int_{t-\varrho}^t [\xi^t M + \dot{e}_f^t W_a] W_a^{-1} [-\xi^t M + \dot{e}_f W_A]^t ds \tag{3.194}$$

Further manipulations of Equation (3.191) result in:

$$
\begin{aligned}
\dot{V}(t) \leq \xi'[\Upsilon_o &+ \varrho M W_a^{-1} M' + \tau(t) N W_c^{-1} N' \\
&+ (\varrho - \tau(t)) N W_c^{-1} N'] \xi(t) + \varrho \dot{x}(t)(W_a + W_c) \dot{e}_f(t) \\
&- \int_{t-\tau(t)}^{t} [\xi' N + \dot{e}_f W_c] W_c^{-1} [\xi' N + \dot{e}_f' W_c]' ds \\
&- \int_{t-\varrho}^{t-\tau(t)} [-\xi' N + \dot{e}_f W_c] W_c^{-1} [-\xi' N + \dot{e}_f' W_c]' ds \\
&- \int_{t-\varrho}^{t} [\xi' M + \dot{e}_f W_a] W_a^{-1} [\xi' M + \dot{e}_f' W_a]' ds \\
\leq \xi'(t)[\Upsilon_o &+ \varrho M W_a^{-1} M' + \varrho N W_c^{-1} N'] \xi(t) \\
&+ \varrho \dot{e}_f'(W_a + W_c) \dot{e}_f(t)
\end{aligned}
\tag{3.195}
$$

In view of Equation (3.181) with $G_o = 0$, $G_d = 0$, $\Gamma_o = 0$, and Schur's complements, it follows from Equation (3.195) that $\dot{V}(t) < 0$ which establishes the internal asymptotic stability.

Consider the performance measure

$$
J = \int_0^\infty (z'(s)z(s) - \gamma^2 w'(s)w(s)) ds
$$

For any $w(t) \in L_2(0, \infty) \neq 0$ and zero initial condition $x(0) = 0$, we have

$$
\begin{aligned}
J &= \int_0^\infty (z'(s)z(s) - \gamma^2 w'(s)w(s) + \dot{V}(x)) ds - \dot{V}(x) \\
&\leq \int_0^\infty (z'(s)z(s) - \gamma^2 w'(s)w(s) + \dot{V}(x)) ds
\end{aligned}
\tag{3.196}
$$

Proceeding further, we get:

$$
\begin{aligned}
z'(s)z(s) - \gamma^2 w'(s)w(s) + \dot{V}(s) &= \bar{\chi}'(s) \bar{\Upsilon} \bar{\chi}'(s), \\
\bar{\chi}(s) = [e_f'(s) \quad e_f'(s - \tau(t)) &\quad e_f'(t - \varrho) \quad w(s)]'
\end{aligned}
\tag{3.197}
$$

where $\bar{\Upsilon}$ corresponds to Υ_o in Equation (3.181) by Schur's complements. It is readily seen from Equation (3.181) that:

$$
z'(s)z(s) - \gamma^2 w'(s)w(s) + \dot{V}(s) < 0
\tag{3.198}
$$

for arbitrary $s \in [t, \infty)$, which implies for any $w(t) \in L_2(0, \infty) \neq 0$ that $J < 0$ leading to $\| z(t) \|_2 < \gamma \| w(t) \|_2$. ∎

3.4.6   Four-tank system simulation

The developed scheme has been evaluated on a four-tank system. In this section, we provide the detailed implementation and simulation of the developed scheme. A laboratory-scale four-tank system was used to collect data at a sampling time of 50 ms. The data sets were generated for a PI-controlled water level control. Different fault scenarios have been considered for the generation of the data sets.

3.4.6.1   Experimental setup

The process data was generated through an experimental setup, as shown in Figure 3.22. A four-tank system was used to collect the data, with the introduction of actuator and sensor

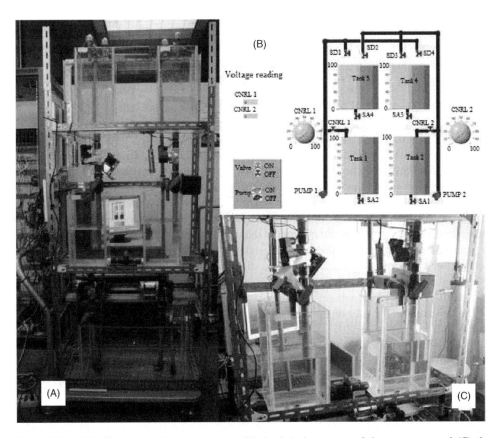

Figure 3.22   (A) The four-tank system setup, (B) the Labview setup of the apparatus and (C) the controlled system

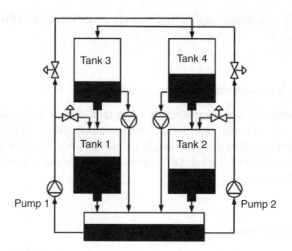

Figure 3.23   Schematic diagram of a four-tank system

faults through the system as can be seen in Figure 3.22(B). An amplified voltage of 20 V was given to each set of tanks to handle the controller effectively for the changes or fluctuation produced in the system. So, fault diagnosis was carried out in a closed-loop setup where the controller is trying to suppress the faults as though they were disturbances.

3.4.6.2   Process data collection and description

The process data was collected at 50 ms sampling time. The main objective of the four-tank system is to reach a reference height in the tanks. The schematic description of the four-tank system can be seen in Figure 3.23. The system has two control inputs (pump throughputs) which can be manipulated to control the water level in the tanks. The two pumps are used to transfer water from a sump into four overhead tanks. By adjusting the bypass valves of the system, the proportion of water pumped into different tanks can be changed to adjust the degree of interaction between the pump throughputs and the water levels. Each pump output goes to two tanks, one lower and one upper, diagonally opposite and the ratio of the split is controlled by the position of the valve. Because of the large water distribution load, the pumps have been supplied 12 V each.

The mathematical model of the four-tank process can be obtained using Bernoulli's law. During this process, several faults are introduced, such as leakage faults, sensor faults, and actuator faults. The leakage faults are introduced by clogging of the pipes of the system, knobs between the tanks etc. Sensor faults are simulated by introducing a gain into the circuit. Similarly, actuator faults are simulated by introducing a gain into the setup that comprises the motor and the pump. A PI controller has been employed in order to reach the desired reference height. Because of the inclusion of faults, the controller was finding it difficult to reach the desired level and so the power of the motor was increased from 10 V to 20 V. This enabled the actuator to perform well in achieving its desired level but led to the controller suppressing the faults injected into the system, which made the fault detection task rather difficult. After the data collection task was completed, techniques such as settling time, steady-state value, and coherence spectra were used to help us get an insight into the faults in the system.

3.4.6.3 Fault model of the four-tank system

Combining all the equations for the interconnected four-tank system we obtain the physical system model. A fault model can then be constructed by adding extra holes to each tank. The mathematical model of the faulty four-tank system can be given as:

$$\frac{dh_1}{dt} = -\frac{a_1}{A_1}\sqrt{2gh_1} + \frac{a_3}{A_1}\sqrt{2gh_3} + \frac{\gamma_1 k_1}{A_1}v_1 + \frac{d}{A_1}$$
$$\qquad -\frac{a_{leak1}}{A_1}\sqrt{2gh_1}$$

$$\frac{dh_2}{dt} = -\frac{a_2}{A_2}\sqrt{2gh_2} + \frac{a_4}{A_2}\sqrt{2gh_4} + \frac{\gamma_2 k_2}{A_2}v_2 - \frac{d}{A_2}$$
$$\qquad -\frac{a_{leak2}}{A_2}\sqrt{2gh_2}$$

$$\frac{dh_3}{dt} = -\frac{a_3}{A_3}\sqrt{2gh_3} + \frac{(1-\gamma_2)k_2}{A_3}v_2$$
$$\qquad -\frac{a_{leak3}}{A_3}\sqrt{2gh_3}$$

$$\frac{dh_4}{dt} = -\frac{a_4}{A_4}\sqrt{2gh_4} + \frac{(1-\gamma_1)k_1}{A_4}v_1$$
$$\qquad -\frac{a_{leak4}}{A_4}\sqrt{2gh_4}$$

$$\frac{dv_1}{dt} = -\frac{v_1}{\tau_1} + \frac{1}{\tau_1}u_1$$

$$\frac{dv_2}{dt} = -\frac{v_2}{\tau_2} + \frac{2}{\tau_2}u_2 \qquad\qquad (3.199)$$

In the discussion that follows, we present simulation results for the developed fault detection scheme. The tasks of our LKF-based, fault-detection filter scheme were executed with increasing precision. Firstly, the data collected from the plant was initialized and the parameters were optimized by the pre-processing and normalization of the data. Then, the LKF-based fault detection filter scheme was implemented to detect the fault profile.

The simulation results for the system under observation are shown in Figures 3.24–3.27 and the error is shown in Figure 3.28. It can be seen that the estimated profiles for States 1–4 (which are presenting Tanks 1–4) are detecting faults, thus pointing clearly to a fault detection filter design.

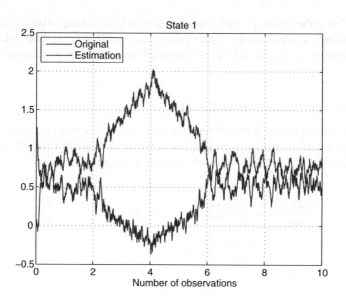

Figure 3.24   Fault detection filter using LKF: State 1

3.5   Robust Adaptive Fault Estimation

In this section, we investigate a fault estimation problem for a class of nonlinear systems subject to multiplicative faults and unknown disturbances. Multiplicative faults mixed with system states and inputs can cause additional complexity in the design of fault estimators because of parameter changes within the process. Especially for nonlinear systems corrupted

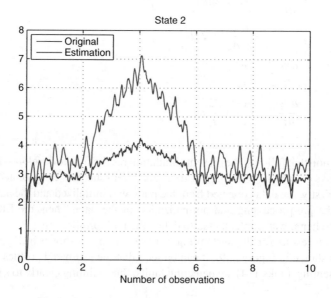

Figure 3.25   Fault detection filter using LKF: State 2

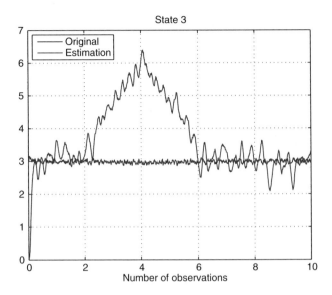

Figure 3.26   Fault detection filter using LKF: State 3

with unknown disturbances, it is not easy to distinguish the real fault factor from the mixed term. Under the nonlinear Lipschitz condition, the proposed robust adaptive fault estimation approach not only estimates the multiplicative faults and system states simultaneously but also extracts the real effect of the faults. Meanwhile, the effect of disturbances is restricted to an L_2 gain performance criterion, which can be formulated into the basic feasibility problem of a linear matrix inequality (LMI). In order to reduce the conservatism of the proposed method,

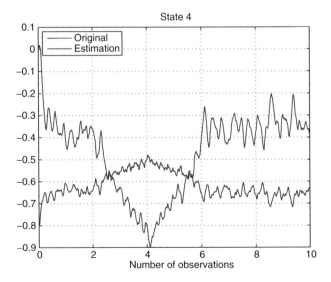

Figure 3.27   Fault detection filter using LKF: State 4

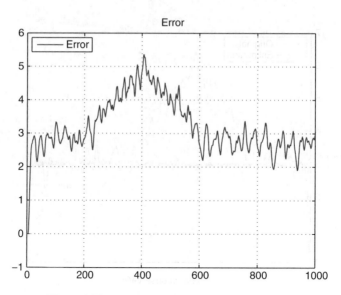

Figure 3.28   Fault detection filter using LKF: Error

a relaxing Lipschitz matrix is introduced. Finally, we illustrate the proposed robust adaptive estimation scheme to verify its efficiency.

3.5.1   Introduction

Issues and concerns about system or process safety and reliability necessitate and foster the development of fault detection and diagnosis for dynamical systems, which has been regarded as one of the most important aspects in seeking effective solutions to guarantee reliable operation of practical control systems at the possible occurrence of system failures or malfunctions.

In order to avoid performance deterioration or system damage, faults have to be found as soon as possible and schemes have to be made to stop propagation of bad effects. Traditional approaches to fault detection and identification are mainly focused on linear systems, which are widely described and well documented in many research articles [6, 58]. However, the majority of practical control systems are nonlinear in nature. Therefore, nonlinear properties cannot be neglected for the purpose of fault diagnosis and identification. This has caused more attention to be given to active research about nonlinear system fault detection and identification [10, 59, 60].

One of the most important tasks in a fault detection and identification scheme, fault estimation is for determining the extent, such as magnitude or frequency, of the faults. Accurate fault estimation can help reconstruct the fault signals so that their effects can be accommodated in the corresponding control reconfiguration. It is not an easy task, especially for nonlinear systems with unknown disturbances. Jiang, Staroswiecki, and Cocquempot [9] transformed a nonlinear system with uncertainties into two subsystems under some geometric conditions, and then established an adaptive observer to obtain the estimations of both states and actuator or sensor faults. Yan and Edwards [61] utilized a sliding-mode observer to realize fault reconstruction. Gao and Ding [62] developed a fault estimator based on a descriptor system formulation that can simultaneously estimate the states and the sensor fault signal superimposed on the output. Hou [63] provided an effective method of estimating the amplitude and frequency of a sinusoidal signal. Other methods have been utilized to tackle fault estimation problems (see, for instance, the work of Vijayaraghavan, Rajamani, and Bokor [64]).

As we know, faults can be classified into additive or multiplicative faults, according to their effects on the system outputs and the system dynamics. Although component faults and some of actuator or sensor faults appear as multiplicative faults (which correspond to parameter changes in the system model), most of the literature concentrates on the effects of additive faults (which result in changes only to the mean value of the system output signal). On the other hand, some studies of multiplicative fault estimation are scattered over some papers [65, 66, 67] and book chapters [5, 25]. As the name implies, faults are mixed with the input and states, in a multiplicative form. This means that the analysis and design for multiplicative fault estimation is not as straightforward as that for additive faults and it is being given more and more attention. In a recent article [17], a good fault detection and isolation scheme was presented, with an unknown fault function that was restricted to a finite set of fault types, each of which was described by the product of an unknown parameter vector characterizing the time-varying magnitude of the fault with a known smooth vector representing the functional structure of the fault.

Taking into account the above conditions, in this section, we focus on a Lipschitz nonlinear system subject to multiplicative faults and unknown disturbances. The goal of our work is to establish a robust adaptive fault estimation scheme which is robust with respect to disturbances and sensitive to faults, to detect and estimate multiplicative faults, and discover the real effect of faults. Based on Lyapunov stability theory and by relaxing a less conservative Lipschitz condition, we develop an estimator to estimate simultaneously the system states and real fault factors. The effect of unknown disturbances is reduced according to an L_2 gain performance criterion. Compared to most of the existing work on fault estimation, the proposed scheme is simple to compute, easy to implement, and capable of estimating the actual size of the faulty parameters in the model.

We demonstrate the effectiveness of proposed fault estimation scheme with an example in a simulation study.

3.5.2   Problem statement

We focus on a class of nonlinear multi-input–multi-output dynamical systems described by

$$\dot{x}(t) = Ax(t) + \phi(x, u, t) + Bu(t) + d(t),$$
$$y(t) = Cx(t), \tag{3.200}$$

where $x(t) \in \mathfrak{R}^n$, $y(t) \in \mathfrak{R}^m$, and $u(t) \in \mathfrak{R}^p$ are the state vector, the output vector, and the input vector, respectively. $d(t)$ represents the system disturbance, and the L_2 norm of the unknown input $d(t)$ is bounded. A, B, and C are the known system matrices of appropriate dimensions. $\phi(x, u, t)$ is a Lipschitz nonlinear vector function with a Lipschitz constant ϕ, such that,

$$\|\phi(x, u, t) - \phi(\hat{x}, u, t)\|_2 \leq \delta \| \left(x(t) - \hat{x}(t) \right) \|_2. \tag{3.201}$$

It should be noted that Equation (3.200) is a general form since most nonlinear functions can be expanded at the equilibrium point. For instance, the nonlinear system $\dot{x} = f(x, u, t)$ is differentiated with respect to x and u, and (x_e, u_e) is the equilibrium point. Applying the Taylor expansion, we get

$$A = \frac{\partial f(x, u, t)}{\partial x}|_{x=x_e, u=u_e}, \quad B = \frac{\partial f(x, u, t)}{\partial u}|_{u=u_e, x=x_e}$$

and $\phi(x, u, t)$ can be assumed to be the remaining term. Further, many nonlinear functions can be assumed as Lipschitz, at least locally. For example, the sinusoidal function $\sin(x)$ appearing in many robotic control systems is globally Lipschitz and the term x^2 can be regarded as locally Lipschitz within a finite range of x.

We now look at a multiplicative fault model. With the assumption that the nonlinear system described by Equation (3.200) is subject to component faults which are parameter changes within the process, the post-fault system is modeled as

$$\dot{x}(t) = Ax(t) + \phi(x, u, t) + Bu(t) + d(t) + \sum_{i=1}^{l} \theta_i(t)g_i(x, u, t), \quad t \geq t_f, \tag{3.202}$$

$$y(t) = Cx(t) \tag{3.203}$$

where $\theta_i(t) \in \Re, i = 1, \ldots, l$, are unknown time functions that are assumed to be zero before the fault occurs and non-zero after. $g_i(x, u, t), i = 1, \ldots, l$, are known functions related to system states and inputs, which also satisfy the Lipschitz condition with a Lipschitz constant δ_i. For simplicity, the time t is dropped from the notation in the following equations.

Assumption 3.9 *The multiplicative fault factors θ_i, $i = 1, \ldots, l$, are unknown and bounded by a constant, that is, $\|\theta_i\| \leq \alpha_i$. The constant α_i is known.*

In Equation (3.202), the term $\sum_{i=1}^{l} \theta_i(t)g_i(x, u, t)$ is generated by multiplicative faults. This representation characterizes a general class of multiplicative faults where θ_i represents the magnitude of the time-varying or constant fault and g_i characterizes the functional structure of the ith fault. Multiplicative faults encountered in a linear system were modeled [25], which can be transformed into this kind of representation. For example, the linear system $\dot{x} = (A + A_f)x$ is subject to multiplicative faults in the form $A_f = \sum_{i=1}^{l} A_i \theta_{Ai}$, so we can get the structure function $g_i = A_i x$. In practice, component faults in the process and some of faults in the sensors and actuators are in the form of multiplicative faults, which change system parameters and usually mix with system states and inputs. Hence such faults result in performance degradation or even instability of the system. Letting $f = \sum_{i=1}^{l} \theta_i(t)g_i(x, u, t)$, we can rewrite Equation (3.202) as

$$\dot{x} = Ax + Bu + \phi(x, u, t) + d(t) + f. \qquad (3.204)$$

It is clear that f is a term induced by the component faults θ_i, $i = 1, \ldots, l$. When the system is in normal operation, $f = 0$. The form in Equation (3.204) has been adopted to treat the additive fault estimation, where the size of f can be estimated. However, it is clear that in modeling the system component faults, the term f is also a function of the system state and input. f alone cannot reflect the real fault sources or sizes. Therefore, it is necessary for us to estimate the real fault factors θ_i, $i = 1, \ldots, l$, instead of the additive fault vector f.

Relaxing the Lipschitz condition, Equation (3.201) can be given in a relaxing matrix form as defined by

$$\|\phi(x, u, t) - \phi(\hat{x}, u, t)\|_2 \leq \left\|H\left(x(t) - \hat{x}(t)\right)\right\|_2. \qquad (3.205)$$

The matrix H could be a sparsely populated matrix. Phanomchoeng and Rajamani illustrate that $\|H(x(t) - \hat{x}(t))\|_2$ is much smaller than $\delta\|x(t) - \hat{x}(t)\|_2$ for the same nonlinear function [68]. The relaxing Lipschitz condition Equation (3.205) is much less conservative.

The objective herein is to design an adaptive estimator with an effective algorithm for the nonlinear system of Equation (3.200)–(3.201) subject to multiplicative faults to estimate the real effect factor θ_i, $i = 1, \ldots, l$, in the post-fault system of Equations (3.202)–(3.203), and make the estimation accurate and insensitive to the unknown disturbances. In order to design an estimator satisfying the above objective, it is assumed that the system states and inputs are all bounded before and after the occurrence of a fault, and the Lipschitz nonlinear functions $\phi(x, u, t)$ and $g_i(x, u, t)$ satisfy the relaxing Lipschitz condition with matrices H and G_i. It should be noted that the feedback control system is capable of making the system bounded even in the presence of a fault. The proposed fault estimation design is independent of the structure of the feedback controller.

3.5.3 Adaptive observer

In this section, an adaptive observer is applied to reconstruct multiplicative fault signals which are mixed with system states and inputs. The designed adaptive observer here for the nonlinear system of Equations (3.200)–(3.201) can be shown to be as follows:

$$\dot{\hat{x}} = A\hat{x} + Bu + \phi(\hat{x}, u) + \sum_{i=1}^{l} \hat{\theta}_i g_i(\hat{x}, u) + L(y - \hat{y}), \qquad (3.206)$$

$$\hat{y} = C\hat{x} \qquad (3.207)$$

$$\dot{\hat{\theta}}_i = \sigma_i g^T \sigma_i(\hat{x}, u)D(y - \hat{y}), \quad i = 1, \ldots, l \qquad (3.208)$$

where $\sigma_i > 0$, $i = 1, \ldots, l$ are constants and $\hat{x}$, $\hat{y}$, and $\hat{\theta}_i$ denote the estimated state, output, and fault variables, respectively. L and D are the design gain matrices. Let $e_x = x - \hat{x}$ and

$e_y = y - \hat{y}$ represent the state and output estimation error; $e_{\theta_i} = \theta_i - \hat{\theta}_i$ denotes fault error. Then we obtain the following estimation error dynamic equations

$$\dot{e}_x = (A - LC)e_x + \phi(x, u) - \phi(\hat{x}, u) + \sum_{i=1}^{l}\left(\theta_i g_i(x, u) - \hat{\theta}_i g_i(\hat{x}, u)\right) + d, \quad (3.209)$$

$$e_y = Ce_x, \quad (3.210)$$

$$\dot{e}_{\theta_i} = -\sigma_i g_i^T(\hat{x}, u)De_y(t). \quad (3.211)$$

The main problem encountered here is that the system is subject to unknown disturbance and the real fault effect factor θ_i which is combined with the system state x and input u. We must design an appropriate estimator which can estimate the fault θ_i effectively and be less sensitive to the disturbance.

First, let us introduce a lemma and a definition which are useful for the analysis of this multiplicative fault estimation problem.

Lemma 3.5   *Barbalat's Lemma: If $\lim_{t\to\infty} \int_0^t f(\tau)d\tau$ exists and is finite, and $f(t)$ is a uniformly continuous function, then $\lim_{t\to\infty} f(\tau) = 0$.*

Definition 3.1   *Persistence of excitation: A piecewise continuous signal vector $\phi : \Re^+ \mapsto \Re^n$ is the persistence of excitation in $\Re^n$ with a level of excitation $\alpha_0 > 0$ if there exist constants $\alpha_1, T_0 > 0$ such that*

$$\alpha_1 I \geq \frac{1}{T_0} \int_t^{t+T_0} \phi(\tau)\phi^T(\tau) \geq \alpha_0 I, \qquad \forall t \geq 0.$$

In the analysis of the estimation error functions Equations (3.209)–(3.211), a sufficient condition for asymptotic stability of the observer is presented and proved in the following theorem.

Theorem 3.13   *Suppose the pair (A, C) is observable, and the matrix C is of full row rank. Assume that $g_i(x, u, t), i = 1, \ldots, l$ are persistence of excitation. If there is a positive definite matrix $P = P^T > 0$ and a matrix D such that*

$$\Pi = \begin{bmatrix} \Lambda + C^T C & P \\ P & -\gamma^2 I \end{bmatrix} < 0, \quad (3.212)$$

$$\Lambda = (A - LC)^T P + P(A - LC) + H^T H + \sum_{i=1}^{l} G_i^T \alpha_i \alpha_i G_i + (l + 1)PP, \quad (3.213)$$

$$DC = P, \quad (3.214)$$

then the observer-based estimator in Equations (3.206)–(3.208) ensures that

- *The estimated x and $\hat{\theta}_i$ asymptotically converge to the nonlinear system state x and the multiplicative fault θ_i, respectively, under the zero disturbance case.*
- *When the unknown disturbance exists, the output error satisfies $\|e_y\|_2^2 < \gamma^2 \|d\|_2^2$.*

Proof. The proof consists of two parts: the internal stability analysis and computation of the robust performance index.

Internal stability analysis Choose $V(t) = e_x^T(t)Pe_x(t) + \sum_{i=1}^l \sigma_i^{-1} e_{\theta_i}^T(t)e_{\theta_i}(t)$ as the Lyapunov function and calculate the derivative of the Lyapunov function $V(t)$. We get

$$
\dot{V} = e_x^T\left((A - LC)^T P + P(A - LC)\right)e_x + 2e_x^T P\left(\phi(x, u) - \phi(\hat{x}, u)\right)
$$

$$
+ \sum_{i=1}^l 2e_x^T P\left(\theta_i g_i(x, u) - \hat{\theta}_i g_i(\hat{x}, u)\right) + \sum_{i=1}^l 2\sigma_i^{-1} e_{\theta_i} \dot{\hat{\theta}}_i + 2e_x^T Pd
$$

$$
= e_x^T\left((A - LC)^T P + P(A - LC)\right)e_x + 2e_x^T P\left(\phi(x, u) - \phi(\hat{x}, u)\right)
$$

$$
+ \sum_{i=1}^l 2e_x^T P\left(\theta_i g_i(x, u) - \hat{\theta}_i g_i(\hat{x}, u)\right) + 2e_x^T Pd.
$$

According to the Lipschitz condition, we have

$$
2e_x^T P\left(\phi(x, u) - \phi(\hat{x}, u)\right) \leq 2\|e_x^T P\|\|\phi(x, u) - \phi(\hat{x}, u)\|
$$

$$
\leq e_x^T P P e_x^T + e_x^T H^T H e_x^T,
$$

$$
2e_x^T P\left(\theta_i g_i(x, u) - \theta_i g_i(\hat{x}, u)\right) \leq 2\|e_x^T P\|\|\theta_i g_i(x, u) - \theta_i g_i(\hat{x}, u)\|
$$

$$
\leq e_x^T P P e_x^T + e_x^T G_i^T \alpha_i \alpha_i G_i e_x^T.
$$

Then the derivative of the Lyapunov function satisfies the following inequality:

$$
\dot{V} \leq e_x^T(\Lambda)e_x + 2e_x^T Pd. \tag{3.215}
$$

In the zero disturbance case, one has

$$
\dot{V} \leq -\lambda_{\min}(-\Lambda)\|e_x\|^2. \tag{3.216}
$$

Based on the Schur complement lemma, the matrix Λ is a negative definite matrix. The inequality in Equation (3.216) indicates $e_x \in L_2$. Because $e_x \in L_\infty$, $\dot{e}_x$ is uniformly bounded. Based on Barbalat's Lemma (3.5), we have $e_x \to 0$ as $t \to 0$. And because of the persistent excitation condition of $g_i(x, u)$, the estimator in Equations (3.206)–(3.208) ensures that $e_{\theta_i} \to 0$ as $t \to 0$.

Robust performance index computation
Defining

$$J = \dot{V} + e_y^T e_y - \gamma^2 d^T d$$

and using Equation (3.215), we can derive that

$$J \le e_x^T \Lambda e_x + 2e_x^T P d + e_x^T C^T C e_x - \gamma^2 d^T d$$
$$\le e^T \Pi e,$$

where

$$e = [e_x^T \quad d^T]^T.$$

It follows that

$$J \le -\lambda_{\min}(-\Pi)\|e\|^2.$$

Under the zero initial condition, we have

$$\int_0^T \left(e_y^T e_y - \gamma^2 d^T d\right) dt = \int_0^T J \, dt - V(T) < 0,$$

which implies

$$\int_0^T e_y^T e_y \, dt \le \gamma^2 \int_0^T d^T d \, dt.$$

This completes the proof of the theorem.   ■

Remark 3.19 *Based on the Schur complement lemma and letting $Y = PL$, $\Pi < 0$ in Equation (3.212) can be rewritten as the following matrix inequality:*

$$\begin{bmatrix} \Xi & P & C^T & P \\ P & -\frac{1}{l+1}I & 0 & 0 \\ C & 0 & -I & 0 \\ P & 0 & 0 & -\gamma^2 I \end{bmatrix} < 0,$$

$$\Xi = A^T P + P A - C^T Y - Y C + H^T H + \sum_{i=1}^l G_i^T \alpha_i \alpha_i G_i.$$

The matrix D can be derived from

$$D = PC^T \left(CC^T\right)^{-1}.$$

Remark 3.20   *A good estimator is designed to make the whole system sensitive to multiplicative faults and insensitive to disturbance. Hence, we can reduce the effect of disturbance d to the formula in Equation (3.212) with a smaller γ using the Matlab LMI toolbox.*

3.6   Notes

The first part of this chapter presented a distributed FDI scheme for a class of interconnected nonlinear uncertain systems. Under certain assumptions, adaptive thresholds were designed for distributed FDI in each subsystem. The important properties of robustness and fault sensitivity (fault detectability and isolability) of the distributed FDI algorithm were investigated.

In the second part, an H_∞ formulation of the SFDC problem in continuous-time switched linear systems using a dynamic-observer detector and a state-feedback controller was presented. It was shown that the structure of dynamic observer can be useful in designing for SFDC problems. An LMI approach for SFDC design in switched systems under switching signals with ADT was introduced; in addition to stabilizing the closed-loop system, it simultaneously guarantees some control and detection objectives. The LMI conditions were presented such that no product of the Lyapunov matrices and the system state matrices are involved, which results in significant conservatism reduction in the SFDC problem. Note that our approach can be easily generalized to discrete-time switched linear systems.

We then proposed a robust fault detection filter with time delay, using Lyapunov–Krasovskii function (LKF). Using an LMI method, the existence conditions and further solution of the optimization problem have been derived and, based on them, an algorithm for the design of the fault detection filters was proposed. The proposed scheme was evaluated on a four-tank system which has a delay when tank filling is in process during leakage, thus demonstrating the effectiveness of the approach.

Finally, we proposed a robust adaptive fault estimation scheme for a kind of Lipschitz nonlinear system subject to multiplicative faults and unknown disturbances. Multiplicative faults are parameter changes within the process which make the design for fault estimation more complicated. The estimator is designed in the context of a trade-off between robustness to disturbances and sensitivity to faults. According to Lyapunov stability theory, the estimator can estimate the real fault factors accurately and simultaneously estimate the system states. The conservatism for the whole fault estimation scheme is reduced by using a relaxing Lipschitz matrix.

References

[1] Spooner, J. T., and Passino, K. M. (1999) "Decentralised adaptive control of nonlinear systems using radial basis neural networks", *IEEE Trans. Automatic Control*, **44**:2050–2057.

[2] Meskin, N., and Khorasani, K. (2009) "Actuator fault detection and isolation for a network of unmanned vehicles", *IEEE Trans. Automatic Control*, **54**:835–840.

[3] Blanke, M., Kinnaert, M., Lunze, J., and Staroswiecki, M. (2006) *Diagnosis and Fault-Tolerant Control*, 2nd Edition, Berlin: Springer.

[4] Gertler, J. J. (1998) *Fault Detection and Diagnosis in Engineering Systems*, New York: CRC Press.

[5] Isermann, R. (2006) *Fault-Diagnosis Systems: An introduction from fault detection to fault tolerance*, Berlin: Springer.

[6]   Frank, P. M. (1990) "Fault diagnosis in dynamic systems using analytical and knowledge-based redundancy: a survey and some new results", *Automatica*, **26**(3):459–474.

[7]   Kabore, P., and Wang, H. (2001) "Design of fault diagnosis filters and fault tolerant control for a class of nonlinear systems", *IEEE Trans. Automatic Control*, **46**:1805–1810.

[8]   Zhang, X., Polycarpou, M. M., and Parisini, T. (2001) "Robust fault isolation of a class of nonlinear input–output systems", *Int. J. Control*, **74**:1295–1310.

[9]   Jiang, B., Staroswiecki, M., and Cocquempot, V. (2004) "Fault diagnosis based on adaptive observer for a class of nonlinear systems with unknown parameters", *Int. J. Control*, **77**(4):367–383.

[10] Xu, A., and Zhang, Q. (2004) "Nonlinear system fault diagnosis based on adaptive estimation", *Automatica*, **40**(7):1181–1193.

[11] Tang, X. D., Tao, G., and Joshi, S. M. (2007) "Adaptive actuator failure compensation for nonlinear MIMO systems with an aircraft control application", *Automatica*, **43**(11):1869–1883.

[12] Shankar, S., Darbha, S., and Datta, A. (2002) "Design of a decentralised detection filter for a large collection of interacting LTI systems", *Math Problems in Engineering*, **8**:233–248.

[13] Patton, R. J., Kambhampati, C., Casavola, A., Zhang, P. *et al.* (2007) "A generic strategy for fault-tolerance in control systems distributed over a network", *European J. Control*, **13**:280–296.

[14] Yan, X., and Edwards, C. (2008) "Robust decentralised actuator fault detection and estimation for large-scale systems using a sliding-mode observer", *Int. J. Control*, **81**:591–606.

[15] Ferrari, R., Parisini, T., and Polycarpou, M. M. (2009), "Distributed fault diagnosis with overlapping decompositions: An adaptive approximation approach", *IEEE Trans. Automatic Control*, **54**:794–799.

[16] Zhang, X., Polycarpou, M. M., and Parisini, T. (2002) "A robust detection and isolation scheme for abrupt and incipient faults in nonlinear systems", *IEEE Trans. Automatic Control*, **47**:576–593.

[17] Zhang, X., Polycarpou, M. M., and Parisini, T. (2010) "Fault diagnosis of a class of nonlinear uncertain systems with Lipschitz nonlinearities using adaptive estimation", *Automatica*, **46**(2):290–299.

[18] Spooner, J. T., and Passino, K. M. (1996) "Adaptive control of a class of decentralised nonlinear systems", *IEEE Trans. Automatic Control*, **41**:280–284.

[19] Hovakimyan, N., Lavretsky, E., Yang, B., and Calise, A. J. (2005) "Coordinated decentralized adaptive output feedback control of interconnected systems", *IEEE Trans. Neural Networks*, **16**:185–194.

[20] Guo, Y., Hill, D. J., and Wang, Y. (2000) "Nonlinear decentralised control of large-scale power systems", *Automatica*, **36**:1275–1289.

[21] Siljak, D. D. (1990) *Decentralisation, Stabilization and Estimation of Large-scale Linear Systems*, Boston, Mass, USA: Academic Press.

[22] Ioannou, P. A., and Sun, J. (1996) *Robust Adaptive Control*, Englewood Cliffs, NJ: Prentice Hall.

[23] Farrell, J., and Polycarpou, M. M. (2006) *Adaptive Approximation Based Control*, Hoboken, NJ: John Wiley & Sons.

[24] Zhang, X., Parisini, T., and Polycarpou, M. M. (2004) "Adaptive fault-tolerant control of nonlinear uncertain systems: an information-based diagnostic approach", *IEEE Trans. Automatic Control*, **49**(8):1259–1274.

[25] Ding, S. X. (2008) *Model-based Fault Diagnosis Techniques: Design Schemes, Algorithms, and Tools*, Berlin, Heidelberg: Springer.

[26] Wan, E. A., der Merwe, R. V., and Nelson, A. T. (2000) "Dual estimation and the unscented transformation", in *Advances in Neural Information Processing Systems*, S. A. Solla, T. K. Leen, and K. R. Miller (Eds), Cambridge, MA: MIT Press, 666–672.

[27] Julier S. J., Uhlmann, J. K., and Durrant-Whyte, H. (1995) "A new approach for filtering nonlinear systems', in *Proc. American Control Conference*, 1628–1632.

[28] Julier, S. J., and Uhlmann, J. K. (1997) "A new extension of the Kalman filter to nonlinear systems", in *Proc. 11th Int. Symposium on AeroSpace/Defense Sensing, Simulation and Controls*, **3068**:182–193.

[29] Moret, E. N. (2001) *Dynamic Modeling and Control of a Car-like Mobile Robot*, MSc Thesis, Virginia Polytechnic Institute and State University, USA.

[30] Wang, J. L., Yang, G.-H., Liu, J. (2007) "An LMI approach to H_- index and mixed H_-/H_∞ fault detection observer design", *Automatica*, **43**(9):1656–1665.

[31] Ding, S. X. (2009) "Integrated design of feedback controllers and fault detectors", *Annual Reviews in Control*, **33**:124–135.

[32] Jacobson, C. A., Nett, C. N. (1991) "An integrated approach to controls and diagnostics using the four parameter controller", *IEEE Control Systems Magazine*, **11**:22–29.

[33] Nett, C. N., Jacobson, C. A., and Miller, A. T. (1988) "An integrated approach to controls and diagnostics", in *Proc. American Control Conference*, Atlanta, GA, USA, 824–835.

[34] Kilsgaard, S., Rank, M. L., Niemann, H. H., and Stoustrup, J. (1996) "Simultaneous design of controller and fault detector", in *Proc. 37th IEEE Conference on Decision Control*, Kobe, Japan, 628–629.

[35] Wang, H., and Yang, G.-H. (2009) "Integrated fault detection and control for LPV systems", *Int. J. Robust and Nonlinear Control*, **19**:341–363.

[36] Hwang, D. S., Peng, S. C., and Hsu, P. L. (1994) "An integrated control/diagnosis system for a hard disk drive", *IEEE Trans. Control Systems Technology*, **2**(4):318–325.

[37] Marcos, A., and Balas, G. J. (2005) "A robust integrated controller/diagnosis aircraft application", *Int. J. Robust and Nonlinear Control*, **15**:531–551.

[38] Casavola, A., Famularo, D., and Franze, G. (2005) "A robust deconvolution scheme for fault detection and isolation of uncertain linear systems: an LMI approach", *Automatica*, **41**:1463–1472.

[39] Boyd, S., Ghaoui, L. E., Fern, E., Balakrishnan, V. (1994) *Linear Matrix Inequalities in System and Control Theory*, Philadelphia, PA: Society for Industrial and Applied Mathematics.

[40] Guo, J., Huang, X., and Cui, Y. (2009) "Design and analysis of robust fault detection filter using LMI tools", *Computers and Mathematics with Applications*, **57**:1743–1747.

[41] Zhong, M. Y., Ding, S., Lam, J., and Wang, H. (2003) "An LMI approach to design robust fault detection filter for uncertain LTI systems", *Automatica*, **39**(3):543–550.

[42] Mahmoudi, A., Momeni, A., Aghdam, A. G., and Gohari, P. (2008) "Switching between finite-time observers", *European J. Control*, **14**(4):297–307.

[43] Orchard, M. E., Vachtsevanos, G. J. (2009) "A particle-filtering approach for on-line fault diagnosis and failure prognosis", *Trans. Institute of Measurement and Control*, **31**(4):221–246.

[44] Baruah, P., and Chinnam, R. B. (2003) "HMMs for diagnostics and prognostics in machining processes", in *Proc. 57th Society for Machine Failure Prevention Technology Conference*, Virginia Beach, VA, 14–18.

[45] Zhang, X. D., Xu, R., Chiman, K., Liang, S. Y., *et al.* (2005) "An integrated approach to bearing fault diagnostics and prognostics, in *Proc. American Control Conference*, Jun 8–10, Portland, OR, **4**:2750–2755.

[46] Camci, F. (2005) *Process monitoring, diagnostics and prognostics using support vector machines and hidden Markov models*, PhD Thesis, Graduate School of Wayne State University, Detroit, Michigan.

[47] Gu, H. Y., Tseng, C. Y., and Lee, L. S. (1991) "Isolated-utterance speech recognition using hidden Markov models with bounded state duration", *IEEE Trans. Signal Processing*, **39**(8):1743–1751.

[48] Falaschi, A. (1992) "Continuously variable transition probability HMM for speech recognition", in *Speech Recognition and Understanding*, P. Laface and R. DeMori (Eds), Heidelberg: Springer-Verlag, 125–130.

[49] Chen, J., and Patton, R. J. (1999) *Robust Model-Based Fault Diagnosis for Dynamic Systems*, Boston, Mass, USA: Kluwer Academic Publishers.

[50] Yoshimura, M., Frank, P. M., and Ding, X. (1997) "Survey of robust residual generation and evaluation methods in observer-based fault detection systems", *J. Process Control*, **7**(6):403–424.

[51] Ding, X. S., Zhong, M. Y., Tang, B. Y., and Zhang, P. (2001) "An LMI approach to the design of fault detection filter for time-delay LTI systems with unknown inputs", in *Proc. American Control Conference*, Arlington, USA, 2137–2142.

[52] Frank, P. M., Ding, S. X., and Seliger, B. K. (2000) "Current developments in the theory of FDI", in *Proc. IFAC Symposium on Fault Detection, Supervision and Safety of Technical Processes*, Budapest, Hungary, **1**:16–27.

[53] Mangoubi, R. S., and Edelmayer, A. M. (2000) "Model based fault detection: the optimal past, the robust present and a few thoughts on the future", in *Proc. IFAC Symposium on Fault Detection, Supervision and Safety of Technical Processes*, Budapest, Hungary, 64–75.

[54] Patton, K. (2008) *Artificial Neural Networks for the Modelling and Fault Diagnosis of Technical Processes* Springer, UK.

[55] Jean-Yves, K., and Woihida, A. (2000) "Robust fault isolation observer for discrete-times systems with time delay", *Proc. IFAC Symposium on Fault Detection, Supervision and Safety of Technical Processes*, Budapest, Hungary, 390–395.

[56] Jiang, B., and Chowdhury, F. N. (2005) "Fault estimation and accommodation for linear MIMO discrete-time systems", *IEEE Trans. Control Systems Technology*, **13**(3):493–499.

[57] Jiang, B., Staroswiecki, M., and Cocquempot, V. (2002) "H_∞ fault detection filter for a class of discrete-time systems with multiple time delays', in *Proc. 15th IFAC World Congress*, Barcelona, Spain.

[58] Duan, G. R., Patton, R. J. (2001) "Robust fault detection using Luenberger-type unknown input observers: a parametric approach", *Int. J. Syst. Sci*, 32(4):533–540.

[59] Boskovic, J. D., Bergstrom, S. E., Mehra, R. K. (2005) "Robust integrated flight control design under failures, damage, and state-dependent disturbances", *J. Guidance, Control and Dynamics*, 28(5):902–917.

[60] De Persis, C., and Isidori, A. 2001 "A geometric approach to nonlinear fault detection and isolation", *IEEE Trans. Automatic Control*, 45(6):853–865.

[61] Yan, X. G., and Edwards, C. (2007) "Nonlinear robust fault reconstruction and estimation using a sliding mode observer", *Automatica*, 43(9):1605–1614.

[62] Gao, Z., and Ding, S. X. (2007) "Actuator fault robust estimation and fault-tolerant control for a class of nonlinear descriptor systems", *Automatica*, 43(5):912–920.

[63] Hou, M. (2007) "Estimation of sinusoidal frequencies and amplitudes using adaptive identifier and observer", *IEEE Trans. Automatic Control*, 52(3):493–499.

[64] Vijayaraghavan, K., Rajamani, R., and Bokor, J. (2007) "Quantitative fault estimation for a class of nonlinear systems", *Int. J. Control*, 80(1):64–74.

[65] Ding, G. R., Frank, P. M., and Ding, E. L. (2003) "An approach to the detection of multiplicative faults in uncertain dynamic systems", in *Proc. 41st IEEE Conference on Decision and Control*, 4371–4376.

[66] Isermann, R. (2005) "Model-based fault-detection and diagnosis: status and applications", *Annual Reviews in Control*, 29(1):71–85.

[67] Tan, C. P., and Edwards, C. (2004) "Multiplicative fault reconstruction using sliding mode observers", in *Proc. 5th Asian Control Conference*, 957–962.

[68] Phanomchoeng, G., and Rajamani, R. (2010) "Observer design for Lipschitz nonlinear systems using Riccati equations", in *Proc. American Control Conference*, 6060–6065.

4

Fault-Tolerant Control Systems

4.1 Model Prediction-Based Design Approach

Fault-tolerant control (FTC) is an integral component in industrial processes as it enables the system to continue robust operation under some conditions. In this chapter, an FTC scheme is proposed for interconnected systems within an integrated design framework to yield a timely monitoring and detection of fault and a reconfiguring of the controller according to those faults. The unscented Kalman filter (UKF) fault detection and diagnosis system is initially run on the main plant and parameter estimation is carried out for the local faults. This critical information is shared through information fusion to the main system where the whole system is decentralized using the overlapping decomposition technique. Using these parameter estimates of decentralized subsystems, a model predictive control (MPC) adjusts its parameters according to the fault scenarios thereby striving to maintain the stability of the system. Experimental results on interconnected, continuous-time, stirred tank reactors(CSTR) with recycling and a four-tank system indicate that the proposed method is capable of correctly identifying various faults and controlling the system under some conditions.

4.1.1 Introduction

In process control industries, failures of some key control and process elements are often encountered. The failure of such major components not only affects the performance of the plant but also leads to critical operation problems giving rise to instability and possible breakdown. For example, a faulty sensor may easily affect the momentum of the production line and, in some cases, may push the other sensors in the plant to work beyond their design configuration, thereby leading to a major control failure in the sensor network-based monitoring of the plant. Other crucial scenarios can be a burned-out thermocouple, a broken transducer or a stuck valve. Therefore, fault tolerance has been one of the major issues in process control and it has become a necessary component for the process industries.

Fault-tolerance, or "graceful degradation", is basically the design property that enables the system to continue operation under some conditions when some part of the system fails. Because of the poor health of the system, the operation continues at a reduced level, rather than

Analysis and Synthesis of Fault-Tolerant Control Systems, First Edition. Magdi S. Mahmoud and Yuanqing Xia.
© 2014 John Wiley & Sons, Ltd. Published 2014 by John Wiley & Sons, Ltd.

failing completely. Moreover, the performance is proportional to the severity of the failure, as compared to a conventionally designed system, in which even a small failure can cause total breakdown. The problem of fault tolerance is of extreme importance when it comes to mission-critical systems and life-critical systems and several approaches have been adopted to implement a reliable fault-tolerant control scheme. Parker, Ng, and Ran report FTC results for direct-drive wind turbines [1] and Dwari and Parsa for five-phase, permanent-magnet motors [2]. A scheme for optimal torque FTC is proposed by Sun *et al.* [3]. Work on a multi-phase power converter drive for fault-tolerant machine development in aerospace applications is presented by de Lillo *et al.* [4]. Recent FTC applications are discussed by Zhao *et al.* [5] and Ceballos *et al.* [6]. Depending upon how the redundancy is being utilized, FTC system design can be classified into two types:

- passive fault-tolerant control (PFTC) systems;
- active fault-tolerant control (AFTC) systems.

In PFTC systems, controllers are fixed and are designed to be robust against a class of *a priori* known faults. This approach needs neither fault diagnosis and detection (FDD) schemes nor controller reconfiguration but it has limited fault-tolerant capabilities. Once the controller is designed in a passive fault-tolerant scheme, it remains fixed during the entire system operation. Even in the event of component failure, the control system should be able to maintain the designed performance. A multiple disjoint decentralized control was proposed, in which redundancy lies in the employment of multiple controllers [7]. Further extensions to control design against actuator failure was developed using a state-feedback controller implementation [8, 9]. PFTC systems are also known in the literature as "reliable control systems" or "control systems with integrity".

In contrast to PFTC systems, AFTC systems react to component failures actively by reconfiguring control actions so that the stability and acceptable performance of the entire system can be maintained. AFTC systems are also referred to as "self-repairing" [10], "reconfigurable" [11], "restructurable" [12], or "self-designing" [13] systems. An AFTC system consists basically of a fault detection and diagnosis scheme, a controller reconfiguration mechanism, and a reconfigurable controller, all working in a systematic manner. AFTC systems have also been called "fault detection, identification (diagnosis) and accommodation schemes" [14, 15]. Xie *et al.* [16] presents a strong tracking-filter-based, generic model control which leads to reconfigurable controllers. Work on AFTC for magnetic levitation systems, in which sensor faults have been considered, has been done by Nazari [17] and Yetendje [18].

We develop an improved active FTC scheme within an integrated fault detection and tolerance design framework. The developed methodology utilizes a fault-tolerant technique based on model prediction to enhance the accuracy and reliability of parametric estimation through UKF in the process fault detection phase.

4.1.2 System description

To achieve effective fault detection and tolerance, we assume that various faults in an interconnected system have been successfully monitored, estimated and protected through tolerance by the encapsulation of the UKF with decentralized control based on model prediction. Figure 4.1

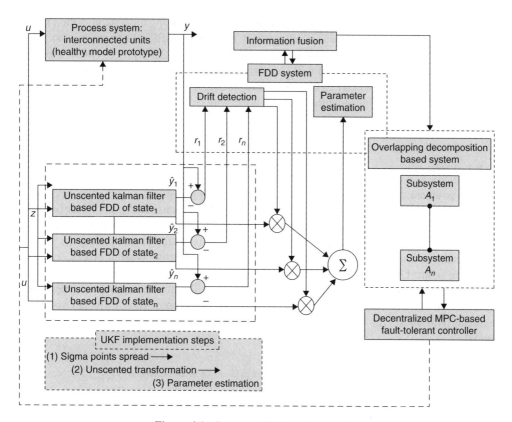

Figure 4.1 Proposed FTC implementation

shows the proposed implementation. Unscented filters are employed in n states of the system to detect faults in a highly dynamic system. The residual compares the output of the unscented filters with the output of n states of a healthy model of the plant (containing no faults); this results in n residuals r_n, which give us the drift detection of the system. The drift detections and output of the unscented filters are added to give us the parameter estimation of the system. The drift detection and parameter estimation are fed into an information unit known as an FDD unit, from where the information is fused, and proceeds to the subsystems of a particular system, made by overlapping decomposition. The fused information tells the strength of the fault and its upper and lower limits, which helps us to build a decentralized, MPC-based, fault-tolerant controller, which adjusts its parameters according to the fault scenarios, thereby striving to maintain the stability of the system.

Assume that a process is monitored by N sensors, described by the following general nonlinear process and measurement models in a discrete time state-space framework:

$$x(k) = f(x(k-1), u(k-1), d(k-1)) + w(k-1)$$
$$z_i = c_i(x(k)) + v_i(k); \quad i = 1, \ldots, N \tag{4.1}$$

where $f(.)$ and $h_i(.)$ are the known nonlinear functions, representing the state transition model and the measurement model, respectively. $x(k) \in R_{n_x}$ is the process state vector, $u(k) \in R_{n_u}$ denotes the manipulated process variables, $d(k) \in R_{n_d}$ represents the process faults modeled by the process disturbances, $z_i(k) \in R_{n_{zi}}$ are the measured variables obtained from the N installed sensors, and $w(k)$ and $v_i(k)$ indicate the stochastic process and measurement disturbances modeled by zero-mean white Gaussian noise with covariance matrices $Q(k)$ and $R_i(k)$, respectively.

4.1.3   Discrete-time UKF

In most practical applications of interest, the process and measurement dynamic models are described by nonlinear equations, as in Equation (4.1). This means that the nonlinear behavior can affect the process operation at least through its process dynamics or measurement equation. In such cases, the standard Kalman filter algorithm is often unsuited to estimating the process states using its linearized, time-invariant, state-space model at the desired process nominal operating point. UKF gives a simple and effective remedy for overcoming such nonlinear estimation problem. Its basic idea is to locally linearize the nonlinear functions, described by Equation (4.1) at each sampling time instant around the most recent process condition estimate. This allows the Kalman filter to be applied to the following linearized time-varying model:

$$x(k) = A(k)x(k-1) + B_u(k)u(k-1) + B_d(k)d(k-1)$$
$$+ w(k-1)$$
$$z_i(k) = C_i(k)x(k) + v_i(k); \quad i = 1, \ldots, N \tag{4.2}$$

where the state transition matrix $A(k)$, the input matrices $B_u(k)$ and $B_d(k)$, and the observation matrix $H_i(k)$ are the Jacobian matrices which are evaluated at the most recent process operating condition in real time, rather than the process fixed nominal values:

$$A(k) = \frac{\partial f}{\partial x}|_{\hat{x}(k)}, \quad B_u(k) = \frac{\partial f}{\partial u}|_{u(k)} \tag{4.3}$$

$$B_d(k) = \frac{\partial f}{\partial d}|_{\hat{d}(k)}, \quad C_i(k) = \frac{\partial h_i}{\partial x}|_{\hat{x}(k)}, \quad i = 1, \ldots, N \tag{4.4}$$

In conventional control, disturbance variables $d(k)$ are treated as known inputs with distinct entry into the process state-space model. This distinction between state and disturbance as non-manipulated variables, however, is not justified from the monitoring perspective using the estimation procedure. Therefore, a new augmented state variable vector $x^*(k) = [d^T(k) \quad x^T(k)]^T$ is developed by considering the process disturbances or faults as additional state variables. To implement this view, the process faults are assumed to be random state variables governed by the following stochastic auto-regressive (AR) model equation:

$$d(k) = d(k-1) + w_d(k-1) \tag{4.5}$$

This assumption changes the linearized model formulations in system of Equation (4.2) to the following augmented state-space model:

$$x^*(k) = A^*(k)x^*(k-1) + B^*(k)u(k-1) + w^*(k-1)$$

$$z_i(k) = C_i^*(k)x^*(k) + v_i(k); \quad i = 1, \ldots, N \tag{4.6}$$

Noting that:

$$A^*(k) = \begin{bmatrix} I^{n_d \times n_d} & 0^{n_d \times n_x} \\ B_d(k)^{n_x \times n_d} & A(k)^{n_x \times n_x} \end{bmatrix}$$

$$B^*(k) = \begin{bmatrix} 0^{n_d \times n_u} & B_u(k)^{n_x \times n_u} \end{bmatrix}^T$$

$$C_i^*(k) = \begin{bmatrix} 0^{1 \times n_d} & C_i(k)^{1 \times n_x} \end{bmatrix}$$

$$W^*(k-1) = \begin{bmatrix} w_d(k-1)^{n_d \times 1} & w(k-1)^{n_x \times 1} \end{bmatrix}^T. \tag{4.7}$$

Assumption 4.1 *There exists a known positive constant L_0 such that for any norm bounded $x_1(t), x_2(t) \in \mathbf{R}^n$, the following inequality holds:*

$$\| f(u(t), y(t), x_1(t)) - f(u(t), y(t), x_2(t)) \|$$

$$\leq L_0 \| x_1(t) - x_2(t) \| \tag{4.8}$$

Assumption 4.2 *The transfer function matrix $C[zI - (A - KC)]^{-1}B$ is strictly positive real, where $K \in \mathbf{R}^{n \times r}$ is chosen such that $A - KC$ is stable.*

Remark 4.1 *For a given positive definite matrix $Q > 0 \in \mathbf{R}^{n \times n}$, there exists matrices $P = P^T > 0 \in \mathbf{R}^{n \times n}$ and a scalar R such that:*

$$(A - KC)^T P(A - KC) = -Q \tag{4.9}$$

$$PB = C^T R \tag{4.10}$$

To detect the fault, the following system is constructed:

$$\hat{x}(k) = A\hat{x}(k) + g(u(k), y(k)) + B\xi_H f(u(k), y(k), \hat{x}(k)) + K(y(k) - \hat{y}(k)) \tag{4.11}$$

$$\hat{y}(k) = C\hat{x}(k) \tag{4.12}$$

where $\hat{x}(k) \in \mathbf{R}^n$ is the state estimate, the input is $u \in \mathbf{R}^m$, and the output is $y \in \mathbf{R}^r$. The pair (A, C) is observable. The nonlinear term $g(u(k), y(k))$ depends on $u(k)$ and $y(k)$, which are directly available. $f(u(k), y(k), x(k)) \in \mathbf{R}^r$ is a nonlinear vector function of $u(k)$, $y(k)$ and

$x(k)$. $\xi(k) \in \mathbf{R}$ is a parameter which changes unexpectedly when a fault occurs. Since it has been assumed that the pair (A, C) is observable, a gain matrix K can be selected such that $A - KC$ is a stable matrix. We define:

$$e_x(k) = x(t) - \hat{x}(k), \quad e_y(k) = y(k) - \hat{y}(k) \tag{4.13}$$

Then, the error equations can be given by:

$$e_x(k + 1) = (A - KC)e_x(k) + B[\xi(k)f(u(k), y(k), x(k))$$
$$-\xi_H f(u(k), y(k), \hat{x}(k))] \tag{4.14}$$

$$e_y(k) = Ce_x(k) \tag{4.15}$$

The convergence of the above filter is guaranteed by Theorem 4.1.

Theorem 4.1   *Under Assumption 4.2, the filter is asymptotically convergent when no fault occurs ($\xi(k) = \xi_H$), i.e., $lim_{k\to\infty}e_y(k) = 0$.*

Proof. Consider the following Lyapunov function

$$V(e(k)) = e_x^T(k)Pe_x(k) \tag{4.16}$$

where P is the solution of Equation (4.9) and Q is chosen such that $\rho_1 = \lambda_{min}(Q) - 2\|C\|.|R|\xi_H L_0 > 0$. Along the trajectory of the fault-free system in Equation (4.14), the corresponding Lyapunov difference along the trajectories $e(k)$ is:

$$\begin{aligned}
\Delta V &= E\{V(e(k + 1)|e_k, p_k)\} - V(e(k))\\
&= E\{e^T(k + 1)P_ie(k + 1)\} - e^T(k)P_ie(k)\\
&= (A_ee_x + B_Lu_e)^T P(A_ee_x + B_Lu_e) - e_x^T(k)Pe_x(k)\\
&= e^T(k)[(P(A - KC) + (A - KC)^T P)\\
&\quad + PB\xi_H[f(u(k), y(k), x(k))\\
&\quad - f(u(k), y(k), \hat{x}(k))]]e(k)
\end{aligned} \tag{4.17}$$

From Assumption 4.1 and Equation (4.9), one can further obtain that

$$\begin{aligned}
\Delta V &\leq -e_x^T(k)Qe_x(t) + 2\|e_y(k)\|.|R|\xi_H L_0\|e_x(k)\|\\
&\leq -\rho_1\|e_x\|^2 < 0
\end{aligned} \tag{4.18}$$

Thus, $lim_{k\to\infty} e_x(k) = 0$ and $lim_{k\to\infty} e_y(k) = 0$. This completes the proof.   ∎

The UKF essentially addresses the approximation issues of the EKF [19, 20, 21]. The basic difference between the EKF and the UKF stems from the manner in which Gaussian random variables (GRV) are presented through system dynamics. In the EKF, the state distribution is approximated by GRV, which are then propagated analytically though the first-order linearization of the nonlinear system. This can introduce large errors in the true posterior mean covariance of the transformed GRV, which may lead to sub-optimal performance and sometimes divergence of the filter. The UKF addresses this problem by using a deterministic sampling approach. The state distribution is approximated by a GRV but is represented using a minimal set of carefully chosen sample points. These sample points completely capture the true mean and covariance of the GRV; when propagated through the *true* nonlinear system, they capture the posterior mean and covariance accurately to second order (Taylor series expansion) for any nonlinearity. The EKF, in contrast, only achieves first-order accuracy.

4.1.4   Unscented Transformation (UT)

The structure of the UKF is elaborated by UT for calculating the statistics of a random variable which undergoes a nonlinear transformation [21]. Consider propagating a random variable x (dimension L) through a nonlinear function, $y = f(x)$. Assume x has mean $\bar{x}$ and covariance P_x. To calculate the statistics of y, we form a matrix $\mathcal{X}$ of $2L + 1$ sigma vectors $\mathcal{X}_i$ according to:

$$\mathcal{X}_0 = \bar{x},$$
$$\mathcal{X}_i = \bar{x} + (\sqrt{(L + \lambda)\mathbf{P}_x})_i, \quad i = 1, \ldots, L$$
$$\mathcal{X}_i = \bar{x} - (\sqrt{(L + \lambda)\mathbf{P}_x})_i - L, \quad i = L + 1, \ldots, 2L \tag{4.19}$$

where $\lambda = \alpha^2(L + \kappa) - L$ is a scaling parameter. The constant α determines the spread of the sigma points around $\bar{x}$ and is usually set to a small positive value ($1 \leq \alpha \leq 10^{-4}$). The constant κ is a secondary scaling parameter, which is usually set to $3 - L$, and β is used to incorporate prior knowledge of the distribution of x (for Gaussian distributions, $\beta = 2$ is optimal). $(\sqrt{(L + \lambda)\mathbf{P}_x})_i$ is the ith column of the matrix square root (that is, the lower-triangular Cholesky factorization). These sigma vectors are propagated through the nonlinear function $\mathcal{Y}_i = f(\mathcal{X}_i), i = 0, \ldots, 2L$. Now the mean and covariance for y are approximated using a weighted sample mean and covariance of the posterior sigma points:

$$\bar{y} \approx \sum_{i=0}^{2L} W_i^{(m)} \mathcal{Y}_i,$$

$$\mathbf{P}_y \approx \sum_{i=0}^{2L} W_i^c (\mathcal{Y}_i - \bar{y})(\mathcal{Y}_i - \bar{y})^T,$$

$$W_0^{(m)} = \frac{\lambda}{L + \lambda},$$

$$W_0^{(c)} = \frac{\lambda}{L + \lambda} + 1 - \alpha^2 + \beta,$$

$$W_i^{(m)} = W_i^{(c)} = \frac{1}{2(L + \lambda)}, i = 1, \ldots, 2L.$$

A block diagram illustrating the steps involved in performing the UT is shown in Figure 4.1.

Remark 4.2 *Note that this method differs substantially from general Monte Carlo sampling methods, which require orders of magnitude more sample points in an attempt to propagate an accurate (possibly non-Gaussian) distribution of the state. The deceptively simple approach taken with the UT results in approximations that are accurate to the third order for Gaussian inputs for all nonlinearities. For non-Gaussian inputs, approximations are accurate to at least the second order, with the accuracy of the third- and higher-order moments being determined by the choice of α and β.*

In view of the foregoing discussion, the UKF is an extension of the UT to the following recursive estimation:

$$\hat{x}_k = x_{k_{prediction}} + \kappa_k \left[y_k - y_{k_{prediction}} \right] \tag{4.20}$$

where the state random variables (RV) are redefined as the concentration of the original state and noise variables: $x_k^a = [x_k^T \quad v_k^T \quad n_k^t]^T$. The UT sigma point selection scheme is then applied to this new augmented state RV to calculate the corresponding sigma matrix, $\mathcal{X}_k^a$. The UKF equations are given below. Note that no explicit Jacobian or Hessian calculations are necessary to implement this algorithm. Initialize with:

$$\hat{x}_0 = \mathsf{E}[x_0],$$
$$P_0 = \mathsf{E}[(x_0 - \hat{x}_0)(x_0 - \hat{x}_0)^T],$$
$$\hat{x}_0^a = \mathsf{E}[x^a] = [\hat{x}_0^T \quad 0 \quad 0]^T. \tag{4.21}$$

For $k \in [1, \ldots, \infty]$, calculate the sigma points:

$$\mathcal{X}_{k-1}^a = [\hat{x}_{k-1}^a \quad \hat{x}_{k-1}^a + \gamma \sqrt{P_{k-1}^a} \quad \hat{x}_{k-1}^a - \gamma \sqrt{P_{k-1}^a}] \tag{4.22}$$

The UKF time-update equations are:

$$\mathcal{X}_{k|k-1}^x = \mathsf{F}(\mathcal{X}_{k-1}^x, \quad u_{k-1}, \quad \mathcal{X}_{k-1}^v),$$

$$\hat{x}_k^- = \sum_{i=0}^{2L} W_i^m \mathcal{X}_{i,k|k-1}^x,$$

$$P_k^- = \sum_{i=0}^{2L} W_i^c (\mathcal{X}_{i,k|k-1}^x - \hat{x}_k^-)(\mathcal{X}_{i,k|k-1}^x - \hat{x}_k^-)^T,$$

$$\mathcal{Y}_{k|k-1} = \mathsf{H}(\mathcal{X}_{k|k-1}^x, \mathcal{X}_{k-1}^n),$$

$$\hat{y}_k^- = \sum_{i=0}^{2L} W_i^m \mathcal{Y}_{i,k|k-1}. \tag{4.23}$$

The UKF measurement-update equations are:

$$P_{\tilde{y}_k \tilde{y}_k} = \sum_{i=0}^{2L} W_i^c (\mathcal{Y}_{i,k|k-1} - \hat{y}_k^-)(\mathcal{Y}_{i,k|k-1} - \hat{y}_k^-)^T,$$

$$P_{x_k y_k} = \sum_{i=0}^{2L} W_i^c (\mathcal{X}_{i,k|k-1} - \hat{x}_k^-)(\mathcal{Y}_{i,k|k-1} - \hat{y}_k^-)^T,$$

$$\kappa_k = P_{x_k y_k} P_{\tilde{y}_k \tilde{y}_k}^{-1},$$

$$\hat{x}_k = \hat{x}_k^- + \kappa_k (y_k - \hat{y}_k^-),$$

$$P_k = P_k^- - \kappa_k P_{\tilde{y}_k \tilde{y}_k} \kappa_k^T \qquad (4.24)$$

where

$$x^a = [x^T \quad v^T \quad n^T]^T,$$

$$\mathcal{X}^a = [(\mathcal{X}^x)^T \quad (\mathcal{X}^v)^T \quad (\mathcal{X}^n)^T]^T, \text{ and}$$

$$\gamma = \sqrt{L + \lambda}. \qquad (4.25)$$

In addition, λ is the composite scaling parameter, L is the dimension of the augmented state, $\mathfrak{R}^v$ is the process-noise covariance, $\mathfrak{R}^n$ is the measurement-noise covariance, and W_i are the weights.

4.1.5   Controller reconfiguration

Controller re-design can be considered by model matching. As the nominal closed-loop system is known, the model of this system can be used as a description of the dynamical properties that the new controller should produce in connection with the faulty plant. That is, the closed-loop system should match the model of the nominal loop. The nominal closed-loop system is composed of the linear nominal plant

$$x(k + 1) = Ax(k) + Bu(k)$$

$$y(k) = Cx(k) \qquad (4.26)$$

and a state-feedback controller $u(k) = -\mathbf{K}x(k)$ both of which yield the model of the closed-loop system:

$$\dot{x}(k) = (A - B\mathbf{K})x(k)$$

$$y(k) = Cx(k) \qquad (4.27)$$

If the controller does not use all the inputs u_f of the input vector u, the matrix $\mathbf{K}$ has zero rows. When the fault f occurs, the faulty plant is given by:

$$x(k+1) = A_f x(k) + B_f u(k)$$
$$y(k) = C_f x(k) \qquad (4.28)$$

where the fault f has changed the system properties, which are now described by the matrices A_f, B_f and C_f. If the set of available inputs and outputs have changed, the matrices B_f and C_f have vanishing columns or rows, respectively. A new state-feedback controller, $u(k) = -K_f x(k)$ should be found such that the closed-loop system

$$x(k+1) = (A_f - B_f \mathbf{K_f}) x(k)$$
$$y(k) = C_f x(k) \qquad (4.29)$$

behaves like the nominal loop. That is, the relation, $A - BK = A_f - B_f K_f$ has to hold, which means that both closed-loop systems have similar dynamics. It cannot be satisfied, unless B and B_f have the same image (e.g., in the case of a redundant actuator). Therefore, the new controller K_f is chosen so as to minimize the difference:

$$\| (A - BK) - (A_f - B_f K_f) \| \qquad (4.30)$$

The solution to this problem is given by:

$$K_f = B_f^+ (A_f - A + BK)$$
$$= (B_f' B_f)^{-1} B_f' (A_f - A + BK) \qquad (4.31)$$

where B_f^+ denotes the pseudo-inverse of B_f. The new controller is adapted to the faulty system and minimizes the difference between the dynamical properties of the nominal loop and the closed-loop system with the faulty plant. In the proposed scheme, this has been done with the help of an overlapping decomposition information set and model predictive control.

4.1.6   Model predictive control

Model predictive control (MPC) is a multi-variable control algorithm that solves, at each sampling instant, a finite horizon optimal control problem. It involves an internal dynamic model of the process using receding horizon control, model assumptions and an optimization cost function J over the receding prediction horizon to calculate the optimum control moves.

4.1.6.1   Receding horizon control

The MPC scheme makes use of the receding horizon principle. At each sample, a finite horizon optimal control problem is solved over a fixed interval of time, the prediction horizon. We assume that the controlled variables, $z(k)$, are to follow some set point trajectory, $r(k)$. A common choice is to use a quadratic cost function that, in combination with a linear system model, yields a finite horizon linear quadratic problem. We assume that a model of the form:

$$x(k+1) = Ax(k) + Bu(k),$$

$$y(k) = C_y x(k),$$

$$z(k) = C_z x(k) + D_z u(k),$$

$$z_c(k) = C_c x(k) + D_c u(k) \tag{4.32}$$

is available. Here $y(k) \epsilon \mathfrak{R}^{p_y}$ is the measured output, $z(k) \epsilon \mathfrak{R}^{p_y}$ the controlled output and $u(k) \epsilon \mathfrak{R}^m$ the input vector. The state vector is $x(k) \epsilon \mathfrak{R}^n$. The MPC controller should also respect constraints on control variables as well as the constrained outputs, $z_c(k) \epsilon \mathfrak{R}^{p_c}$.

$$\triangle u_{min} \leq \triangle u(k) \leq \triangle u_{max},$$

$$u_{min} \leq u(k) \leq u_{max}, \quad z_{min} \leq z_c(k) \leq z_{max} \tag{4.33}$$

where $\triangle u(k) = u(k) - u(k-1)$ are the control increments. The distinction between controlled and constrained variables is natural, since only the controlled variables have specified reference values.

4.1.6.2 An optimal control problem

The optimal control problem is the core element of the MPC algorithm. Consider the following quadratic cost function:

$$J(k) = \sum_{i=Hw}^{H_p+H_w+1} \| \hat{z}(k+i|k) - r(k+i|k) \|_Q^2$$

$$+ \sum_{i=0}^{H_u-1} \| \triangle \hat{u}(k+i|k), -r(k+i|k) \|_R^2 \tag{4.34}$$

where $\hat{z}(k+i|k)$ are the predicted controlled outputs at time k and $\triangle \hat{u}(k+i|k)$ are the predicted control increments. The matrices $Q \geq 0$ and $R > 0$ are weighting matrices that are assumed to be constant over the prediction horizon. The length of the prediction horizon is H_p and the first sample to be included in the horizon is H_w. H_w may be used to shift the control horizon, but we assume that $H_w = 0$. The control horizon is given by H_u. The cost function, Equation (4.34), may be rewritten as:

$$J(k) = \| Z(k) - \tau(k) \|_Q^2 + \| \triangle U \|_R^2 \tag{4.35}$$

where

$$
Z(k) = \begin{bmatrix} \hat{z}(k|k) \\ \vdots \\ \hat{z}(k + H_p - 1|k) \end{bmatrix}, \tau(k) = \begin{bmatrix} r(k|k) \\ \vdots \\ r(k + H_p - 1|k) \end{bmatrix},
$$

$$
\Delta U(k) = \begin{bmatrix} \Delta u(k|k) \\ \vdots \\ \Delta u(k + H_u - 1|k) \end{bmatrix},
$$

$$
Q = diag \begin{bmatrix} Q & Q & \ldots, & Q \end{bmatrix}, R = diag \begin{bmatrix} R & R & \ldots, & R \end{bmatrix}
$$

By deriving the prediction expressions, we can write

$$
Z(k) = \Psi x(k) + \Gamma u(k - 1) + \Theta \Delta U(k) \tag{4.36}
$$

where

$$
\Psi = \begin{bmatrix} C_z \\ C_z A \\ C_z A^2 \\ \vdots \\ C_z A^{H_p - 1} \end{bmatrix}, \Gamma = \begin{bmatrix} D_z \\ C_z B + D_z \\ C_z AB + C_z B + D_z \\ \vdots \\ C_z \sum_{i=0}^{H_p - 2} A^i B + D_z \end{bmatrix}
$$

$$
\Theta = \begin{bmatrix} D_z & 0 & \cdots \\ C_z B + D_z & D_z & 0 \\ C_z AB + C_z B + D_z & \ddots & \vdots \\ \vdots & \ddots & 0 \\ C_z \sum_{i=0}^{H_p - 2} A^i B + D_z & \cdots & D_z \\ \vdots & \ddots & \vdots \\ C_z \sum_{i=0}^{H_p - 2} A^i B + D_z & \cdots & \cdots, \end{bmatrix}
$$

Also, let

$$
E(k) = \tau(k) - \Psi x(k) - \Gamma u(k - 1) \tag{4.37}
$$

This quantity could be interpreted as the free response of the system, if all the decision variables at $t = k$, $\Delta \mathcal{U}(k)$, were set to zero.

Remark 4.3 *An efficient technique for interconnected systems is the overlapping decomposition [22]. The model predictive control has been derived here by subsystems from decentralized overlapping decomposition. Consider the two systems.*

$$\mathbf{S} : \dot{x} = Ax, \tilde{\mathbf{S}} = \dot{\tilde{x}} = \bar{A}\tilde{x} \tag{4.38}$$

where $x(t) \in \mathbf{R}^n$ is the state of $\mathbf{S}$ and $\tilde{x}(t) \in \mathbf{R}^{\tilde{n}}$. According to the expansion–contraction or "Inclusion Principle", $\tilde{\mathbf{S}}$ includes $\mathbf{S}$ or $\mathbf{S}$ is included by $\tilde{\mathbf{S}}$ iff

$$\tilde{x} = Vx, \quad \tilde{A} = VAU + M, \quad UV = I_n \tag{4.39}$$

It is obvious that the stability of $\tilde{\mathbf{S}}$ implies the stability of $\mathbf{S}$ and further details about computing matrices U, V, M and other related issues can be found in.

Remark 4.4 *The problem of minimizing the cost function is a quadratic programming (QP) problem. The algorithm for obtaining the optimal control signal at each sample assumes that the present state vector is available. Since this is often not the case, state estimation is required. The celebrated separation principle, stating that the optimal control and optimal estimation problems are solved independently and yields a globally optimal controller for linear systems, suggests an attractive approach. We let the solution of the optimization problem be based on an estimate of the state vector, $\hat{x}(k)$ instead of on the true state vector $x(k)$. For this purpose, a Kalman filter can be used. Apart from estimating the state of the system, an estimator could be used to estimate disturbances, assuming that a disturbance model is available. For example, error-free tracking may be achieved by including a particular disturbance model in the observer.*

Remark 4.5 *In our case of UKF-based fault detection (comprising of drift detection and parameter estimation), the following is constructed:*

$$\hat{x}(k) = A\hat{x}(k) + g(u(k), y(k)) + B\hat{\xi}(k)f(u(k), y(k), \hat{x}(k)) + K(y(k) - \hat{y}(k)) \tag{4.40}$$

$$\hat{y}(k) = C\hat{x}(k) \tag{4.41}$$

where $\hat{x}(k) \in \mathbf{R}^n$ is the estimated vector and $\hat{\xi}(k)$ is an estimate of $\xi(k)$. The value of $\hat{\xi}(k)$ is set to ξ_H until a fault is detected. It is assumed that after a fault occurs, $\xi(k) = \xi_f = constant \neq \xi_H, |\xi_f| \leq \xi_0$. We introduce:

$$e_x(k) = x(k) - \hat{x}(k),$$
$$e_y(k) = y(k) - \hat{y}(k),$$
$$e_0(k) = \xi_f - \hat{\xi}(k). \tag{4.42}$$

Then, the reconfigurable fault control can be obtained such that:

$$e_x(k+1) = (A - KC)e_x(k) + B[\xi_f f(u(k), y(k), x(k)) - \hat{\xi}(k)f(u(k), y(k), \hat{x}(k))], \quad (4.43)$$

$$e_y(k) = Ce_x(k). \tag{4.44}$$

The convergence of the above adaptive reconfiguration is guaranteed by Theorem 4.2.

Theorem 4.2 *Under Assumptions 4.1 and 4.2, the system in Equation (4.43) and the following diagnostic algorithm:*

$$\Delta\xi = \Gamma f^T(u(k), y(k), \hat{x}(k))Re_y(k) \tag{4.45}$$

can realize $lim_{t\to\infty}e_x(k) = 0$ and a bounded $e_0(k) \in L^2$. Furthermore, $lim_{k\to\infty}e_\xi(k) = 0$ under a persistent excitation, where R is given by Equation (4.9) and $\Gamma > 0$ is a weighting scalar.

Proof. Consider the following Lyapunov function:

$$V(e(k)) = e_x^T(k)Pe_x(k) + \Gamma^{-1}e_\xi^2(t) \tag{4.46}$$

From Equation (4.43) and Equation (4.45), its first forward difference is:

$$\begin{aligned}
\Delta V &= E\{V(e(k+1)|e_k, p_k)\} - V(e(k)) \\
&= E\{e^T(k+1)P_ie(k+1)\} - e^T(k)P_ie(k) \\
&= (A_e e_x + B_L u_e)^T P(A_e e_x + B_L u_e) - e_x^T(k)Pe_x(k) \\
&= e^T(k)(P(A - KC) + (A - KC)^T P) \\
&\quad + PB[\xi_f f(u(k), y(k), x(k)) \\
&\quad - \hat{\xi}(k)f(u\xi(k), y(k), \hat{x}(k))]e(k) \\
&\quad - 2e_\xi(k)f^T(u(k), y(k), \hat{x}(k))Re_y(k)
\end{aligned} \tag{4.47}$$

According to Assumptions 4.1 and 4.2, one can further obtain that

$$\begin{aligned}
\Delta V &\leq -e_x^T(k)Qe_x(k) - 2e_\xi(k)f^T(u(k), y(k), \hat{x}(k))Re_y(k) \\
&\quad 2e_x^T(k)C^T R\{e_\xi f(u(k), y(k), x(k)) - \\
&\quad \hat{\xi}(k)f(u(k), y(k), \hat{x}(k))\}
\end{aligned} \tag{4.48}$$

where $\rho_2 = \lambda_{min}(Q) - 2\|C\|.|R|\xi_0 L_0, |\xi_f| \leq \xi_0$, $Q > 0$, is chosen such that $\rho_2 > 0$. The inequality of Equation (4.48) implies the stability of the origin $e_x = 0$, $e_\xi = 0$ and the uniform boundedness of e_x and e_ξ with $e_x \in L_2$. On the other hand, from Equation (4.43), $\dot{e}_x$ is uniformly bounded as well. According to Barbalat's Lemma A.10, one can get

$$\lim_{k\to\infty} e_x(k) = 0 \tag{4.49}$$

The persistent excitation condition means that there exist two positive constants σ and t_0 such that for all t the following inequality holds:

$$\sum_{m=k}^{k+k_0} f^T(y(m), u(m), x(m)) B^T B f^T(y(m), u(m), x(m))$$

$$\geq \sigma I. \tag{4.50}$$

Then from Equation (4.43), Equation (4.45), Equation (4.49), and Equation (4.50), one can conclude that $\lim_{t \to \infty} e_\xi(k) = 0$. This completes the proof. ∎

The following sections show the detailed implementation and simulation of the proposed scheme.

4.1.7   Interconnected CSTR units

In this section, we introduce a benchmark example of a plant composed of interconnected units with recycling [23]. We consider a plant composed of two well-mixed, non-isothermal, continuous stirred-tank reactors (CSTRs) with interconnections, where three parallel irreversible elementary exothermic reactions of the form $A \xrightarrow{k_1} B$, $A \xrightarrow{k_2} U$, $A \xrightarrow{k_3} R$ take place. As shown in Figure 4.2, the feed to CSTR 1 consists of two streams, one containing fresh A at flow rate F_0, temperature T_0, and molar concentration C_{A0}, and another containing recycled A from

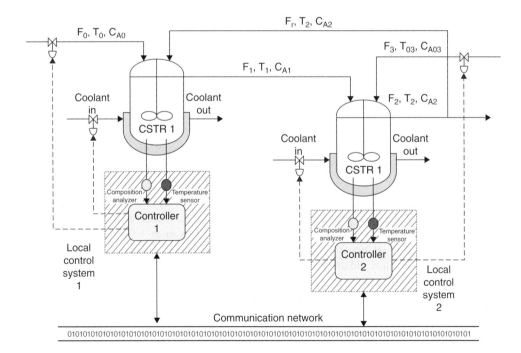

Figure 4.2   Process flow diagram for two interconnected CSTR units

the second reactor at flow rate F_r, temperature T_2, and molar concentration C_{A_2}. The feed to CSTR 2 consists of the output of CSTR 1, and an additional fresh stream feeding pure A at flow rate F_3, temperature T_{0_3}, and molar concentration C_{A0_3}. The output of CSTR 2 is passed through a separator that removes the products and recycles unreacted A to CSTR 1. Due to the non-isothermal nature of the reactions, a jacket is used to remove heat from and provide heat to both reactors.

A leak in both flux may be simulated by means of the new coefficients α_1 and α_2 whose values are $0 < \alpha_1 \leq 1$ and $1 < \alpha_2 \leq 1$. The mathematical model of the faulty CSTR can be given as:

$$\dot{T}_1 = \frac{\alpha_1 \times F_0}{V_1}(T_0 - T_1) + \frac{\alpha_1 \times F_R}{V_1}(T_2 - T_1)$$
$$+ \sum_{i=1}^{3} G_i(T_i)C_{A1} + \frac{Q_1}{\rho C_p V_1} \tag{4.51}$$

$$\dot{C}_{A1} = \frac{\alpha_1 \times F_R}{V_1}(C_{A0} - C_{A1}) + \frac{F_R}{V_1}(C_{A2} - C_{A1})$$
$$- \sum_{i=1}^{3} R_i(T_i)C_{A1}V_1 \tag{4.52}$$

$$\dot{T}_2 = \frac{\alpha_2 \times F_1}{V_2}(T_1 - T_2) + \frac{\alpha_2 \times F_3}{V_2}(T_{03} - T_2)$$
$$+ \sum_{i=1}^{3} G_i(T_2)C_{A2} + \frac{Q_2}{\rho C_p V_2} \tag{4.53}$$

$$\dot{C}_{A2} = \frac{\alpha_2 \times F_1}{V_2}(C_{A1} - C_{A2}) + \frac{\alpha_2 \times F_3}{V_2}(C_{A03} - C_{A2})$$
$$- \sum_{i=1}^{3} R_i(T_2)C_{A2} \tag{4.54}$$

where $R_i(T_j) = k_{i0}exp(-E_i/RT_j)$, $G_i(T_j) = (-\Delta H_i/\rho C_P)$ for $j = 1, 2$. ΔH_i, k_i, E_i, $i = 1, 2, 3$, denote the enthalpy, pre-exponential constants and activation energies of the three reactions, respectively. The control objective is to stabilize the plant at the (open-loop) unstable steady state using the heat input rate Q_1 and the inlet reactant concentration C_{A0} as manipulated inputs for the first reactor, and using the heat input rate Q_2 and the inlet reactant concentration C_{A03} as manipulated inputs for the second reactor. Operation at the unstable point is typically sought to avoid high temperatures, while simultaneously achieving reasonable conversion. Under nominal conditions, $\alpha_i = 1$; the performance may change suddenly or gradually when α_i is a function of time. The fault is modeled as a step function. The heat exchange surface normally has a transient degradation caused by the dirt on both sides of the wall.

4.1.8 Four-tank system

A four-tank system consists of four interconnected water tanks and two pumps. Its manipulated variables are the voltages to the pumps and the controlled variables are the water levels in the two lower tanks. The quadruple-tank process is built by considering the concept of two two-tank processes. The four-tank system presents a multi-input–multi-output (MIMO) system. This system is a real-life control problem prototyped for experimentation to try to solve it in the most efficient way, since it deals with multiple variables, thus it gives a reflection of large systems in industry. The schematic description of the four tank system can be seen in Figure 4.3. The system has two control inputs (pump throughputs) that can be manipulated to control the water level in the tanks. The two pumps are used to transfer water from a sump into four overhead tanks. By adjusting the bypass valves of the system, the proportion of water pumped into different tanks can be changed to adjust the degree of interaction between the pump throughputs and the water levels. Thus each pump output goes to two tanks, one lower and another upper and diagonally opposite, and the ratio of the split is controlled by the position of the valve. Because of the large water distribution load, the pumps have been supplied 12 V each. The mathematical modeling of the four-tank process can be obtained using Bernoulli's law. Combining all the equations for the interconnected four-tank system, we obtain a model of the physical system. A fault model can then be constructed by adding extra holes to each tank.

The mathematical model of the faulty four-tank system can be given as:

$$\frac{dh_1}{dt} = -\frac{a_1}{A_1}\sqrt{2gh_1} + \frac{a_3}{A_1}\sqrt{2gh_3}$$

$$+ \frac{\gamma_1 k_1}{A_1}v_1 + \frac{d}{A_1} - \frac{a_{leak1}}{A_1}\sqrt{2gh_1} \tag{4.55}$$

$$\frac{dh_2}{dt} = -\frac{a_2}{A_2}\sqrt{2gh_2} + \frac{a_4}{A_2}\sqrt{2gh_4}$$

$$+ \frac{\gamma_2 k_2}{A_2}v_2 - \frac{d}{A_2} - \frac{a_{leak2}}{A_2}\sqrt{2gh_2} \tag{4.56}$$

$$\frac{dh_3}{dt} = -\frac{a_3}{A_3}\sqrt{2gh_3} + \frac{(1-\gamma_2)k_2}{A_3}v_2$$

$$- \frac{a_{leak3}}{A_3}\sqrt{2gh_3} \tag{4.57}$$

$$\frac{dh_4}{dt} = -\frac{a_4}{A_4}\sqrt{2gh_4} + \frac{(1-\gamma_1)k_1}{A_4}v_1$$

$$- \frac{a_{leak4}}{A_4}\sqrt{2gh_4} \tag{4.58}$$

$$\frac{dv_1}{dt} = -\frac{v_1}{\tau_1} + \frac{1}{\tau_1}u_1 \tag{4.59}$$

$$\frac{dv_2}{dt} = -\frac{v_2}{\tau_2} + \frac{2}{\tau_2}u_2 \tag{4.60}$$

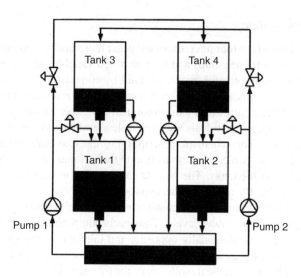

Figure 4.3 A four-tank system

4.1.9   Simulation results

In the following discussion, we present the simulation results for UKF for a fault detection and fault tolerance scheme with model-prediction-based decentralized control. A series of simulation runs was conducted on the interconnected CSTR units to evaluate the effectiveness of the proposed scheme. The same fault scenarios were used to perform different sets of experiments. The details of the algorithm can be seen in the work of Mahmoud and Khalid [24].

4.1.10   Drift detection in the interconnected CSTRs

A fault may occur in any phase or in any part of the plant. Critical faults not detected on time can lead to adverse effects. In the following discussion, the drift detection of the faults using UKF is clarified. It can be seen from Figure 4.4 that, despite an offset, the signature of the fault is the same as the nominal case. This is due to the closed loop which is performing the job with feedback. This may be also intrinsic in the physical mechanism of a real-time system generating the data, thus making life difficult for a fault detection mechanism and suppressing the deviation and drifts. Thus, here, the UKF-based drift detection can give us a better picture for the fault scenario as shown in Figure 4.5. The kinks seen in the middle of the flux leak profile of State 1 can alert the engineer to unusual practice in the process.

4.1.11   Information fusion from UKF

Once the UKF is carrying out parametric estimation, we can get system states for the leakage. For example, if we have a flux leak in State 1, the faulty state matrix from UKF can be extracted and fed into the MPC for appropriate control of faults. The results are compared for the expanded version of system control and for overlapping control; it can be seen

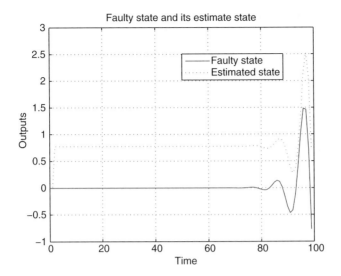

Figure 4.4   Interconnected CSTR units: leak estimate and fault estimate of State 1

(Figures 4.6–4.9) that the overlapping control is also performing its job by controlling the faults, which gives a better response time towards the control of the system states, with fewer over-shoots and less excitation.

Further, when the information is fused in the model-predictive control, the results show that the system somehow recovers from the faults rather than there being a complete breakdown of the system (see Figure 4.10). This explains how with the help of an unscented filter, the

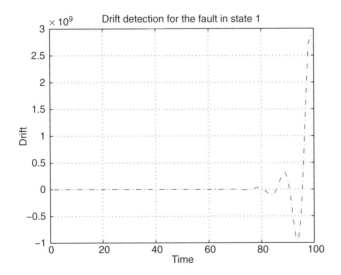

Figure 4.5   Interconnected CSTR units: drift detection for the leak in State 1

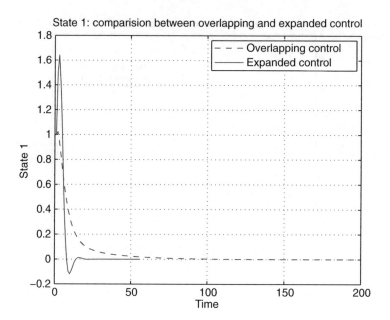

Figure 4.6   Interconnected CSTR units, State 1: overlapping decentralized control

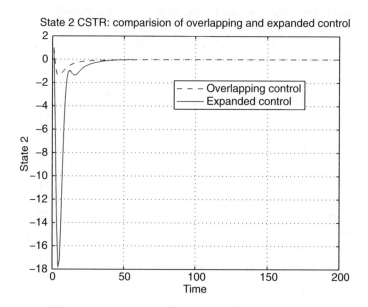

Figure 4.7   Interconnected CSTR units, State 2: overlapping decentralized control

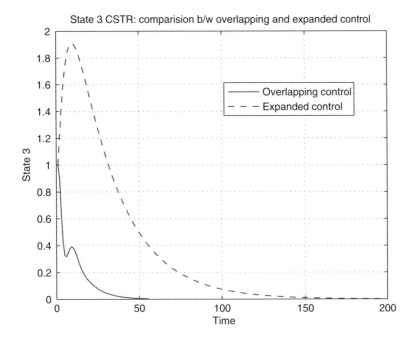

Figure 4.8   Interconnected CSTR units, State 3: overlapping decentralized control

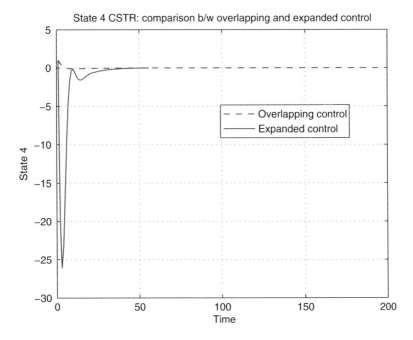

Figure 4.9   Interconnected CSTR units State, 4: overlapping decentralized control

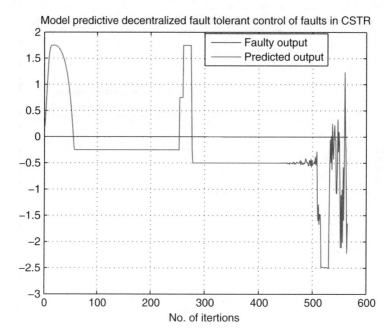

Figure 4.10   Interconnected CSTR units: MPC-based decentralized fault-tolerant control

fault detection part results in measuring the drift and parameter potency of fault in each state of CSTR, which is then fed into the information fusion unit where, with the help of overlapping decomposition and MPC, we are able to control the fault, given the bounds of fault uncertainties.

4.1.12   Drift detection in the four-tank system

In a similar way, we present simulation results for UKF for fault detection and fault tolerance with model-prediction-based decentralized control in the four-tank system. A series of simulation runs was conducted on the four-tank system to evaluate the effectiveness of the proposed scheme. The fault scenarios defined in Section 3.4.6 have been used to perform different sets of experiments.

A fault may occur in any phase or in any part of the plant. Critical faults not detected on time can lead to adverse effects. It can be seen from Figure 4.11 that the fault is so incipient that, apart from in the beginning, the level of water is achieving the same height. This is mainly due to the closed-loop system, where the controller is performing its job of achieving the desired set-point of water level in the tanks, thus suppressing any kind of deviations and drifts caused by leakage faults. Considering this situation, UKF-based drift detection can give us a better picture for the fault scenario as shown in Figure 4.12. The kinks shown in the middle of the height achievement of water in tank 1 can alert the engineer to unusual practice in the process, thus the need to prepare for action or constant monitoring.

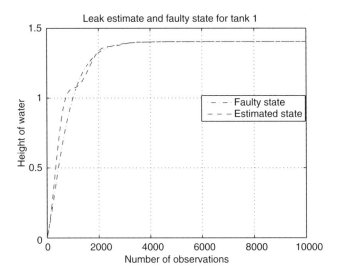

Figure 4.11 Four-tank system: leak estimate and fault estimate of Tank 1

Once, the parametric estimation is carried out by the UKF, we can get system states for the leakage fault. For example, if we have leakage in State 1, the UKF gives us the following A matrix:

$$A_{leak} = \begin{bmatrix} 1.0806 & 0.0034 & 0.0009 & -0.0877 \\ 1 & 0 & 0 & 0 \\ 0 & 1 & 0 & 0 \\ 0 & 0 & 1 & 0 \end{bmatrix} \tag{4.61}$$

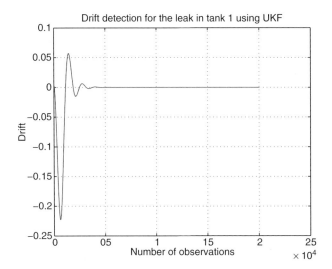

Figure 4.12 Four-tank system: drift detection for the leak in Tank 1

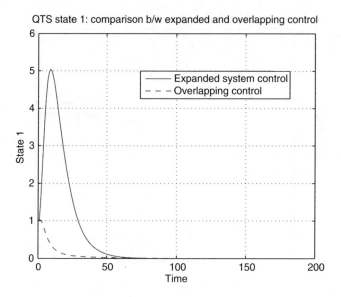

Figure 4.13   Four-tank system State 1: overlapping decentralized control

This matrix is fed into the model-predictive control algorithm in order to upgrade it according to the fault at hand. The results are compared for the expanded version of the system control and the overlapping control. It can be seen from Figures 4.13–4.16 that the overlapping control is performing its job by controlling the faults and giving a better response time towards the control of the system states, with fewer over-shoots and less excitation i.e. it makes the system stable in less time.

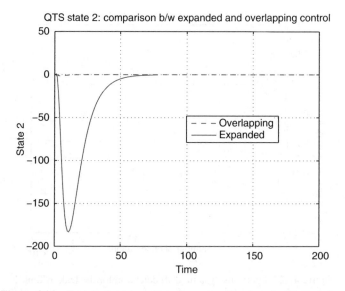

Figure 4.14   Four-tank system State 2: overlapping decentralized control

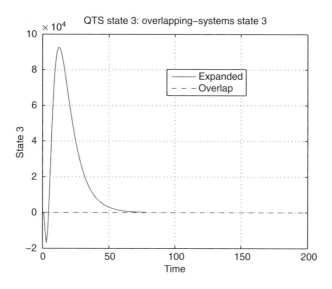

Figure 4.15 Four-tank system State 3: overlapping decentralized control

Further, the information is fused in the model-prediction control and the results show that the system somehow recovers itself from the faults rather than resulting in a complete breakdown of the system (see Figure 4.17). This explains how with the help of an unscented filter, the fault detection part measures the drift and parameter potency of leakage fault in each tank state, which data is then fed into the information fusion unit in the form of a leakage matrix. With the help of overlapping decomposition and MPC, we are able to achieve the desired set-point water-level of tank despite leakage faults, given the bounds of the leakage fault.

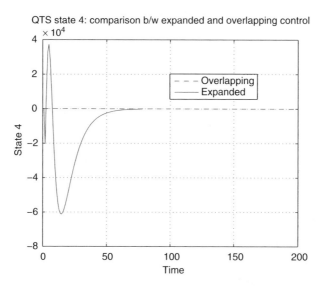

Figure 4.16 Four-tank system State 4: overlapping decentralized control

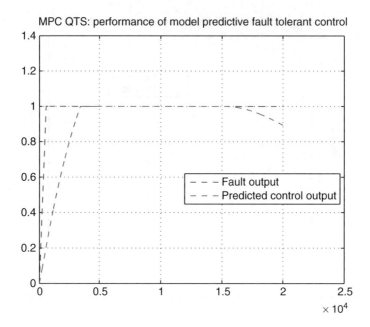

Figure 4.17 Four-tank system: MPC-based decentralized fault-tolerant control

4.2 Observer-Based Active Structures

Active fault-tolerant control (AFTC) strategies are meant to manage faulty situations by maintaining overall system stability and acceptable performance, after the fault detection by an FDI unit [25]. Although some stability and local performance proofs exist, very few papers consider the effect of FDI performance (e.g., detection delay) and reconfiguration mechanisms on the global stability and performance levels of the overall fault-tolerant scheme. Usually, these properties are checked by means of a Monte-Carlo campaign through nonlinear simulations. Some more formal solutions have been also published in the past five years. The effect of the FDI delay can be analyzed for linear systems [26]. Yang, Jiang, and Staroswiecki [27] studied an FTC scheme that uses a switching algorithm for fault isolation: a sequence of controllers is switched, until the appropriate one is found. However, as outlined by the authors, a delay in setting the correct controller still exists. Some works combine a fault-tolerant controller with a diagnostic filter (see, for instance, [28, 29, 30]). However, the structure and parameters of the existing control certification rules generally need to be modified to use such a combination. A fault-tolerant controller with a diagnostic filter would require a new expensive certification campaign for fault-free situations and could be a major concern to safety-critical systems (such as aeronautic systems).

4.2.1 Problem statement

Consider the standard feedback configuration shown in Figure 4.18, where G is an uncertain, linear, time-invariant plant subjected to faults and K_o is the nominal controller. K_o is validated

Figure 4.18   Standard feedback configuration

and certified for fault-free situations. K_o must not be modified, to avoid a costly additional certification campaign in already validated situations.

Following the basic ideas presented by Cieslak *et al.* [31], we propose to tackle the design of FTC loops as in the block diagram shown in Figure 4.19. The principle can be summarized as follows: when a fault is detected by the FDI unit, a signal $\tilde{u}$ is added to the control signal in order to compensate for the fault. This loop is activated using a switching logic. Hence, the overall scheme ensures (validated and certified) nominal performance in fault-free situations. This is one major advantage of the proposed method.

The FTC design problem of the architecture depicted in Figure 4.19 can be formulated as follows: suppose that the faulty system is stabilizable and the fault is detectable. The goal is to design a fault-tolerant controller $\tilde{K}$ and the dynamical filters H_u and H_y to produce the two signals

$$u(s) = u_o(s) + \tilde{K}(s)r(s) \tag{4.62}$$

$$r(s) = H(s)u(s) + H_y(s)y(s) \tag{4.63}$$

such that the stability of the feedback system and the required control objectives are guaranteed for the considered fault and such that r meets some robustness level against model perturbations and disturbances, and sensitivity to fault.

Using an H_∞/H_2 formulation, this means that $\tilde{K}$, H_u and H_y should satisfy the constraints

$$\|T_z\|_\infty < \gamma \tag{4.64}$$

$$\|T_{dr}\|_\infty < \gamma_1, \quad \|T_{fr}\|_2 < \gamma_2 \tag{4.65}$$

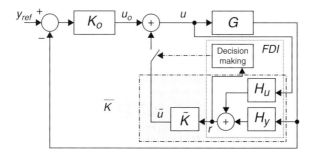

Figure 4.19   FTC setup

T_z is related to the transfer of control objectives to be achieved in the presence of fault. T_{dr} and T_{fr} denote the transfers between r and disturbances and between r and fault, respectively. γ, γ_1 and γ_2 represent some performance level to achieve and r is the fault-indicating signal (the residual).

Remark 4.6 *In the setup of Figure 4.19, to avoid algebraic loop problems, it is necessary that the D-matrix of the H_u or $\tilde{K}$ filter must be null, i.e. H_u or $\tilde{K}$ must be a strict proper transfer. In this book, we consider only strictly proper systems. The task of diagnosis will be ensured by using Luenberger observers. Hence, the previous requirement will always be achieved with this framework since the H_u filter will be a strict proper transfer.*

Before proceeding to the design of the FTC strategy, we highlight some requirements and interesting features of the interaction between the FDI and FTC units.

4.2.2 A separation principle

Assume that G is a strict proper transfer function. Suppose that the FDI unit is based on a Luenberger formulation, that is:

$$H_u(s) = C(sI - A - LC)^{-1}B \tag{4.66}$$

$$H_y(s) = -C(sI - A - LC)^{-1}L - I \tag{4.67}$$

where (A, B, C) is the state-space representation of G, and L denotes the observer gain. Let (A_o, B_o, C_o, D_o) and $(\tilde{A}, \tilde{B}, \tilde{C}, \tilde{D})$ be the state-space representations of K_o and $\tilde{K}$ respectively. The overall state-space representation of Figure 4.19 when the FTC loop is activated is given by

$$
\begin{bmatrix} \dot{x}(t) \\ \dot{x}_o(t) \\ \dot{\tilde{x}}(t) \\ \dot{\zeta}(t) \end{bmatrix} =
\left[
\begin{array}{cc:c:c}
A - BD_oC & BC_o & B\tilde{C} & B\tilde{D}C \\
-B_oC & A_o & 0 & 0 \\
\hdashline
0 & 0 & \tilde{A} & \tilde{B}C \\
\hdashline
0 & 0 & 0 & A - LC
\end{array}
\right]
\begin{bmatrix} x(t) \\ x_o(t) \\ \tilde{x}(t) \\ \zeta(t) \end{bmatrix}
$$

$$
+ \begin{bmatrix} BD_o \\ B_o \\ 0 \\ 0 \end{bmatrix} y_{ref}(t)
$$

$$
y(t) = \begin{pmatrix} C & 0 & 0 & 0 \end{pmatrix}
\begin{bmatrix} x(t) \\ x_o(t) \\ \tilde{x}(t) \\ \zeta(t) \end{bmatrix}
\tag{4.68}
$$

where x, x_o, $\tilde{x}$, $\hat{x}$ and ζ correspond to the states of G, K_o, $\tilde{K}$, the FDI filter, and the estimation error $x - \hat{x}$ respectively. From Equation (4.68), it can be seen that the A-matrix is upper block triangular. It follows that the asymptotical stability depends on the stability of the FTC loop composed of $\tilde{K}$, H_u and H_y, and depends on the stability of the nominal control loop. In other words, the separation principle given by Cieslak *et al.* [31] is preserved in spite of the modification of the FTC architecture. Another interesting feature is related to the second separation principle that appears on the FTC loop, which highlights a well-known result in linear-quadratic-Gaussian (LQG) control theory: the control part (here, the FTC filter $\tilde{K}$) and the observation gain L (which impacts directly on the performance of FDI part) can be designed separately.

In the following discussion, we propose to design a unique filter $\overline{K}$ (see Figure 4.19) under control specifications. Based on observer-based structures [32], $\overline{K}$ is reformulated as:

$$\overline{K} : \begin{cases} \dot{\hat{x}}(t) = A\hat{x}(t) + Bu(t) + L(y(t) - C\hat{x}(t)) \\ \dot{x}_q(t) = A_q x_q(t) + B_q(y(t) - C\hat{x}(t)) \\ \tilde{u}(t) = C_q x_q(t) + D_q(y(t) - C\hat{x}(t)) - K\hat{x}(t) \end{cases} \tag{4.69}$$

where (A_q, B_q, C_q, D_q) is the state-space representation of an extra Youla parameter Q that can be either static or dynamic. x_q is related to the state of Q, K is a static control gain, and L is an estimation gain. The overall state-space representation of Figure 4.20 when the FTC loop is activated can thus be written as follows:

$$\begin{bmatrix} \dot{x}(t) \\ \dot{x}_o(t) \\ \dot{x}_q(t) \\ \dot{\zeta}(t) \end{bmatrix} = \left[\begin{array}{cccc} A - BD_oC - BK & BC_o & \vdots & BC_q & \vdots & BD_qC + BK \\ \hline -B_oC & A_o & \vdots & 0 & \vdots & 0 \\ \hline 0 & 0 & \vdots & A_q & \vdots & B_qC \\ \hline 0 & 0 & \vdots & 0 & \vdots & A - LC \end{array} \right] \begin{bmatrix} x(t) \\ x_o(t) \\ x_q(t) \\ \zeta(t) \end{bmatrix}$$

$$+ \begin{bmatrix} BD_o \\ B_o \\ 0 \\ 0 \end{bmatrix} y_{ref}(t)$$

$$y(t) = \begin{bmatrix} C & 0 & 0 & 0 \end{bmatrix} \begin{bmatrix} x^T(t) & x_o^T(t) & x_q^T(t) & \zeta^T(t) \end{bmatrix}^T \tag{4.70}$$

Equation (4.70) shows clearly the separation principles. In addition, an FDI part can be distinguished (via the presence of the term $A - LC$ in Equation (4.70)) in spite of a design based on control specifications. This nice feature permits us to define $\overline{K}$ as the set of admissible FDI or FTC units since $\overline{K}$ provides a fault compensation signal $\tilde{u}$ (the FTC part) and an FDI part (able to take some decisions) can be isolated by using the formulation of Equation (4.69). This last point underlines an interesting feature: a specific move to the design of an FDI filter is not always necessary to active FTC strategies.

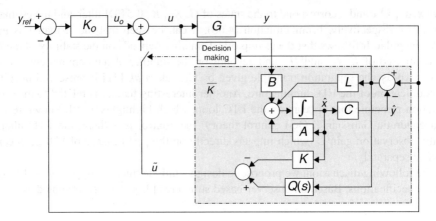

Figure 4.20 Proposed FTC setup with observer-based structure

4.2.3 FDI residuals

Using the formulation of Equation (4.69), two signals can be interpreted as the fault-indicating signals. The first is related to the signal r_1 that enters into the estimation gain L. Without loss of generality, the state-space representation of the FDI part is defined as

$$\begin{cases} \dot{\hat{x}}(t) = (A - LC)\hat{x}(t) + Bu(t) + Ly(t) \\ r_1(t) = -C\hat{x}(t) + y(t) \end{cases} \tag{4.71}$$

From Equation (4.71), the obtained FDI part possesses the classic observer features and its FDI performance level depends of the value of L.

The other possible fault-indicating signal corresponds to the output of the Youla parameter Q. The residuals r_2 correspond to r_1 filtered by the Youla parameter Q. The state-space representation of the FDI part is given by:

$$\begin{cases} \dot{\hat{x}}(t) = (A - LC)\hat{x}(t) + Bu(t) + Ly(t) \\ \dot{x}_q(t) = -B_q C\hat{x}(t) + A_q x_q(t) + B_q y(t) \\ r_2(t) = -D_q C\hat{x}(t) + C_q x_q(t) + D_q y(t) \end{cases} \tag{4.72}$$

From Equation (4.72), it can be difficult to determine the impact of the gain L and the Youla parameter Q on the diagnosis performances. To deal with this problem, we propose to use a generic approach that is able to compute a robustness indicator γ_1 and a sensitivity indicator γ_2. The selection of the appropriate residuals (r_1 or r_2) is done by selecting the FDI filter possessing the best performance levels.

4.2.4 Control of integrity

FTC strategies mainly take place in safety-critical systems. An important requirement of these systems is achieving integrity control [33]. In addition, as can be seen in Figure 4.20, the

FTC scheme can work in open-loop (fault-free) situations. Therefore, an important design requirement is that the FTC loop ($\overline{K}$ in Figure 4.20) must be stable. This point is discussed by Cieslak, Henry, and Zolghadri [34].

4.2.5   Overall stability

Looking at Figure 4.20, it is natural to ask about the overall stability of the FTC architecture due to the presence of the switch. Here, the common quadratic Lyapunov function existence is used [35]. If all modes of a switched system have a common Lyapunov function, then the switching sequence subjected to imperfections of the FDI part (detection delay) or the FTC part (reconfiguration delay) does not affect the stability of the switched system.

The overall stability of the FTC architecture, see Figure 4.20, is thus guaranteed if the following assumption is achieved.

Assumption 4.3   *To preserve the overall system stability of Figure 4.20 under any switching sequence (with any detection and reconfiguration delay) it is necessary to find a common Lyapunov function for all possible operating modes. That is, it is necessary to calculate a matrix $P = P^T > 0$ such that:*

$$A_i^T P + PA_i < 0, \quad i = 1, \ldots, \tau \tag{4.73}$$

where τ is the number of possible operating modes and A_i is the evolution matrix of Figure 4.20 in different operating modes.

Remark 4.7   *It is obvious that undesirable transients can appear when using switching logic. A practical solution to the study of the rate and the amplitude of these bumps is presented by Cieslak et al. [31].*

4.2.6   Design outline

The preceding analysis allows consideration of an outline for the design of an active FTC scheme where fault diagnosis and fault tolerance performances can be managed. The proposed scheme consists firstly in designing the set of admissible FDI and FTC units $\overline{K}$ under control objectives. Once $\overline{K}$ is designed, the observer-based structures [32] are used to create a fault-diagnosis scheme able to activate, in a timely fashion, the fault-tolerant compensation signal $\tilde{u}$, see Equation (4.69). There exist several configurations of Equation (4.69) that guarantee the control objectives ensured by $\overline{K}$ [32]. An analysis campaign based on the generalization of the structured singular value μ_g [36] is thus used to compute the FDI performance levels (robustness, sensitivity, etc.) of each configuration. The results of this analysis campaign is finally used to select the best FDI–FTC pair extracted from $\overline{K}$.

The FTC design problem can thus be formulated as: Suppose that the faulty system is stabilizable and the fault is detectable. The goal is to select from the set of all admissible

FDI–FTC units $\overline{K}$, a static gain $\overline{K}$, an observer gain L and a dynamic Youla parameter that comply with Assumption 4.3 and produce the two signals (see Figure 4.20):

$$r_1(t) = y(t) - \hat{y}(t) \quad \text{or } r_2(t)$$

$$= Q(s)(y(t) - \hat{y}(t)) \tag{4.74}$$

$$\tilde{u}(t) = C_q x_q(t) - K\hat{x}(t) + D_q(y(t) - \hat{y}(t)) \tag{4.75}$$

such that the stability of the feedback system and the required control objectives are guaranteed for the considered fault and such that the selected residuals meet some robustness level against disturbances and sensitivity to fault. Using an H_∞/H_2 formulation, this means that $\overline{K}$ in Equation (4.69) should satisfy the constraints:

$$\|T_{zK}\|_\infty < \gamma, \quad Re\{\lambda_i\{\overline{A}\}\} < 0 \tag{4.76}$$

$$\|T_{dri}\|_\infty < \gamma_1 \quad \|T_{fri}\|_2 > \gamma_2, i = 1, 2 \tag{4.77}$$

where T_{zK} is related to the transfer of control objectives to be achieved in the presence of faults. $T_{d\Re\bullet}$ and $T_{f\Re\bullet}$ denote the transfers between residuals (r_1 or r_2) and disturbances and between residuals (r_1 or r_2) and faults, respectively. γ, γ_1 and γ_2 represent some performance level to achieve. $Re\{\lambda_i\{\overline{A}\}\} < 0$ indicates that $\overline{K}$ must be stable.

4.2.7   Design of an active FTC scheme

The proposed solution is based on four steps:

1. Design $\overline{K}$.
2. Extract from $\overline{K}$ all configurations of K, L and Q achieving Assumption 4.3.
3. Compute the FDI performance indicators.
4. Select the best solution from the analysis campaign.

In the following discussion, we focus on Step 2, extracting the FDI and FTC parts from $\overline{K}$. Interested readers can refer to the work of Cieslak, Henry, and Zolghadri [34] for the mathematical development of the other steps.

4.2.8   Extraction of FDI–FTC pairs

The aim of this subsection is to extract from $\overline{K}$ the state-space representation of the dynamical transfer Q (since $\overline{K}$ is an augmented filter with respect to the controlled system G of Figure 4.18), the estimation gain L and the static control gain K. The derived observer-based controller of Equation (4.69) is the input–output equivalent to $\overline{K}$ [32].

Let us consider the state-space representations of $\overline{K}$:

$$\dot{\overline{x}}(t) = \overline{A}\overline{x}(t) + \overline{B}_u u(t) + \overline{B}_y y(t) \tag{4.78}$$

$$\tilde{u}(t) = \overline{C}\overline{x}(t) + \overline{D}_y y(t) \tag{4.79}$$

The state equation of $\overline{K}$, Equation (4.78), can be expressed as a Luenberger observer of the variable $z(t) = Tx(t)$. The state vector of $\overline{K}$ is thus defined such that:

$$\overline{x}(t) = \hat{z}(t) \tag{4.80}$$

According to Luenberger's formulation [37], it follows that the state equation can be given by

$$\dot{\hat{z}}(t) = F\hat{z}(t) + Gy(t) + TBu(t) \tag{4.81}$$

where F, G and T are matrices of appropriate dimensions achieving

$$TA - FT = GC, \qquad F \text{ is stable} \tag{4.82}$$

in order to ensure the asymptotic stability of the observer. Note that G in Figure 4.18 represents the system but G in Equation (4.81) is the observer gain.

Using the output Equation (4.79) and $u = \tilde{u} + u_o$ (see Figure 4.20), Equation (4.81) becomes

$$\dot{\hat{z}}(t) = F\hat{z}(t) + Gy(t) + TB\overline{C}\overline{x}(t) + TBC_o x_o(t) + TB(\overline{D}_y - D_o)y(t) \tag{4.83}$$

With Equation (4.79), Equation (4.80), and $u = \tilde{u} + u_o$, Equation (4.78) can be reformulated as:

$$\dot{\hat{z}}(t) = (\overline{A} + \overline{B}_u\overline{C})\hat{z}(t) + (\overline{B}_y + \overline{B}_u(\overline{D}_y - D_o))y(t) + \overline{B}_u C_o x_o(t) \tag{4.84}$$

The identification of Equation (4.83) and Equation (4.84) can lead to the following algebraic relations:

$$F = \overline{A}$$
$$G = \overline{B}_y \tag{4.85}$$

These two equations with Equation (4.82) guarantee that we are dealing with an observer-based controller. Substituting Equation (4.85) in Equation (4.82), we get:

$$\overline{A}T - TA + \overline{B}_y C = 0 \tag{4.86}$$

Hence, the stable filter $\overline{K}$ can be formulated as an observer-based controller by solving in T, the non-symmetric Sylvester Equation (4.86) that is a particular case of the non-symmetric algebraic Riccati equation. T can be computed by standard invariant subspace techniques which consist in finding an invariant subspace of the closed-loop system matrix and partitioning the obtained vectors. (This part has been omitted – interested readers can refer to the work of

Alazard and Apkarian [32] for more details.) Now, consider the following Schur decomposition of the matrix F

$$F = V\tilde{F}V^* = (V_1 \quad V_2) \begin{bmatrix} \tilde{F}_{11} & \tilde{F}_{12} \\ 0 & \tilde{F}_{22} \end{bmatrix} \begin{bmatrix} V_1^* \\ V_2^* \end{bmatrix} \tag{4.87}$$

where V_1^* is the conjugate-transpose of V_1, $VV^* = I$ and $\tilde{F}_{22}$ is a square matrix with the same dimensions as the A matrix of G. Let us perform the following variable change

$$\hat{z}(t) = (V_1 \quad V_2) \begin{bmatrix} w_1 \\ w_2 \end{bmatrix} \tag{4.88}$$

and introduce the notation:

$$\begin{bmatrix} \tilde{G}_1 \\ \tilde{G}_2 \end{bmatrix} \begin{bmatrix} V_1^* \\ V_2^* \end{bmatrix} G, \begin{bmatrix} \tilde{T}_1 \\ \tilde{T}_2 \end{bmatrix} = \begin{bmatrix} V_1^* \\ V_2^* \end{bmatrix} T \tag{4.89}$$

Using the same calculations as those presented by Alazard and Apkarian [32], it is possible to reformulate Equation (4.69):

$$\dot{\hat{x}}(t) = A\hat{x}(t) + Bu(t) + \tilde{T}_2^{-1}\tilde{G}_2(y(t) - C\hat{x}(t)) \tag{4.90}$$

$$\dot{x}_q(t) = \tilde{F}_{11}x_q(t) + (\tilde{G}_1 - \tilde{T}_1\tilde{T}_2^{-1}\tilde{G}_2)(y(t) - C\hat{x}(t)) \tag{4.91}$$

$$\tilde{u}(t) = (\overline{C}T + \overline{D}_yC)\hat{x}(t) + \overline{C}V_1x_q(t) + \overline{D}_y(y(t) - C\hat{x}(t)) \tag{4.92}$$

These theoretical developments permit us to provide the following theorem.

Theorem 4.3   *Assume that a solution to the non-symmetric algebraic Riccati Equation (4.86) exists. That is, there is a matrix T such that Equation (4.82) holds. From Equation (4.78), Equation (4.79), Equation (4.85), Equation (4.87) and Equation (4.89), it is possible to extract from $\overline{K}$ all parameters of the observer-based controller structure shown in Figure 4.20. The state-space representation of Q, the estimation gain L and the static control gain K are defined according to:*

$$A_q = \tilde{F}_{11} = V_1^*FV_1 \tag{4.93}$$

$$B_q = \tilde{G}_1 - \tilde{T}_1\tilde{T}_2^{-1}$$

$$\tilde{G}_2 = V_1^*(I - T(V_2^*T)^{-1}V_2^*)G \tag{4.94}$$

$$C_q = \overline{C}V_1 \tag{4.95}$$

$$D_q = \overline{D}_y \tag{4.96}$$

$$L = \tilde{T}_2^{-1}\tilde{G}_2 = (V_2^*T)^{-1}V_2^*G \tag{4.97}$$

$$K = -(\overline{C}T + \overline{D}_yC) \tag{4.98}$$

Proof. The proof follows from immediate application of the set identification of Equations (4.90)–(4.92) and Equation (4.69).   ∎

Remark 4.8   *As mentioned by Alazard and Apkarian [32] (see also Section 8.4), there are many solutions for T. Each solution corresponds to a particular choice of the eigenvalues among the set of closed-loop eigenvalues. This facet permits several configurations of FTC and FDI parts.*

4.2.9   Simulation

The overall procedure presented in the preceding sections is now illustrated by a basic academic example that details the steps of the design. The example is a tank process linearized around an equilibrium point. The input model u is a parameter controlling the flow rate to the tank via a valve position. The output is the tank level h measured by a sensor in the tank. This sensor provides high-frequency measurement noise, denoted n. The fault scenario consists of a component fault that leads to instability. The model that describes the system in both fault-free and faulty situations, is given by

$$\begin{cases} \dot{x}(t) = A(I + \rho_1)x(t) + Bu(t) \\ h(t) = Cx(t) + n(t) \end{cases} \tag{4.99}$$

where $A = [-1]$, $B = [4]$ and $C = [1]$. ρ_1 represents the fault; that is, $\rho_1 = 0$ denotes no fault and $\rho_1 = -2$ is the considered fault. The process is controlled by a regulator K_o. Its state-space representation is defined according to:

$$A_o = \begin{bmatrix} -22.38 & 101.9 & 30.85 \\ -717.16 & -582.4 & 1034.87 \\ 0 & 0 & -0.002 \end{bmatrix}, B_o = \begin{bmatrix} 0 \\ 0 \\ 2.99 \end{bmatrix} \tag{4.100}$$

$$C_o = \begin{pmatrix} -2.39 & 11.39 & 3.45 \end{pmatrix}, D_o = (0) \tag{4.101}$$

Remark 4.9   *Note that a preliminary study revealed that K_o maintains the stability of the nominal control loop despite the fault.*

4.2.9.1   Designing an active FTC scheme

We can now design an active FTC scheme based on the four steps in Section 4.2.7:

1. Design $\overline{K}$:
   A μ-analysis procedure reveals that the obtained stable filter $\overline{K}$ of seventh order achieves the desired control performance.
2. Extract FDI–FTC units:
   In this example, the static control gain K and the estimation gain L are scalar matrices since the system is a first-order process. Hence, the Youla parameter Q is a dynamic transfer function of fifth order. It follows that 42 combinations of K, L and Q can be obtained. However, only 20 configurations achieve Assumption 4.3.

Table 4.1   FDI performance levels of residuals r_1

N^o	$\lambda\{A - LC\}$	γ_1	γ_2	$\Omega = (0; \omega_{max})$	J
1	-5.8×10^5	0.96	6.5×10^{-9}	$(0; 3 \times 10^8]$	7×10^7
2	-1.6×10^5	0.96	2.3×10^{-8}	$(0; 8.7 \times 10^7]$	2×10^7
3	-4×10^3	0.96	9×10^{-7}	$(0; 2.3 \times 10^6]$	5.3×10^5
4	-498	0.96	8×10^{-6}	$(0; 2.6 \times 10^5]$	6×10^5
5	-58	0.96	6.5×10^{-5}	$(0; 3 \times 10^4]$	7.3×10^3
6	-4×10^{-4}	0.077	0.1	$(0; 3 \times 10^{-2}]$	$-$

From the term $(A - BD_oC - BK)$ of Equation (4.70), it can be seen that K can affect the eigenvalues of the nominal control loop. We propose not modifying the dynamics of the nominal control loop (see Section 4.2.1). Hence, the eigenvalue of $(A - BD_oC - BK)$ is fixed to its nominal value, -1. It follows that only six of the 20 possible configurations are viable. The assignment procedure of the six viable configurations operates such that one eigenvalue is assigned to $A - LC$ and the last five eigenvalues to Q.

3. Compute the FDI performance indicators.
   According to Henry and Zolghadri [36], the robustness performance level γ_1, the fault sensitivity performance level γ_2 and the frequency range Ω (where γ_2 is ensured) are computed. Note that a high value of Ω involves fast dynamics of the observation error. From Table 4.1, it can be seen that the obtained results coincide with the classic observer features when the considered residuals are r_1 (i.e. a high gain of L leads to a high value of Ω which outlines fast dynamics).
   The same analysis is done considering r_2, see Table 4.2. It can be seen that the six configurations ensure the same level of FDI performance. It is a rather obvious result, since the eigenvalues of the FDI part given by Equation (4.72) are always the same in spite of the six considered configurations of $A - LC$ and Q.

4. Select the best solution from the analysis.
   We propose to select the configuration that minimizes the following problem:

$$(L, K, Q(s)) = arg \min(J_i(\gamma_1, \gamma_2, \Omega)), i = 1, \ldots, 6 \qquad (4.102)$$

$$s.t. \omega_{max} > 1 rd/s, \Omega = (0; \omega_{max}),$$
$$J_i(\gamma_1, \gamma_2, \Omega) = \alpha_1(\gamma_1/\gamma_2) + \alpha_2(1/\max(\Omega)) \qquad (4.103)$$

where i is an index indicating a viable configuration, and α_1 and α_2 are the design parameters. Here, we take $\alpha_1 = 0.5$ and $\alpha_2 = 0.5$ (no privileged objectives). Equation (4.102) defines the optimization criteria and Equation (4.103) gives a constraint to avoid slow dynamics in observing the error (which lead to an overly long convergence time). From Table 4.1 and Table 4.2, it can be seen that the best results are obtained when r_2 are used as fault-indicating signals. Configuration 1 is now tested for simulation results.

Table 4.2   FDI performance levels of residuals r_2

N^o	$\lambda\{A - LC\}$	γ_1	γ_2	$\Omega = (0; \omega_{max})$	J
$-$	$-$	0.077	9×10^{-4}	$(0; 2.7 \times 10^3]$	42

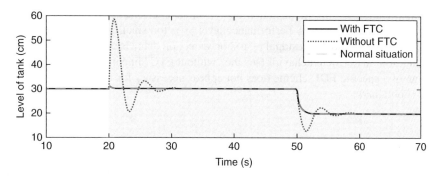

Figure 4.21 Simulation results: Output signals

4.2.9.2 Simulation results

The FTC architecture depicted in Figure 4.20 is implemented. The decision-making rule consists of a simple threshold on the norm of the residual vector. Figures 4.21 and 4.22 show the simulation results. To highlight the method, the same simulations are run in nominal operating mode (fault-free), in a faulty situation without FTC, and in a faulty situation with FTC. The initial operating point is fixed at $h_{ref} = 30$ cm. At $t = 50$ s, h_{ref} is changed to $h_{ref} = 20$ cm.

The fault is simulated at $t = 20$ s. Figure 4.22 highlights that fault detection cannot be achieved if we only have access to the residuals r_1 (since the norm of residuals r_1 stays under

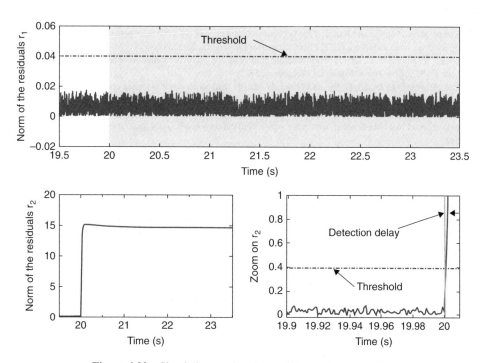

Figure 4.22 Simulation results: Norm of the residuals r_1 and r_2

the threshold leading to infinite detection delay). Indeed, according to the results presented in Table 4.1, the fault sensitivity performance level γ_2 is too small against the attenuation of noise measurement. Using the residual r_2, a short detection delay is obtained (see Figure 4.22) and the undesirable transient behavior (see the "without FTC" line in Figure 4.21) is avoided. The design of a specific FDI scheme does not appear necessary for this example since a safe situation is obtained.

4.3 Notes

In this chapter, an effective integrated fault detection and fault-tolerant control technique is developed for a class of interconnected process systems characterized by actuators and sensors that may undergo several different types of failure. Typical faults in interconnected process systems were considered and a UKF-based approach was employed for fault detection and parameter estimation. The fault detection system was decentralized using the overlapping decomposition technique. Then model-prediction-based fault-tolerant controllers were designed using the parameter estimates from the fault detection decentralized subsystem. It has been demonstrated that all the signals in the system are bounded and that the tracking error converges to zero asymptotically despite multiple actuator and sensor faults. The proposed scheme has been successfully evaluated on interconnected CSTR units with recycling and a four-tank system, thus underpinning the proposed scheme with its practical implementation.

References

[1] Parker, M. A., Ng, C., and Ran, L. (2011) "Fault-tolerant control for a modular generator converter scheme for direct-drive wind turbines", *IEEE Trans. Industrial Electronics*, **58**(1):305–315.

[2] Dwari, S., and Parsa, L. (2011) "Fault-tolerant control of five-phase permanent-magnet motors with trapezoidal back EMF", *IEEE Trans. Industrial Electronics*, **58**(2):476–485.

[3] Sun, Z., Wang, J., Jewell, G. W., and Howe, D. (2010) "Enhanced optimal torque control of fault-tolerant PM machine under flux-weakening operation", *IEEE Trans. Industrial Electronics*, **57**(1):344–353.

[4] de Lillo, L., Empringham, L., Wheeler, P. W., Khwan-On, S. et al. (2010) "Multiphase power converter drive for fault-tolerant machine development in aerospace application", *IEEE Trans. Industrial Electronics*, **57**(2):575–583.

[5] Zhao, W., Chau, K. T., Cheng, M., Ji, J., and Zhu, X. (2010) "Remedial brushless AC operation of fault-tolerant doubly salient permanent-magnet motor drives", *IEEE Trans. Industrial Electronics*, **57**(6):2134–2141.

[6] Ceballos, S., Pou, J., Robles, E., Zaragoza, J., and Martin, J. L. (2010) "Performance evaluation of fault-tolerant neutral-point-clamped converters", *IEEE Trans. Industrial Electronics*, **57**(8):2709–2718.

[7] Siljak, D. D. (1980) "Reliable control using multiple control systems", *Int. J. Control*, **31**(2):303–329.

[8] Shimemura, E., and Fujita, M. (1985) "A design method for linear state feedback systems processing integrity based on a solution of a Riccati-type equation", *Int. J. Control*, **42**(4):887–899.

[9] Zhao, Q., and Jiang, J. (1998) "Reliable state feedback control systems design against actuator failures", *Automatica*, **34**(10):1267–1272.

[10] Chandler, P. R. (1984) "Self-repairing flight control system reliability and maintainability program: Executive overview", in *Proc. IEEE National Aerospace and Electronics Conference*, 586–590.

[11] Moerder, D. D., Halyo, N., Broussard, J. R., and Caglayan, A. K. (1989) "Application of precomputed control laws in a reconfigurable aircraft flight control system", *J. Guidance, Control, and Dynamics*, **12**(3):325–333.

[12] Montoya, R. J., Howell, W. E., Bundick, W. T., Ostroff, A. J., et al. (1982) "Restructurable controls", *Tech. Rep. NASA CP-2277*, Proc. of a Workshop held at NASA Langley Research Center, Hampton, VA, September 21–22.

[13] Monaco, J., Ward, D., Barron, R., and Bird, R. (1997) "Implementation and flight test assessment of an adaptive, reconfigurable flight control system", in *Proc. 1997 AIAA Guidance, Navigation, and Control Conference*, 1443–1454.

[14] Maybeck, P. S., and Stevens, R. D. (1991) "Reconfigurable flight control via multiple model adaptive control methods", *IEEE Trans. Aerospace and Electronic Systems*, 27(3):470–479.

[15] Rothenhagen, K., and Fuchs, F. W. (2009) "Current sensor fault detection, isolation, and reconfiguration for doubly fed induction generator", *IEEE Trans. Industrial Electronics*, 56(10):4239–4245.

[16] Xie, X., Zhou, D., and Jin, Y. (1999) "Strong tracking filter based adaptive generic model control", *J. Process Control*, 9:337–350.

[17] Nazari, R. (2010) "Fault-tolerant control of a magnetic levitation system using virtual-sensor-based reconfiguration", *Proc. 18th Mediterranean Conf on Control and Automation (MED)*, 1067–1072.

[18] Yetendje, A. (2010) "Multisensor fusion fault-tolerant control of a magnetic levitation system", *Proc. 18th Mediterranean Conf on Control and Automation (MED)*, 1055–1060.

[19] Wan, E. A., der Merwe, R. V., and Nelson, A. T. (2000) "Dual estimation and the unscented transformation", in *Advances in Neural Information Processing Systems*, S. A. Solla, T. K. Leen, and K. R. Miller (Eds), Cambridge, MA: MIT Press, 666–672.

[20] Julier S. J., Uhlmann, J. K., and Durrant-Whyte, H. (1995) "A new approach for filtering nonlinear systems", in *Proc. American Control Conference*, 1628–1632.

[21] Julier, S. J., and Uhlmann, J. K. (1997) "A new extension of the Kalman filter to nonlinear systems", in *Proc. 11th Int. Symposium on AeroSpace/Defense Sensing, Simulation and Controls*, 3068:182–193.

[22] Mahmoud, M. S. (2010) *Decentralized Control and Filtering in Interconnected Dynamical Systems*, New York: CRC Press, Taylor and Francis Group.

[23] Sun, Y., and El-Farra, N. H. (2008) "Quasi-decentralized model-based networked control of process systems", *Computers and Chemical Engineering*, 2016–2029.

[24] Mahmoud, M. S., and Khalid, H. M. (2010) "Detail Evaluation of Fault Tolerant Schemes", *MsM-KFUPM-Tolerant-509*, Internal Report, Saudia Arabia: KFUPM.

[25] Zhang, Y., and Jiang, J. (2008) "Bibliographical review on reconfigurable fault-tolerant control systems", *Annual Reviews in Control*, 32:229–252.

[26] Shin, J. Y., and Belcastro, C. M. (2006) "Performance analysis on fault tolerant control system", *IEEE Trans. Control System Technology*, 14(9):1283–1294.

[27] Yang, H., Jiang, B., and Staroswiecki, M. (2009) "Supervisory fault tolerant control for a class of uncertain nonlinear systems", *Automatica*, 45:2319–2324.

[28] Marcos, A., and Balas, G. J. (2005) "A robust integrated controller/diagnosis aircraft application", *Int. J. Robust and Nonlinear Control*, 15:531–551.

[29] Oudghiri, M., Chadli, M., and El Hajjaji, A. (2008) "Robust observer-based fault tolerant control for vehicle lateral dynamics", *Int. J. Vehicle Design*, 48:173–189.

[30] Weng, Z., Patton, R., and Cui, P. (2008) "Integrated design of robust controller and fault estimator for linear parameter varying systems", in *Proc. 17th World Congress IFAC*, Seoul Korea, 117–124.

[31] Cieslak, J., Henry, D., Zolghadri, A., and Goupil, P. (2008) "Development of an active fault tolerant flight control strategy", *AIAA J. Guidance, Control and Dynamics*, 31(1):135–147.

[32] Alazard, D., and Apkarian, P. (1999) "Exact observer-based structures for arbitrary compensators", *Int. J. Robust and Nonlinear Control*, 9:101–118.

[33] Maciejowski, J. M. (1989) *Multivariable Feedback Design*, New York: Addison Wesley.

[34] Cieslak, J., Henry, D., and Zolghadri, A. 2009 "Design of Fault Tolerant Control Systems: a Flight Simulator Experiment", in *Proc. 7th IFAC Symposium SafeProcess*, Barcelona, Spain, 138–143.

[35] Liberzon, D. (2003) *Switching in Systems and Control*, Boston, Mass, USA: Birkhuser.

[36] Henry, D., and Zolghadri, A. (2005) "Design and analysis of robust residual generators for systems under feedback control", *Automatica*, 41(2):251–264.

[37] Blanke, M., Staroswiecki, M., and Wu, N. E. (2001) "Concepts and methods in fault-tolerant control", in *Proc. American Control Conference*, 2606–2620.

5

Fault-Tolerant Nonlinear Control Systems

5.1 Comparison of Fault Detection Schemes

It is quite difficult to compare fault detection (FD) schemes. The choice of a particular FD scheme depends upon several factors, for instance, information about the process itself (linear or nonlinear, time-varying or time-invariant), the availability of the process model, the types of unknown inputs, open loop or closed loop, and the application sensitivity, for example how much safety is required. Typical applications include nuclear power plants and aircraft systems. In situations where a process model is available or it is easy to model the process (for example, electrical and mechanical systems), model-based approaches are preferred. It is difficult to obtain an analytical model for chemical or industrial processes and sometimes a mathematical model is too complex to handle. In such situations, qualitative model-based approaches or signal-based approaches are employed. It is worth noting that analytical model-based approaches are efficient in detecting faults and their online implementation is easier.

Looking at analytical model-based approaches, an interesting comparison is given by Isermann [1], who reports the advantages and disadvantages of each approach. For instance, parameter estimation requires the structure of the model to be known whereas parity space and observer-based approaches assume that the model and its parameters are known. Parameter estimation approaches require input excitation while the other two schemes do not admit such limitations. For the detection of multiplicative faults, the parameter estimation approach is a better choice. A further essential difference is that fault isolation can based on the single output information in the parameter estimation approach, which is not possible in the other two approaches. Parity space approaches are more sensitive to measurement noises than observer-based and parameter estimation approaches. To cut a long story short, it can be concluded that different methods work well in different scenarios. Hence, it is possible to combine different approaches to enjoy the advantages of each approach. Some combining strategies are given by Isermann [1].

Analysis and Synthesis of Fault-Tolerant Control Systems, First Edition. Magdi S. Mahmoud and Yuanqing Xia.
© 2014 John Wiley & Sons, Ltd. Published 2014 by John Wiley & Sons, Ltd.

5.2   Fault Detection in Nonlinear Systems

Most real-time systems are nonlinear in nature. If these systems are running at a fixed operating point, the well-established techniques for linear FDI can be utilized by considering their linearized model. However, when the nonlinear system is not running at a fixed operating point, linear FDI techniques cannot be employed. In particular, if the operating point deviates too much because of the fault, the linear FDI scheme is submitted to increased modeling errors and may run out of its valid range of operation, which may generate false alarms instead of detecting faults [2]. For this reason, a great deal of attention has been attracted to the research in FD for nonlinear systems.

Several approaches to fault detection in nonlinear systems have been proposed. These include nonlinear observer-based approaches [3, 4, 5], parity space approaches [6, 7, 8], neural networks [9, 10] and fuzzy nets [6, 11, 12]. Since nonlinear observer-based fault detection is the major interest of this chapter, an overview of some state-of-the art methods for observer-based residual generation in nonlinear systems is given in the next section.

5.3   Nonlinear Observer-Based Residual Generation Schemes

Nonlinear observer-based residual generation has been a topic of intensive research over the past three decades. Various approaches have been proposed in literature.

5.3.1   General considerations

Due to the complexity of modern engineering systems, it is increasingly important to ensure their reliability. This has motivated researchers to concentrate on FTC, which is primarily meant to ensure safety, that is, the stability of a system after the occurrence of a fault. There are two approaches to synthesize controllers that are tolerant to system faults. One approach, known as passive FTC, aims at designing a controller which is *a priori* robust to some given expected faults. Another approach, known as active FTC, relies on the availability of a fault detection and diagnosis (FDD) block that gives, in real-time, information about the nature and intensity of the fault. This information is then used by a control reconfiguration block to adjust online the control effort in such a way as to maintain stability and to optimize the performance of the faulty system [13].

Passive FTC has the drawback of being reliable only for the expected class of faults that have been taken into account in the design. Furthermore, the performance of the closed loop are not optimized for each fault scenario. However, it has the advantage of avoiding a time delay because of online diagnosis of faults and reconfiguration of the controller, required by active FTC [14, 15]; time-delay avoidance is very important in practical situations where the time window during which the system stays stabilizable is very short. In practical applications, passive FTCs complement active FTC schemes. Indeed, passive FTCs are necessary during the fault detection and estimation phases [16], where they are used to ensure the stability of the faulty system, before switching to active FTCs that recover some performance after the fault is detected and estimated. Passive FTC is also used as a complement of active FTC in switching-based active FTC, where the active FTC switches between different passive FTC, each controller being designed off-line to cope with a finite number of expected faults and stored in a controller bank [17].

Nonlinear regulation theory has been used to solve the nonlinear fault-tolerant control (NFTC) problem for particular practical examples, that is, induction motors [18] and robot manipulators [19]. The faults treated were modeled as additive actuator faults. Benosman and Lum [20] used Lyapunov reconstruction techniques to solve the problem of loss of actuator effectiveness for nonlinear models affine in the control. Note that the main drawback of this scheme is that it is based on *a priori* knowledge of a stabilizing feedback for the nominal safe model and knowledge of the associated Lyapunov function in closed form.

After the occurrence of a fault, the faulty system is expected to be unable to perform the tasks required and planned initially for the safe system. Therefore, new, less demanding, tasks have to be generated online for the faulty system. This is based on two main stages. The first stage concerns online trajectory planning or reshaping, using an online optimization scheme that generates the closest trajectory to the nominal one, but without violating the new constraints of the faulty system. The second stage concerns the control reallocation problem, using nonlinear model predictive control (NMPC). This scheme deals with nonminimum-phase, nonlinear models affine in the control.

An important part of FTC specializes in actuator faults. Indeed, FTCs dealing with actuator faults are relevant to practical applications and have already been the subject of many publications [20, 21, 22, 23, 24, 25].

Mhaskar *et al.* considered uncertain nonlinear models with constrained inputs [26, 27] and active FTC with respect to additive actuator faults for nonlinear systems affine in the control [24]. Constrained actuators were considered, and state-feedback as well as output-feedback FDDs/FTCs were proposed. Zhang *et al.* proposed an active NFTC for the class of SISO nonlinear systems, with incipient faults [16, 28]. The structure of the FTC was based on three controllers: a nominal controller for the safe system guarantees the system trajectory's boundedness until the fault was detected; a second controller that recovered some control performance before the fault was isolated; after isolation, a third controller, based on the fault model, was used to improve the control performance. The reconfiguration of the controllers was based on adaptive backstepping approaches.

5.3.2   Extended Luenberger observer

The basic idea is to linearize the nonlinear model around the current estimated state, instead of an extended point, such as the origin. The detailed study of this class of observers was made by Zeitz [29]. For the purpose of fault detection, a similar approach was used by Adjallah, Maquin, and Ragot [30]. The discrete-time version of this observer was studied by Witczak [5]. Sometimes, because of the time-varying nature and the linearization errors, it is difficult to obtain a proper filter gain matrix, which limits its use in practice.

5.3.3   Nonlinear identity observer approach

The idea of nonlinear identity, observer-based fault detection was initiated by Hengy and Frank [31]. Later,it was presented with more details in various survey papers [4, 32, 33, 34]. The basic idea behind the design is to linearize the observer error dynamics at the estimated state and neglect the higher-order terms. A filter gain is determined in such a way that the equilibrium point of the error dynamics, that is, $e = 0$ is asymptotically stable. A solution to this problem was first proposed by Adjallah, Maquin, and Ragot [30] by assuming that the measurements are linear.

5.3.4 Unknown input observer approach

The intuitive idea of the unknown input observer (UIO) is to design an observer for the purpose of fault detection in such a way that the effect of the unknown inputs on the residual signal is completely eliminated. This approach takes advantage of the structure of the system model which is in observable form while admitting the difficulty of transforming general nonlinear systems into the form suitable for this approach. Further, even for the linear case, the existence conditions for UIO are restrictive.

5.3.5 The disturbance decoupling nonlinear observer approach

The disturbance decoupling nonlinear observer for fault detection was addressed by Seliger and Frank [35], who relaxed the existence conditions of nonlinear UIO. The basic idea of a disturbance decoupling nonlinear observer is the same as a UIO but a more general class of nonlinear systems is used. Further, a nonlinear transformation is used instead of a linear one. The transformation is required to be a state transformation and is used in such a way that the disturbances are decoupled from the faults. A difficulty arises when the disturbance distribution matrix depends explicitly on the input. Some solutions were proposed to avoid this difficulty [35]. After achieving the required transformation, the observer can be designed using any observer design method, for instance, the nonlinear UIO method.

5.3.6 Adaptive nonlinear observer approach

Conventional observer-based methods have poor performance in detecting slowly developing faults in nonlinear uncertain systems, especially when the uncertainties are dominant [36]. To overcome this difficulty, one solution is the use of an adaptive observer. This class of observers has been widely utilized for detecting faults in nonlinear systems in the presence of model uncertainties. Another advantage of this approach is that some uncertain parameters may be estimated online, which can be used for improving robustness against model uncertainties.

5.3.7 High-gain observer approach

High-gain observers [37] are usually designed with the objective of handling model uncertainties. They have also been applied to fault detection in nonlinear systems [38, 39, 40]. An intrinsic feature of any high-gain observer is the peaking phenomenon, which is also addressed in the literature.

5.3.8 Sliding-mode observer approach

Sliding-mode observers have been widely applied to fault detection in nonlinear systems [41, 42, 43]. The inherent property of sliding-mode observers of being robust to uncertainties and disturbances makes them suitable for state estimation and fault detection.

5.3.9   Geometric approach

The nonlinear geometric approach to fault detection [44] is based on the detection filter design using a geometric approach for linear systems. The residual generator is designed in such a way that the obtained residual signal depends trivially on fault and non-trivially on disturbances in its decoupled form. The shortcoming of this method is the requirement for a deep understanding of geometric theory. Further, the disturbance decoupling process may lead to faults that lie in the same subspace as the disturbances being undetectable.

5.3.10   Game-theoretic approach

Game theory has been utilized in designing observers for linear systems [45]–[47] and non-linear systems [48]. It has also been utilized in the context of fault detection [49, 50]. The advantages of game-theoretic design are that extreme case scenarios for disturbances and filter gains can be treated very easily and that it can be applied to the more general class of nonlinear systems. [50] used the game-theoretic approach for fault detection in continuous-time nonlinear systems. They proposed a fault detection filter which attenuates the effect of measurement noises while keeping the minimal effect of fault on the "innovation signal" (also known as the "residual signal"). They exploited the tools of geometric techniques in order to get a special form of nonlinear system in which the disturbance subspace is decoupled. The residual generator thus designed depends non-trivially on fault and trivially on disturbances (in decoupled form). It can be noted from the proposed design scheme that the filter provides disturbance attenuation to the desired level in the presence of fault which has minimal effect on the innovation signal. However, it does not address the problem of $\mathcal{H}_\infty$ index-based design which is often required in the design. In addition, the proposed approach considered the innovation signal as a residual signal. However, in real FD problems, the residual signal is the difference between the actual output of the process and the estimate. For the optimal design of FD systems, the residual signal (not the innovation signal) should be considered.

5.3.11   Observers for Lipschitz nonlinear systems

In Lipschitz nonlinear systems, the nonlinear function is assumed to be bounded by some constant. Nonlinear systems involving sinusoidal nonlinearities are always bounded and can be considered as globally Lipschitz nonlinear systems. Systems having nonlinearities other than sinusoidal can be bounded for a particular range of process operation and hence are regarded as locally Lipschitz nonlinear systems. Lipschitz nonlinear systems is an active field of research in the context of control, observers and FDI. Some of the reasons are that:

- Any type of nonlinearity can be transformed into a Lipschitz nonlinear system, at least locally.
- Analytical solutions can be presented for nonlinear problems involving Lipschitz nonlinearities.
- Well-established LMI techniques can be utilized to formulate the nonlinear problems involving Lipschitz nonlinearities. These LMIs can then be solved using standard computer software, such as MATLAB®.

5.3.12   Lyapunov-reconstruction-based passive scheme

Let us start with some passive NFTC algorithms. As we stressed before, these types of FTC are not expected to "do the job alone", since in practice they have to be associated with some active FTCs to obtain an efficient controller tolerant to faults.

We first consider nonlinear systems of the form

$$\dot{x} = f(x) + g(x)u, \tag{5.1}$$

where $x \in \mathfrak{R}^n$ and $u \in \mathfrak{R}^m$ represent, respectively, the state and the input vectors. The vector fields f and columns of g are supposed to satisfy the classical smoothness assumptions, with $f(0) = 0$. We also assume the system to be locally reachable [51]. Adding to the previous classical assumptions, we also need the following assumptions to hold.

Assumption 5.1   *We assume the existence of a nominal closed-loop control $u_{nom}(t, x)$, such that the solutions of the closed-loop system*

$$\dot{x} = f(x) + g(x)u_n(t, x) \tag{5.2}$$

satisfy

$$|x(t)| \leq \beta(|x(t_0)|, t - t_0), \, \forall t_0 \in D, \, \forall t \geq t_0$$

where

$$D = \{x \in \mathfrak{R}^n \, | \, |x| < r_0\}, r_0 > 0$$

and $\beta(.)$ is a class KL function.

Assumption 5.2   *We assume two types of actuator fault:*

- *faults that enter the system in an additive way; that is, the faulty model is*

$$\dot{x} = f(x) + g(x)(u + F(t, x)), \tag{5.3}$$

  where F represents the actuator fault and such that

$$|F(t, x)| \leq b(t, x)$$

  where

$$b : [0, \infty) \times D \to R$$

  is a nonnegative continuous function.

- *loss of actuator effectiveness, represented by a multiplicative matrix α as*

$$\dot{x} = f(x) + g(x)\alpha u, \tag{5.4}$$

where $\alpha \in \Re^{m \times m}$ is a diagonal continuous time-variant matrix, with the diagonal elements $\alpha_{ii}(t), i = 1, \ldots, m$ such that $0 < \varepsilon_1 \le \alpha_{ii}(t) \le 1$.

The following propositions are quite standard:

Proposition 5.1   *Consider the control law*

$$u(t, x) = u_n(t, x) - sgn\left(\left(\frac{\partial V}{\partial x}g\right)^T\right)(b(t, x) + \epsilon), \quad \epsilon > 0, \tag{5.5}$$

where $u_n(t, x)$ is such that Assumption 5.1 is satisfied; V is the associated Lyapunov function; $b(t, x)$ is defined in Assumption 5.2; and sgn(v) denotes the vector sign function, such that sgn(v)(i) = sgn(v(i)). It ensures that the equilibrium point $x = 0$ is locally uniformly asymptotically stable (UAS) in D for the closed-loop system of Equations (5.3) and (5.5).

Proposition 5.2   *Consider the control law*

$$u(t, x) = u_{nom}(t, x) - sgn\left(\left(\frac{\partial V}{\partial x}g\right)^T\right)\left(|u_{nom}| + \frac{|u_{nom}|}{\epsilon_1}\beta_1\right), \beta_1 \ge 1, \tag{5.6}$$

where $u_{nom}(t, x)$ is such that Assumption 5.1 is satisfied, V is the associated Lyapunov function, and sgn($\cdot$) denotes the sign function. It ensures that the equilibrium point $x = 0$ is locally UAS in D for the closed-loop system of Equations (5.4) and (5.6).

Proposition 5.1 and Proposition 5.2 ensure robust stabilization with respect to additive as well as multiplicative actuator faults. However, they are discontinuous because of the sign function. We now present "continuous" versions of the propositions.

Proposition 5.3   *The control law*

$$u(t, x) = u_n(t, x) - sat\left(\left(\frac{\partial V}{\partial x}g\right)^T\right)(b(t, x) + \epsilon), \quad \epsilon > 0 \tag{5.7}$$

ensures that the solutions of the closed-loop system of Equations (5.3) and (5.7) satisfy

$$\forall x(t_0) \quad such\ that\ |x(t_0)| \le \alpha_2^{-1}(\alpha_1(r_0)),$$

$$\exists T \ge 0, \quad such\ that \begin{cases} |x(t)| \le \beta(|x(t_0)|, t - t_0), & \forall t_0 \le t \le t_0 + T, \\ |x(t)| \le \alpha_2^{-1}(\alpha_1(\tilde{x})), & \forall t \ge t_0 + T, \end{cases} \tag{5.8}$$

where, for a vector v,

$$sat(v) = \begin{cases} \frac{v(i)}{\tilde{\epsilon}}, & \text{if } |v(i)| \leq \tilde{\epsilon}, \\ sgn(v(i)) & \text{if } |v(i)| > \tilde{\epsilon}, \end{cases}$$

$$\tilde{x} = \alpha_3^{-1}(2m\tilde{\epsilon}b_{\max}) \leq \alpha_2^{-1}(\alpha_1(r_0)),$$

$$b(t, x) \leq b_{\max}, \quad \forall t, \forall x \in D, \tag{5.9}$$

and α_1, α_2, and α_3 are class K functions in D and β is class KL.

Proposition 5.4   *The control law*

$$u(t, x) = u_{nom}(t, x) - sat\left(\left(\frac{\partial V}{\partial x}g\right)^T\right)\left(|u_{nom}| + \frac{|u_{nom}|}{\epsilon_1}\beta_1\right), \quad \beta_1 \geq 1 \tag{5.10}$$

ensures that the solutions of the closed-loop system of Equations (5.4) and (5.10) satisfy

$$\forall x(t_0) \quad \text{such that } |x(t_0)| \leq \alpha_2^{-1}(\alpha_1(r_0)),$$

$$\exists T \geq 0, \quad \text{such that } \begin{cases} |x(t)| \leq \beta(|x(t_0)|, t - t_0), & \forall t_0 \leq t \leq t_0 + T, \\ |x(t)| \leq \alpha_2^{-1}(\alpha_1(\tilde{x})), & \forall t \geq t_0 + T, \end{cases} \tag{5.11}$$

where, for a vector v,

$$sat(v) = \begin{cases} \frac{v(i)}{\tilde{\epsilon}}, & \text{if } |v(i)| \leq \tilde{\epsilon}, \\ sgn(v(i)) & \text{if } |v(i)| > \tilde{\epsilon}, \end{cases}$$

$$\tilde{x} = \alpha_3^{-1}(2m\tilde{\epsilon}u_{nom-\max}) \leq \alpha_2^{-1}(\alpha_1(r_0)),$$

$$|u_{nom}| \leq u_{nom-\max}, \quad \forall t, \tag{5.12}$$

and α_1, α_2, and α_3 are class K functions in D and β is class KL.

The continuous controllers in Equation (5.7) and Equation (5.10) do not guarantee the local UAS. However, they guarantee that the closed-loop trajectories are bounded by a class K function, and that this bound can be made as small as desired by choosing a small $\tilde{\epsilon}$ in the definition of the function sat. The passive NFTC recalled above is in closed form and thus is easy to implement. However, it has two main drawbacks: it is based on the availability of the closed-form expression of the Lyapunov function associated with the nominal stabilizing law; and it does not consider input saturations in the control design. In an attempt to overcome those limitations, we consider other controllers.

Theorem 5.1   *Consider the closed-loop system that consists of the faulty system of Equation (5.4), with constant unknown matrix α, and the dynamic state feedback:*

$$\dot{u} = -L_g W(x)^T - k\xi, \quad u(0) = 0,$$

$$\xi = \epsilon_1\left(-(L_g W(x))^T - k\xi\right), \xi(0) = 0, \tag{5.13}$$

where W is a C^1 radially unbounded, positive semidefinite function, such that $L_f W \leq 0$, and $k > 0$. Consider the fictitious system:

$$\dot{x} = f(x) + g(x)\xi,$$

$$\dot{\xi} = \epsilon_1 \left(-(L_g W)^T + \tilde{v} \right), \tag{5.14}$$

$$y = h(\xi) = \xi.$$

If Equation (5.14) is (G)ZSD with the input $\tilde{v}$ and the output y, then the closed-loop system of Equation (5.4) with Equation (5.13) admits the origin $(x, \xi) = (0, 0)$ as a (globally) asymptotically stable ((G)AS) equilibrium point.

In Theorem 5.1, one of the necessary conditions is the existence of $W \geq 0$, such that the uncontrolled part of Equation (5.3) satisfies $L_f W \leq 0$. To avoid this condition, which may not be satisfied for some practical systems, the following theorem provides an answer.

Theorem 5.2   *Consider the closed-loop system that consists of the faulty system of Equation (5.4), with constant unknown matrix α, and the dynamic state feedback:*

$$\dot{u} = \frac{1}{\epsilon_1} \left(-k(\xi - \beta K(x)) \right) - \beta L_g W^T + \beta \frac{\partial K}{\partial x}(f + g\xi) \right),$$

$$\beta = diag(\beta_{11}, \ldots, \beta_{mm}), \ 0 \leq \frac{\tilde{\epsilon}_1}{\epsilon_1} \leq \beta_{ii} \leq 1, \tag{5.15}$$

$$\xi = -k(\xi - \beta K(x)) - \beta L_g W^T + \beta \frac{\partial K}{\partial x}(f + g\xi), \quad \xi(0) = 0, \quad u(0) = 0,$$

where $k > 0$ and the C^1 function $K(x)$ is such that a C^1 radially unbounded, positive semidefinite function W satisfies

$$\frac{\partial W}{\partial x}(f(x) + g(x)\beta K(x)) \leq 0, \forall x \in \Re^n, \forall \beta = diag(\beta_{11}, \ldots, \beta_{mm}), o < \tilde{\epsilon}_1 \leq \beta_{ii} \leq 1.$$

$$\tag{5.16}$$

Consider the fictitious system:

$$\dot{x} = f(x) + g(x)\xi,$$

$$\dot{\xi} = \beta \frac{\partial K}{\partial x}(f + g\xi) - \beta L_g W^T + \tilde{\tilde{v}}, \tag{5.17}$$

$$\tilde{y}\beta = \xi - \beta K(x).$$

If Equation (5.17) is (G)ZSD with the input $\tilde{\tilde{v}}$ and the output $\tilde{y}$, for all β such that β_{ii}, $i = 1, \ldots, m, 0 < \tilde{\epsilon}_1 \leq \beta_{ii} \leq 1$. Then, the closed-loop system of Equation (5.4) with Equation (5.15) admits the origin $(x, \xi) = (0, 0)$ as a GAS equilibrium point.

Theorems 5.1 and 5.2 may guarantee global asymptotic stability. However, the conditions required may be difficult to satisfy for some systems. The following control law ensures, under less-demanding conditions, semiglobal stability instead of global stability.

Theorem 5.3   *Consider the closed-loop system that consists of the faulty system of Equation (5.4), with constant matrix α and the dynamic state feedback:*

$$\dot{u} = -k(\xi - u_{unom}(x)), \quad k > 0,$$

$$\dot{\xi} = -k\epsilon_1(\xi - u_{unom}(x)), \quad \xi(0) = 0, \quad u(0) = 0, \tag{5.18}$$

where the nominal controller $u_{nom}(x)$ achieves semiglobal asymptotic and local exponential stability of $x = 0$ for the safe system in Equation (5.1). Then, the closed-loop Equation (5.4) with Equation (5.18) admits the origin $(x, \xi) = (0, 0)$ as asemiglobal asymptotically stable equilibrium point.

The following result is for general nonlinear models, nonnecessarily affine on u, with input saturation.

Theorem 5.4   *Consider the closed-loop system that consists of the faulty system:*

$$\dot{x} = f(x) + g(x, \alpha u)\alpha u, \tag{5.19}$$

for $\alpha \in [\epsilon_1, 1]$, and the static state feedback:

$$u(x) = -\lambda(x)G(x, 0)^T,$$

$$G(x, 0) = \frac{\partial W(x)}{\partial x}\epsilon_1 g(x, 0),$$

$$\lambda(x) = \frac{2\overline{u}}{\left(1 + \gamma_1(|x|^2 + 4\overline{u}^2|G(x, 0)|^2)\right)(1 + |G(x, 0)|^2)} > 0,$$

$$\gamma_1 = \int_0^{2s} \frac{\overline{\gamma}_1(s)}{1 + \overline{\gamma}_1(1)}ds, \tag{5.20}$$

$$\overline{\gamma}_1(s) = \frac{1}{s}\int_s^{2s} (\overline{\gamma}_1(t) - 1)dt + s$$

$$\tilde{\gamma}_1(s) = \max_{\{(x,u)|x|^2+|u|^2 \leq s\}} \left\{1 + \int_0^1 \frac{\partial W(x)}{\partial x}\frac{\partial g(x, \tau\epsilon_1 u)}{\partial u}d\tau\right\},$$

where W is a C^2 radially unbounded, positive semidefinite function, such that $L_f W \leq 0$. Consider the fictitious system:

$$\dot{x} = f(x) + g(x, \epsilon_1 u)\epsilon_1 u,$$

$$y = \frac{\partial W(x)}{\partial x}\epsilon_1 g(x, \epsilon_1 u). \tag{5.21}$$

If Equation (5.21) is (G)ZSD, then the closed-loop system of Equation (5.19) with Equation (5.20) admits the origin as a (G)AS equilibrium point. Furthermore $|u(x)| \leq \bar{u}$, for all x.

For the particular case of affine nonlinear systems, that is, $g(x, u) = g(x)$, we have the following proposition, which is a direct consequence of Theorem 5.4.

Proposition 5.5 *Consider the closed-loop system that consists of the faulty system of Equation (5.4), with constant unknown matrix α, and the static state feedback:*

$$u(x) = -\lambda(x)G(x)^T,$$

$$G(x) = \frac{\partial W(x)}{\partial x}\epsilon_1 g(x), \qquad\qquad (5.22)$$

$$\lambda(x) = \frac{2\bar{u}}{|1 + G(x)|^2}.$$

where W is a C^2 radially unbounded, positive semidefinite function, such that $L_f W \leq 0$. Consider the fictitious system:

$$\dot{x} = f(x) + g(x)\epsilon_1 u, \qquad\qquad (5.23)$$

$$y = \frac{\partial W(x)}{\partial x}\epsilon_1 g(x).$$

If Equation (5.23) is (G)ZSD, then the closed-loop system of Equation (5.4) with Equation (5.22) admits the origin as a (G)AS equilibrium point. Furthermore $|u(x)| \leq \bar{u}, \forall x$.

5.3.13 Time-varying results

This section presents the results of the previous section for time-varying faults.

Theorem 5.5 *Consider the closed-loop system that consists of the faulty system of Equation (5.4) with the dynamic state feedback:*

$$\dot{u} = -L_g W(x)^T - k\xi, \quad k > 0, \quad u(0) = 0,$$

$$\dot{\xi} = \tilde{\alpha}(t)\left(-(L_g W(x))^T - k\xi\right), \quad \xi(0) = 0, \qquad\qquad (5.24)$$

where $\tilde{\alpha}(t)$ is a C^1 function, such that $0 < \epsilon_1 \leq \tilde{\alpha}(t) \leq 1, \forall t$, and W is a C^1 positive semidefinite function, such that

- $L_f W \leq 0$;
- *the system $\dot{x} = f(x)$ is AS conditionally to the set $M = \{x | W(x) = 0\}$;*

- *for all $(\overline{x}, \overline{\xi})$-limiting solutions for the system:*

$$\dot{x} = f(x) + g(x)\xi,$$

$$\dot{\xi} = \alpha(t)\left(-(L_g W(x))^T - k\xi\right) \qquad (5.25)$$

$$y = h(x, \xi) = \xi,$$

with respect to unbounded sequence $\{t_n\}$ in $[0, \infty)$, then if $h(\overline{x}, \overline{\xi}) = 0$, a.e., then either $(\overline{x}, \overline{\xi})(t_0) = (0, 0)$ for some $t_0 \geq 0$ or $(0, 0)$ is an ω-limit point of $(\overline{x}, \overline{\xi})$, that is, $\lim_{t \to \infty}(\overline{x}, \overline{\xi})(t) \to (0, 0)$.

Then the closed-loop system of Equation (5.4) with Equation (5.24) admits the origin $(x, \xi) = (0, 0)$ as a UAS equilibrium point.

Theorem 5.6   *Consider the closed-loop system that consists of the faulty system:*

$$\dot{x} = f(x) + g(x, \alpha(t)u)\alpha(t)u, \qquad (5.26)$$

for $\alpha \in [\epsilon_1, 1]$, $\forall t$, with the static state feedback:

$$u(x) = -\lambda(x)G(x, 0)^T,$$

$$G(x, 0) = \frac{\partial W(x)}{\partial x} g(x, 0),$$

$$\lambda(x) = \frac{2\overline{u}}{\left(1 + \gamma_1(|x|^2 + 4\overline{u}^2|G(x, 0)|^2)\right)(1 + |G(x, 0)|^2)} > 0,$$

$$\gamma_1 = \int_0^{2s} \frac{\overline{\gamma}_1(s)}{1 + \overline{\gamma}_1(1)} ds, \qquad (5.27)$$

$$\overline{\gamma}_1(s) = \frac{1}{s} \int_s^{2s} (\overline{\gamma}_1(t) - 1)dt + s,$$

$$\tilde{\gamma}_1(s) = \max_{\{(x,u)|x|^2 + |u|^2 \leq s\}} \left\{1 + \int_0^1 \frac{\partial W(x)}{\partial x} \frac{\partial g(x, \tau\epsilon_1 u)}{\partial u} d\tau\right\},$$

where W is a C^2, positive semidefinite function, such that

- $L_f W \leq 0$;
- *the system $\dot{x} = f(x)$ is AS conditionally to the set $M = \{x | W(x) = 0\}$;*
- *for all $\overline{x}$-limiting solutions for the system*

$$\dot{x} = f(x) + g(x, \epsilon_1 u(x))\left(-\lambda(x)\alpha(t)\frac{\partial W}{\partial x}(x)g(x, 0)\right)^T, \qquad (5.28)$$

$$y = h(x) = \lambda(x)^{0.5}|\frac{\partial W}{\partial x}(x)g(x, 0)|,$$

with respect to unbounded sequence $\{t_n\}$ in $[0, \infty)$, then if $h(\overline{x}) = 0$, a.e., either $\overline{x}(t_0) = 0$ for some $t_0 \geq 0$ or 0 is an ω-limit point of x.

Then the closed-loop system of Equation (5.26) with Equation (5.27) admits the origin $x = 0$ as a UAS equilibrium point. Furthermore $|u(x)| \leq \overline{u}, \forall x$.

Proposition 5.6   *Consider the closed-loop system that consists of the faulty system of Equation (5.4) with the static state feedback:*

$$u(x) = -\lambda(x)G(x)^T,$$

$$G(x) = \frac{\partial W(x)}{\partial x}g(x),$$ (5.29)

$$\lambda(x) = \frac{2\overline{u}}{1 + |G(x)^2|}.$$

where W is a C^2, positive semidefinite function, such that

- *$L_f W \leq 0$;*
- *The system $\dot{x} = f(x)$ is AS conditionally to the set $M = \{x | W(x) = 0\}$;*
- *for all $\overline{x}$-limiting solutions for the system*

$$\dot{x} = f(x) + g(x)\left(-\lambda(x)\alpha(t)\frac{\partial W}{\partial x}(x)g(x)\right)^T,$$

$$y = h(x) = \lambda(x)^{0.5}|\frac{\partial W}{\partial x}(x)g(x)|,$$ (5.30)

with respect to unbounded sequence $\{t_n\}$ in $[0, \infty)$, then if $h(\overline{x}) = 0$, a.e., then either $\overline{x}(t_0) = 0$ for some $t_0 \geq 0$ or 0 is an ω-limit point of x.

Then the closed-loop system of Equation (5.4) with Equation (5.29) admits the origin $x = 0$ as a UAS equilibrium point. Furthermore $|u(x)| \leq \overline{u}, \forall x$.

Passive NFTC schemes are valid for a large class of nonlinear systems, not necessarily affine in the control, and take into account input saturations; however, the conditions to satisfy might be difficult to check when dealing with models having a large number of states. Needless to stress, passive FTCs cannot cope with faults alone – they have to be associated with active FTCs.

5.3.14   *Optimization-based active scheme*

Passive FTCs ensure the stability of the faulty system during the time period when the FDD is estimating the fault; active FTCs take over from the passive FTCs and, using the estimated faulty model, try to optimize the performance of the faulty system. We present some active NFTC schemes, in the rest of this section.

Benosman and Lum [52] studied the problem of graceful performance degradation for affine nonlinear systems. The method is an optimization-based scheme that gives a constructive way to reshape online the output reference for the post-fault system, and explicitly take into account the actuator and state saturations. Online output reference reshaping is associated with an online, MPC-based, controller reconfiguration that forces the post-fault system to track the new output reference.

The models considered are affine in the control:

$$\dot{x} = f(x) + g(x)u,$$
$$y = h(x), \tag{5.31}$$

where $x \in \mathfrak{R}^n$, $u \in \mathfrak{R}^{n_a}$, and $y \in \mathfrak{R}^m$ represent the state, input and controlled output vectors, respectively. The vector fields f, columns of g, and function h are supposed to satisfy the following classical assumptions.

Assumption 5.3 $f : \mathfrak{R}^n \to \mathfrak{R}^n$ and the columns of $g : \mathfrak{R}^n \to \mathfrak{R}^{n \times n_a}$ are smooth vector fields on a compact set X of $\mathfrak{R}^n$, and $h(x)$ is a smooth function on X with $f(0) = 0$, $h(0) = 0$.

Assumption 5.4   Equation (5.1) has a well-defined (vector) relative degree $\{r_1, \ldots, \mathfrak{R}^m\}$ at each point $x^0 \in X$ [53].

Assumption 5.5   The system is fully or over-actuated, in the sense that the number of actuators is at least equal to the number of controlled outputs, that is, $n_a \geq m$.

Assumption 5.6   Assumptions 5.3–5.5 are preserved after the occurrence of a fault in the system.

Assumption 5.7   The desired nominal trajectory is feasible by the nominal (safe) system, within its input and state limits.

The control objective is, then, to find a controller u such that the nominal as well as the faulty system's output vector y tracks asymptotically a desired smooth feasible trajectory $y_d(t)$, while satisfying the constraints on the actuators and states:

$$u \in \Omega \triangleq \{u = (u_1, u_2, \ldots, u_{n_a})^T | u_i^- \leq u_i \leq u_i^+, i = 1, 2, \ldots, n_a\},$$
$$x \in X \triangleq \{x = (x_1, x_2, \ldots, x_n)^T | x_i^- \leq x_i \leq x_i^+, i = 1, 2, \ldots, n\}, \tag{5.32}$$

where

$$u^- = (u_1^-, u_2^-, \ldots, u_{n_a}^-)^T, \quad u^+ = (u_1^+, u_2^+, \ldots, u_{n_a}^+)^T$$
$$x^- = (_1^-, x_2^-, \ldots, x_n^-)^T, \quad x^+ = (x_1^+, x_2^+, \ldots, x_n^+)^T$$

are vectors of the lower and upper actuator limits and state limits, respectively. To find such a controller, the problem is formulated in terms of the following optimization:

$$\min_{(a,t_{2F})} J = \min_{(a,t_{2F})} \int_{t_{1F}}^{t_{2F}} (y_n(t) - y_a(t))^T Q_1 (y_n(t)$$

$$- y_d(t))dt + \int_{t_{1F}}^{t_{2F}} u(t)^T Q_2 u(t)dt, \tag{5.33}$$

under the constraints

$$\dot{x} = f_F(x) + g_F(x)u,$$

$$y_d(t, a, t_{2F}) = h(x),$$

$$u^- \le u \le u^+, \quad x^- \le x \le x^+,$$

$$y^{(k)}(t_{1F}) \triangleq (y_1^{(k)}(t_{1F}), \ldots, y_m^{(k)}(t_{1F}))^T$$

$$y_n^{(k)}(t_{1F}) = (y_{n_1}^{(k)}(t_{1F}), \ldots, y_{n_m}^{(k)}(t_{1F}))^T,$$

$$y^{(k)}(t_{2F}) = (y_1^{(k)}(t_{2F}), \ldots, y_m^{(k)}(t_{2F}))^T$$

$$y_n^{(k)}(t_{2n}) = (y_{nom_1}^{(k)}(t_{2n}), \ldots, y_{n_m}^{(k)}(t_{2nom}))^T,$$

$$k = 0, \ldots, s, \quad t_{2F} \ge t_{2n}, \tag{5.34}$$

where

$$y_d(t) = \left(\sum_{i=1}^{i=l+1} a_{i1}((t - t_{1F})/(t_{2F} - t_{1F}))^{i-1}, \ldots, \sum_{i=1}^{i=l+1} a_{im}((t - t_{1F})/(t_{2F} - t_{1F}))^{i-1} \right)^T$$

and $s \in N^+$, $Q_1 \in \Re^{m \times m}$, and $Q_2 \in \Re^{n_a \times n_a}$ are positive definite weight matrices; $a = (a_{(1)1}, \ldots, a_{(l+1)1}, \ldots, a_{(1)m}, \ldots, a_{(l+1)m})^T \in \Re^{m(l+1)}$ is the vector of the polynomial coefficients; t_{2F} is the final motion time for the optimal trajectory vector $y_d(t)$; t_{2n} is the final motion time for the nominal trajectory vector $y_n(t) \triangleq (y_{n_1}(t), \ldots, y_{n_m}(t))^T$; and where f_F, g_F hold for the modified vector field f and matrix g after the occurrence of the fault. The existence of solutions and the computation scheme was studied for different cases: without internal dynamics; with internal dynamics for minimum phase systems; and with internal dynamics for nonminimum phase systems. In the following discussion, based on the availability of an FDD module, two cases are studied:

- where FDD provides a precise post-fault model;
- the realistic case, where FDD gives a delayed imprecise post-fault model.

This optimization-based scheme can deal with the general class of nonlinear models affine in the control, with state and input constraints and can include a stable inversion part to deal with nonminimum phase systems. However, the necessary online computation can be time-consuming for large models.

5.3.15   Learning-based active scheme

We report here the results presented by Polycarpou [54], who used a learning scheme to modify
the feedback control so as to stabilize the system in the presence of a fault.

The following class of systems is considered

$$\dot{x} = f(x) + G(x)[u + \eta(x, t) + \beta(t - T)\xi(x)] \tag{5.35}$$

where, $x \in \Re^n$, and $u \in \Re^m$ are the state and control vectors, respectively; $G = [g_1, g_2, \ldots, g_m]$ is an $n \times m$ matrix function; $f, g_i : \Re^n \to \Re^n, i = 1, \ldots, m$ are known smooth
vector fields representing the nominal system dynamics; $\beta(t - T)$ is a step function repre-
senting an abrupt fault occurring at an unknown time T; $\eta(x, t)$ represents the time-varying
model uncertainties; and $\xi(x)$ is the vector of state-dependent faults. The author assumes the
existence of a nominal controller $u_N(x)$ that guarantees uniform stabilization of the nominal
system:

$$\dot{x} = f(x) + G(x)u \tag{5.36}$$

The scheme also assumes the availability of the closed form Lyapunov function V_N associated
with the nominal stable feedback system:

$$\dot{x} = f(x) + G(x)u_N(x). \tag{5.37}$$

Polycarpou proposes the following NFTC:

$$
\begin{aligned}
u &= u_N(x) + \phi(x, \hat{\theta}, \bar{\theta}), \\
\phi(x, \hat{\theta}, \bar{\theta}) &= -\Omega(x)^T \hat{\theta} - \bar{\theta}\omega(x), \\
\dot{\hat{\theta}} &= \Gamma\Omega(x)p(x), \\
\dot{\bar{\theta}} &= \gamma\omega(x)^T p(x), \\
p(x) &= \left(\frac{\partial V_N}{\partial x}G(x)\right)^T, \\
\omega_i(x) &= tanh\left(\frac{p_i(x)}{\epsilon}\right), \quad \epsilon > 0, i = 1, \ldots, m,
\end{aligned}
\tag{5.38}
$$

where $\Omega(x)$ is a $q \times m$ and represents the basis function for the neural network approximation
of the fault f by $\hat{\Omega}f(x, \hat{\theta}) = \Omega(x)^T\hat{\theta}, \hat{\theta} \in \Re^q$. Then, under the assumption of matching
conditions, that is, η and ξ are in the range space of G, Polycarpou proves that the feedback
controller Equation (5.38) stabilizes the faulty system of Equation (5.35). However, this control
law, is based on knowledge of the full state vector and might lead to chattering effects if the
parameter ϵ is chosen too small.

5.3.16 Adaptive backstepping-based active scheme

The scheme presented here is based on the results of Zhang, Parisini, and Polycarpou [16, 28]. The systems studied are of the form

$$\dot{x}_i = x_{i+1} + \phi_i(\overline{x}_i) + \eta_i(x, u, t) + \beta_i(t - T_0)\xi_i(\overline{x}_i), i = 1, \ldots, n - 1,$$

$$\dot{x}_n = \phi_0(x)u + \phi_n(x) + \eta_n(x, u, t) + \beta_n(t - T_0)\xi_n(x),$$

$$y = x_1$$

(5.39)

where $x \in \Re^n$ is the state vector, $\overline{x}_i = (x_1, \ldots, x_i)^T$, $u \in R$, and $y \in R$ are the input and the output, respectively. The function ϕ_0 is a nonzero smooth function, and $\phi_i, \eta_i, f_i, i = 1, \ldots, n$ are smooth functions. The control goal is to force the output y to track a desired trajectory $y_r(t)$, where $y_r^{(l)}, l = 0, \ldots, n$ are known, piecewise continuous and bounded. As in Section 5.3.15, η_i, ξ_i, and $\beta_i, i = 1, \ldots, n$ represent the model uncertainties, the expected faults, and the time profile of the faults, respectively. Then, based on an assumption of the availability of an FDD module that detects and estimates the fault, we propose the following three-stage controller:

$$u = \begin{cases} u_0(x, y_d, t), & t < T_d, \\ u_D(x, y_d, t), & T_d \leq t < T_{isol}, \\ u_1(x, y_d, t), & T \geq T_{isol}, \end{cases}$$

(5.40)

where T_d and T_{isol} are the time of the fault detection and fault isolation, respectively. Based on the adaptive-backstepping approach, we propose the following expressions for the three controllers.

- For $t < T_d$

$$u_0(t) = \frac{\alpha_n + y_r^{(n)}}{\phi_0(x)}$$

(5.41)

with

$$\alpha_0 = 0,$$

$$\alpha_1 = -c_1 z_1 - c_2 z_2 - \phi_1,$$

$$\alpha_i = -c_1 z_i - z_{i-1} - \phi_i - c_2 z_i \sum_{j=1}^{i-1} \left(\frac{\partial \alpha_{i-1}}{\partial x_j}\right) + \sum_{j=0}^{i-2} \frac{\partial \alpha_{i-1}}{\partial y_r^{(j)}} y_r^{(j+1)}$$

(5.42)

$$+ \sum_{j=1}^{i-1} \frac{\partial \alpha_{i-1}}{\partial x_j}(x_{j+1} + \phi_j), \quad i = 2, \ldots, n,$$

$$z_i = x_i - \alpha_{i-1} - y_r^{(i-1)}, \quad i = 1, \ldots, n.$$

- For $T_d \leq t < T_{isol}$

$$u_D(t) = \frac{\alpha_n + y_r^{(n)}}{\phi_0(x)} \tag{5.43}$$

with

$$\alpha_0 = 0,$$

$$\alpha_1 = -c_1 z_1 - \phi_1 - \hat{\theta}_1^T \varphi_1 + p_1(y, \hat{\theta}_1, \hat{\Psi}, y_r),$$

$$\alpha_i = -z_{i-1} - c_i z_i - \phi_i - \hat{\theta}_1^T \varphi_1(\overline{x}_i)$$

$$+ \sum_{k=1}^{i-1} \left[\frac{\partial \alpha_{i-1}}{\partial x_k} (x_{k+1} + \phi_k, \hat{\theta}_k^T \varphi_k(\overline{x}_k)) \right] \tag{5.44}$$

$$+ \sum_{k=1}^{i-1} \left[\frac{\partial \alpha_{i-1}}{\partial y_r^{(k-1)}} y^{(k)_r} + \frac{\partial \alpha_{i-1}}{\partial \hat{\theta}_k} \tau_{ki} \right]$$

$$+ \sum_{k=1}^{i-1} \frac{\partial \alpha_{i-1}}{\partial x_k} \varphi_k(\overline{x}_k)^T \Gamma_k \sum_{l=k}^{i-2} \left(\frac{\partial \alpha_l}{\partial \hat{\theta}_k} \right)^T z_{l+1},$$

$$+ \rho_i(\overline{x}_i, \overline{\theta}_1, \hat{\Psi}, \overline{y}_r^{(i-1)}), \quad i = 2, \ldots, n.$$

The parameter adaptive laws are:

$$\dot{\hat{\theta}}_k(t) = \tau_{kn}, \quad 1 \leq k \leq n,$$

$$\dot{\hat{\Psi}} = \Gamma_\Psi \left[\sum_{k=1}^{n} z_k \omega_k - \sigma(\hat{\Psi} - \Psi^0) \right], \quad \Psi^0 \geq 0, \Gamma_\Psi > 0, \sigma > 0 \tag{5.45}$$

with

$$z_i = x_i - \alpha_{i-1} - y_r^{(i-1)}, i = 1, \ldots, n,$$

$$\tau_{11} = \Gamma_1 \left[\varphi_1 \left(x_1 z_1 - \sigma \left(\hat{\theta}_1 - \theta_1^0 \right) \right) \right], \sigma > 0, \Gamma_1 > 0, \tag{5.46}$$

$$\tau_{ki} = \tau_{k(i-1)} - \Gamma_k z_i \frac{\partial \alpha_{i-1}}{\partial x_k} \varphi_k(\overline{x}_k), \Gamma_k > 0, 1 \leq k \leq i-1, i = 2, \ldots, n,$$

$$\tau_{ii} = \Gamma_i \left[\varphi_i(\overline{x}_k) z_i - \sigma \left(\hat{\theta}_i - \theta_i^0 \right) \right], \Gamma_1 > 0, i = 2, \ldots, n.$$

where $\varphi_i, i = 1, \ldots, n$ are the basis functions of the linear approximation for the unknown fault function, that is, $\hat{\xi}_i(\overline{x}_i, \hat{\theta}_i) = (\hat{\theta}_i)^T \varphi_i(\overline{x}_i)$, and $\varphi_i, i = 1, \ldots, n$ are the given bounding control functions.

- For $t \geq T_{isol}$

$$u_I(t) = \frac{a_n + y_r^{(n)}}{\phi_0(x)},$$

$$\dot{\hat{\theta}}_k(t) = \tau_{kn}^l, \quad 1 \leq k \leq n, \tag{5.47}$$

$$\dot{\hat{\Psi}}(t) = \Gamma_\Psi \left[\sum_{k=1}^{n} z_k \omega_k - \sigma(\hat{\Psi} - \Psi^0) \right], \quad \Psi^0 \geq 0, \Gamma_\Psi > 0, \sigma > 0,$$

associated with the update laws in Equation (5.46), except that the basis functions and the bounding control functions are different for u_D, since in this case they are specific to the isolated fault. We have proved that, under the assumption of bounded uncertainties $\eta_i(x, u, t)$ and bounded fault approximation-error $\xi_i(\overline{x}_i) - \hat{\xi}_i(\overline{x}_i, \hat{\theta}_i)$ bounded $\forall i$, all the signals and parameter estimates are uniformly bounded; that is, $z(t)$, $\hat{\theta}(t)$, $\hat{\Psi}(t)$, and $x(t)$ are bounded $\forall t$. However, this approach is based on the special structure of the faulty model of Equation (5.39) and assumes the availability of state vector measurements for the feedback control. Finally, the FDD and FTC presented here are based on the assumption of the fault being part of an *a priori* known set of expected faults' models.

5.3.17   Switched control-based active scheme

We report here on the schemes introduced by Mhaskar *et al.* [24, 55], which consider problems of FDD and FTC for a class of nonlinear systems with input constraints. The models studied are of the form

$$\dot{x} = f(x) + G_{k(t)}(x)(u_{k(t)}(y) + \tilde{u}_{k(t)(t)}),$$

$$y(x) = h(x), u_k(t) \in U_k, u_{k(t)}(y) + \tilde{u}_{k(t)(t)} \in U_k, \tag{5.48}$$

$$k(t) \in k = \{1, \ldots, N\}, N < \infty, U_k = \{u \in \Re^m : |u| \leq u_k^{\max}\}, u_k^{\max} > 0, \forall k,$$

where $x \in \Re^n$ is the vector of state variables, $y \in \Re^m$ is the vector of measurable variables, and $u_k(y) \in \Re^m$ denotes the control vector under the kth configuration. The additive actuator faults are modeled by $\tilde{u}_k$. The vector function f and the matrices $G_k(x)$, $\forall k$ are assumed to be sufficiently smooth on their domains of definition. For each value of $k \in K$, the system is controlled via a different set of manipulated inputs, which defines a given control configuration. The nonlinear model of Equation (5.48) is associated with the following assumption.

Assumption 5.8   *Consider Equation (5.48) in configuration k under state feedback. Then for every input $u_{j,k}$, $j = 1, \ldots, m$, there exists a unique state $x_{i,k}$, $i = 1, \ldots, n$, such that with $x_{i,k}$ as output, the relative degree of $x_{i,k}$ with respect to $u_{j,k}$ and only with respect to $u_{j,k}$ is equal to 1.*

This assumption means that each actuator is the only one influencing at least some state. This implies that the effect of a specific actuator on the system evolution is completely distinguishable, which allows fault isolation in that specific actuator. This sufficient condition for fault detection or isolation can be relaxed if the input enters the model in an "upper-triangular" or "lower-triangular" form ([24], Remark 3). The authors introduced a nonlinear FDD in the following theorem.

Theorem 5.7   *Consider the model of Equation (5.48) in configuration k which satisfies Assumption 5.8 under the control law:*

$$u_k = -\omega_k(x, u_k^{\max})(L_{G_k} V_k(x))^T,$$

$$\omega_k(x, u_k^{\max}) = \begin{cases} \dfrac{\alpha_k(x) + \sqrt{b_k^2(x) + (u_k^{\max} |b_k^T(x)|)^4}}{|b_k^T(x)|^2 \left(1 + \sqrt{1 + \left(u_k^{\max} |b_k^T(x)|^2\right)^2}\right)}, & b_k^T(x) \neq 0, \\ \\ 0, & b_k^T(x) = 0, \end{cases}$$

$$\alpha_k(x) = L_{f_k} V_k(x) + \rho_k V_k(x), \qquad \rho_k > 0,$$

$$b_k(x) = L_{G_k} V_k(x).$$

assuming that the set
$$\Phi_k(u_k^{\max}) = \{x, \text{ such that } L_{f_k} V_k(x) + \rho_k V_k(x) \leq u_k^{\max} |(L_{G_k} V_k(x))^T|\},$$
contains the origin and a neighborhood of the origin.

Let the fault detection and isolation filter for the *j*th manipulated input in the *k*th configuration be described by

$$\dot{\tilde{x}}_{i,k} = f_i(x_1, \ldots, \tilde{x}_{i,k}, \ldots, x_n)$$

$$+ g_{j,k}[i](x_1, \ldots, \tilde{x}_{i,k}, \ldots, x_n) \times u_{j,k}(x_1, \ldots, \tilde{x}_{i,k}, \ldots, x_n)$$

$$e_{i,k} = \tilde{x}_{i,k} - x_i, \tag{5.50}$$

where $g_{j,k}[i]$ denotes the *i*th element of the vector $g_{j,k}$, $\tilde{x}_{i,k}(0) = x_i(0)$ and the subscripts *i* and *k* refer to the *i*th state under the *k*th control configuration. Let $T_{j,k}^f$ be the earliest time for which $\tilde{u}_{j,k} \neq 0$, then the fault detection and isolation filter of Equation (5.50) ensures that $\lim_{t \to T_{j,k}^f} e_{i,k}(t) \neq 0$. Also, $e_{i,k}(t) \neq 0$ only if $\tilde{u}_{j,k} \neq 0, 0 \leq s < t$.
The NFTC is introduced in the following theorem.

Theorem 5.8   *Consider the closed-loop system of Equations (5.48) and (5.49), and let* $x(0) \in \Omega_{k_0}$ *for some* $k_0 \in K$, *with* Ω_k *being defined as:* $\Omega_k(u_k^{\max}) = \{x \in \Re^n : V_k(x) \leq c_k^{\max}\} \subset \Phi_k, c_k^{\max} > 0$ *is a level set of* V_k. *Let* T_{j,k_0} *be the earliest time such that* $e_{i,k_0} \neq 0$ *for some i corresponding to a manipulated input* u_{j,k_0} *in Equation (5.50). Then, the following switching rule:*

$$K(t) = \begin{cases} k_0, & 0 \leq t < T_{j,k_0}, \\ q \neq k_0, & t \geq T_{j,k_0}, \quad x(T_{j,k_0}) \in \Omega_q, \quad u\Omega_{j,k_0} \notin u_q, \end{cases} \tag{5.51}$$

guarantees asymptotic stability of the origin of the closed-loop system of Equations (5.48) and (5.49).

This active NFTC is applicable for the general class of nonlinear models affine in the control and is based on a state feedback (in the same papers, Mhaskar *et al.* proposed an extension to the case of output feedback). However, they require Assumption 5.8 to hold to be able to detect and isolate the actuator fault. Another point is that this scheme does not consider multiplicative actuator faults.

5.3.18 Predictive control-based active scheme

Mhaskar *et al.* [26, 27] studied the problem of NFTC for nonlinear models affine in the control, with input constraints and uncertainties. Actuator faults are treated, under the assumption of controllability of the faulty system. Let us recall below the main result of this work.

The models considered are of the form

$$\dot{x} = f(x) + G_k(x)u_k + W_k(x)\theta_k(t), \quad u_k \in U_k, \theta_k \in \Theta_k,$$

$$k \in \{1, \ldots, N\}, \quad N < \infty, \tag{5.52}$$

where $x \in \Re^n$ denotes the vector of state variables, $u \in U_k \subset \Re^m$, $U_k = \{u \in \Re^m$, such that $|u| \leq u_k^{max}\}$, and $u_k^{max} > 0, \forall k$ denotes the vector of constrained inputs. The vector $\theta_k(t) = [\theta_k^1 \ldots, \theta_k^q]^T \in \Theta_k \subset \Re^q$ denotes the vector of time-varying uncertainties but bounded variables taking values in a nonempty compact convex subset of $\Re^q$. The vector $f(x)$ (such that $f(0) = 0$), the matrices $G_k(x) = [g_k^1(x) \ldots, g_k^m(x)]$, $g_k^i \in \Re^n$, $i = 1, \ldots, m$, and $W(x) = [w_k^1(x) \ldots, w_k^q(x)]$, $w_k^i \in \Re^n$, $i = 1, \ldots, q$, are assumed to be sufficiently smooth in their domain of definition. For each value of the index k, the process is controlled via a different manipulated input, which defines a given control configuration. Switching between the available N control configurations is controlled by a higher-level supervisor, which ensures that only one control configuration is active at any given time, and allows only a finite number of switches over any finite interval of time. The main idea of this work is that, after the occurrence of a fault, the system is associated with one of the N configurations. They build off-line a bank of N nonlinear model-predictive stabilizing controllers and, based on the value of the state vector at the time of fault occurrence, they switch among these controllers to ensure the stability of the faulty system. To make the presentation of the NFTC clear, we follow the structure of Mhaskar *et al.* [26] and present first a Lyapunov-based switched controller, then the associated nonlinear model predictive controller (NMPC) and finally the NFTC based on this algorithm.

5.3.18.1 Lyapunov-based switched controller

The Lyapunov-based controller associated with the system of Equation (5.52) is given by the bounded state feedback:

$$u_k^b = \frac{\alpha_k(x) + \sqrt{\alpha_{1,k}(x)^2 + (u_k^{max}\beta_k(x))^4}}{\beta_k(x)^2\left(1 + \sqrt{1 + (u_k^{max}\beta_k(x))^2}\right)}(L_{G_k V_k})^T, \tag{5.53}$$

where V_k is a robust control Lyapunov function (RCLF) (as defined by Freeman and Kokotović [56], page 49: $\alpha_k(x) = L_f V_k + (\rho_k\|x\| + \chi_k\theta_k^b\|L_{W_k} V_k\|)(\|x + \rho_k\|x\| + \chi_k\theta_k^b\|L_{W_k} V_k\|)$,

$\beta_k(x) = \|L_{G_k} V_k\|$, $L_{G_k} V_k = [L_{g_k^1} V_k \dots L_{g_k^m} V_k]$, $L_{W_k} V_k = [L_{W_k^1} V_k \dots L_{W_k^q} V_k]$, $\theta_k^b > 0$, such that $\|\theta_k(t)\| \leq \theta_k^b$, $\forall t$, and $\rho_k > 0$, $k > 1$, $\phi_k > 0$.

The following convergence result was reported by Mhaskar *et al.* [26] and proven by El-Farra and Christofides [57]:

Let $\Gamma_k(\theta_k^b, u_k^{max}) = \{x \in \Re^n : \alpha_{1,k}(x) \leq u_k^{max}\beta_k(x)\}$ and assume that $\Omega_k = \{x \in \Re^n : V_k(x) \leq c_k^{max}\} \subseteq \Pi_k(\theta_k^b, u_k^{max})$, for some $c_k^{max} > 0$. Then, given any positive real number, d_k^r, such that: $D_k^r = \{x \in \Re^n : \|x\| \leq d_k^r\} \subset \Omega_k$ and $\forall x_0 \in \Omega_k$, $\exists \epsilon_k^{r*} > 0$, such that if $\phi_k/(\chi_k - 1) < \epsilon_k^{r*}$ the solutions of the closed-loop system of Equations (5.52) and (5.53) satisfy $x(t) \in \Omega_k$, $\forall t$ and $\limsup_{t\to\infty} \|x(t)\| \leq d_k^r$.

We also need to recall a convergence result [26], that characterizes the behavior of the solutions of Equations (5.52) and (5.53), when the continuous controller Equation (5.53) is implemented in discrete time. The result is as follows:

Consider the system of Equation (5.52) for a fixed k with $\theta_k = 0$, $\forall t$, associated with the controller of Equation (5.53). Let $u_k(t) = u_k^b(j\Delta_k)$, $j\Delta_k \leq t < (j+1)\Delta_k$, $j = 0, \dots, \infty$. Then, $\forall d_k > 0$, there exists $\Delta_k^* > 0$, $\delta_k' > 0$, $\epsilon_k^* > 0$ such that if $\Delta_k \in (0, \Delta_k^*]$ and $x(0) \in \Omega_k$ then $x(t) \in \Omega_k \forall t$ and $\limsup_{t\to\infty} \|x(t)\| \leq d_k$. Also, if $V_k(x(0)) \leq \delta_k'$ then $V_k \leq \delta_k' \forall \tau \in [0, \Delta_k)$ and if $\delta_k' < V_k(x(0)) \leq c_k^{max}$, then $V_k(x(\tau)) \leq -\epsilon_k^* \forall \tau [0, \Delta_k)$.

5.3.18.2 Nonlinear model predictive controller

Next we report the Lyapunov-based predictive control associated with the Lyapunov-based controller of Equation (5.53). Mhaskar *et al.* reported [26] and proved [58] the following result:

Consider the system of Equation (5.52), for a fixed value of k, with $\theta_k(t) = 0$, $\forall t$, associated with the following NMPC controller:

$$\min\{J(x, t, u_k), u_k \in S_k\},$$

$$J(x, t, u_k) = \int_t^{t+T} \left(\|x^u(s, x, t)\|_{Q_k}^2 + \|u_k\|_{R_k}^2\right)ds, \quad Q_k \geq 0, R_k > 0,$$

$$\text{such that } \dot{x} = f_k(x) + G_k(x)u_k, \tag{5.54}$$

$$\dot{V}_k(x(\tau)) \leq -\epsilon_k \quad \text{if } V_k(x(t)) > \delta_k', \quad \tau[t, t + \Delta k),$$

$$V_k(x(\tau)) \leq \delta_k' \quad \text{if } V_k(x(t)) \leq \delta_k', \quad \tau \in [t, t + \Delta k),$$

where ϵ_k, δ_k' are as defined above, S_k is the family of piecewise continuous functions with period Δ_k mapping $[t, t + T]$ into U_k, $T > 0$ is the horizon of the optimization, and V_k is RCLF that yields a stability region Ω_k, under continuous implementation of the controller Equation (5.53), with a fixed $\rho_k > 0$. Then, $\forall d_k > 0$, $\exists \Delta_k^* > 0$, and $\delta_k' > 0$, such that, if $x(0) \in \Omega_k$ and $\Delta \in \Delta(0, \Delta_k^*]$, then $x(t) \in \Omega_k$, $\forall t$ and $\limsup_{t\to\infty} \|x(t)\| \leq d_k$.

5.3.18.3 Predictive control-based NFTC

Finally, we can report the predictive control-based NFTC:

Consider the system of Equation (5.52), for which the bounded controllers of Equation (5.53) and Lyapunov-based MPCs of Equation (5.54) have been designed and for which the stability regions Ω_j, $j = 1, \ldots, N$ under the MPCs have been explicitly characterized. Let $d_{\max} = \max_{j=1,\ldots,N} d_j$, d_j as defined above, and let $\Omega_U = \cup_{j=1}^{j=N} \Omega_j$. Define $J_j(t) = \int_t^{t+T_j} (\|x^u(s, x, t)\|_{Q_k}^2 + \|u_k^b\|_{R_k}^2) ds$, where $t + T_j \geq t$ is the earliest time at which the state of the closed-loop system under the bounded controller enters the level set defined by $V_j(x) = \delta'_j$. Then, let $k(0) = i$ for some index $i \in \{1, \ldots, N\}$ and $x(0) \in \Omega_i$. Let T_i^f be the earliest time at which a fault occurs. Furthermore, let $f = \{j : \text{such that } j \neq i, x(T_i^f) \in \Omega_j\}$, and let l be such that $J_l = \min_{j \in f} J_j$. Then, the following switching rule

$$k(t) \begin{cases} i, & 0 \leq t < T_i^f, \\ l, & t \geq T_i^f \end{cases} \tag{5.55}$$

guarantees that $x(t) \in \Omega_U, \forall t \geq 0$ and $\limsup_{t \to \infty} \|x(t)\| \leq d_{\max}$.

Remark 5.1 *An extension of the preceding results to the uncertain case $\theta_k(t) \neq 0$, $\forall k$, $\forall t$ can be found in the work of Mhaskar et al. [26, 27]. The active NFTC, based on the computation off-line of a bank of robust nonlinear controllers, is valid for general nonlinear models affine in the control, however, it is based on the availability of a robust control Lyapunov function in closed-form, which is usually not easily accessible [56]. We can also point out that Mhaskar et al. [26, 27] assumed the availability of an FDD bloc and did not consider the problems of fault isolation and estimation delays as well as FDD uncertainties.*

5.4   Integrated Control Reconfiguration Scheme

This section describes the design of fault diagnosis and active fault-tolerant control schemes that can be developed for nonlinear systems. The methodology is based on a fault detection and diagnosis procedure relying on adaptive filters designed via the nonlinear geometric approach, which allows one access to the disturbance de-coupling property. The controller reconfiguration exploits directly the online estimate of the fault signal.

We consider the classical model of an inverted pendulum on a cart as an application example, in order to highlight the complete design procedure, including the mathematical aspects of the nonlinear disturbance de-coupling, as well as the feasibility and efficiency of the proposed approach. Extensive simulations of the benchmark process and Monte Carlo analysis are practical tools for assessing experimentally the robustness and stability properties of the developed fault-tolerant control scheme, in the presence of modeling and measurement errors. The fault-tolerant control method is also compared with an approach relying on sliding-mode control, in order to evaluate the benefits and drawbacks of both techniques. This comparison highlights that the proposed design methodology can constitute a reliable and robust approach for application to real, nonlinear processes.

5.4.1   Introduction

It is well-known that feedback control systems for practical engineering applications strongly rely on actuators, sensors and data acquisition or interface components to enable proper functioning of the physical controlled system. Faulty conditions in those system components may lead to instability or a drastic reduction in performance. On some occasions, it may even

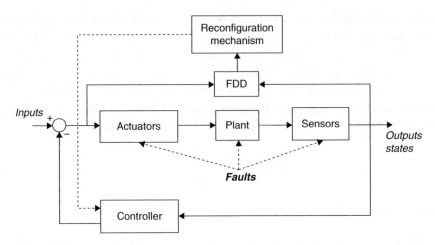

Figure 5.1   An active fault-tolerant control scheme

cause damage to the physical system. There has been a growing demand for reliability, safety and fault tolerance in control system applications. It is therefore necessary to design control systems that are capable of tolerating potential faults in order to improve their reliability and availability, while providing desirable performance. In this book, these types of control system are called "fault-tolerant control systems"; they possess the ability to accommodate component faults automatically. They are capable of maintaining the overall system stability and accept-able performance in the event of such faults. Briefly stated, a closed-loop control system which can tolerate component malfunctions while maintaining desirable performance and stability properties is said to be a fault-tolerant control system. As shown in Figure 5.1, fault-tolerant control system design is based on a fault detection and diagnosis (FDD) scheme. Since fault identification is essential, FDD is mainly used to highlight the requirement of fault estimation.

5.4.2   Basic features

We learned in Chapter 1 that fault-tolerant control methods are classified into two types:

- passive fault-tolerant control schemes (PFTCSs);
- active fault-tolerant control schemes (AFTCSs).

In PFTCSs, controllers are fixed and designed to be robust against a class of presumed faults. This scheme does not require FDD schemes or controller reconfiguration but it has limited fault-tolerant capabilities. On the other hand, AFTCSs react to system component failures actively by reconfiguring control actions so that the stability and acceptable performance of the entire system can be maintained.

Needless to stress, a successful AFTCS design relies heavily on real-time FDD schemes to provide the most up-to-date information about the actual status of the system. Looked at in this light, the ultimate goal in a fault-tolerant control system is to design a controller with a suitable structure for achieving stability and satisfactory performance, not only when all control components are functioning normally, but also in cases when there are faults in sensors, actuators, or other system components. This calls for a good FDD scheme for timely and accurate detection and location of the fault by using, for example, a residual generation approach with dynamic observers or filters.

In the following discussion, we focus on the implementation of an AFTCS approach that integrates a reliable and robust fault diagnosis scheme with the design of a controller reconfiguration system. The approach is based on a nonlinear FDD procedure which provides fault detection and isolation and fault size estimation. It utilizes disturbance-decoupled adaptive nonlinear filters designed by the nonlinear geometric method to provide feedback of the estimated fault signal. The controller reconfiguration exploits a second control loop, relying on the online estimate of the fault signal. This eventually improves the final performances of the overall system.

5.4.3 Nonlinear model of a pendulum on a cart

We consider the dynamic model of a pendulum on a cart, as shown in Figure 5.2.

Assuming that the cart has mass M, the pendulum mass m is concentrated at the tip of a pole of length L, and that there are no friction effects, the dynamic model obtained using Hamilton's principle is

$$(M+m)\ddot{x} + mL\ddot{\theta}cos\theta - mL\dot{\theta}^2 sin\theta = F$$

$$m\ddot{x}cos\theta + mL\ddot{\theta}cos\theta - mg sin\theta = \tau \qquad (5.56)$$

where g is the gravity constant, F is the linear force acting on the cart, and τ the torque acting directly at the base of the pole. In terms of

$$\mathbf{x} = \begin{bmatrix} x_1 & x_2 & x_3 & x_4 \end{bmatrix}^t$$

$$= \begin{bmatrix} x & \dot{x} & \theta & \dot{\theta} \end{bmatrix}^t$$

$$u = F, \ d = \tau \qquad (5.57)$$

where u is the control input and d is the disturbance term, the model of Equation (5.56) can be expressed in the compact form

$$\dot{\mathbf{x}} = N(\mathbf{x}) + M(\mathbf{x})u + D(\mathbf{x})d \qquad (5.58)$$

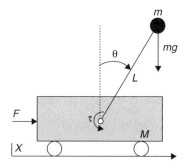

Figure 5.2   Inverted pendulm on a cart

where

$$
N(\mathbf{x}) = \begin{bmatrix} x_2 \\ \frac{mLx_4^2 sinx_3 - mg sinx_3 cosx_3}{M+m sin^2 x_3} \\ x_4 \\ \frac{(M+m)g sinx_3 - mLx_4^2 sinx_3 cosx_3}{(M+m sin^2 x_3)L} \end{bmatrix}
$$

$$
M(\mathbf{x}) = \begin{bmatrix} 0 \\ \frac{1}{M+m sin^2 x_3} \\ 0 \\ \frac{-cosx_3}{(M+m sin^2 x_3)L} \end{bmatrix}
$$

$$
D(\mathbf{x}) = \begin{bmatrix} 0 \\ \frac{cosx_3}{(M+m sin^2 x_3)L} \\ 0 \\ \frac{1}{(M+m sin^2 x_3)L^2} \end{bmatrix} \tag{5.59}
$$

where $N(\mathbf{x})$, $M(\mathbf{x})$ and $D(\mathbf{x})$ are smooth vector fields. The presented FDD scheme belongs to the nonlinear geometric approach (NGA) [44], where a coordinate transformation highlighting a subsystem affected by the fault and decoupled by the disturbances is the starting point for designing a set of adaptive filters. They are able both to detect an additive fault acting on a single actuator and to estimate the magnitude of the fault. It is worth observing that, by means of this NGA, the fault estimate is decoupled from disturbance d. In the faulty situation, the nonlinear model of the system including an additive fault f is considered in the form

$$
\dot{\mathbf{x}} = N(\mathbf{x}) + M(\mathbf{x})(u + f) + D(\mathbf{x})d
$$

$$
\mathbf{y} = H(\mathbf{x}) \tag{5.60}
$$

The design of the strategy for diagnosis of fault f with disturbance decoupling, by means of the NGA approach, is organized as follows:

1. Compute Σ_*^P, the minimal conditioned invariant distribution containing P, where P is the distribution spanned by the columns of $D(\mathbf{x})$.
2. Compute Ω^*, the maximal observability codistribution contained in $(\Sigma_*^P)^\perp$.
3. Consider the fault detectability condition: if $M(\mathbf{x}) \notin (\Omega^*)^\perp$, the fault is detectable and a suitable change of coordinates can be determined.

The invariant distribution Σ_*^P is recursively processed by

$$
S_o = \bar{P} = D(\mathbf{x})
$$

$$
S_{k+1} = \bar{S} + \sum_{j=0}^{m} [M(\mathbf{x})_j, \ \bar{S} \cap \mathbf{ker}\{dH(\mathbf{x})\}] \tag{5.61}
$$

where m is the number of inputs, $\bar{S}(\bar{P})$ represents the involutive closure of $S(P)$, $[M, \Xi]$ is the distribution spanned by all vector fields $[M, \sigma]$, $\sigma \in \Xi$ and $[M, \sigma]$ is the Lie bracket of M, σ.

It has been shown [44] that, if there exists $k \geq 0$ such that $S_{k+1} \longrightarrow S_k$, the recursion stops and $\Sigma_*^P = S_k$.

Next, Ω^* can be determined by the following algorithm:

$$Q_o = (\Sigma_*^P)^\perp \cap \mathbf{span}\{d H(\mathbf{x})\}$$

$$Q_{k+1} = (\Sigma_*^P)^\perp \cap \sum_{j=0}^{m} [L_{M(\mathbf{x})_j} Q_k + \mathbf{span}\{d H(\mathbf{x})\}] \tag{5.62}$$

where $L_g \Lambda$ denotes the codistribution spanned by all covector fields $L_g \lambda$, $\lambda \in \Lambda$ and $L_g \lambda$ is the derivative of λ along g.

In a similar way, there exists an integer k^* such that $Q_{k^*+1} \longrightarrow Q_{k^*}$, in which case Q_{k^*} is indicated as the observability codistribution algorithm of $((\Sigma_*^P)^\perp)$ and represents the maximal observability codistribution contained in $P^\perp \equiv \Omega^*$. Consequently, if $M(\mathbf{x}) \notin (\Omega^*)^\perp$, the disturbance d can be decoupled and the fault f is detectable. Further details can be found in the work of De Persis and Isidori [44].

5.4.4   NGA adaptive filter design

It can be shown that $Q_{k^*} = o.c.a.\,((\sum_*^P)^\perp)$ represents the maximal observability codistribution contained in $P^\perp$, that is, Ω^* [44]. Therefore, with reference to the model in Equation (5.60), when $l(x) \notin (\Omega^*)^\perp$, the disturbance d can be de-coupled and the fault f is detectable.

The examined NGA for the fault diagnosis problem [44] is based on a coordinate change in the state space, $\Phi(x)$, and in the output space, $\Psi(y)$. They consist of a surjection Ψ_1 and a function Φ_1 such that $\Omega^* \cap span\{dh\} = span\{d(\Psi_1)\}$ and $\Omega^* = span\{d\Phi_1\}$, where

$$\begin{cases} \Phi(x) = \begin{pmatrix} \bar{x}_1 \\ \bar{x}_2 \\ \bar{x}_3 \end{pmatrix} = \begin{pmatrix} \Phi_1(x) \\ H_2 h(x) \\ \Phi_3(x) \end{pmatrix}, \\ \\ \Psi(y) = \begin{pmatrix} \bar{y}_1 \\ \bar{y}_2 \end{pmatrix} = \begin{pmatrix} \Psi_1(y) \\ H_2 y \end{pmatrix} \end{cases} \tag{5.63}$$

are (local) diffeomorphisms and H_2 is a selection matrix, that is, its rows are a subset of the rows of the identity matrix. By using the new (local) state and output coordinates $(\bar{x}, \bar{y})$, the system of Equation (5.60) is transformed as follows:

$$\begin{cases} \dot{\bar{x}}_1 = n_1(\bar{x}_1, \bar{x}_2) + g_1(\bar{x}_1, \bar{x}_2)c + l_1(\bar{x}_1, \bar{x}_2, \bar{x}_3)f, \\ \dot{\bar{x}}_2 = n_2(\bar{x}_1, \bar{x}_2, \bar{x}_3) + g_2(\bar{x}_1, \bar{x}_2, \bar{x}_3)c \\ \qquad + l_2(\bar{x}_1, \bar{x}_2, \bar{x}_3)f + p_2(\bar{x}_1, \bar{x}_2, \bar{x}_3)d, \\ \dot{\bar{x}}_3 = n_3(\bar{x}_1, \bar{x}_2, \bar{x}_3) + g_3(\bar{x}_1, \bar{x}_2, \bar{x}_3)c \\ \qquad + l_3(\bar{x}_1, \bar{x}_2, \bar{x}_3)f + p_3(\bar{x}_1, \bar{x}_2, \bar{x}_3)d, \\ \bar{y}_1 = h(\bar{x}_1), \\ \bar{y}_2 = \bar{x}_2, \end{cases} \tag{5.64}$$

with $l_1(\bar{x}_1, \bar{x}_2, \bar{x}_3)$ not being identically zero. In this way [44], we obtain the observable subsystem of Equation (5.64), which, if it exists, is affected by the fault and not by disturbances and the other faults to be de-coupled.

This transformation can be applied to the system of Equation (5.60) if and only if the fault detectability condition is satisfied. In the new reference frame, Equation (5.60) can be decomposed into three subsystems, Equation (5.64), where the first one (the $\bar{x}_1$-subsystem) is always de-coupled from the disturbance vector and affected by the fault as follows:

$$\begin{cases} \dot{\bar{x}}_1 = n_1(\bar{x}_1, \bar{y}_2) + g_1(\bar{x}_1, \bar{y}_2)c + l_1(\bar{x}_1, \bar{y}_2, \bar{x}_3)f, \\ \bar{y}_1 = h(\bar{x}_1), \end{cases} \tag{5.65}$$

where, as the state $\bar{x}_2$ in Equation (5.64) is assumed to be measured, the variable $\bar{x}_2$ in Equation (5.65) is considered an independent input denoted by $\bar{y}_2$.

Further manipulations lead to the following:

$$S_0 = \bar{P} = cl(p_d(x))$$

$$= cl\left(\begin{bmatrix} 0 \\ -\dfrac{\cos x_3}{L(M+m\sin^2 x_3)} \\ 0 \\ \dfrac{1}{L^2(M+\sin^2 x_3)} \end{bmatrix}\right) \equiv p_d(x). \tag{5.66}$$

By recalling that ker $\{dh\} = \emptyset$, it follows that $\sum_{*}^{P} = \bar{P}$ as $\bar{S}_0 \cap ker\{dh\} = \emptyset$. Thus, the algorithm in Equations (5.61)–(5.62) stop with $S_1 = S_0 = \sum_{*}^{P}$.

On the other hand, in order to solve Equation (5.62), it is necessary to compute the expression $(\sum_{*}^{P})^{\perp} = (\bar{P})^{\perp}$. It is worth noting that, for the case under investigation, the determination of the codistribution $(\sum_{*}^{P})^{\perp} = (\bar{P})^{\perp}$ is enhanced due to the sparse structure of Equation (5.66). Moreover, by means of Equation (5.61), the computation of $(\sum_{*}^{P})^{\perp} = (\bar{P})^{\perp}$ leads to a codistribution $\Omega^* = o.c.a.((\sum_{*}^{P})^{\perp})$ spanned by exact differentials.

Finally, any codistribution Ω which is a conditioned invariant contained in $\bar{P}^{\perp}$ spanned by exact differentials, with $\Omega = o.c.a.((\Omega)$ and $l(x) \notin (\Omega)^{\perp}$, can be used to define the coordinate change in Equation (5.63). Therefore, the computation of the maximal observability codistribution is not required.

By observing that

$$(\bar{P})^{\perp} = \left(\begin{bmatrix} 0 \\ -\dfrac{\cos x_3}{L(M+m\sin^2 x_3)} \\ 0 \\ \dfrac{1}{L^2(M+\sin^2 x_3)} \end{bmatrix}\right)^{\perp}$$

$$= \begin{bmatrix} 1 & 0 & 0 & 0 \\ 0 & 0 & 1 & 0 \\ 0 & 0 & 1 & 0 \\ 0 & 1 & -Lx_4\sin x_3 & L\cos x_3 \end{bmatrix} \tag{5.67}$$

and noting that $span\{dh\} = I_4$, from Equation (5.62) it follows that $\Omega = (\sum_*^P)^\perp = (\bar{P})^\perp$ and $(\Omega^*)^\perp = \sum_*^P = \bar{P}$. The fault in the model of Equation (5.60) is detectable if $l(x) \notin (\Omega^*)^\perp = \sum_*^P = \bar{P}$. This condition is fulfilled because of the expression of $l(x)$ in Equation (5.59).

As $dim\{\Omega^*\} = 3$ and $dim\{\Omega^* \cap span\{dh\}\} = 3$, it follows that $\Phi_1(y) : \Re^4 \to \Re^3$. Moreover, as $\Omega^* \cap span\{dh\} = span\{d(\Psi_1 o h)\}$, $H_2 y : \Re^4 \to \Re^1 o$. Thus, as $h(x) = I_4 x$, the surjection $\Psi(y(x))$ is given by

$$\Psi(y(x))$$
$$= \begin{pmatrix} \Psi_1(x) \\ H_2 x \end{pmatrix} = \left(\begin{bmatrix} x_2 + L x_4 \cos x_3 \\ x_1 \\ x_3 \\ [x_4] \end{bmatrix} \right), \tag{5.68}$$

where $H_2 = \begin{bmatrix} 0 & 0 & 0 & 1 \end{bmatrix}$. Note that, since $dh = I_4$, the diffeomorphism $\Phi_1(x)$ such that $\Omega^* = span\{d(\Phi_1)\}$ is given by

$$\Phi_1(x) = \Psi_1(y(x)) = \Psi_1(x). \tag{5.69}$$

Hence, the $\bar{x}_1$-subsystem state variable is

$$\bar{x}_1 = \begin{bmatrix} \bar{x}_{11} \\ \bar{x}_{12} \\ \bar{x}_{13} \end{bmatrix} = \begin{bmatrix} x_2 + L x_4 \cos x_3 \\ x_1 \\ x_3 \end{bmatrix}. \tag{5.70}$$

It is worth observing that only $\bar{x}_{11}$ is affected by the faults, and that the differentials of $x_2 + L x_4 \cos x_3$ span an observability codistribution Ω contained in $P^\perp$ with $\Omega = o.c.a. (\Omega)$. Hence, as previously remarked, in order to estimate the fault, it is possible to use the scalar subsystem defined by the coordinate $\bar{x}_{11} = x_2 + L x_4 \cos x_3$, whose dynamics are defined by

$$\dot{\bar{x}}_{11} = \frac{d(x_2 + L x_4 \cos x_3)}{dt}, \tag{5.71}$$

from which, assuming that the whole state is measured, the NGA adaptive filter (NGA-AF) can be computed.

With reference to Equation (5.65), the NGA-AF can be designed if the condition of De Persis and Isidori [44] and the following new constraints are satisfied:

- The $\bar{x}_1$-subsystem is independent of the $\bar{x}_3$ state components.
- The fault is a step function of the time, and hence the parameter f is a constant to be estimated.
- There exists a proper scalar component $\bar{x}_{1s}$ of the state vector $\bar{x}_1$ such that the corresponding scalar component of the output vector is $\bar{y}_{1s} = \bar{x}_{1s}$ and the following relation holds [59]:

$$\dot{\bar{y}}_{1s}(t) = M_1(t) \cdot f + M_2(t), \tag{5.72}$$

where $M_1(t) \neq 0, \forall t \geq 0$. Moreover, $M_1(t)$ and $M_2(t)$ can be computed for each time instant, since they are functions just of the input and output measurements. Equation (5.72) describes the general form of the system under diagnosis.

Under these conditions, the design of the adaptive filter is achieved, with reference to the system model Equation (5.72), in order to provide a fault estimation $\hat{f}(t)$, which asymptotically converges to the magnitude of the fault f.

The proposed adaptive filter is based on the least squares algorithm with a forgetting factor and it is described by the following adaptation law:

$$\begin{cases} \dot{P} = \beta P - \frac{1}{N^2} P^2 \check{M}_1^2, & P(0) = P_0 > 0, \\ \dot{\hat{f}} = P \epsilon \check{M}_1, & \hat{f}(0) = 0, \end{cases} \tag{5.73}$$

with the following equations representing the output estimation and the corresponding normalized estimation error:

$$\begin{cases} \hat{\bar{y}}_{1s} = \check{M}_1 \hat{f} + \check{M}_2 + \lambda \check{\bar{y}}_{1s}, \\ \epsilon = \frac{1}{N^2} (\bar{y}_{1s} - \hat{\bar{y}}_{1s}), \end{cases} \tag{5.74}$$

where all the involved variables of the adaptive filter are scalar. In particular, $\lambda > 0$ is a parameter related to the bandwidth of the filter, $\beta \geq 0$ is the forgetting factor and $N^2 = 1 + \check{M}_1^2$ is the normalization factor of the least squares algorithm. Moreover, the proposed adaptive filter adopts the signals $\check{M}_1, \check{M}_2, \check{\bar{y}}_{1s}$, which are obtained by means of a low-pass fitering of the signals $M_1, M_2, \bar{y}_{1s}$ as follows:

$$\begin{cases} \dot{\check{M}}_1 = -\lambda \check{M}_1 + M_1, & \check{M}_1(0) = 0, \\ \dot{\check{M}}_2 = -\lambda \check{M}_2 + M_2, & \check{M}_2(0) = 0, \\ \dot{\check{\bar{y}}}_{1s} = -\lambda \check{\bar{y}}_{1s} + \bar{y}_{1s}, & \check{\bar{y}}_{1s}(0) = 0. \end{cases} \tag{5.75}$$

Thus, the adaptive filter is described by the system of Equations (5.73)–(5.75).

It can be proved that the asymptotic relation between the normalized output estimation error $\epsilon(t)$ and the fault estimation error $f - \hat{f}(t)$ is the following:

$$\lim_{t \to \infty} \epsilon(t) = \lim_{t \to \infty} \frac{\check{M}_1(t)}{N^2(t)} \left(f - \hat{f}(t) \right). \tag{5.76}$$

Moreover, it can be proved that the adaptive filter described by the relations Equation (5.73)–(5.75) provides an estimate $\hat{f}(t)$ that asymptotically converges to the magnitude of the step fault f. The proofs are quite standard and left to the reader.

In order to design the NGA-AF scheme, it is possible to design a four-NGA adaptive filter in the form of Equations (5.73)–(5.75), allowing for estimating the magnitude of a step fault acting on the linear force actuator of the inverted pendulum. In order to decouple the effect of the disturbance d from the fault estimator, it is necessary to select from the $\bar{x}_1$-subsystem the following state component

$$\bar{x}_{1s} = \bar{x}_{11} = x_2 + L x_4 \cos x_3. \tag{5.77}$$

Hence, it is possible to describe the specific expression of the fault dynamics, Equation (5.72). The design of the NGA-AF for f is based on

$$
\begin{cases}
\dot{\bar{y}}_{1s} = M_1 \cdot f + M_2, \\[2mm]
M_1 = \dfrac{1 - \cos^2 x_3}{M + m \sin^2 x_3}, \\[4mm]
M_2 = \dfrac{m L x_4^2 \sin x_3 - mg \sin x_3 \cos x_3}{M + m \sin^2 x_3} \\[4mm]
\quad + \dfrac{(M + m)g \sin x_3 \cos^2 x_3}{M + m \sin^2 x_3} \\[4mm]
\quad - \dfrac{m L x_4^2 \sin x_3 \cos^2 x_3}{M + m \sin^2 x_3} \\[4mm]
\quad - x_4^2 \sin x_3 + \dfrac{1 - \cos^2 x_3}{M + m \sin^2 x_3} u.
\end{cases}
\tag{5.78}
$$

It is worth noting that, if the disturbance model Equation (5.60) had been considered, the matrices p_d and $(\bar{P})^\perp$ would have been computed as follows:

$$
p(x) = p_d(x) =
\begin{bmatrix}
1 & 0 \\
0 & 0 \\
0 & 1 \\
0 & 0
\end{bmatrix}
\tag{5.79}
$$

and

$$
(\bar{P})^\perp =
\begin{bmatrix}
1 & 0 \\
0 & 0 \\
0 & 1 \\
0 & 0
\end{bmatrix}^\perp
=
\begin{bmatrix}
0 & 1 & 0 & 0 \\
0 & 0 & 0 & 1
\end{bmatrix}.
\tag{5.80}
$$

It is easy to verify that the condition $l(x) \notin (\Omega^*)^\perp = (\bar{P}) = (\sum_*^P)$ still holds. Thus, when both longitudinal and angular velocity disturbance are present, the design of the NGA-AF is based on the following terms:

$$
\begin{cases}
M_1 = \dfrac{1}{M + m \sin^2 x_3}, \\[4mm]
M_2 = \dfrac{m L x_4^2 \sin x_3 - mg \sin x_3 \cos x_3}{M + m \sin^2 x_3} \\[4mm]
\quad + \dfrac{u}{M + m \sin^2 x_3}.
\end{cases}
\tag{5.81}
$$

However, only the NGA-AF described by the system of Equation (5.78) is considered in the following discussion.

In order to compute the simulation results described in next section, the AFTC scheme has been completed by means of an optimal state-feedback control law, designed on the basis of the linear approximation of the model in Equation (5.58) in a neighborhood of

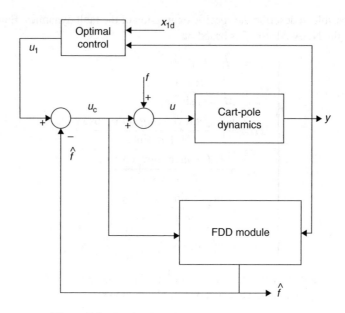

Figure 5.3 Logic of the integrated AFTCS strategy

$X_o = \begin{bmatrix} x_{1d} & 0 & 0 & 0 \end{bmatrix}^T$, in which x_{1d} can be any value. In fact, the linear approximation is independent of x_1, so that the input vector of the optimal controller can be calculated as $\tilde{X} = [(x_1 - x_{1d}) x_2 x_3 x_4]$, and the cart–pole system will be stabilized in the upright position at any linear position reference.

The logic scheme of the integrated adaptive fault-tolerant approach is shown in Figure 5.3, which uses the following nomenclature and symbols:

x_{1d}: desired value of the linear position;

u: actuated input;

u_c: controlled input;

u_l: output signal from the optimal controller;

y: measured outputs;

f: actuator fault;

$\hat{f}$: estimated actuator fault.

The logic scheme depicted in Figure 5.3 shows that the AFTCS strategy is implemented by integrating the FDD module with the existing control system. From the controlled input and output signals, the FDD module provides the correct estimation $\hat{f}$ of the f actuator fault, which is injected to the control loop, in order to compensate the effect of the actuator fault. After this correction, the optimal controller provides the exact tracking of the reference signal, x_{1d}.

5.4.5 Simulation results

The simulation results highlight that the model state variables remain bounded in a set, which ensures control performance, even in the presence of large faults. Moreover, the assumed fault conditions do not modify the system structure, thus guaranteeing the global stability.

To show the diagnostic characteristics brought by the application of the proposed AFTCS and FDD schemes to the scenario of the inverted pendulum on a cart, the nonlinear dynamic model of the mechanical system was implemented in MATLAB®/Simulink.

The following values of the system parameters were assumed: $M = 1$ kg, $m = 0.1$ kg, $L = 0.3$ m, $g = 9.81$ m/s². The optimal controller was designed using the LQR approach in order to minimize the cost function:

$$J = \int_0^{+\infty} (\tilde{X}^T Q \tilde{X} + u R u) dt \tag{5.82}$$

with $Q = 10 I_4$ and $R = 1$.

In order to show the capabilities of the proposed AFTCS strategy, the system was tested, setting as a reference x_{1d} a square-wave with a 0.2 m amplitude and a 50 s period. A random disturbance d modeled as a zero-mean, band-limited noise was applied. It is worth noting that the filter is structurally de-coupled from this disturbance torque, while the measurement noise and the modeling errors may affect the fault estimation.

The following results refer to the simulation of a fault f modeled as a step signal with a size of 0.1 N, commencing at $t = 66$ s. Figure 5.4 shows the estimate of the actuator fault f (solid line), when compared with the simulated actuator fault (dashed line). The fault estimate was achieved by using the logic scheme represented in Figure 5.3 and from the FDD module described earlier.

As will be shown in the performance evaluation, after a suitable choice of the parameters for the filter in Equations (5.73)–(5.75), the FDD module provides quite an accurate estimate of the fault size, with minimal detection delay. Residual errors are caused by the measurement noise and the mismatch between the parameters in the plant and those in the NGA adaptive filters.

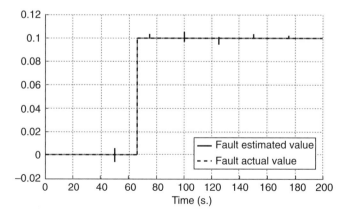

Figure 5.4 Real-time estimate $\hat{f}$ of the actuator fault f

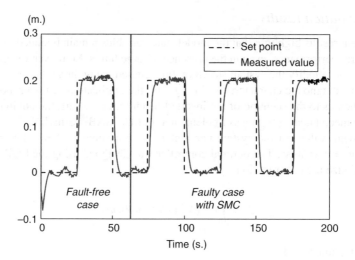

Figure 5.5   Patterns of the linear position x_1 of the cart

Figure 5.5 shows the cart position x_1 compared with its desired value x_{1d}. When the fault is not acting on the system, the position error is quite small and is affected mainly by the disturbance torque d. The fault commences at $t = 66$ s but the fault estimate feedback is applied after $t = 110$ s.

As highlighted in Figure 5.5, during the time interval 66 s $< t <$ 110 s, the steady-state error cannot be eliminated by the optimal controller without the AFTCS. On the other hand, when the proposed AFTCS scheme is switched on, the steady-state error caused by the fault is almost zero.

The simulation results summarized in Figs. 5.4 and 5.5 show the effectiveness of the integrated FDD and AFTCS strategy presented here; it is able to improve the control objective recovery and the reference tracking in the presence of actuator fault. However, the asymptotic fault accommodation, the transient and asymptotic stability of the controlled system, which in this paper are assessed in simulation, require further theoretical studies and investigations.

It is worth observing that the suggested NGA-AF provides not only the fault detection and isolation, but also the fault estimate. The proposed NGA-AF is less sensitive to measurement noise, which allows for a smaller minimal detectable fault.

Note finally that a fault modeled as an additive step function has been considered, since it represents the most common fault situation in connection with the FDD scheme, as reported [60]. Moreover, the FDD module can easily be generalized to estimate, for example, a polynomial function of time or a generic fault signal belonging to a given class of faults if the NGA-AF contains the internal model of the fault itself. Generalization to more general fault functions is beyond the scope of this book; it will be investigated in further works.

A more realistic fault scenario for the inverted pendulum, an intermittent fault, is considered in the following discussion. The results refer to the simulation of a fault f modeled as a sequence of rectangular pulses with variable sizes, commencing at $t = 66$ s. Figure 5.6 shows the estimate of the intermittent actuator fault f (solid line) compared with the simulated actuator fault (dashed line). Also in this case, after a suitable choice of parameters for the NGA-AF, the FDD module provides quite a good estimate of the fault signal.

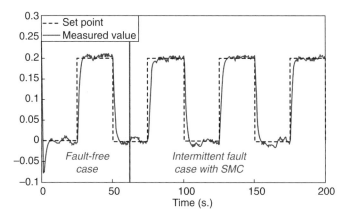

Figure 5.6   Real-time estimate $\hat{f}$ of the intermittent fault f

Under this condition, Figure 5.7 shows the cart position x_1 compared with its desired value x_{1d}. The intermittent fault commences at $t = 66$ s, but the fault estimate feedback is applied after $t = 110$ s. During the time interval 66 s $< t <$ 110 s a steady-state error is present, without applying the AFTCS strategy. However, when the AFTCS scheme is working, the steady-state error due to the intermittent fault is almost eliminated.

The achieved simulation results highlight that the presented FDD and AFTCS integrated strategy is also effective for the case of intermittent faults with variable amplitudes.

5.4.6   Performance evaluation

In this section, further experimental results are reported. They refer to the performance evaluation of the developed AFTCS scheme with respect to modeling errors and measurement uncertainty. In particular, the simulation of different fault-free and faulty data sequences was

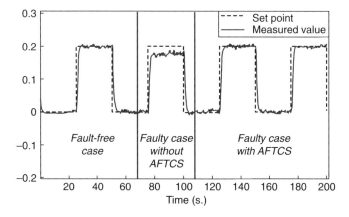

Figure 5.7   Linear position x_1 of the cart with an intermittent fault, with and without the AFTCS

Table 5.1   Simulated cart–pole parameter uncertainties

Variable	Nominal value	Min. error	Max. error
M	1 kg	$\pm\,0.1\,\%$	$\pm\,50\,\%$
m	0.1 kg	$\pm\,0.1\,\%$	$\pm\,50\,\%$
u	u_0	$\pm\,0.1\,\%$	$\pm\,25\,\%$
y	y_0	$\pm\,0.1\,\%$	$\pm\,25\,\%$
f	0	0.1 N	1 N

performed by exploiting the cart–pole benchmark simulator and a MATLAB® Monte Carlo analysis. The Monte Carlo tool is useful at this stage as the AFTCS strategy performance depends on the residual error magnitude from the model approximation as well as on the input–output measurement errors.

In particular, the nonlinear cart–pole simulator developed in Simulink is able to vary the statistical properties of the signals used for modeling possible process parameter uncertainty and measurement errors. Monte Carlo analysis represents a viable method for analyzing some properties of the developed AFTCS scheme when applied to the process under consideration. Under this assumption, Table 5.1 reports the nominal values of the examined cart–pole model parameters with their uncertainty.

Monte Carlo analysis was performed by describing the variables as Gaussian stochastic processes, with zero-mean and standard deviations corresponding to the minimal and maximal error values in Table 5.1. Moreover, it is assumed that the input and the output signals, u and y, are affected by measurement errors, expressed as percentage standard deviations of the corresponding nominal values u_o and y_o.

To evaluate the performance and robustness of the AFTCS scheme, the following indices were used on 500 Monte Carlo runs:

- **Fault reconstruction relative error, ϵ_{mf}:** the mean value of the relative error between the size of the examined fault f and the size of the estimated fault $\hat{f}$;
- **Reference tracking relative error, ϵ_{mr}:** the mean value of the relative error between the reference signal x_{1d} and the controlled output x_1;
- **Settling time, T_{ms}:** the mean value of the settling time of the controlled output x_1 with respect to the set-point signal x_{1d};
- **Fault detection delay, τ_{mf}:** the mean value of the time delay required for detecting the considered fault signal;
- **Computation time, τ_c:** the mean value of the time required to perform the computational process;
- **Control signal energy, E_{mu}:** the mean value of energy of the control input signal u.

These criteria were computed for several possible combinations of the parameter values in Table 5.1. Table 5.2 summarizes the results obtained by considering the described FDD module and the FTC strategy. In particular, it shows the values of the performance indices according to the best, worst and average cases, with reference to the possible combinations of parameters from Table 5.1.

Table 5.2 Monte Carlo analysis of the AFTCS scheme

Index	Best case	Average case	Worst case
ϵ_{mf}	3 %	9 %	18 %
ϵ_{mr}	13 %	14 %	15 %
T_{ms}	18.04 s	18.20 s	18.60 s
τ_{mf}	0.65 s	1.20 s	2.16 s
E_{mu}	0.05	0.06	0.07
τ_c	0.015 s	0.017 s	0.019 s

Table 5.2 shows that, with the proper design of the FDD logic in connection with the FTC scheme, it is possible to achieve reference tracking errors less than 15 %, settling times smaller than 19 s, with minimal control input energy of 0.07, small detection delay, and computation time less than 0.017 s. The simulations showed that these values are almost independent for the actuator fault size, which varies from 0.1 N to 1.0 N.

The results demonstrate that Monte Carlo simulation is an effective tool for experimentally testing the design robustness, stability and reliability of the proposed AFTCS method with respect to modeling uncertainty. The simulation technique facilitates an assessment of the performances of the developed FDD and FTC strategies.

Finally, it is worth noting that, when the case of step fault signals is considered, perfect knowledge of the system parameters allows the engineer to settle the optimal threshold logic and therefore provide an analytical estimate of the performance indices (detection time, false alarm rate etc.). On the other hand, when the parameters of the nonlinear model are uncertain, the performance indices are represented by stochastic variables whose probability distributions can be evaluated only by means of approximations, due to the nonlinearity of the model itself. However, these approximations are not required if the Monte Carlo tool is exploited, since nonlinear maps of stochastic variables do not have to be managed in an analytical way.

5.4.7   Comparative studies

This section provides some comparative results with respect to other FDD and FTC schemes emphasizing the advantages and drawbacks of the AFTCS method with respect to the sliding-mode control (SMC) approach.

The SMC can be designed on the basis of a linear or a nonlinear model. In both cases, the design procedure is based on the selection of an appropriate switching manifold, and then on the determination of a control law, including a discontinuous term, which ensures the sliding motion in this manifold. However, SMC design for the nonlinear case is generally applied to systems in the "regular" form, which consists of two blocks: one depending on the control, with the same dimension as the control vector, and the other independent. Such a regular form may be obtained by means of a nonlinear coordinate transformation. On the other hand, if a linear model is used, the transformation into the regular form and the design of the sliding-mode dynamics are simpler, since known results from linear control techniques (i.e., pole placement, eigenstructure assignment, and optimal quadratic) are applicable.

Even if the cart–pole benchmark has been studied for SMC design for both its nonlinear model and its linear approximation ([61], Par. 4.2 and 5.4), the proposed AFTCS strategy is compared here with an SMC design based on Ackermann's formula ([61], Par. 5.4). This approach allows the determining of a state-space discontinuity plane equation in an explicit form, without transforming the original system into a regular form, such that the sliding motions on that plane are governed by linear dynamics with desired eigenvalue placement, independent of disturbances.

This design procedure is therefore inherently tolerant of faults, since the disturbance torque and actuator faults are both de-coupled from the sliding motion. The sliding surface based on Ackermann's formula for a linear system $\dot{x} = Ax + bu$ (with $A \in \Re^{n \times n}$) is designed as follows:

$$
\begin{aligned}
s &= c^T x = 0, \\
c^T &= e^T P(A), \\
e^T &= [0 \ldots, 0\,1]\begin{bmatrix} b & Ab \ldots, A^{n-1}b \end{bmatrix}^{-1}, \\
P(A) &= (A - \lambda_1 I)(A - \lambda_2 I) \ldots, (A - \lambda_{n-1} I),
\end{aligned}
\tag{5.83}
$$

where $\lambda_1, \lambda_2, \ldots, \lambda_{n-1}$ are the desired eigenvalues of the sliding mode. The control law defined as

$$
u = -M sign(s) \tag{5.84}
$$

enforces a sliding motion in the plane $s = 0$ if

$$
M > |c^T Ax| + f_M, \tag{5.85}
$$

where f_M is an upper bound on an additive disturbance on the control input. In the case of the linearized model of the cart–pole system, such disturbance plays the role of both the disturbance torque and the actuator fault.

In order to provide a brief but clear insight into the aforementioned FDD techniques, the comparison has been performed under the same working conditions and based on the indices suggested earlier. It is worth noting that the FTC scheme implemented via the SMC does not exploit the FDD module for fault estimation shown in Figure 5.3. Ss previously remarked, the SMC strategy is inherently fault-tolerant, as disturbance and fault are de-coupled via the sliding motion. Moreover, the discontinuous control action generated from the SMC, and used by the FDD module, would inevitably worsen the fault reconstruction.

Figure 5.8 shows the linear position x_1 of the cart in the fault-free and faulty cases, when the SMC is used for a step fault situation. Figure 5.9 depicts x_1 when the SMC is exploited in connection with the intermittent fault case. Figures 5.5, 5.8, and 5.9 highlight the fact that FTC schemes with different controllers are approximately equally able to accommodate the actuator fault cases considered.

However, a comparison of Tables 5.2 and 5.3 shows that the FTC scheme using the SMC achieves better performance in terms of tracking errors ϵ_{mr} and settling times T_{ms}. However,

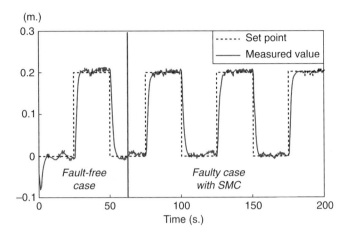

Figure 5.8 Linear position x_1 of the cart in the fault-free and faulty cases, with the SMC and for the step fault case

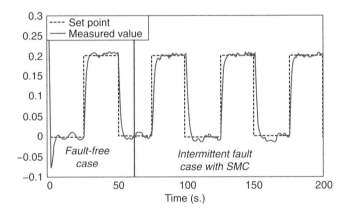

Figure 5.9 Linear position x_1 with the SMC and an intermittent fault situation

Table 5.3 Monte Carlo analysis with the sliding mode controller

Indices	Best case	Average case	Worst case
ϵ_{mr}	9 %	10 %	11 %
T_{ms}	20 s	21 s	22 s
E_{mu}	20	20.5	20.8
τ_c	0.082 s	0.084 s	0.087 s

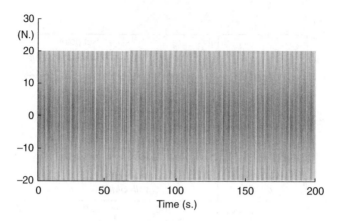

Figure 5.10   Control signal activity of the SMC

as reported in Table 5.3 and depicted in Figure 5.10, the control input energy E_{mu} required by the SMC is much greater than in the case of the suggested AFTCS (see Table 5.2). Moreover, the SMC increases the computational time τ_c considerably.

Figure 5.10 shows the control signal activity generated by the SMC and Figure 5.11 shows the control signal activity from the LQR.

Remark 5.2   *A few comments can finally be drawn here. When the modeling of the dynamic system can be perfectly obtained, the suggested AFTCS achieves good performance. This scheme showed also interesting robustness properties in the presence of unmodeled disturbance, modeling mismatch and measurement errors. However, with a scheme relying, e.g., on an SMC, the tracking relative error can fall below the value for the suggested AFTCS scheme. The SMC strategy takes advantage of its intrinsic robustness capabilities, even if it requires an increased control effort and an average computation time about 3.59 times bigger than the*

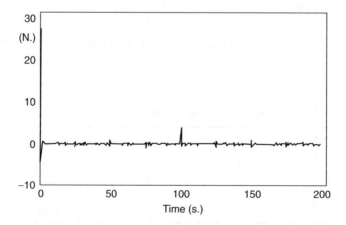

Figure 5.11   Control signal activity of the LQR

one required by the suggested AFTCS scheme. However, it represents the time required for computing both the controller and the process simulations. It is worth noting also that the FDD unit was not modified while changing the control strategy. Moreover, one of the advantages of the suggested AFTCS scheme consists in improving the fault-tolerance characteristics of those controllers that are not intrinsically fault tolerant. However, for control schemes that are already fault tolerant, the proposed AFTCS method enhances the behaviour of the complete system in transient conditions.

5.5 Notes

Many of the early published work on FTC concentrated initially on linear FTC. More and more research has started focussing on nonlinear FTC problems. Indeed, the latter is more challenging because of the difficulties intrinsic to nonlinear systems. Some encouraging results have been obtained. In this chapter, effort has focused on summarizing recently obtained results on NFTC with particular attention to the degree of the "model-nonlinearities". Although several interesting results have been obtained so far, it is generally considered that work treating the problems of both nonlinear FDD and nonlinear FTC together in an effective applicable methods is still required. Real-life applications of NFTC theories are also missing from recent work. In particular, the case of infinite-dimension, nonlinear models, that is, nonlinear partial derivative equations-based models, has yet to be studied; recent results in this direction are presented by [62]–[65].

The chapter also discussed the development of an active fault-tolerant control scheme that integrates a robust fault diagnosis method with the design of a controller reconfiguration system. The methodology was based on a fault detection and diagnosis procedure relying on disturbance de-coupled adaptive filters designed via the nonlinear geometric approach. With reference to the achieved performance of the overall fault-tolerant control scheme, the advantages and drawbacks of the complete design scheme applied to a nonlinear inverted pendulum were discussed and compared with reference to widely used control strategies. The stability and robustness of the developed fault-tolerant control scheme, together with the evaluation of the achievable performance, were estimated in simulation methods including the Monte Carlo tool.

References

[1] Isermann, R. (2006) *Fault-diagnosis Systems: An Introduction from Fault Detection to Fault Tolerance*, Berlin: Springer.

[2] Frank, P. M., and Ding, X. (1997) "Survey of robust residual generation and evaluation methods in observer-based fault detection systems", *J. Process Control*, 7(6):403–424.

[3] Chen, J., and Patton, R. J. (1999) *Robust Model-Based Fault Diagnosis for Dynamic Systems*, Boston, Mass, USA: Kluwer Academic Publishers.

[4] Garcia, E. A., and Frank, P. M. (1997) "Deterministic nonlinear observer-based approaches to fault diagnosis", *Control Engineering Practice*, 5:663–670.

[5] Witczak, M. (2007) *Modelling and Estimation Strategies for Fault Diagnosis of Nonlinear Systems: From Analytical to Soft Computing Approaches*, Lecture notes in Control and Information Sciences, **354**, Berlin, Germany: Springer.

[6] Nguang, S. K., Zhang, P., and Ding, S. X. (2007) "Parity relation based fault estimation for nonlinear systems: An LMI approach", *Int. J. Automation & Computing*, 4:164–168.

[7] Halder, B., and Sarkar, N. (2007) "Robust fault detection of robotic manipulator", *Int. J. Robotic Research*, **26**:273–285.

[8] Halder, B., and Sarkar, N. (2007) "Robust nonlinear analytic redundancy for fault detection and isolation in mobile robot", *Int. J. Automation & Computing*, **2**:177–182.

[9] Talebi, H. A., Abdollahi, F., Patel, R. V., and Khorasani, K. (2009) *Neural Network-Based State Estimation of Nonlinear Systems: Application to Fault Detection and Isolation*, Berlin: Springer.

[10] Talebi, H. A., and Khorasani, K. (2007) "An intelligent sensor and actuator fault detection and isolation scheme for nonlinear systems", in *Proc. 46th IEEE Conf. Decis. Contr. (CDC)*, New Orleans, LA, USA, 2620–2625.

[11] Frank, P. M. 1996 "Analytical and qualitative model-based fault diagnosis: a survey and some new results", *European J. Control*, **2**(1):6–28.

[12] El-Ghatwary, M. E. M. (2007) *Robust Fuzzy Observer-Based Fault Detection for Nonlinear Systems*, PhD thesis, University of Duisburg-Essen, Germany.

[13] Benosman, M. (2010) "A survey of some recent results on nonlinear fault tolerant control", *Math Problems in Engineering*, Article ID 586169, 1–25.

[14] Zhang, Y., and Jiang, J. (2006) "Issues on integration of fault diagnosis and reconfigurable control in active fault-tolerant control systems", in *Proc. 6th IFAC Symposium of Fault Detection Supervision and Safety for Technical Processes*, Beijing, China, 1513–1524.

[15] Staroswiecki, M., Yang, H., and Jiang, B. (2006) "Progressive accommodation of aircraft actuator faults", in *Proc. 6th IFAC Symposium of Fault Detection Supervision and Safety for Technical Processes*, Beijing, China, 877–882.

[16] Zhang, X., Parisini, T., and Polycarpou, M. M. (2004) "Adaptive fault-tolerant control of nonlinear uncertain systems: an information-based diagnostic approach", *IEEE Trans. Automatic Control*, **49**(8):1259–1274.

[17] Ingimundarson, A., and Sanchez-Pena, R. (2008) "Using the unfalsified control concept to achieve fault tolerance", in *Proc. 17th IFAC World Congress*, Seoul, Korea, 1236–1242.

[18] Bonivento, C., Gentili, L., and Paoli, A. (2004) "Internal model based fault tolerant control of a robot manipulator", in *Proc. 43rd IEEE Conference on Decision and Control*, **5**:5260–5265.

[19] Bonivento, C., Isidori, A., Marconi, L., and Paoli, A. (2004) "Implicit fault-tolerant control: application to induction motors", *Automatica*, **40**(3):355–371.

[20] Benosman, M., and Lum, K. Y. (2008) "Passive actuators' fault-tolerant control for affine nonlinear systems", in *Proc. 17th IFAC World Congress*, Seoul, Korea, 14229–14234.

[21] Tao, G., Chen, S., and Joshi, S. M. (2002) "An adaptive actuator failure compensation controller using output feedback", *IEEE Trans. Automatic Control*, **47**(3):506–511.

[22] Tao, G., Chen, S., and Joshi, S. M. (2002) "An adaptive actuator failure compensation controller using output feedback", *IEEE Trans. on Automatic Control*, **47**(3):506–511.

[23] Wu, N. E., Zhang, Y., and Zhou, K. (2000) "Detection, estimation, and accommodation of loss of control effectiveness", *Int. J. Adaptive Control and Signal Processing*, **14**(7):775–795.

[24] Mhaskar, P., McFall, C., Gani, A., Christofides, P. D., and Davis, J. F. (2008) "Isolation and handling of actuator faults in nonlinear systems", *Automatica*, **44**(1):53–62.

[25] Richter, J., and Lunze, J. (2008) "Reconfigurable control of Hammerstein systems after actuator faults", in *Proc. 17th IFAC World Congress*, Seoul, Korea, 3210–3215.

[26] Mhaskar, P., Gani, A., and Christofides, P. D. (2006) "Fault-tolerant control of nonlinear processes: performance-based reconfiguration and robustness", *Int. J. Robust and Nonlinear Control*, **16**(3):91–111.

[27] Mhaskar, P. (2006) "Robust model predictive control design for fault-tolerant control of process systems", *Industrial and Engineering Chemistry Research*, **45**(25):8565–8574.

[28] Zhang, X., Polycarpou, M. M., and Parisini, T. (2001) "Integrated design of fault diagnosis and accommodation schemes for a class of nonlinear systems", in *Proc. 40th IEEE Conference on Decision and Control*, Orlando, FL, USA, **2**:1448–1453.

[29] Zeitz, M. (1987) "The extended Luenberger observer for nonlinear systems", *Systems and Control Letters*, **9**:149–156.

[30] Adjallah, K., Maquin, D., and Ragot, J. (1994) "Nonlinear observer-based fault detection", in *Proc. 3rd IEEE Conference of Control Applications*, Glasgow, UK, 1115–1120.

[31] Hengy, D., and Frank, P. M. (1986) "Component failure detection via nonlinear observers", in *Proc. IFAC Workshop on Fault Detection and Safety of Chemical Plants*, Kyoto, Japan, 153–157.

[32] Frank, P. M. (1986) "Enhancement of robustness in observer-based fault detection", *Int. J. Control*, **59**:955–981.

[33] Frank, P. M., Schreier, G., and Garcia, E. A. (1999) "New direction in nonlinear observer design", in *Nonlinear Observers for Fault Detection and Isolation*, 401–422, Berlin: Springer-Verlag.

[34] Seliger, R., and Frank, P. M. (2000) "Issues of fault diagnosis for dynamic systems", in *Robust Observer-based Fault Diagnosis in Nonlinear Uncertain Systems*, 145–185, London: Springer-Verlag.

[35] Seliger, R., and Frank, P. M. (1991) "Robust component fault detection and isolation in nonlinear dynamic systems", in *Proc. IFAC Symposium on Fault Detection, Supervision and Safety of Technical Process*, Baden-Baden, Germany, 313–318.

[36] Frank, P. M. (1994) "On-line fault detection in uncertain nonlinear systems using diagnostic observers: a survey", *Int. J. Systems Science*, **25**(12):2129–2154.

[37] Gauthier, J. P., Hammouri, H., and Othman, S. (1992) "A simple observer for nonlinear systems: application to bioreactors", *IEEE Trans. Automatic Control*, **37**(6):875–880.

[38] Guerra, R. M., Garrido, R., and Miron, A. O. (2005) "The fault detection problem in nonlinear systems using residual generators", *IMA J. Math. Control and Information*, **22**:119–136.

[39] Hammouri, H., Kinnaert, M., and El Yaagoubi, E. H. (1999) "Observer-based approach to fault detection and isolation for nonlinear systems", *IEEE Trans. Automatic Control*, **44**(10):1879–1884.

[40] Besancon, G. (2003) "High-gain observation with disturbance attenuation and application to fault detection and isolation for nonlinear systems", *Automatica*, **39**:1095–1102.

[41] Yan, X. G., and Edwards, C. (2007) "Sensor fault detection and isolation for nonlinear systems based on sliding mode observer", *Int. J. Adaptive Control and Signal Processing*, **21**(8–9):657–673.

[42] Yan, X. G., and Edwards, C. (2007) "Nonlinear robust fault reconstruction and estimation using a sliding mode observer", *Automatica*, **43**(9):1605–1614.

[43] Yan, X. G., and Edwards, C. (2008) "Robust sliding mode observer-based actuator fault detection and isolation for a class for nonlinear systems", *Int. J. Systems Science*, **39**(4):1605–1614.

[44] De Persis, C., and Isidori, A. 2001 "A geometric approach to nonlinear fault detection and isolation", *IEEE Trans. Automatic Control*, **45**(6):853–865.

[45] de Souza, C. E., Shaked, U., and Fu, M. (1995) "Robust $\mathcal{H}_\infty$ filtering for continuous-time varying uncertain systems with deterministic input signals", *IEEE Trans. Automatic Control*, **43**:709–719.

[46] Limebeer D. J. N, Anderson, B. D. O., and Handel, B. (1994) "A Nash game approach to mixed $\mathcal{H}_2/\mathcal{H}_\infty$ control", *IEEE Trans. Automatic Control*, **39**(1):69–82.

[47] Theodor, Y., and Shaked, U. (1996) "A dynamic game approach to mixed $\mathcal{H}_2/\mathcal{H}_\infty$ estimation", *Int. J. Robust Nonlinear Control*, **6**:331–345.

[48] Aliyu, M. D. S., and Boukas, E. K. (2009) "Discrete-time mixed $\mathcal{H}_2/\mathcal{H}_\infty$ nonlinear filtering", *Int. J. Robust Nonlinear Control*, **19**(4):394-417, 2009.

[49] Chung, W. H., and Speyer, J. L. (1998) "A game theoretic fault detection filter", *IEEE Trans. Automatic Control*, **43**:143–161.

[50] De Persis, C., and Isidori, A. (2002) "On the design of fault detection filter with game theoretic-optimal sensitivity", *Int. J. Robust Nonlinear Control*, **12**:729–747.

[51] Vidyasagar, M. (1993) *Nonlinear Systems Analysis*, 2nd edition, NJ, USA: Prentice-Hall.

[52] Benosman, M., and Lum, K. Y. (2009) "Online references reshaping and control reallocation for nonlinear fault tolerant control", *IEEE Trans. Control Systems Technology*, **17**(2):366–379.

[53] Isidori, A. (1989) *Nonlinear Control Systems*, 2nd edition, Communications and Control Engineering Series, Berlin, Germany: Springer.

[54] Polycarpou, M. M. (2001) "Fault accommodation of a class of multivariable nonlinear dynamical systems using a learning approach", *IEEE Trans. Automatic Control*, **46**(5):736–742.

[55] Mhaskar, P., Gani, A., El-Farra, N. H., McFall, C., et al. (2006) "Integrated fault detection and fault-tolerant control of process systems", *AIChE Journal*, **52**(6):2129–2148.

[56] Freeman, R. A., and Kokotović, P. V. (1996) *Robust Nonlinear Control Design: State-Space and Lyapunov Techniques*, Boston, Mass, USA: Birkhäuser.

[57] El-Farra, N. H., and Christofides, P. D. (2003) "Coordinating feedback and switching for control of hybrid nonlinear processes", *AIChE Journal*, **49**(8):2079–2098.

[58] Mhaskar, P., El-Farra, N. H., and Christofides, P. D. (2005) "Predictive control of switched nonlinear systems with scheduled mode transitions", *IEEE Trans. Automatic Control*, **50**(11):1670–1680.

[59] Bonfe, M., Castaldi, P., Geri, W., and Simani, S. (2007) "Nonlinear actuator fault detection and isolation for a general aviation aircraft", *Space Technology: Space Engineering, Telecommunication, Systems Engineering and Control*, **27**(23):107–113.

[60] Edwards, C., Lombaerts, T., and Smail, H., (Eds) (2010) *Fault Tolerant Flight Control: A Benchmark Challenge*, 1st Edition, Lecture Notes in Control and Information Sciences, 399, UK: Springer.

[61] Utkin, V. I., Guldner, J., and Shi, J. (1999) *Sliding Mode Control in Electromechanical Systems*, 1st Edition, Series in Systems & Control Engineering, London: Taylor & Francis.

[62] Armaou, A., and Demetriou, M. A. (2008) "Robust detection and accommodation of incipient component and actuator faults in nonlinear distributed processes", *AIChE Journal*, **54**(10):2651–2662.

[63] El-Farra, N. H., Demetriou, M. A., and Christofides, P. D. (2008) "Actuator and controller scheduling in nonlinear transport-reaction processes", *Chemical Engineering Science*, **63**(13):3537–3550.

[64] El-Farra, N. H. (2006) "Integrated fault detection and fault-tolerant control architectures for distributed processes", *Industrial and Engineering Chemistry Research*, **45**(25):8338–8351.

[65] El-Farra, N. H., and Ghantasala, S. (2007) "Actuator fault isolation and reconfiguration in transport-reaction processes", *AIChE Journal*, **53**(6):1518–1537.

6

Robust Fault Estimation

This chapter will focus on a framework of robust fault estimation observer design for continuous-time and discrete-time systems. Increased productivity requirements and stringent performance specification of engineering systems have motivated intensive investigations into the design of fault detection and isolation and fault-tolerant control algorithms. Observer-based fault detection and isolation techniques have received considerable attention and have been successfully applied to practical systems. Residual generation approaches using observers, where the difference between the system output and observer output is processed to form residuals, have been widely used.

6.1 Introduction

Fault detection and isolation (FDI) is used to monitor the system and to determine the location of the fault. After that, a fault estimation (FE) module is activated to determine online the magnitude of the fault, which may be used for fault-tolerant control. As for the issue of FE for continuous-time systems, fruitful results have been obtained during the past two decades. Common FE methods use adaptive observer [1]–[3] or sliding-mode observer [4]–[7], and generally apply to minimum-phase systems, which is a restrictive condition. Shafai *et al.* [8] and Koenig and Mammar [9] used a proportional–integral observer under the constant fault assumption, but FE performance was not considered, and no general approach was given.

On the other hand, it is well known that discrete-time observers are more practical and challenging than continuous-time cases because most continuous-time control systems are implemented digitally. Compared with continuous-time systems, only a few results have been reported for FE in discrete-time systems; the topic of fault detection has attracted considerable attention [10] but FE was not included in these works. For a class of uncertain linear systems, Saif [11] and Jiang and Chowdhury [12] studied an FE method based on a special coordinate transformation but the online fault estimate at time k needs the output vector at time $k + 1$. It is seen that the method does not satisfy the causality constraint. Gao, Breikin, and Wang [13]

Analysis and Synthesis of Fault-Tolerant Control Systems, First Edition. Magdi S. Mahmoud and Yuanqing Xia.
© 2014 John Wiley & Sons, Ltd. Published 2014 by John Wiley & Sons, Ltd.

constructed an augmented system, in which the designed discrete-time observer can realize unknown input estimation. However, it was assumed that the fault difference between two successive fault values is sufficiently small to be neglected, that is, only the constant fault was considered. This is a severe restriction.

In the following discussion, the objective is to analyze and develop a general framework of the observer-based robust FE scheme for both continuous-time and discrete-time systems. To achieve robust FE, we propose a multiconstrained FE observer (MFEO), containing an H_∞ performance and a regional pole placement. We also introduce existence conditions for both continuous-time and discrete-time systems; these conditions are less restrictive than those of adaptive and sliding-mode observers because our approach can deal with nonminimum-phase systems. The conventional multiconstrained design method always generates unavoidable conservatism. By extending the slack variable technique [14, 15] to the MFEO design, we obtain less conservative results.

6.2 System Description

Consider the following system

$$\delta[x(t)] = Ax(t) + B_u(t) + B_f f(t) + B_d d(t), \tag{6.1}$$

$$y(t) = Cx(t) + D_u(t) + D_f f(t) + D_d d(t), \tag{6.2}$$

where $\delta[\cdot]$ denotes the derivative operator $\dot{x}(t)$ for continuous-time systems and the shift operator $x(t+1)$ for discrete-time cases. $x(t) \in \Re^n$ is the state, $u(t) \in \Re^m$ is the input, $y(t) \in \Re^p$ is the output, and $f(t) \in \Re^{n_f}$ represents possible actuator and sensor faults. $d(t) \in \Re^{n_d}$ is disturbance vector that belongs to $L_2[0, \infty)$ for continuous-time systems and $l_2[0, \infty)$ for discrete-time cases. A, B, B_f, B_d, C, D, D_f, and D_d are constant real matrices of appropriate dimensions. The pair (A, C) is observable. Without loss of generality, it is supposed that the fault distribution matrices B_f and D_f are of full-column rank.

Remark 6.1 *From the system description of Equations (6.1) and (6.2), we can see that a general system is considered in this chapter, including possible actuator and sensor faults. The purpose of this chapter is to establish a new unified FE framework for both continuous-time and discrete-time systems. The issue of fault detection has received considerable attention over the past two decades, but there are only a few results on FE. The topic of FE is more interesting and challenging and has not been fully solved yet.*

Remark 6.2 *In this section, only linear time-invariant systems are considered. As we know, most practical systems are nonlinear in nature, but in practical situations, they are usually linearized to design controllers and observers, such as flight control systems. The FE problem of linear systems has not yet been fully solved, so research on linear systems is still very meaningful and valuable. Moreover, note that the MFEO design can be extended to linear time-delay systems, Lipschitz nonlinear systems, TS fuzzy model-based nonlinear systems, and so on.*

Before ending this section, we recall the following lemma, which will be used to prove the main results:

Lemma 6.1 *[16] For a given matrix $A \in \Re^{n \times n}$, the eigenvalues of A belong to the circular region $D(\alpha, r)$ with center $\alpha + j0$ and radius r if and only if there exists a symmetric positive definite matrix $P \in \Re^{n \times n}$ such that*

$$\begin{bmatrix} -P & PA_r \\ \bullet & -P \end{bmatrix} < 0, \tag{6.3}$$

where $A_r = (A - \alpha I_n)/r$. Here and in the rest of this chapter, $*$ denotes the symmetric elements in a symmetric matrix.

Remark 6.3 *Lemma 6.1 is used to place the eigenvalues of matrix A into a desired circular region to improve the system's transient performance. For the sake of calculation convenience, Equation (6.3) can be changed into the following equivalent form:*

$$\begin{bmatrix} -P & P(A - \alpha I_n) \\ \bullet & -\Re^2 P \end{bmatrix} < 0. \tag{6.4}$$

6.3 Multiconstrained Fault Estimation

In this section, we achieve robust fault estimation with a multi-constrained fault estimation observer under the H_∞ performance specification with the regional pole constraint. The existence conditions for both continuous-time and discrete-time systems are then derived explicitly. By introducing slack variables, improved results on the multiconstrained fault estimation observer design are established such that different Lyapunov functions can be separately designed for each constraint.

6.3.1 Observer design

To estimate the size of the fault vector, we construct the FE observer

$$\delta[\hat{x}(t)] = A\hat{x}(t) + B_u(t) + B_f \hat{f}(t) - L(\hat{y}(t) - y(t)), \tag{6.5}$$

$$\hat{y}(t) = C\hat{x}(t) + D_u(t) + D_f \hat{f}(t) \tag{6.6}$$

and FE algorithm

$$\delta[\hat{f}(t)] = \begin{cases} -F(\hat{y}(t) - y(t)) & \text{for continuous-time systems} \\ \hat{f}(t) - F(\hat{y}(t) - y(t)) & \text{for discrete-time systems} \end{cases}, \tag{6.7}$$

where $\hat{x}(t) \in \Re^n$ is the observer state, $\hat{y}(t) \in \Re^p$ is the observer output, and $\hat{f}(t) \in \Re^{n_f}$ is an estimate of $f(t)$. $L \in \Re^{n \times p}$ and $F \in \Re^{n_f \times p}$ are FE observer gain matrices to be designed.

From Equations (6.5)–(6.7), we can see that apart from matrices L and F, the others are all known. Therefore, the problem of the FE observer design is determined as needing to look for gain matrices L and F to achieve robust FE.

Let

$$e_x(t) = \hat{x}(t) - x(t), \quad e_y(t) = \hat{y}(t) - y(t),$$
$$e_f(t) = \hat{f}(t) - f(t),$$

then the error dynamics is given by

$$\delta[e_x(t)] = (A - LC)e_x(t) + (B_f - LD_f)e_f(t) + (LD_d - B_d)d(t), \quad (6.8)$$
$$e_y(t) = Ce_x(t) + D_f e_f(t) - D_d d(t) \quad (6.9)$$

For continuous-time systems, the FE error is

$$\dot{e}_f(t) = -F(\hat{y}(t) - y(t)) - \dot{f}(t)$$
$$= -FCe_x(t) - FD_f e_f(t) + FD_d d(t) - \dot{f}(t). \quad (6.10)$$

For discrete-time cases, it is shown that

$$e_f(t+1) = \hat{f} - F(\hat{y}(t) - y(t)) - f(t+1)$$
$$= \hat{f}(t) - f(t) + f(t) - F(\hat{y}(t) - y(t)) - f(t+1)$$
$$= \hat{f}(t) - f(t) + f(t) - FCe_x(t) - FD_f e_f(t) + FD_d d(t) - f(t+1)$$
$$= e_f(t) - FCe_x(t) - FD_f e_f(t) + FD_d d(t) + f(t) - f(t+1)$$
$$= -FCe_x(t) + (I_{n_f} - FD_f)e_f(t) + FD_d d(t) - \Delta f(t), \quad (6.11)$$

where $\Delta f(t) = f(t+1) - f(t)$.

Finally, it is concluded that

$$\delta[e_f(t)] = \begin{cases} -FCe_x(t) - FD_f e_f(t) + FD_d d(t) - \dot{f}(t) \\ \quad \text{for continuous-time systems} \\ -FCe_x(t) + (I_{n_f} - FD_f)e_f(t) + FD_d d(t) - \Delta f(t) \\ \quad \text{for discrete-time systems.} \end{cases} \quad (6.12)$$

To calculate gain matrices L and F conveniently, after analyzing the error dynamics in detail, we can obtain the following augmented system:

$$\delta \begin{bmatrix} e_x(t) \\ e_f(t) \end{bmatrix} = \begin{cases} \begin{bmatrix} A - LC & B_f - LD_f \\ -FC & -FD_f \end{bmatrix} \begin{bmatrix} e_x(t) \\ e_f(t) \end{bmatrix} + \\ \begin{bmatrix} LD_d - B_d & 0 \\ FD_d & -I_{n_f} \end{bmatrix} \begin{bmatrix} d(t) \\ \dot{f}(t) \end{bmatrix} \\ \text{for continuous-time systems} \\ \begin{bmatrix} A - LC & B_f - LD_f \\ -FC & I_{n_f} - FD_f \end{bmatrix} \begin{bmatrix} e_x(t) \\ e_f(t) \end{bmatrix} + \\ \begin{bmatrix} LD_d - B_d & 0 \\ FD_d & -I_{n_f} \end{bmatrix} \begin{bmatrix} d(t) \\ \Delta f(t) \end{bmatrix} \\ \text{for discrete-time systems} \end{cases} . \tag{6.13}$$

Let

$$\bar{e}(t) = \begin{bmatrix} e_x(t) \\ e_f(t) \end{bmatrix}, \quad v_C(t) = \begin{bmatrix} d(t) \\ \dot{f}(t) \end{bmatrix}, \quad v_D(t) = \begin{bmatrix} d(t) \\ \Delta f(t) \end{bmatrix},$$

$$\bar{A}_C = \begin{bmatrix} A & B_f \\ 0_{n_f \times n} & 0_{n_f} \end{bmatrix}, \bar{A}_D = \begin{bmatrix} A & B_f \\ 0_{n_f \times n} & I_{n_f} \end{bmatrix}, \bar{L} = \begin{bmatrix} L \\ F \end{bmatrix},$$

$$\bar{C} = \begin{bmatrix} C & D_f \end{bmatrix}, \quad \bar{B}_d = \begin{bmatrix} B_d & 0_{n \times n_f} \\ 0_{n_f \times n_d} & I_{n_f} \end{bmatrix}, \quad \bar{D}_d = \begin{bmatrix} D_d & 0_{p \times n_f} \end{bmatrix},$$

then it follows that

$$\delta[\bar{e}(t)] = (\bar{A} - \bar{L}\bar{C})\bar{e}(t) + (\bar{L}\bar{D}_d - \bar{B}_d)v(t), \tag{6.14}$$

where

$$\bar{A} = \begin{cases} \bar{A}_C & \text{for continuous-time systems} \\ \bar{A}_D & \text{for discrete-time systems} \end{cases},$$

$$v(t) = \begin{cases} v_C(t) & \text{for continuous-time systems} \\ v_D(t) & \text{for discrete-time systems} \end{cases}.$$

Remark 6.4   *From error dynamics Equation (6.14), we can see that the new matrices $\bar{A}, \bar{C}, \bar{B}_d$, and $\bar{D}_d$ are known matrices, whereas the matrix $\bar{L}$ contains the two matrices L and F that have to be designed. Therefore, the proposed FE observer design is converted to the problem of seeking the gain matrix $\bar{L}$ such that $(\bar{A} - \bar{L}\bar{C})$ is stable and the FE error $e_f(t)$ is insensitive to the term $v(t)$ as much as possible (i.e., $e_f(t)$ is as small as possible).*

Assumption 6.1   *$\dot{f}(t)$ belongs to $L_2[0, \infty)$ for continuous-time systems and $\Delta f(t)$ belongs to $l_2[0, \infty)$ for discrete-time cases.*

Remark 6.5   *For continuous-time systems, an FE method based on a sliding-mode observer requires preliminary knowledge of the upper bound of faults [4]–[7] and the adaptive observer-based FE design assumes that $\dot{f}(t) = 0$ after the fault occurrence [1, 2]. For discrete-time systems, it was supposed [13] that the fault difference item $\Delta f(t)$ was sufficiently small to be neglected, that is, constant faults were considered. At the same time, the FE filter needs the assumption that $f(t) \in L_2[0, \infty)$ for continuous-time systems [17] and $f(t) \in l_2[0, \infty)$ for discrete-time cases [18]. Compared with the earlier design methods, it is readily seen that Assumption 6.1 is more general.*

In Theorem 6.1, an MFEO design method under an H_∞ performance specification with a regional pole constraint is proposed to achieve robust FE.

Theorem 6.1   *Let a prescribed H_∞ performance level γ and a circular region $D(\alpha, r)$ be given. Error dynamics Equation (6.14) satisfies the H_∞ performance index $\|e_f(t)\|_2 < \gamma \|v(t)\|_2$, and the eigenvalues of $(\bar{A} - \bar{L}\bar{C})$ belong to $D(\alpha, r)$ if and only if there exist two symmetric positive definite matrices $\bar{P}_1, \bar{P}_2 \in \Re^{(n+n_f)\times(n+n_f)}$ and a matrix $\bar{L} \in \Re^{(n+n_f)\times p}$ such that the following conditions hold:*

$$
\begin{cases}
\begin{bmatrix}
\bar{P}_1(\bar{A}_C - \bar{L}\bar{C}) + (\bar{A}_C - \bar{L}\bar{C})^T & \bar{P}_1(\bar{L}\bar{D}_d - \bar{B}_d) & \bar{I}_{n_f} \\
\bullet & -\gamma I_{(n_d+n_f)} & 0 \\
\bullet & \bullet & -\gamma I_{n_f}
\end{bmatrix} < 0, \\
\qquad\qquad\qquad \textit{for continuous-time systems} \\[4pt]
\begin{bmatrix}
-\bar{P}_1 & \bar{P}_1(\bar{A}_D - \bar{L}\bar{C}) & \bar{P}_1(\bar{L}\bar{D}_d - \bar{B}_d) & 0 \\
\bullet & -\bar{P}_1 & 0 & \bar{I}_{n_f} \\
\bullet & \bullet & -\gamma I_{(n_d+n_f)} & 0 \\
\bullet & \bullet & \bullet & -\gamma I_{n_f}
\end{bmatrix} < 0, \\
\qquad\qquad\qquad \textit{for discrete-time systems}
\end{cases}
\tag{6.15}
$$

and

$$
\begin{cases}
\begin{bmatrix}
-\bar{P}_2 & \bar{P}_2(\bar{A}_C - \bar{L}\bar{C}) - \alpha\bar{P}_2 \\
\bullet & -\Re^2\bar{P}_2
\end{bmatrix} < 0, \textit{ for continuous-time systems} \\[4pt]
\begin{bmatrix}
-\bar{P}_2 & \bar{P}_2(\bar{A}_D - \bar{L}\bar{C}) - \alpha\bar{P}_2 \\
\bullet & -\Re^2\bar{P}_2
\end{bmatrix} < 0, \textit{ for discrete-time systems}
\end{cases}
\tag{6.16}
$$

where $\bar{I}_{n_f} = \begin{bmatrix} 0_{n\times n_f} \\ I_{n_f} \end{bmatrix}$.

Proof. Constraint (6.15): Using the bounded real lemmas for continuous-time/discrete-time systems [19], one obtains Equation (6.15) directly.

   Constraint (6.16): For error dynamics Equation (6.14), by setting $(\bar{A} - \bar{L}\bar{C}) \to A$ and $\bar{P}_2 \to P$ in Lemma 3.5, one obtains Equation (6.16).   ∎

Remark 6.6   *In Theorem 6.1, considering H_∞ performance specification allows the restraint of the effect of the term $v(k)$ as much as possible; introducing the regional pole constraint aims at improving the performance of FE. However, the two conditions might affect the ultimate H_∞ attenuation level. Therefore, the two constraints of the MFEO are partially conflicting with each other, and a trade-off between the estimation performance and the ultimate H_∞ attenuation level must be reached.*

Remark 6.7   *Introducing two Lyapunov matrices $\bar{P}_1$, $\bar{P}_2$ rather than a single one provides some degrees of design freedom and allows Theorem 6.1 to be a necessary and sufficient condition. However, the coupling of the observer gain matrix $\bar{L}$ between Equation (6.15) and Equation (6.16) results in a nonconvex problem, which cannot be handled by linear optimization procedures. To recover convexity, we must require that all specifications are enforced by a common Lyapunov matrix, that is, $\bar{P} = \bar{P}_1 = \bar{P}_2$, resulting in Corollary 6.1.*

Corollary 6.1   *Let a prescribed H_∞ performance level γ and a circular region $D(\alpha, r)$ be given. If there exists a symmetric positive definite matrix $\bar{P} \in \Re^{(n+n_f)\times(n+n_f)}$ / and a matrix $\bar{Y} \in \Re^{(n+n_f)\times p}$ such that the following conditions hold:*

$$
\begin{cases}
\begin{bmatrix} \bar{P}\bar{A}_C + \bar{A}_C^T\bar{P} - \bar{Y}\bar{C} - \bar{C}^T\bar{Y}^T & \bar{Y}\bar{D}_d - \bar{P}\bar{B}_d & \bar{I}_{n_f} \\ \bullet & -\gamma I_{(n_d+n_f)} & 0 \\ \bullet & \bullet & -\gamma I_{n_f} \end{bmatrix} < 0, \\
\qquad\qquad \textit{for continuous-time systems} \\
\begin{bmatrix} -\bar{P} & \bar{P}\bar{A}_D - \bar{Y}\bar{C} & \bar{Y}\bar{D}_d - \bar{P}\bar{B}_d & 0 \\ \bullet & -\bar{P} & 0 & \bar{I}_{n_f} \\ \bullet & \bullet & -\gamma I_{(n_d+n_f)} & 0 \\ \bullet & \bullet & \bullet & -\gamma I_{n_f} \end{bmatrix} < 0, \\
\qquad\qquad \textit{for discrete-time systems}
\end{cases}
\tag{6.17}
$$

and

$$
\begin{cases}
\begin{bmatrix} -\bar{P} & \bar{P}\bar{A}_C - \bar{Y}\bar{C} - \alpha\bar{P} \\ \bullet & -\Re^2\bar{P} \end{bmatrix} < 0, \textit{ for continuous-time systems} \\
\begin{bmatrix} -\bar{P} & \bar{P}\bar{A}_D - \bar{Y}\bar{C} - \alpha\bar{P} \\ \bullet & -\Re^2\bar{P} \end{bmatrix} < 0, \textit{ for discrete-time systems}
\end{cases}
,
\tag{6.18}
$$

where $\bar{I}_{n_f} = \begin{bmatrix} 0_{n\times n_f} \\ I_{n_f} \end{bmatrix}$, then error dynamics Equation (6.14) satisfies the H_∞ performance index $\|e_f(t)\|_2 < \gamma\|v(t)\|_2$, the eigenvalues of $(\bar{A} - \bar{L}\bar{C})$ belong to $D(\alpha, r)$, and the MFEO gain matrix is given by $\bar{L} = \bar{P}^{-1}\bar{Y}$.

Remark 6.8   *Under a given regional pole constraint, the minimum H_∞ attenuation level in Corollary 6.1 can be obtained by solving the following optimization problem:*

$$\textit{minimize } \gamma \textit{ subject to Equation (6.17) and Equation (6.18).}$$

This remark also applies to the other H_∞ performance optimization calculations.

6.3.2   Existence conditions

In this section, we discuss the existence condition of the MFEO for continuous-time and discrete-time systems. From Equation (6.14), the task of Theorem 6.1 is to look for $\bar{L}$ such that $(\bar{A} - \bar{L}\bar{C})$ is robustly stable under the H_∞ performance specification with a regional pole constraint. So the existence condition of the MFEO is that the pair

$$(\bar{A}, \bar{C}) = \begin{cases} (\bar{A}_C, \bar{C}) = \left(\begin{bmatrix} A & B_f \\ 0_{n_f \times n} & 0_{n_f} \end{bmatrix}, \begin{bmatrix} C & D_f \end{bmatrix} \right), & \text{for continuous-time systems} \\[3mm] (\bar{A}_D, \bar{C}) = \left(\begin{bmatrix} A & B_f \\ 0_{n_f \times n} & I_{n_f} \end{bmatrix}, \begin{bmatrix} C & D_f \end{bmatrix} \right), & \text{for discrete-time systems} \end{cases}$$

is observable.

6.3.2.1   Existence conditions for continuous-time systems

With linear systems theory, the observability of the pair $(\bar{A}_C, \bar{C})$ is equivalent to

$$\text{rank} \left(\begin{bmatrix} \bar{A}_C - \lambda I_{n+n_f} \\ \bar{C} \end{bmatrix} \right) = n + n_f, \quad \forall \lambda \in C, \tag{6.19}$$

and one has

$$\text{rank} \left(\begin{bmatrix} \bar{A}_C - \lambda I_{n+n_f} \\ \bar{C} \end{bmatrix} \right) = \text{rank} \left(\begin{bmatrix} A - \lambda I_n & B_f \\ 0 & -\lambda I_{n_f} \\ C & D_f \end{bmatrix} \right). \tag{6.20}$$

When $\lambda = 0$, it is shown that

$$\text{rank} \left(\begin{bmatrix} A - \lambda I_n & B_f \\ 0 & -\lambda I_{n_f} \\ C & D_f \end{bmatrix} \right) = \text{rank} \left(\begin{bmatrix} A & B_f \\ C & D_f \end{bmatrix} \right). \tag{6.21}$$

When $\lambda \neq 0$, one obtains

$$\begin{aligned} &\text{rank} \left(\begin{bmatrix} A - \lambda I_n & B_f \\ 0 & -\lambda I_{n_f} \\ C & D_f \end{bmatrix} \right) \\ &= \text{rank} \left(\begin{bmatrix} I_n & -\lambda^{-1} B_f & 0 \\ 0 & I_{n_f} & 0 \\ 0 & -\lambda^{-1} D_f & I_p \end{bmatrix} \cdot \begin{bmatrix} A - \lambda I_n & 0 \\ 0 & -\lambda I_{n_f} \\ C & 0 \end{bmatrix} \right) \\ &= \text{rank} \left(\begin{bmatrix} A - \lambda I_n & 0 \\ 0 & -\lambda I_{n_f} \\ C & 0 \end{bmatrix} \right) \\ &= n_f + \text{rank} \left(\begin{bmatrix} A - \lambda I_n \\ C \end{bmatrix} \right) \end{aligned} \tag{6.22}$$

Therefore, it is concluded that the pair $(\bar{A}_C, \bar{C})$ is observable if

$$\text{rank}\left(\begin{bmatrix} A & B_f \\ C & D_f \end{bmatrix}\right) = n + n_f$$

in Equation (6.21) and

$$\text{rank}\left(\begin{bmatrix} A - \lambda I_n \\ C \end{bmatrix}\right) = n$$

in Equation (6.22) hold simultaneously.

6.3.2.2 Existence conditions for discrete-time systems

The observability of the pair $(\bar{A}_D, \bar{C})$ is equivalent to

$$\text{rank}\left(\begin{bmatrix} \bar{A}_D - \lambda I_{n+n_f} \\ \bar{C} \end{bmatrix}\right) = n + n_f, \quad \forall \lambda \in C, \tag{6.23}$$

and one obtains

$$\text{rank}\left(\begin{bmatrix} \bar{A}_D - \lambda I_{n+n_f} \\ \bar{C} \end{bmatrix}\right) = \text{rank}\left(\begin{bmatrix} A - \lambda I_n & B_f \\ 0 & (1-\lambda)I_{n_f} \\ C & D_f \end{bmatrix}\right). \tag{6.24}$$

When $\lambda = 1$, one obtains

$$\text{rank}\left(\begin{bmatrix} A - \lambda I_n & B_f \\ 0 & (1-\lambda)I_{n_f} \\ C & D_f \end{bmatrix}\right) = \text{rank}\left(\begin{bmatrix} A - I_n & B_f \\ C & D_f \end{bmatrix}\right). \tag{6.25}$$

When $\lambda \neq 1$, one obtains

$$\text{rank}\left(\begin{bmatrix} A - \lambda I_n & B_f \\ 0 & (1-\lambda)I_{n_f} \\ C & D_f \end{bmatrix}\right)$$

$$= \text{rank}\left(\begin{bmatrix} I_n & (1-\lambda)^{-1}B_f & 0 \\ 0 & I_{n_f} & 0 \\ 0 & (1-\lambda)^{-1}D_f & I_p \end{bmatrix} \cdot \begin{bmatrix} A - \lambda I_n & 0 \\ 0 & (1-\lambda)I_{n_f} \\ C & 0 \end{bmatrix}\right)$$

$$= \text{rank}\left(\begin{bmatrix} A - \lambda I_n & 0 \\ 0 & (1-\lambda)I_{n_f} \\ C & 0 \end{bmatrix}\right)$$

$$= n_f + \text{rank}\left(\begin{bmatrix} A - \lambda I_n \\ C \end{bmatrix}\right). \tag{6.26}$$

Therefore, it is concluded that the pair $(\bar{A}_D, \bar{C})$ is observable if

$$\text{rank}\left(\begin{bmatrix} A - I_n & B_f \\ C & D_f \end{bmatrix}\right) = n + n_f$$

in Equation (6.25) and

$$\text{rank}\left(\begin{bmatrix} A - \lambda I_n \\ C \end{bmatrix}\right) = n$$

in Equation (6.26) hold simultaneously.

Remark 6.9 *For continuous-time and discrete-time systems, the feasibility of*

$$rank\left(\begin{bmatrix} A - \lambda I_n \\ C \end{bmatrix}\right) = n$$

is guaranteed because the pair (A, C) is supposed to be observable, which is commonly used for observer design. On the other hand, in Equation (6.21) and Equation (6.25), conditions

$$rank\left(\begin{bmatrix} A & B_f \\ C & D_f \end{bmatrix}\right) = n + n_f, \quad and \quad rank\left(\begin{bmatrix} A - I_n & B_f \\ C & D_f \end{bmatrix}\right) = n + n_f,$$

imply that there are no invariant zeros at zero and one, which allows the MFEO to deal with nonminimum-phase systems. Therefore, compared with adaptive observer and sliding-mode observer approaches, the MFEO possesses wider application ranges, which can be suitable to more general systems.

6.3.3 Improved results

To guarantee the convergence performance of the MFEO, multiconstrained design is involved, leading to possible conservatism, because of the common Lyapunov matrix $\bar{P}$ in Corollary 6.1. Multiconstrained design problems may have no solution because of the increased number of constraints that are considered. In this section, we extend the idea introduced by de Oliveira, Geromel, and Bernussou [14] and Xie [15] to the MFEO design so that less restrictive conclusions are obtained, by introducing slack variables such that different Lyapunov matrices $\bar{P}_1$ and $\bar{P}_2$ can be designed. Theorems 6.2 and 6.3 yield an MFEO design with improved performance.

Theorem 6.2 *Continuous-time case: Let a prescribed H_∞ performance level γ, a circular region $D(\alpha, r)$, and a sufficiently small positive scalar ε be given. If there exist two symmetric* *positive definite matrices* $\bar{P}_1, \bar{P}_2 \in \mathfrak{R}^{(n+n_f)\times(n+n_f)}$ *and two matrices* $\bar{S} \in \mathfrak{R}^{(n+n_f)\times(n+n_f)}$, $\bar{Y} \in \mathfrak{R}^{(n+n_f)\times p}$ *such that the following conditions hold:*

$$
\begin{bmatrix}
\Pi & \bar{P}_1 - \bar{S} + \varepsilon(\bar{A}_C^T \bar{S}^T - \bar{C}^T Y^T) & \bar{Y}\bar{D}_d - \bar{S}\bar{B}_d & \bar{I}_{n_f} \\
\bullet & -\varepsilon(\bar{S} + \bar{S}^T) & \varepsilon(\bar{Y}\bar{D}_d - \bar{S}\bar{B}_d) & 0 \\
\bullet & \bullet & -\gamma I_{(n_d+n_f)} & 0 \\
\bullet & \bullet & \bullet & -\gamma I_{n_f}
\end{bmatrix} < 0, \qquad (6.27)
$$

and

$$
\begin{bmatrix}
-\bar{S} - \bar{S}^T + \bar{P}_2 & \bar{S}\bar{A}_C - \bar{Y}\bar{C} - \alpha\bar{S} \\
* & -\mathfrak{R}^2 \bar{P}_2
\end{bmatrix} < 0, \qquad (6.28)
$$

where

$$
\bar{I}_{n_f} = \begin{bmatrix} 0_{n\times n_f} \\ I_{n_f} \end{bmatrix}, \quad \Pi = \bar{S}\bar{A}_C + \bar{A}_C^T \bar{S}^T - \bar{Y}\bar{C} - \bar{C}^T \bar{Y}^T
$$

then error dynamics Equation (6.14) satisfies the H_∞ performance index $\|e_f(t)\|_2 < \gamma \|v(t)\|_2$, the eigenvalues of $(\bar{A}_C - \bar{L}\bar{C})$ belong to $D(\alpha, r)$, and the MFEO gain matrix is given by $\bar{L} = \bar{S}^{-1}\bar{Y}$.

Proof. By introducing the slack matrices $\bar{S}$ and ε, different Lyapunov matrices $\bar{P}_1$ and $\bar{P}_2$ can be designed. On the basis of the proof of Theorem 6.1, here, we need to prove the equivalence of Equations (6.15) and (6.27) and the equivalence of Equations (6.16) and Equation (6.28). ∎

6.3.3.1 Equation (6.15) → Equation (6.27)

When choosing $\bar{P}_1 = \bar{S} = \bar{S}^T$ in Equation (6.27), we can obtain

$$
\begin{bmatrix}
\Theta & \varepsilon(\bar{A}_C - \bar{L}\bar{C})^T \bar{P}_1 & \bar{P}_1(\bar{L}\bar{D}_d - \bar{B}_d) & \bar{I}_{n_f} \\
\bullet & -2\varepsilon\bar{P}_1 & \varepsilon\bar{P}_1(\bar{L}\bar{D}_d - \bar{B}_d) & 0 \\
\bullet & \bullet & -\gamma I_{(n_d+n_f)} & 0 \\
\bullet & \bullet & \bullet & -\gamma I_{n_f}
\end{bmatrix} < 0.
$$

$$
\Theta = \bar{P}_1(\bar{A}_C - \bar{L}\bar{C}) + (\bar{A}_C - \bar{L}\bar{C})^T \bar{P}_1 \qquad (6.29)
$$

Then, using the Schur complement, Equation (6.29) is equivalent to

$$
\begin{bmatrix}
\bar{P}_1(\bar{A}_C - \bar{L}\bar{C}) + (\bar{A}_C - \bar{L}\bar{C})^T \bar{P}_1 & \bar{P}_1(\bar{L}\bar{D}_d - \bar{B}_d) & \bar{I}_{n_f} \\
\bullet & -\gamma I_{(n_d+n_f)} & 0 \\
\bullet & \bullet & -\gamma I_{n_f}
\end{bmatrix} +
$$

$$
\frac{\varepsilon}{2}
\begin{bmatrix}
(\bar{A}_C - \bar{L}\bar{C})^T \bar{P}_1(\bar{A}_C - \bar{L}\bar{C}) & (\bar{A}_C - \bar{L}\bar{C})^T \bar{P}_1(\bar{L}\bar{D}_d - \bar{B}_d) & 0 \\
\bullet & (\bar{L}\bar{D}_d - \bar{B}_d)\bar{P}_1(\bar{L}\bar{D}_d - \bar{B}_d) & 0 \\
\bullet & \bullet & 0
\end{bmatrix} < 0. \qquad (6.30)
$$

We can conclude that if Equation (6.15) holds, Equation (6.30) is always satisfied for a sufficiently small positive scalar ε, so Equation (6.15) implies Equation (6.27).

6.3.3.2 Equation (6.27) → Equation (6.15)

If Equation (6.27) is feasible, then premultiplying and postmultiplying by

$$
\begin{bmatrix}
I_{(n+n_f)} & (\bar{A}_C - \bar{L}\bar{C})^T & 0 & 0 \\
0 & (\bar{L}\bar{D}_d - \bar{B}_d)^T & I_{(n_d+n_f)} & 0 \\
0 & 0 & 0 & I_{n_f}
\end{bmatrix}
$$

and its transpose, we can obtain Equation (6.15) directly.

6.3.3.3 Equation (6.16) → Equation (6.28)

When choosing $\bar{P}_2 = \bar{S} = \bar{S}^T$ in Equation (6.28), we can obtain Equation (6.16), and then Equation (6.16) implies Equation (6.28).

6.3.3.4 Equation (6.28) → Equation (6.16)

In Equation (6.28), we can obtain $\bar{S} + \bar{S}^T > \bar{P}_2 > 0$, which implies $\bar{S}$ is nonsingular. Because $\bar{P}_2$ is symmetric positive definite, the inequality $(\bar{P}_2 - \bar{S})\bar{P}_2^{-1}(\bar{P}_2 - \bar{S})^T \geq 0$ holds, and it can also be expressed as $-\bar{S} - \bar{S}^T + \bar{P}_2 \geq -\bar{S}\bar{P}_2^{-1}\bar{S}^T$. Therefore, it follows from Equation (6.28) that

$$
\begin{bmatrix}
-\bar{S}\bar{P}_2^{-1}\bar{S}^T & \bar{S}(\bar{A}_C - \bar{L}\bar{C}) - \alpha\bar{S} \\
\bullet & -\Re^2 \bar{P}_2
\end{bmatrix} < 0. \tag{6.31}
$$

Premultiplying and postmultiplying by diag $\bar{P}_2\bar{S}^{-1}$, I_{n+n_f} and its transpose, we obtain Equation (6.16), which establishes that Equation (6.28) implies Equation (6.16).

Theorem 6.3　Discrete-time case　*Let a prescribed H_∞ performance level γ and a circular region $D(\alpha, r)$ be given. If there exist two symmetric positive definite matrices $\bar{P}_1, \bar{P}_2 \in \Re^{(n+n_f)\times(n+n_f)}$ and two matrices $\bar{S} \in \Re^{(n+n_f)\times(n+n_f)}$, $\bar{Y} \in \Re^{(n+n_f)\times p}$ such that:*

$$
\begin{bmatrix}
-\bar{S} - \bar{S}^T + \bar{P}_1 & \bar{S}\bar{A}_D - \bar{Y}\bar{C} - \bar{Y}\bar{D}_d - \bar{S}\bar{B}_d & 0 & \\
\bullet & -\bar{P}_1 & 0 & \bar{I}_{n_f} \\
\bullet & \bullet & -\gamma I_{(n_d+n_f)} & 0 \\
\bullet & \bullet & \bullet & -\gamma I_{n_f}
\end{bmatrix} < 0, \tag{6.32}
$$

and

$$
\begin{bmatrix}
-\bar{S}\bar{S}^T + \bar{P}_2 & \bar{S}\bar{A}_D - \bar{Y}\bar{C} - \alpha\bar{S} \\
\bullet & -\Re^2 \bar{P}_2
\end{bmatrix} < 0, \tag{6.33}
$$

where $\bar{I}_{n_f} = \begin{bmatrix} 0_{n\times n_f} \\ I_{n_f} \end{bmatrix}$, then error dynamics Equation (6.14) satisfies the H_∞ performance index $\|e_f(t)\|_2 < \gamma\|v(t)\|_2$, the eigenvalues of $(\bar{A}_D - \bar{L}\bar{C})$ belong to $D(\alpha, r)$, and the MFEO gain matrix is given by $\bar{L} = \bar{S}^{-1}\bar{Y}$.

Proof. Similar to Theorem 6.2, by introducing a slack matrix $\bar{S}$, different Lyapunov matrices $\bar{P}_1$ and $\bar{P}_2$ can be designed, for discrete-time systems. The equivalence of Equation (6.16) and Equation (6.33) is shown by reference to Theorem 6.2 and thus is omitted here. Here we only prove the equivalence of Equation (6.15) and Equation (6.32). ∎

6.3.3.5 Equation (6.15) → Equation (6.32)

When choosing $\bar{P}_1 = \bar{S} = \bar{S}^T$ in Equation (6.32), we can obtain Equation (6.15), and then Equation (6.15) implies Equation (6.32).

6.3.3.6 Equation (6.32) → Equation (6.15)

In Equation (6.32), we can obtain $\bar{S} + \bar{S}^T > \bar{P}_1 > 0$, which implies that $\bar{S}$ is nonsingular. Because $\bar{P}_1$ is symmetric positive definite, the inequality $(\bar{P}_1 - \bar{S})\bar{P}_1^{-1}(\bar{P}_1 - \bar{S})^T \geq 0$ holds, and it can also be expressed as $-\bar{S} - \bar{S}^T + \bar{P}_1 \geq -\bar{S}\bar{P}_1^{-1}\bar{S}^T$. Therefore, it follows from Equation (6.32) that

$$\begin{bmatrix} -\bar{S}\bar{P}_1^{-1}\bar{S}^T & \bar{S}(\bar{A}_D - \bar{L}\bar{C}) & \bar{S}(\bar{L}\bar{D}_d - \bar{B}_d) & 0 \\ \bullet & -\bar{P}_1 & 0 & \bar{I}_{n_f} \\ \bullet & \bullet & -\gamma I_{(n_d + n_f)} & 0 \\ \bullet & \bullet & \bullet & -\gamma I_{n_f} \end{bmatrix} < 0, \tag{6.34}$$

and then premultiplying and postmultiplying by diag $(\bar{P}_1\bar{S}^{-1}, I_{n+n_f}, I_{n_d+n_f}, I_{n_f})$ and its transpose, we obtain Equation (6.15), which establishes that Equation (6.32) implies Equation (6.15).

Remark 6.10 *In Theorems 6.2 and 6.3, thanks to the slack variables, the different Lyapunov matrices $\bar{P}_1$ and $\bar{P}_2$ can be designed, so that Theorems 6.2 and 6.3 are more flexible than Corollary 6.1. Under the same regional pole constraint, one may obtain a smaller H_∞ attenuation level, which is at least equal to that obtained from Corollary 6.1.*

Remark 6.11 *On the basis of Corollary 6.1, an improved MFEO design of continuous-time systems is obtained by introducing slack variables, that is, Theorem 6.2. Because of the conjunction of the parameter ε and matrix S, the conditions of Theorem 6.2 are nonconvex. Therefore, when solving the conditions of Theorem 6.2, the parameter ε needs to be given in advance such that the conditions of Theorem 6.2 become a convex optimization problem that can be solved by convex optimization tools, such as CVX and YALMIP. It should be noted that the optimal value γ in Theorem 6.2 does not monotonously decrease with ε. A global search of the parameter ε is needed with the purpose of looking for the minimum value γ.*

Remark 6.12 *For conditions of Theorem 6.2, to calculate the minimum value γ, it seems that a global search of the parameter ε would impose a great computation burden. In fact, we generally need to look for the minimum value γ within a small region of the parameter ε, that is, compared with Corollary 6.1, the computation complexity would not increase much. Moreover, under the same regional pole constraint, it is possible to obtain a smaller γ by solving conditions of Theorem 6.2.*

6.3.4 Simulation results

In this section, a VTOL aircraft is taken to illustrate the effectiveness of our presented method. A continuous-time dynamic model of the VTOL model in the vertical plane borrowed from Saif and Guan [20] is given in the state space formulation, and the model parameters are as follows:

$$
A = \begin{bmatrix} -9.9477 & -0.7476 & 0.2632 & 5.0337 \\ 52.1659 & 2.7452 & 5.5532 & -24.4221 \\ 26.0922 & 2.6361 & -4.1975 & -19.2774 \end{bmatrix}, \; B = \begin{bmatrix} 0.4422 & 0.1761 \\ 3.5446 & -7.5922 \\ -5.5200 & 4.4900 \\ 0 & 0 \end{bmatrix},
$$

$$
C = \begin{bmatrix} 1 & 0 & 0 & 0 \\ 0 & 1 & 0 & 0 \end{bmatrix}, \quad D = \begin{bmatrix} 0 & 0 \\ 0 & 0 \end{bmatrix},
$$

where $x(t) = [V_h, V_v, q, \theta]$, $u(t) = [\delta_c, \delta_l]$. The states are horizontal velocity $V_h(kts)$, vertical velocity V_v (kts), pitch rate $q(deg/s)$, and pitch angle $\theta(deg)$. The inputs are collective pitch control $\delta_c(deg)$ and longitudinal cyclic pitch control δ_l (deg). The outputs are the first and second states. Here, we consider actuator faults. Such faults usually occur in the input channel, so we assume $B_f = B$ and $D_f = 0$. Here, it is assumed that disturbance distribution matrices are $B_d = [0.1\ 0.1\ 0.1\ 0.1]^T$ and $D_d = [0.1\ 0.1]^T$.

We consider the continuous-time case and the discrete time case separately.

6.3.4.1 Continuous-time plant

Because the dynamics (A, B_f, C, D_f) has two unstable invariant zeros $0.8230 \pm 1.9693j$ (which means the system (A, B_f, C, D_f) is nonminimum phase), the adaptive and sliding-mode observer methods cannot be used; the MFEO design is possible because the pair (A, C) is observable and there is no invariant zero of the system (A, B_f, C, D_f) at zero.

Solving the conditions in Theorem 6.2 with a regional pole constraint $D(-8, 7.5)$ and a small positive scalar $\varepsilon = 0.04$, one can obtain the minimum H_∞ attenuation level $\gamma = 4.3090$ with

$$
\bar{P}_1 = 10^3 \begin{bmatrix} 8.4274 & 0.4165 & 0.7971 & -4.1413 & -0.2453 & -0.2940 \\ 0.4165 & 0.0893 & 0.0681 & -0.1680 & -0.0257 & -0.0003 \\ 0.7971 & 0.0681 & 0.1232 & -0.3337 & -0.0152 & -0.0367 \\ -4.1413 & -0.1680 & -0.3337 & 2.8463 & 0.1727 & 0.1252 \\ -0.2453 & -0.0257 & -0.0152 & 0.1727 & 0.0221 & -0.0056 \\ -0.2940 & -0.0003 & -0.0367 & 0.1252 & -0.0056 & 0.0254 \end{bmatrix},
$$

$$
\bar{P}_2 = 10^3 \begin{bmatrix} 6.4865 & 0.5283 & 0.8198 & -3.2074 & -0.1831 & -0.2461 \\ 0.5283 & 0.0773 & 0.0851 & -0.2013 & -0.0214 & -0.0112 \\ 0.8198 & 0.0851 & 0.1399 & -0.3286 & -0.0126 & -0.0413 \\ -3.2074 & -0.2013 & -0.3286 & 2.2906 & 0.1387 & 0.1013 \\ -0.1831 & -0.0214 & -0.0126 & 0.1387 & 0.0201 & -0.0079 \\ -0.2461 & -0.0112 & -0.0413 & 0.1013 & -0.0079 & 0.0255 \end{bmatrix},
$$

$$\bar{S} = 10^3 \begin{bmatrix} 5.6789 & 0.2245 & 0.3763 & -3.2876 & -0.2443 & -0.1278 \\ 0.4657 & 0.0714 & 0.0653 & -0.2202 & -0.0233 & -0.0068 \\ 0.8101 & 0.0772 & 0.1298 & -0.3209 & -0.0161 & -0.0360 \\ -2.7558 & -0.0621 & -0.1248 & 2.3466 & 0.1618 & 0.0502 \\ -0.1304 & -0.0072 & 0.0082 & 0.1627 & 0.0207 & -0.0104 \\ -0.2543 & -0.0087 & -0.0373 & 0.0797 & -0.0048 & 0.0210 \end{bmatrix}.$$

The MFEO gain matrix is

$$\bar{L} = \begin{bmatrix} 16.4181 & -1.4565 \\ 4.6466 & 12.3602 \\ 22.7405 & -5.1994 \\ -21.7315 & -0.6850 \\ 482.7330 & -6.9702 \\ 423.2193 & -16.3978 \end{bmatrix}.$$

Assume that an actuator fault is created as

$$f_1(t) = 0, \quad f_2(t) = \begin{cases} 0, & 0 \leq t < 10\,\text{s} \\ 2(1 - e^{-1(t-10)}), & 10\,\text{s} \leq t < 20\,\text{s} \\ 2 - 4(1 - e^{-(t-20)}), & 20\,\text{s} \leq t < 40\,\text{s} \end{cases}.$$

Under the input $u(t) = [5 \quad 5]^T$, we obtain the following simulation results. The eigenvalue distribution situation of $(\bar{A}_C - \bar{L}\bar{C})$ is shown in Figure 6.1. Figure 6.2 illustrates the result of

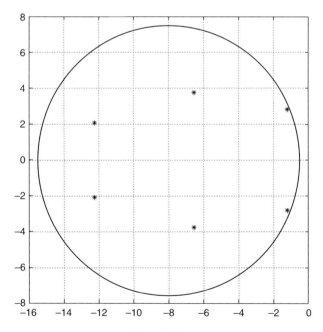

Figure 6.1 $(\bar{A}_C - \bar{L}\bar{C})$ pole distribution

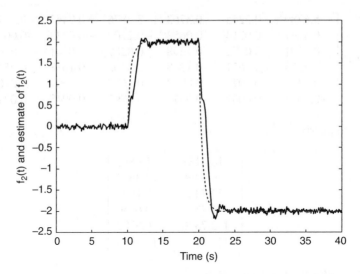

Figure 6.2   Fault $f_2(t)$ (dotted line) and its estimate $\hat{f}_2(t)$ (solid line)

FE using the MFEO method. Even if there exist unstable invariant zeros, it is seen that the proposed MFEO still can achieve FE.

Under the common regional pole constraint, it can be checked that Corollary 6.1 (for continuous-time systems) gives the minimum H_∞ attenuation value, $\gamma = 4.3091$, which is larger than 4.3106, thus illustrating the fact that Theorem 6.2 is less conservative than Corollary 6.1.

6.3.4.2   Discrete-time plant

We now deal with the discretization of the continuous-time VTOL plant. By using the zero-order hold with a sampling period $T = 0.2$ s, the discrete-time VTOL model is given by

$$
A = \begin{bmatrix} -0.0717 & -0.0681 & 0.0205 & 0.5076 \\ 6.7971 & 1.2917 & 0.7634 & -3.5029 \\ 2.1633 & 0.2191 & 0.4345 & -1.8614 \\ 0.2966 & 0.0300 & 0.1360 & 0.7606 \end{bmatrix}, \quad B = \begin{bmatrix} -0.0172 & 0.0927 \\ 0.7044 & -1.2657 \\ -0.5131 & 0.4350 \\ -0.0670 & 0.0559 \end{bmatrix},
$$

$$
B_f = B, \quad B_d = \begin{bmatrix} 0.0126 \\ 0.0720 \\ 0.0223 \\ 0.0223 \end{bmatrix},
$$

and matrices C, D, D_f, and D_d do not change.

It can be shown that the discrete-time dynamics (A, B_f, C, D_f) has two unstable invariant zeros, $1.0834 \pm 0.4740j$. Solving the conditions in Theorem 6.3 with a regional pole constraint $D(0.5, 0.5)$, one obtains the minimum H_∞ attenuation level $\gamma = 6.5625$ with

$$\bar{P}_1 = 10^3 \begin{bmatrix} 1.7938 & 0.1188 & 0.0989 & -0.7312 & -0.0784 & -0.0260 \\ 0.1188 & 0.0116 & 0.0135 & -0.0321 & -0.0041 & -0.0026 \\ 0.0989 & 0.0135 & 0.0262 & -0.0294 & -0.0018 & -0.0050 \\ -0.7312 & -0.0321 & -0.0294 & 0.5013 & 0.0457 & 0.0054 \\ -0.0784 & -0.0041 & -0.0018 & 0.0457 & 0.0052 & -0.0001 \\ -0.0260 & -0.0026 & -0.0050 & 0.0054 & -0.0001 & 0.0019 \end{bmatrix},$$

$$\bar{P}_2 = 10^3 \begin{bmatrix} 3.6239 & 0.1943 & 0.1801 & -1.8613 & -0.1598 & -0.0710 \\ 0.1943 & 0.0146 & 0.0191 & -0.0854 & -0.0072 & -0.0051 \\ 0.1801 & 0.0191 & 0.0375 & -0.0875 & -0.0059 & -0.0080 \\ -1.8613 & -0.0854 & -0.0875 & 1.1729 & 0.0987 & 0.0323 \\ -0.1598 & -0.0072 & -0.0059 & 0.0987 & 0.0089 & 0.0020 \\ -0.0710 & -0.0051 & -0.0080 & 0.0323 & 0.0020 & 0.0028 \end{bmatrix},$$

$$\bar{S} = 10^3 \begin{bmatrix} 2.6063 & 0.1384 & 0.0792 & -1.1632 & -0.1120 & -0.0373 \\ 0.1241 & 0.0113 & 0.0128 & -0.0345 & -0.0040 & -0.0029 \\ 0.1312 & 0.0162 & 0.0307 & -0.0486 & -0.0036 & -0.0061 \\ -1.2881 & -0.0538 & -0.0352 & 0.7974 & 0.0716 & 0.0140 \\ -0.0813 & -0.0040 & -0.0012 & 0.0461 & 0.0053 & -0.0000 \\ -0.0845 & -0.0048 & -0.0058 & 0.0379 & 0.0026 & 0.0027 \end{bmatrix}.$$

The MFEO gain matrix is

$$\bar{L} = \begin{bmatrix} 0.9658 & -0.0597 \\ 0.0476 & 1.1449 \\ 2.1543 & 0.0214 \\ -0.8647 & -0.1893 \\ 24.6443 & 2.0312 \\ 22.7798 & 1.3677 \end{bmatrix}$$

Figure 6.3 shows the eigenvalue distribution situation of $(\bar{A}_D - \bar{L}\bar{C})$. Figure 6.4 shows the simulation result of FE. We can see that, even if there exist unstable invariant zeros in the discrete-time dynamics (A, B_f, C, D_f), FE can still be achieved using the proposed design method.

Similarly, we compare the minimum H_∞ attenuation value between Corollary 6.1 (for discrete-time systems) and Theorem 6.3 under the common regional pole constraint. The minimum attenuation values are $\gamma = 7.0298$ and $\gamma = 6.5625$, respectively, so it is seen that by introducing the slack matrix $\bar{S}$, Theorem 6.3 is less conservative than Corollary 6.1.

6.4 Adaptive Fault Estimation

Issues and concerns about system or process safety and reliability have necessitated and fostered the development of fault detection and diagnosis for dynamical systems. FDD has been regarded as one of the most important aspects in seeking effective solutions to guarantee

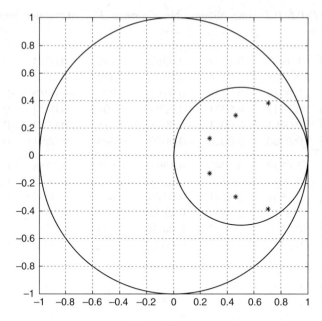

Figure 6.3 $(\bar{A}_D - \bar{L}\bar{C})$ pole distribution

reliable operation of practical control systems at the possible occurrence of system failures or malfunctions. During the past two decades, significant research results in the area of fault detection and accommodation have been seen [21, 22].

6.4.1   Introduction

In order to avoid performance deterioration or system damage, faults have to be found as soon as possible and schemes have to be made to stop propagation of bad effects. Traditional

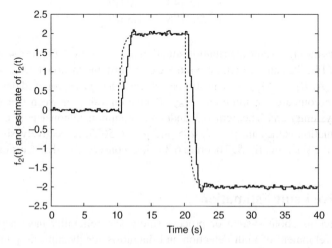

Figure 6.4   Fault $f_2(t)$ (dotted line) and its estimate $\hat{f}_2(t)$ (solid line)

approaches to fault detection and identification are mainly focused on linear systems, which are widely described and well documented in many research articles [23, 24]. However, the majority of practical control systems are nonlinear in nature. Nonlinear properties cannot be neglected for the purpose of fault diagnosis and identification, therefore active research into nonlinear system fault detection and identification has been given more and more attention [25, 26, 27].

One of the most important tasks in a fault detection and identification scheme, fault estimation is for determining the extent, such as magnitude or frequency, of the faults. Accurate fault estimation can help reconstruct the fault signals so that their effects can be accommodated in the corresponding control reconfiguration. Nevertheless, it is not an easy task, especially for nonlinear systems with unknown disturbance. Jiang, Staroswiecki, and Cocquempot [28] transformed a nonlinear system with uncertainties into two subsystems under some geometric conditions, and then established an adaptive observer to obtain the estimations of both states and actuator or sensor faults. Yan and Edwards [29] utilized sliding-mode observer to realize fault reconstruction. Gao and Ding [30] developed a fault estimator based on a descriptor system formulation for the sensor fault estimation problem. It can simultaneously estimate the states and the sensor fault signal superimposed on the output. Hou [31] provided an effective method to estimate the amplitude and frequency of a sinusoidal signal. Some other methods have been utilized to tackle fault estimation problems; see, for instance, the work of Vijayaraghavan, Rajamani, and Bokor [32].

As we learned earlier, faults can be classified into additive and multiplicative faults according to their effects on the system outputs and the system dynamics. Although component faults and some actuator or sensor faults appear in the form of multiplicative faults that correspond to parameter changes in the system model, most of the literature about fault estimation has paid attention to the effects of additive faults, which result in changes only in the mean value of the system output signal. On the other hand, some studies about multiplicative fault estimation can be found [33, 21, 34].

Generally, it is relatively harder to separate the effects of the faults from the input and states because they are mixed in a multiplicative form. Therefore, the analysis and design for multiplicative fault estimation is not as straightforward as that for additive faults. Though it is more difficult, the estimation of the real effect of multiplicative faults has been given more and more attention. In a recent article [35], a good fault detection and isolation scheme was presented for a system developing an unknown fault function that was restricted to a finite set of fault types, and each type was described by the product of an unknown parameter vector characterizing the time-varying magnitude of the fault with a known smooth vector representing the functional structure of the fault.

Taking into account the conditions mentioned above, we now focus on a Lipschitz nonlinear system subject to multiplicative faults and unknown disturbances. Our aim is to establish a robust adaptive fault estimation scheme which is robust with respect to the disturbance and sensitive to the faults, to detect and estimate multiplicative faults and get the real effect of faults. Based on Lyapunov stability theory and by relaxing a less conservative Lipschitz condition, an estimator is developed to estimate both the system states and real fault factors simultaneously. The effect of unknown disturbances is reduced according to an L_2 gain performance criterion. Compared to most of the existing work on fault estimation, the proposed scheme is simple to compute, easy to implement, and capable of estimating the actual size of the faulty parameters in the model.

6.4.2 Problem statement

In this section, we focus on a class of nonlinear multi-input–multi-output dynamical systems described by

$$\dot{x}(t) = Ax(t) + \phi(x, u, t) + Bu(t) + d(t), \tag{6.35}$$

$$y(t) = Cx(t) \tag{6.36}$$

where $x(t) \in \Re^n$, $y(t) \in \Re^m$, and $u(t) \in \Re^p$ are the state vector, the output vector, and the input vector, respectively. $d(t)$ represents the system disturbance and the L_2 norm of the unknown input $d(t)$ is bounded. A, B, and C are the known system matrices of appropriate dimensions. $\phi(x, u, t)$ is a Lipschitz nonlinear vector function with a Lipschitz constant ϕ, that is,

$$\|\phi(x, u, t) - \phi(\hat{x}, u, t)\|_2 \leq \delta \| \left(x(t) - \hat{x}(t) \right) \|_2. \tag{6.37}$$

It should be noted that Equation (6.35) is a general form since most nonlinear functions can be expanded at the equilibrium point. For instance, the nonlinear system $\dot{x} = f(x, u, t)$ is differentiated with respect to x and u, and (x_e, u_e) is the equilibrium point. Applying the Taylor expansion, we can get $A = \frac{\partial f(x, u, t)}{\partial x}\big|_{x=x_e, u=u_e}$, $B = \frac{\partial f(x, u, t)}{\partial u}\big|_{u=u_e, x=x_e}$, and $\phi(x, u, t)$ can be assumed to be the remaining term. Further, many nonlinear functions can be assumed as Lipschitz, at least locally. For example, the sinusoidal function $\sin(x)$ that appears in many robotic control systems is globally Lipschitz. And the term x^2 can be regarded as locally Lipschitz within a finite range of x.

With the assumption that the nonlinear system of Equations (6.35) and (6.36) is subject to component faults that are parameter changes within the process, the post-fault system is modeled as

$$\dot{x}(t) = Ax(t) + \phi(x, u, t) + Bu(t) + d(t) + \sum_{i=1}^{l} \theta_i(t) g_i(x, u, t), \quad t \geq t_f, \tag{6.38}$$

$$y(t) = Cx(t) \tag{6.39}$$

where $\theta_i(t) \in \Re, i = 1, \ldots, \ell$, are unknown time functions which are assumed to be zero before the fault occurrence, and nonzero after the fault occurrence. $g_i(x, u, t), i = 1, \ldots, l$, are known functions related to system states and inputs, which also satisfy the Lipschitz conditions with a Lipschitz constant δ_i. For simplicity, the time t is dropped from the notation in the following equations.

Assumption 6.2 *The multiplicative fault factors $\theta_i, i = 1, \ldots, l$, are unknown and bounded by a constant, that is, $\|\theta_i\| \leq \alpha_i$. The constant α_i is known.*

In Equation (6.38), the term $\sum_{i=1}^{l} \theta_i(t) g_i(x, u, t)$ is generated by the multiplicative faults. This representation characterizes a general class of multiplicative faults where θ_i represents the magnitude of the time-varying or constant fault and g_i characterizes the functional structure of the ith fault. Multiplicative faults encountered in a linear system can be transformed into this kind of representation. For example, the linear system $\dot{x} = (A + A_f)x$ is subject to multiplicative faults in the form $A_f = \sum_{i=1}^{l} A_i \theta_{Ai}$, so we can get the structure function $g_i = A_i x$. In practice, component faults in the process and some faults in the sensors and actuators are in the form of multiplicative faults, which changes system parameters and usually mixes with system states and inputs. Hence such faults result in performance degradation or even instability of the system. Letting $f = \sum_{i=1}^{l} \theta_i(t) g_i(x, u, t)$, we can rewrite Equation (6.38) as

$$\dot{x} = Ax + Bu + \phi(x, u, t) + d(t) + f. \tag{6.40}$$

It is clear that f is a term induced by the component faults $\theta_i, i = 1, \ldots, l$. When the system is in normal operation, $f = 0$. The form of Equation (6.40) has been adopted to treat the additive fault estimation, where the size of f can be estimated. However, it is clear that in modeling the system component faults, the term f is also a function of the system state and input. f alone cannot reflect the real fault sources or size. Therefore, it is necessary for us to estimate the real fault factors $\theta_i, i = 1, \ldots, l$, instead of the additive fault vector f.

Relaxing the Lipschitz condition, Equation (6.37), can result in a relaxing matrix form which is defined as

$$\|\phi(x, u, t) - \phi(\hat{x}, u, t)\|_2 \leq \left\| H\left(x(t) - \hat{x}(t)\right)\right\|_2. \tag{6.41}$$

The matrix H could be a sparsely populated matrix. There is an example to illustrate that $\|H(x(t) - \hat{x}(t))\|_2$ is much smaller than $\delta\|x(t) - \hat{x}(t)\|_2$ for the same nonlinear function in the work of Phanomchoeng and Rajamani [36]. The relaxing Lipschitz condition Equation (6.41) is much less conservative.

The objective here is to design an adaptive estimator with an effective algorithm for the nonlinear system of Equations (6.35) and (6.36) subject to multiplicative faults that will estimate the real effect factor $\theta_i, i = 1, \ldots, \ell$, in the post-fault system of Equations (6.38) and (6.39). The adaptive estimator should also make the estimation accurate and insensitive to the unknown disturbances. In order to design an estimator satisfying the above objective, it is assumed that the system states and inputs are all bounded before and after the occurrence of a fault and that the Lipschitz nonlinear functions $\phi(x, u, t)$ and $g_i(x, u, t)$ satisfy the relaxing Lipschitz condition with matrices H and G_i. It should be noted that the feedback control system is capable of making the system bounded even in the presence of a fault. The proposed fault estimation design is independent of the structure of the feedback controller.

6.4.3 Robust adaptive estimation

In this section, an adaptive observer is applied to reconstruct multiplicative fault signals which are mixed with system states and inputs. The adaptive observer designed for the nonlinear system of Equations (6.35) and (6.36) can be shown to be as follows:

$$\dot{\hat{x}} = A\hat{x} + Bu + \phi(\hat{x}, u) + \sum_{i=1}^{l} \hat{\theta}_i g_i(\hat{x}, u) + L(y - \hat{y}), \tag{6.42}$$

$$\hat{y} = C\hat{x}, \tag{6.43}$$

$$\dot{\hat{\theta}}_i = \sigma_i g^T \sigma_i(\hat{x}, u) D(y - \hat{y}), \quad i = 1, \ldots, \ell \tag{6.44}$$

where $\sigma_i > 0, i = 1, \ldots, \ell$ are constants and $\hat{x}$, $\hat{y}$, and $\hat{\theta}_i$ denote the estimated state, output, and fault variables, respectively. L and D are the design gain matrices. Let $e_x = x - \hat{x}$ and $e_y = y - \hat{y}$ represent the state and output estimation error; $e_{\theta_i} = \theta_i - \hat{\theta}_i$ denotes fault error. Then we obtain the following estimation error dynamic equations:

$$\dot{e}_x = (A - LC)e_x + \phi(x, u) - \phi(\hat{x}, u) + \sum_{i=1}^{l}\left(\theta_i g_i(x, u) - \hat{\theta}_i g_i(\hat{x}, u)\right) + d, \quad (6.45)$$

$$e_y = Ce_x, \quad (6.46)$$

$$\dot{e}_{\theta_i} = -\sigma_i g_i^T(\hat{x}, u)De_y(t) \quad (6.47)$$

The main problem encountered here is that the system is subject to unknown disturbance and the real fault effect factor θ_i, which is combined with the system state x and input u. We must design an appropriate estimator which can estimate the fault θ_i effectively and be less sensitive to the disturbance.

In the analysis of the estimation error functions, Equations (6.45)–(6.47), a sufficient condition for asymptotic stability of the observer is presented and proved in the following theorem.

Theorem 6.4   *Suppose the pair (A, C) is observable, and the matrix C is of full row rank. Assume that $g_i(x, u, t), i = 1, \ldots, l$, are persistence of excitation. If there is a positive definite matrix $P = P^T > 0$ and a matrix D such that*

$$\Pi = \begin{bmatrix} \Lambda + C^T C & P \\ P & -\gamma^2 I \end{bmatrix} < 0, \quad (6.48)$$

$$\Lambda = (A - LC)^T P + P(A - LC) + H^T H + \sum_{i=1}^{l} G_i^T \alpha_i \alpha_i G_i + (l+1)PP, \quad (6.49)$$

$$DC = P \quad (6.50)$$

then the observer-based estimator in Equations (6.42)–(6.44) ensures that

- *The estimated x and $\hat{\theta}_i$ asymptotically converge to the nonlinear system state x and the multiplicative fault θ_i respectively under the zero disturbance case.*
- *When the unknown disturbance exists, the output error satisfies $\|e_y\|_2^2 < \gamma^2 \|d\|_2^2$.*

Proof. The proof consists of two parts: the internal stability analysis and computing the robust performance index.   ∎

6.4.4   Internal stability analysis

We choose the Lyapunov function candidate as follows:

$$V(t) = e_x^T(t)Pe_x(t) + \sum_{i=1}^{l} \sigma_i^{-1} e_{\theta_i}^T(t)e_{\theta_i}(t)$$

and calculate the derivative of the Lyapunov function $V(t)$. We get

$$\dot{V} = e_x^T\left((A - LC)^T P + P(A - LC)\right)e_x + 2e_x^T P\left(\phi(x, u) - \phi(\hat{x}, u)\right)$$

$$+ \sum_{i=1}^{l} 2e_x^T P\left(\theta_i g_i(x, u) - \hat{\theta}_i g_i(\hat{x}, u)\right) + \sum_{i=1}^{l} 2\sigma_i^{-1} e_{\theta_i} \dot{\hat{\theta}}_i + 2e_x^T Pd$$

$$= e_x^T\left((A - LC)^T P + P(A - LC)\right)e_x + 2e_x^T P\left(\phi(x, u) - \phi(\hat{x}, u)\right)$$

$$+ \sum_{i=1}^{l} 2e_x^T P\left(\theta_i g_i(x, u) - \hat{\theta}_i g_i(\hat{x}, u)\right) + 2e_x^T Pd.$$

According to the Lipschitz condition, we have

$$2e_x^T P\left(\phi(x, u) - \phi(\hat{x}, u)\right) \leq 2\|e_x^T P\| \|\phi(x, u) - \phi(\hat{x}, u)\|$$

$$\leq e_x^T PP e_x^T + e_x^T H^T H e_x^T,$$

$$2e_x^T P\left(\theta_i g_i(x, u) - \theta_i g_i(\hat{x}, u)\right) \leq 2\|e_x^T P\| \|\theta_i g_i(x, u) - \theta_i g_i(\hat{x}, u)\|$$

$$\leq e_x^T PP e_x^T + e_x^T G_i^T \alpha_i \alpha_i G_i e_x^T.$$

Then the derivative of the Lyapunov function satisfies the following inequality:

$$\dot{V} \leq e_x^T(\Lambda)e_x + 2e_x^T Pd \qquad (6.51)$$

In the zero disturbance case, one has

$$\dot{V} \leq -\lambda_{\min}(-\Lambda)\|e_x\|^2 \qquad (6.52)$$

By the Schur complements, matrix Λ is a negative definite matrix. Inequality (6.52) indicates $e_x \in L_2$. Because $e_x \in L_\infty$, $\dot{e}_x$ is uniformly bounded. Based on Barbalat's Lemma A.10, we have $e_x \to 0$ as $t \to 0$. Because of the persistent excitation condition of $g_i(x, u)$, the estimator of Equations (6.42)–(6.44) ensures that $e_{\theta_i} \to 0$ as $t \to 0$.

6.4.5 Robust performance index

We define

$$J = \dot{V} + e_y^T e_y - \gamma^2 d^T d.$$

Using Equation (6.51), we can derive that

$$J \leq e_x^T \Lambda e_x + 2e_x^T Pd + e_x^T C^T C e_x - \gamma^2 d^T d$$

$$\leq e^T \Pi e,$$

where

$$e = [e_x^T \quad d^T]^T.$$

It follows that

$$J \leq -\lambda_{\min}(-\Pi)\|e\|^2.$$

Under the zero initial condition, we have

$$\int_0^T \left(e_y^T e_y - \gamma^2 d^T d\right) dt = \int_0^T J dt - V(T) < 0,$$

which implies

$$\int_0^T e_y^T e_y dt \leq \gamma^2 \int_0^T d^T d dt.$$

This completes the proof of the theorem.

Remark 6.13 *Based on the Schur Complement Lemma and letting $Y = PL$, $\Pi < 0$, Equation (6.48) can be rewritten as the following matrix inequality:*

$$\begin{bmatrix} \Xi & P & C^T & P \\ P & -\frac{1}{l+1}I & 0 & 0 \\ C & 0 & -I & 0 \\ P & 0 & 0 & -\gamma^2 I \end{bmatrix} < 0,$$

$$\Xi = A^T P + PA - C^T Y - YC + H^T H + \sum_{i=1}^l G_i^T \alpha_i \alpha_i G_i.$$

The matrix D can be derived from

$$D = PC^T \left(CC^T\right)^{-1}.$$

Remark 6.14 *A good estimator is designed to make the whole system sensitive to the multiplicative fault and insensitive to the disturbance. Hence, we can reduce the effect of disturbance d on the Equation (6.48) with a smaller γ using the Matlab LMI toolbox.*

6.4.6 Simulation

We consider a one-link manipulator with revolute joints actuated by a DC motor, which is an excellent example used to verify design schemes in many works [36, 35]. The corresponding state-space model with no faults and disturbance is

$$\dot{q}_m = \omega_m,$$

$$\dot{\omega}_m = \frac{k}{J_m}(q_1 - q_m) - \frac{B}{J_m}\omega_m + \frac{k_\tau}{J_m}u,$$

$$\dot{q}_1 = \omega_1,$$

$$\dot{\omega}_1 = -\frac{k}{J_1}(q_1 - q_m) - \frac{mgh}{J_1}\sin(q_1),$$

where q_1 and q_m are the angular position of the link and the motor, respectively; ω_1 is the angular velocity of the link and ω_m is the angular velocity of the motor; J_1 and J_m are the inertia of the link and the motor; and the control u is the torque of the motor. The nonlinear system with multiplicative fault and disturbance is shown in the following:

$$\dot{x} = Ax + Bu + \phi(x) + d + \sum_{i=1}^{l} \theta_i g_i(x, u),$$

$$y = Cx,$$

with

$$A = \begin{bmatrix} 0 & 1 & 0 & 0 \\ -48.6 & -1.25 & 48.6 & 0 \\ 0 & 0 & 0 & 1 \\ 19.5 & 0 & -19.5 & 0 \end{bmatrix}, \quad B = \begin{bmatrix} 0 \\ 21.6 \\ 0 \\ 0 \end{bmatrix},$$

$$\phi = \begin{bmatrix} 0 \\ 0 \\ 0 \\ -3.33\sin(x_3) \end{bmatrix}, \quad C = \begin{bmatrix} 1 & 0 & 0 & 0 \\ 0 & 1 & 0 & 0 \end{bmatrix}.$$

The unknown disturbance is $d = \begin{bmatrix} 0 & d_1 & 0 & d_2 \end{bmatrix}^T$, where the L_2 norms of d_1 and d_2 are assumed to be bounded. Two types of component fault are considered here:

- An abnormal friction appears in the motor which leads to parameter changes in the system state matrix. Suppose that the viscous friction constant B increases by 20 % at $t = 5$ s. In this case, $\theta_1 \in [0, 1]$ represents the real multiplicative fault parameter. When $\theta_1 = 0$, the system is in normal operation. $\theta_1 = 0.2$ at $t = 5$ s and the viscous friction fault structure function $g_1 = \begin{bmatrix} 0 & -1.25x_2 & 0 & 0 \end{bmatrix}^T$.
- The actuator causes a multiplicative fault which is in the form of $u = (1 + \theta_2)\bar{u}$. $\theta_2 \in [-1, 0]$ represents the magnitude of the fault. When $\theta_2 = 0$, the actuator is in normal operation; $\theta_2 = -1$ represents the complete failure of the actuator and the fault structure function is $g_2 = \begin{bmatrix} 0 & 21.6u & 0 & 0 \end{bmatrix}^T$. Here we suppose that the actuator efficiency decreases by 30 % at $t = 15$ s.

The nonlinear term ϕ is a Lipschitz nonlinear function with a global Lipschitz constant $\delta = 3.33$ and the relaxing Lipschitz matrix is

$$H = \begin{bmatrix} 0 & 0 & 0 & 0 \\ 0 & 0 & 0 & 0 \\ 0 & 0 & 0 & 0 \\ 0 & -3.33 & 0 & 0 \end{bmatrix}.$$

Based on the multiplicative fault estimation strategy, the robust adaptive multiplicative fault estimator is established. The simulation of the robust adaptive fault estimation is performed in Simulink. A sinusoidal wave input of this system is given by $u = \sin(t)$. The initial condition is $x(0) = 0$. According to Theorem 6.4, the observer gain is obtained with a robust performance $\gamma^2 = 0.3$ shown here

$$L = \begin{bmatrix} 5.9554 & -46.0479 \\ 0.5437 & 30.1714 \\ 20.7891 & 70.8924 \\ 24.3609 & 72.2822 \end{bmatrix}.$$

Figure 6.5 shows the output estimation error for the system subject to an extra abnormal friction fault and Figure 6.6 illustrates the estimated multiplicative fault θ_1. The fault is accurately estimated compared to the desired trajectory. Figures 6.7 and 6.8 depict the output estimation error and the estimation of multiplicative fault θ_2 when the actuator efficiency degradation occurs in the system. Again the fault is successfully estimated. From these simulation results, we can see that the proposed robust adaptive estimation scheme not only guarantees the state estimation and multiplicative fault estimation accurately, but also makes the estimator insensitive to unknown disturbances.

6.5 Adaptive Tracking Control Scheme

Unmanned aerial vehicles (UAV) has become one of the most active research areas. A major challenge in this area is the design of flight control systems. Path tracking is one of the most challenging aspects. Wang and Lin [37] present an adaptive dynamic surface control scheme for a class of linear multivariable systems. We use the scheme in the following discussion to design an active fault-tolerant control scheme for a hypersonic UAV with actuator loss-of-effectiveness faults. A nonlinear fault detection observer is designed to detect the actuator fault occurring in the attitude systems of UAV, which determines the switching time from the normal controller to the FTC one. When an actuator fault occurs, we design a dynamic surface control-based active FTC scheme which guarantees the attitude of the faulty UAV asymptotically tracking the desired command signal.

6.5.1 Attitude dynamics

A hypersonic UAV with uncertainty in parameter and external disturbance input has attitude dynamics [38] as

$$(M + \Delta M)\dot{w} = -w^{\times}(M + \Delta M)w + u + d \tag{6.53}$$

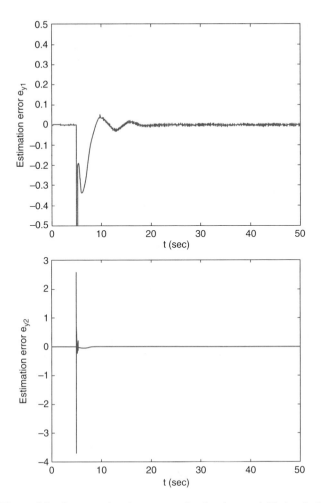

Figure 6.5 Output estimation error under the abnormal friction fault

where $M \in \Re^{3 \times 3}$ is the nominal inertia matrix, $\Delta M \in \Re^{3 \times 3}$ is an uncertain part of the inertia matrix, which is caused by fuel consumption and variations of particular payloads from a nominal one. Introduce

$$w = \begin{bmatrix} p \\ q \\ r \end{bmatrix}$$

$$u = \begin{bmatrix} u_1 \\ u_2 \\ u_3 \end{bmatrix}$$

$$d = \begin{bmatrix} d_1 \\ d_2 \\ d_3 \end{bmatrix}$$

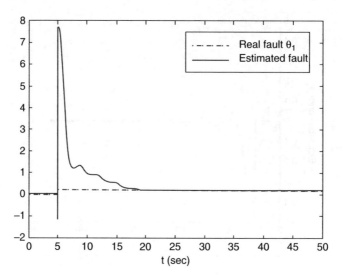

Figure 6.6   Estimation of viscous friction

as the angular rate vector, the control torque vector and the external disturbance vector, respectively. Let the operator $w^\times$ denote a skew-symmetric matrix acting on the vector $w = [\, w_1 \quad w_2 \quad w_3 \,]^T$ that has the following form:

$$w^\times = \begin{bmatrix} 0 & -w_3 & w_2 \\ w_3 & 0 & -w_1 \\ -w_2 & w_1 & 0 \end{bmatrix}$$

For simplicity, the hypersonic UAV attitude dynamics [38] is rewritten as

$$M\dot{w} = -w^\times M w + u + \eta(w, d) \tag{6.54}$$

where $\eta(w, d) = -\Delta M \dot{w} - w^\times \Delta M w + d$ represents the combination of parameter uncertainty and external disturbance. The following assumption provides a condition for $\eta(w, d)$.

Assumption 6.3   *The combination of parameter uncertainty and external disturbance represented by $\eta(w, d)$ in (6.54) is the unknown nonlinear function of w and d, but bounded by the known constant $\bar{\eta}$. Specifically, it is assumed that $\forall w \in \mathfrak{R}^3, d \in \mathfrak{R}^3, |\eta_i(w, d)| < \bar{\eta}_i$ ($i = 1, 2, 3$), $\bar{\eta}_i$ is the known constant.*

It follows [39] that the attitude kinematics of a hypersonic UAV is described by

$$\dot{\gamma} = R(\gamma)w$$

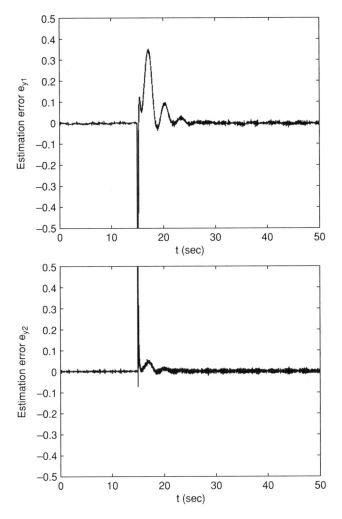

Figure 6.7 Output estimation error under actuator fault

where the rotational matrix $R(\gamma) \in \{R_1(\gamma), R_2(\gamma)\}$ is given by

$$R_1(\gamma) = \begin{bmatrix} 1 & tan\theta sin\Phi_b & tan\theta cos\Phi_b \\ 0 & cos\Phi_b & -sin\Phi_b \\ 0 & \frac{sin\Phi_b}{cos\theta} & \frac{cos\Phi_b}{cos\theta} \end{bmatrix}, \quad \gamma = \begin{bmatrix} \Phi_b \\ \theta \\ \Psi \end{bmatrix}$$

$$R_2(\gamma) = \begin{bmatrix} cos\alpha & 0 & sin\alpha \\ 0 & 1 & 0 \\ sin\alpha & 0 & -cos\alpha \end{bmatrix}, \quad \gamma = \begin{bmatrix} \Phi \\ \alpha \\ \beta \end{bmatrix}$$

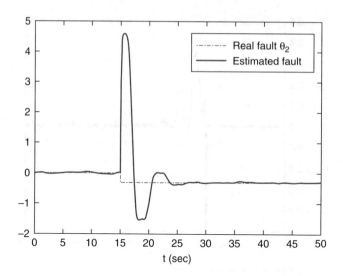

Figure 6.8   Estimation of actuator fault

Remark 6.15   *It shold be noted that $R_1(\gamma)$, $R_2(\gamma)$ are often used in the ascent and reentry phases, repectively. In the following discussion, only the reentry phase is considered. The rotation matrix is a square matrix with real entries that is used to perform a rotation in Euclidean space. More specifically, its elements can be characterized as orthogonal matrices with unity determinant.*

The command torque u is known to be related to the deflection command vector δ, namely,

$$u = L\delta$$

where $L \in \mathfrak{R}^{3 \times m}$ and m is the number of the control-surface deflection variables. Figure 6.9 shows the configuration of the X-33 hypersonic unmanned aerial vehicle. It has four sets of control surfaces: rudders, body flaps, inboard elevons, and outboard elevons, with left and right sides for each set. Each of the control surfaces can be actuated independently. The control-surface deflection variables, collectively known as the effector vector, are given [39] by

$$\delta = \begin{bmatrix} \delta_{rei}, & \delta_{lei}, & \delta_{rft}. & \delta_{lft}, & \delta_{rvr}, & \delta_{lvr}, & \delta_{reo}, & \delta_{leo} \end{bmatrix}^T$$

6.5.2   Fault detection scheme

The hypersonic UAV attitude dynamics described by Equation (6.54) is for a fault-free actuator. To put the fault-tolerant control problem in perspective, the faulty attitude dynamics of a

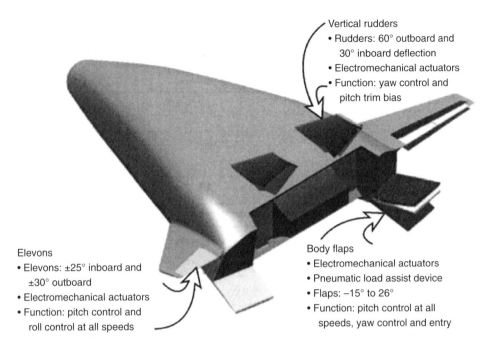

Vertical rudders
- Rudders: 60° outboard and
  30° inboard deflection
- Electromechanical actuators
- Function: yaw control and
  pitch trim bias

Elevons
- Elevons: ±25° inboard and
  ±30° outboard
- Electromechanical actuators
- Function: pitch control and
  roll control at all speeds

Body flaps
- Electromechanical actuators
- Pneumatic load assist device
- Flaps: −15° to 26°
- Function: pitch control at all
  speeds, yaw control and entry

Figure 6.9   Configuration of the X-33 hypersonic unmanned aerial vehicle

hypersonic UAV must be established in terms of the loss of control effectiveness. Therefore the hypersonic UAV attitude dynamics under an actuator fault is given by

$$M\dot{w} = -w^{\times}Mw + L(I - F)\delta + \eta(w, d) \tag{6.55}$$

where $F = diag\{fi\}(i = 1, 2, \ldots 8)$ and $f_i \in [0, \epsilon_i]$, f_i is an unknown constant, ϵ_i represents the maximum percentage of the admissible loss of control effectiveness satisfying $0 \le \epsilon_i < 1$.
The model of Equation (6.55) can also be represented as

$$\dot{w} = -M^{-1}w^{\times}Mw + M^{-1}L\delta + M^{-1}\eta(w, d) - M^{-1}LF\delta. \tag{6.56}$$

To proceed further, we seek to design a nonlinear fault detection observer of the form

$$\dot{\hat{w}} = -\lambda(\hat{w} - w) - M^{-1}w^{\times}Mw + M^{-1}L\delta \tag{6.57}$$

where $\hat{w}$ is the estimated angular rate and $\lambda = diag\{\lambda_1, \lambda_2, \lambda_3\}$, $-\lambda_i < 0$ $(i = 1, 2, 3)$ are the poles of fault detection observer Equation (6.56), which are determined *a priori*. Note that $M^{-1}\eta = [M^{-1}\eta_1, M^{-1}\eta_2, M^{-1}\eta_3]^T$.
Let $\tilde{w} \triangleq w - \hat{w}$ be the observer error vector, then the adaptive threshold $\bar{\Upsilon}$ for fault detection can be chosen as

$$\bar{\Upsilon}_i \triangleq e^{-\lambda_i t}\tilde{w}_i(0) + \int_0^t e^{-\lambda_i(t-\tau)}(M^{-1}\bar{\eta})_i\, d\tau, \quad (i = 1, 2, 3) \tag{6.58}$$

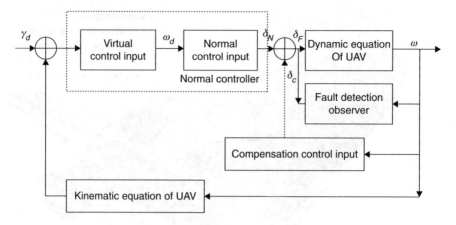

Figure 6.10 UAV using dynamic surface control

Note that the second term in $\bar{\Upsilon}_i$ can be obtained as the output of a linear filter (with transfer function $\frac{1}{s+\lambda_i}$, for example $\frac{1}{s+10}$ and zero initial conditions) with input given by $(M^{-1}\bar{\eta})_i$. Fault detection is therefore made when the modulus of at least one of the observer error components $|\tilde{w}_i|$ exceeds its corresponding adaptive threshold. T_d (the fault detection time) is defined as $|\tilde{w}_i| > \bar{\Upsilon}_i > (i = 1, \ i = 2, \ i = 3)$ for $t > T_0$; that is

$$T_d := \inf \cup_{i=1,2,3} \{t > T_o : |\tilde{w}_i| > \bar{\Upsilon}_i\} \tag{6.59}$$

6.5.3 Fault-tolerant tracking scheme

The task now is to design a fault-tolerant controller for the faulty system in order to guarantee that the closed-loop signals are bounded and the attitude of UAV γ asymptotically tracks a reference command γ_d. A fault-tolerant control scheme for UAV using the dynamic surface control configuration scheme is depicted in Figure 6.10.

It must be observed that during normal operational mode (hypersonic UAV with a fault-free actuator), control input δ_N is designed for the attitude dynamics of the fault-free case using both adaptive and dynamic surface control techniques [40]. When an actuator fault occurs, a compensation control input δ_C is added to the normal control input δ_N to reduce the effects of actuator fault. For the purpose of designing a normal control input N, the following new variables are defined:

$$p_1 = \gamma \in \Re^3, \quad p_2 = w \in \Re^3. \tag{6.60}$$

The attitude dynamics of UAV Equation (6.53) and Equation (6.55) are transformed into the following form:

$$\dot{p}_1 = R(p_1)p_2 \tag{6.61}$$

$$\dot{p}_2 = -M^{-1}p_2^\times M p_2 + M^{-1}L\delta + M^{-1}\eta(p_2, d). \tag{6.62}$$

Theorem 6.5   *Consider a fault-free UAV attitude dynamics given by Equations (6.61) and (6.62) under Assumption 6.3. The normal control input δ_N and adaptive control law $\dot{\hat{\eta}}$ guarantee the asymptotic output tracking of the UAV attitude control system:*

$$\delta_N = L^T (LL^T)^{-1}[-K_2 Z_2 + p_2^\times M p_2 - sign(Z_2)\hat{\eta} + M\dot{\alpha}_2] \qquad (6.63)$$

$$\dot{\hat{\eta}} = C_1 Z_2 \qquad (6.64)$$

Proof. Similar to [40], we assume that p_2 in Equation (6.61) is a virtual control input. Let

$$z_1 = p_1 - \gamma_d \qquad (6.65)$$

which is called the first error surface, $\gamma_d \in \mathfrak{R}^3$ is the desired attitude angle of UAV. Taking the time derivative of z_1, one has

$$\dot{Z}_1 = \dot{p}_1 - \dot{\gamma}_d = R(p_1)p_2 - \dot{\gamma}_d = R(p_1)[p_2 - \mathfrak{R}^{-1}(p_1)\dot{\gamma}_d] \qquad (6.66)$$

We select an appropriate virtual control p_{2d} as

$$p_{2d} = -K_1 Z_1 + \mathfrak{R}^{-1}(p_1)\dot{\gamma}_d \qquad (6.67)$$

By introducing a new state variable α_2 and passing p_{2d} through a first-order filter with time constant ε_2 to obtain α_2:

$$\varepsilon_2 \dot{\alpha}_2 + \alpha_2 = p_{2d} \qquad \alpha_2(0) = p_2(0) \qquad (6.68)$$

Next, consider the system of Equation (6.62) and let

$$Z_2 = p_2 - \alpha_2 \qquad (6.69)$$

which is called the second error surface. Taking the time derivative of Z_2, we have

$$\dot{Z}_2 = -M^{-1} p_2^\times M p_2 + M^{-1} L\delta + M^{-1}\eta(p_2, d) - \dot{\alpha}_2 \qquad (6.70)$$

From Equation (6.70), the normal control input and the parameter updating law are designed as

$$\delta_N = L^T (L.L^T)^{-1}[-K_2 Z_2 + p_2^\times M p_2 - sign(Z_2)\hat{\eta} + M\dot{\alpha}_2] \qquad (6.71)$$

$$\dot{\hat{\eta}} = C_1 Z_2 \qquad (6.72)$$

To examine the stability analysis of the closed-loop control system, we define the filter error as $\phi = \alpha_2 - x_{2d}$. Mathematical manipulations yield

$$\dot{\phi} = \dot{\alpha}_2 - \dot{p}_2 = \frac{-\phi}{\varepsilon_2} + K_1 \dot{Z}_1 - \mathfrak{R}^{-1}(p_1)\dot{\gamma}_d - \mathfrak{R}^{-1}(p_1)\ddot{\gamma}_d \qquad (6.73)$$

Considering a Lyapunov function candidate as follows:

$$V = V_1 + V_2 = \frac{1}{2}(Z_1^2 + \phi^2) + \frac{1}{2}(Z_2^2 + \frac{1.\tilde{\eta}^2}{C_1}) \tag{6.74}$$

we take the time derivative of V_1 and obtain

$$\dot{V}_1 = Z_1^T \dot{Z}_1 + \phi^T \dot{\phi}$$

$$= Z_1^T R(p_1)[p_2 - \Re^{-1}(p_1)\dot{\gamma}_d] + \phi^T [\frac{-\phi}{\varepsilon_2}$$

$$+ K_1 \dot{Z}_1 - \dot{\Re}^{-1}(p_1)\dot{\gamma}_d - \Re^{-1}(p_1)\ddot{\gamma}_d] \tag{6.75}$$

By subtracting Equation (6.67) from Equation (6.75) and manipulating, it yields

$$\dot{V}_1 \leq -K_1 \|R(p_1)\| \|Z_1\|^2 - \frac{1}{\varepsilon_2}\|\phi\|^2 + K \tag{6.76}$$

Similarly, one can obtain the following:

$$\dot{V}_2 = Z_2^T \dot{Z}_2 + \tilde{\eta}^T \dot{\tilde{\eta}}$$

$$= Z_2^T[-M^{-1}p_2^\times Mp_2 + M^{-1}L\delta + M^{-1}\eta(p_2,d) - \dot{\alpha}] + \tilde{\eta}^T \dot{\tilde{\eta}} \tag{6.77}$$

Subtracting Equation (6.71) from Equation (6.77) and rearranging, we obtain:

$$\dot{V}_2 \leq -K_2\|Z_2\|^2 + \frac{\tilde{\eta}^T}{C_1}(C_1 Z_2 - \dot{\tilde{\eta}}) = -K_2\|Z_2\|^2 \tag{6.78}$$

From Equation (6.75) and Equation (6.78), it can easily be found that

$$\dot{V} = \dot{V}_1 + \dot{V}_2 \leq -K_1\|R(p_1)\|\|Z_1\|^2 - \frac{1}{\varepsilon_2}\|\phi\|^2 \tag{6.79}$$

$$-K_2\|Z_2\|^2 + k$$

Selecting an appropriate $\varepsilon_2 > 0$, such that $\dot{V} < 0$, then the Lyapunov stability theory guarantees the global uniform boundedness of Z_1 and Z_2. It follows that $Z_1 \to 0$ as $t \to \infty$. Since $Z_1 = p_1 - \gamma_d$, p_1 is also bounded and $\lim_{t \to 0} p_1 = \gamma_d$. This completes the proof. ∎

In the following discussion, we extend the result to deal with the fault-tolerant control problem of UAV attitude control systems by developing a compensation control input δ_C on the basis of the nominal controller δ_N to compensate for the effects of actuator fault. Therefore, the fault-tolerant control input δ_F of the faulty attitude control system of Equation (6.56) consists of two parts:

$$\delta_F = \delta_N + \delta_C \tag{6.80}$$

For the fault-tolerant controller, the design of the nominal control input δ_N remains the same. In terms of Equation (6.56) and Equation (6.70), we can obtain

$$
\begin{aligned}
\dot{Z}_2 &= -M^{-1}p_2^\times M p_2 + M^{-1}L(I-F)\delta + M^{-1}\eta(p_2,d) - \dot{\alpha}_2 \\
&= -M^{-1}p_2 M p_2 + M^{-1}L(\delta_N + \delta_C) - M^{-1}LF(\delta_N + \delta_C) \\
&\quad + M^{-1}\eta(p_2,d) - \dot{\alpha}_2 \\
&= -M^{-1}p_2 M p_2 + M^{-1}L\delta_N - M^{-1}LF\delta_N + M^{-1}L\zeta\delta_C \\
&\quad + M^{-1}\eta(p_2,d) - \dot{\alpha}_2
\end{aligned}
\tag{6.81}
$$

where $\zeta = I_3 F$. It can be seen easily that $\|\zeta\| < 1$.

Now a compensation control input δ_C is designed as

$$
\delta_C = -sign(Z_2)L^T(L.L^T)^{-1}\Big[\|L\delta_N\| + \frac{\lambda\|L\delta_N\|}{\bar{\zeta}}\Big]
\tag{6.82}
$$

where $\zeta = 1 - max\{fi\}$, $(i = 1, 2, 3)$, $\lambda > 0$ is a positive constant.

From Equations (6.80)–(6.82), the time derivative of V_2 is given by

$$
\begin{aligned}
\dot{V}_2 &= Z_2^T \dot{Z}_2 + \tilde{\eta}^T \dot{\tilde{\eta}} \\
&= Z_2^T[-M^{-1}p_2^\times M p_2 + M^{-1}L\delta - M^{-1}LF\delta_N \\
&\quad + M^{-1}L\zeta\delta_C + M^{-1}\eta(p_2,d) - \dot{\alpha}_2] + \tilde{\eta}^T \dot{\tilde{\eta}} \\
&\leq K_2\|Z_2\|^2 - \|Z_2^T\|\|M^{-1}\|\|L\delta_N\| - \lambda\|Z_2^T\|\|M^{-1}\|\|L\delta_N\| \\
&\leq -K_2\|Z_2\|^2 - (1+\lambda)\|Z_2^T\|\|\|M^{-1}\|\|L\delta_N\| \\
&\leq -K_2\|Z_2\|^2
\end{aligned}
\tag{6.83}
$$

Note that the inequality $0 < \bar{\zeta} < \|\zeta\|_{min} \leq 1$ is used in the operation of Equation (6.75), from which (with a stability analysis similar to that of Theorem 6.5), the following result can be obtained.

Theorem 6.6 *Consider the faulty UAV attitude control system of Equations (6.55) and (6.56); the fault-tolerant control input δ_F described in Equation (6.80) can guarantee the asymptotic output tracking of the UAV attitude control system.*

Remark 6.16 *The chattering effect caused by the sign function can be overcome by employing boundary layers. Therefore the control signal δ_N can be modified as*

$$
\delta_N = L^T(L.L^T)^{-1}\Big[- K_2 Z_2 + p_2^\times M p_2 - \frac{Z_2}{\|Z_2\| + \rho}\hat{\eta} + M\dot{\alpha}_2\Big]
$$

$$
\dot{\hat{\eta}} = C_1 Z_2
\tag{6.84}
$$

and the compensated control input δ_C can be modified as

$$\delta_C = -L^T(L.L^T)^{-1} \frac{Z_2}{\|Z_2\| + \rho_2} \left[\|L\delta_N\| + \lambda \frac{\|L\delta_N\|}{\bar{\zeta}}\right] \tag{6.85}$$

where $\rho_1 > 0$ and $\rho_2 > 0$ are two positive constant scalars.

Remark 6.17 *Dynamics surface is an improved backstepping control method, the primary advantage of which is that it can avoid the problem of explosion of terms inherent in the backstepping design procedure. For further details, the reader is referred to the work of Qian, Jiang, and Xu [40].*

6.6 Notes

In this chapter, a design framework of robust fault estimation observer for both continuous-time and discrete-time systems was investigated, and their feasibility were established. The degree of design conservatism was further reduced by introducing slack variables. Simulation results for a VTOL aircraft were presented to illustrate the effectiveness of the proposed method. Next, a robust adaptive fault estimation scheme was proposed for a type of Lipschitz nonlinear system subject to multiplicative faults and unknown disturbances. Multiplicative faults are parameter changes within the process that make the design for fault estimation more complicated. The estimator was designed in the context of a trade-off between robustness to disturbances and sensitivity to faults. According to Lyapunov stability theory, the estimator can estimate the real fault factors accurately, and simultaneously estimate the system states. The conservatism for the whole fault estimation scheme was reduced by using a relaxing Lipschitz matrix. The adaptive fault estimation algorithms have been verified by a one-link manipulator control system. Finally, an adaptive fault-tolerant control approach for a class of unmanned aerial vehicle attitude dynamical systems with actuator loss-of-effectiveness fault was studied in detail. A nonlinear fault detection observer was designed to detect the faulty condition in the UAV attitude control system. By utilizing the dynamic surface control technique, a fault-tolerant control strategy was developed for the faulty UAV attitude control systems. A compensatory control input δ_C was designed and added to the normal control signal δ_N in faulty condition. Stability of the closed-loop control system is proved on the basis of Lyapunov theory.

References

[1] Wang, H., and Daley, S. (1996) "Actuator fault diagnosis: an adaptive observer-based technique", *IEEE Trans. Automatic Control*, **41**(7):1073–1078.

[2] Jiang, B., Staroswiecki, M., and Cocquempot, V. (2002) "Fault identification for a class of time-delay systems", in *Proc. American Control Conference*, Anchorage, 2239–2244.

[3] Zhang, K., Jiang, B., and Cocquempot, V. (2009) "Fast adaptive fault estimation and accommodation for nonlinear time-varying delay systems", *Asian J. Control*, **11**(6):643–652.

[4] Edwards, C., Spurgeon, S. K., Patton, R. J. (2000) "Sliding mode observers for fault detection and isolation", *Automatica*, **36**(4):541–553.

[5] Yan, X. G., Edwards, C., and Spurgeon, S. K. (2004) "Output feedback sliding mode control for non-minimum phase systems with nonlinear disturbances", *Int. J. Control*, **77**(15):1353–1361.

[6] Chen, W., and Saif, M. (2007) "Observer-based strategies for actuator fault detection, isolation and estimation for certain class of uncertain nonlinear systems", *IET Control Theory and Applications*, **6**(1):1672–1680.

[7] Chen, W., and Saif, M. (2007) "A sliding mode observer-based strategy for fault detection, isolation, and estimation in a class of Lipschitz nonlinear systems", *Int. J. Systems Science*, **38**(12):943–955.

[8] Shafai, B., Pi, C. T., and Nork, S. (2002) "Simultaneous disturbance attenuation and fault detection using proportional integral observers", in *Proc. American Control Conference*, Anchorage, 1647–1649.

[9] Koenig, D., and Mammar, S. (2002) "Design of proportional-integral observer for unknown input descriptor systems", *IEEE Trans. Automatic Control*, **47**(12):2057–2062.

[10] Wang, H., and Lam, J. (2002) "Robust fault detection for uncertain discrete-time systems", *J. Guidance, Control, and Dynamics*, **25**(2):291–301.

[11] Saif, M. (1998) "Robust discrete time observer with application to fault diagnosis", *IEE Proc. Control Theory and Applications*, **145**(3):353–357.

[12] Jiang, B., and Chowdhury, F. N. (2005) "Fault estimation and accommodation for linear MIMO discrete-time systems", *IEEE Trans. Control Systems Technology*, **13**(3):493–499.

[13] Gao, Z., Breikin, T., and Wang, H. (2008) "Discrete-time proportional and integral observer and observer-based controller for systems with both unknown input and output disturbances", *Optimal Control Applications and Methods*, **29**(3):171–189.

[14] de Oliveira, M. C., Geromel, J. C., and Bernussou, J. (2002) "Extended H_2 and H_∞ norm characterizations and controller parameterizations for discrete-time systems", *Int. J. Control*, **75**(9):666–679.

[15] Xie, W. (2008) "Multi-objective $H_\infty \alpha$-stability controller synthesis of LTI systems. *IET Control Theory and Applications*, **2**(1):51–55.

[16] Garcia, G., and Bernussou, J. (1995) "Pole assignment for uncertain systems in a specified disk by state feedback", *IEEE Trans. Automatic Control*, **40**(1):184–190.

[17] Nguang, S. K., Shi, P., and Ding, S. (2006) "Delay-dependent fault estimation for uncertain time-delay nonlinear systems: an LMI approach", *Int. J. Robust and Nonlinear Control*, **16**(18):913–933.

[18] Li, X.-J., and Yang, G.-H. (2009) "Fault estimation for discrete-time delay systems in finite frequency domain", in *Proc. American Control Conference*, St Louis, 4328–4333.

[19] Gahinet, P., and Apkarian, P. (1994) "A linear matrix inequality pproach to H_∞ control", *Int. J. Robust and Nonlinear Control*, **4**(4):421–448.

[20] Saif, M., and Guan, Y. (1993) "A new approach to robust fault detection and identification", *IEEE Trans. Aerospace and Electronic Systems*, **29**(3):685–695.

[21] Isermann, R. (2005) "Model-based fault-detection and diagnosis: status and applications", *Annual Reviews in Control*, **29**(1):71–85.

[22] Zhang, Y., and Jiang, J. (2008) "Bibliographical review on reconfigurable fault-tolerant control systems", *Ann. Review Control*, **32**(2):229–252.

[23] Duan, G. R., Patton, R. J. (2001) "Robust fault detection using Luenberger-type unknown input observers: a parametric approach", *Int. J. Syst. Sci*, 32(4):533–540.

[24] Frank, P. M. (1990) "Fault diagnosis in dynamic systems using analytical and knowledge-based redundancy: a survey and some new results", *Automatica*, **26**(3):459–474.

[25] Boskovic, J. D., Bergstrom, S. E., Mehra, R. K. (2005) "Robust integrated flight control design under failures, damage, and state-dependent disturbances", *J. Guidance, Control and Dynamics*, **28**(5):902–917.

[26] De Persis, C., and Isidori, A. (2001) "A geometric approach to nonlinear fault detection and isolation", *IEEE Trans. Automatic Control*, **45**(6):853–865.

[27] Xu, A., and Zhang, Q. (2004) "Nonlinear system fault diagnosis based on adaptive estimation", *Automatica*, **40**(7):1181–1193.

[28] Jiang, B., Staroswiecki, M., and Cocquempot, V. (2004) "Fault diagnosis based on adaptive observer for a class of nonlinear systems with unknown parameters", *Int. J. Control*, **77**(4):367–383.

[29] Yan, X. G., and Edwards, C. (2007) "Nonlinear robust fault reconstruction and estimation using a sliding mode observer", *Automatica*, **43**(9):1605–1614.

[30] Gao, Z., and Ding, S. X. (2007) "Actuator fault robust estimation and fault-tolerant control for a class of nonlinear descriptor systems", *Automatica*, **43**(5):912–920.

[31] Hou, M. (2007) "Estimation of sinusoidal frequencies and amplitudes using adaptive identifier and observer", *IEEE Trans. Automatic Control*, **52**(3):493–499.

[32] Vijayaraghavan, K., Rajamani, R., and Bokor, J. (2007) "Quantitative fault estimation for a class of nonlinear systems", *Int. J. Control*, **80**(1):64–74.

[33] Ding, G. R., Frank, P. M., and Ding, E. L. (2003) "An approach to the detection of multiplicative faults in uncertain dynamic systems", in *Proc. 41st IEEE Conference on Decision and Control*, 4371–4376.

[34] Tan, C. P., and Edwards, C. (2004) "Multiplicative fault reconstruction using sliding mode observers", in *Proc. 5th Asian Control Conference*, 957–962.

[35] Zhang, X., Polycarpou, M. M., and Parisini, T. (2010) "Fault diagnosis of a class of nonlinear uncertain system with Lipschitz nonlinearities using adaptive estimation", *Automatica*, **46**(2):290–299.

[36] Phanomchoeng, G., and Rajamani, R. (2010) "Observer design for Lipschitz nonlinear systems using Riccati equations", in *Proc. American Control Conference*, 6060–6065.

[37] Wang, C. L., and Lin, Y. (2009) "Adaptive dynamic surface control for linear multivariable systems", *Automatica*, **46**(10):1703–1711.

[38] Tong, S. C., Li, Y. M., and Wang, T. (2009) "Adaptive fuzzy backstepping fault-tolerant control for uncertain nonlinear systems based on dynamic surface", *Int. J. Innov. Comput. Inf. and Control*, **5**(10A):3249–3262.

[39] Benallegue, A., Mokhtari, A., and Fridman, L. (2008) "High-order sliding-mode observer for a quadrotor UAV", *Int. J. Robust Nonlinear Control*, **18**(4–5):427–440.

[40] Qian, M., Jiang, B., and Xu, D. (2012) "Fault tolerant tracking control scheme for UAV using dynamic surface control technique", *Circuits, Systems and Signal Processing*, **31**(012):1713–1729.

7

Fault Detection of Networked Control Systems

7.1 Introduction

During the last three decades, model-based fault detection (FD) technology has attracted much attention [1, 2, 3]. From the rich theoretical results and increasing applications in industrial practice, it is well understood that

- The model-based fault detection problem is an output estimation problem.
- The fault detection problem is a multi-objective design problem. In order to ensure a quick and reliable detection of faults, both the robustness of the FD system to model uncertainty or unknown disturbances and its sensitivity to faults should be taken into consideration.

In the context of linear time-invariant (LTI) systems, a number of approaches have been proposed for the design of FD systems.

The rapid development of communication techniques promote, because of their flexibility, easy maintenance and low cost, the widespread application of networks in building complex large-scale or teleautomatic systems. Sensors, actuators, processes, and control and supervision stations now exchange information through networks, instead of point-to-point connections. The networks may also be used to achieve other aims. On the other side, networks introduce the following problems:

- data dropout;
- limited bandwidth;
- time delay caused by data transmission;
- information loss because of encoding and quantization.

These problems have been intensively studied by the control community in the last several years [4, 5, 6, 7, 8, 9, 10, 11], including

- analysis of the impact of the network on control performance;

Analysis and Synthesis of Fault-Tolerant Control Systems, First Edition. Magdi S. Mahmoud and Yuanqing Xia.
© 2014 John Wiley & Sons, Ltd. Published 2014 by John Wiley & Sons, Ltd.

- design of control algorithms taking into account the above factors;
- proposal of new network protocols suitable for control.

The purpose of this chapter is to discuss the fault detection problem of networked control systems (NCS). Attention is focused on the analysis and design of FD systems in the case of missing measurements caused by data dropout. The main tasks are to:

- propose to modify the structure of te residual generator and schedule network resource dynamically to cope with missing measurements;
- show that the residual dynamics can be modeled by a discrete-time Markovian jump linear system;
- give a residual evaluation scheme aiming to reduce false alarm rate;
- give a co-design approach for the time-variant residual generator and threshold.

In addition, attention is paid to

- filtering in the case of missing measurements; a detailed up-to-date summary of research in this direction can be found in the work of Wang, Ho, and Liu [12].
- threshold selection for residual evaluation.

Residual evaluation is an important step in model-based FD approaches [1, 2, 3]. The concept of a threshold selector was first systematically introduced by Emami-Naeini, Akhter, and Rock [13] to cope with the residual evaluation problem of LTI systems with model uncertainty. More recently, Stoustrup, Niemann, and la Cour-Harbo [14] determined an optimal threshold value aiming to achieve balance between the false alarm rate and the miss detection rate. Ding *et al.* unified the problem of threshold calculation of residual limit monitoring and trend analysis schemes as a standard optimization problem that can be solved with the LMI technique [15]. They also proposed an approach [16] to determining the threshold which satisfies a given false alarm rate using probabilistic robustness technique for systems with norm-bounded uncertainty. In this chapter, the threshold will be used to reduce the false alarm rate caused by missing measurements.

7.2   Problem Formulation

Assume that the process to be monitored is an LTI system described by

$$x(k+1) = Ax(k) + Bu(k) + E_f f(k)$$
$$y(k) = Cx(k) + F_f f(k) \tag{7.1}$$

where $x \in \Re^n$ denotes the state vector, $u \in \Re^p$ the control input vector, $y \in \Re^m$ the measurable output vector and $f \in \Re^{k_f}$ the vector of faults to be detected, A, B, C, E_f, F_f are known matrices of compatible dimensions. For the sake of clarity, A is assumed to be stable.

For the purpose of fault detection, the necessary information is control input u and measurement output y. In the case of perfect communication, the observer-based residual generator can be constructed as

$$\hat{x}(k+1) = A\hat{x}(k) + Bu(k) + L(y(k) - \hat{y}(k)) \tag{7.2}$$

$$r(k) = W(y(k) - \hat{y}(k)), \hat{y}(k) = C\hat{x}(k)$$

where $r \in \Re^{k_r}$ is the so-called residual signal and L and W are parameters that ensure the stability and dynamics of the residual signal. A number of approaches can be used to select L and W to meet the design specifications [1, 2, 3]. The existence of fault is clearly indicated by the deviation of the residual signal from zero.

In the following discussion, it is assumed that the control and supervision stations for the process of Equation (7.1) are located together. While the residual generator can directly obtain the control input u from the controller, it must get access to measurement y through the network. Thus, in cases of data dropout because of a network jam, some measurements may get lost during the communication between the sensors and the supervision station. As a result, the standard residual generator Equation (7.2) designed under the assumption of perfect communication cannot be directly applied. It should be either modified or completely re-designed to accommodate the unusual situation of missing measurements.

7.3 Modified Residual Generator Scheme

In this section, we propose a way to modify the standard residual generator Equation (7.2). After the analysis of residual dynamics, a threshold is introduced to evaluate the residual, in order to ensure reliable detection of the fault.

7.3.1 Modified residual generator and dynamic analysis

Because of the stochastic character of data dropout, we introduce a stochastic variable γ to represent data communication status. $\gamma(k) = 1$ means that the measurement at time point k arrives correctly, while $\gamma(k) = 0$ means that this measurement is lost. Note that the problem of modifying the residual generator is one of which signal we should take if $\gamma(k) = 0$ and $y(k)$ is not at hand. One natural choice, if the current measurement does not arrive on time, is to simultaneously

- use the last available measurement to generate the residual signal
- inform (warn) the network coordination (resource allocation) center to reduce the data dropout by, for instance, dynamically adjusting the bandwidth assigned to this process.

The two tasks can be implemented with the aid of Figure 7.1, in which the individual blocks interact with the process via banks of actuators and sensors through the network and under the influence of the network coordination center.

The measurement management block is added before the residual generator to embody the above idea. The modified residual generator can be modeled by

$$\hat{x}(k+1) = A\hat{x}(k) + Bu(k) + L(y^a(k) - \hat{y}(k))$$

$$r(k) = W(y^a(k) - \hat{y}(k)), \hat{y}(k) = C\hat{x}(k) \tag{7.3}$$

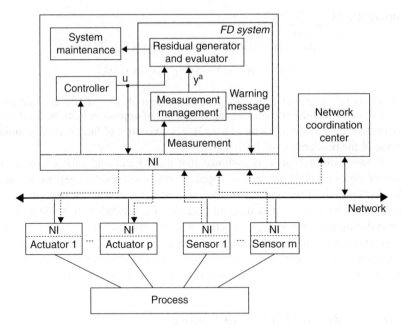

Figure 7.1   Scheme description (NI means "network interface")

where

$$y^a(k) = \begin{cases} y(k), & \textit{if } \gamma(k) = 1 \\ \text{the last available measurement,} & \textit{if } \gamma(k) = 0 \end{cases} \qquad (7.4)$$

It can be seen that the dynamics of residual generated by Equation (7.3) depends on

- the value of $\gamma(k)$;
- the difference between real value of $y(k)$ and the used value $y^a(k)$.

Before beginning with the quantitative analysis of residual dynamics, we make the assumption that the 2-norm of the sequence

$$\theta(k) := y(k) - y^a(k)$$

is bounded by δ.

In the case of $\gamma(k) = 0$, the residual dynamics is governed by

$$e(k + 1) = (A - LC)e(k) + (E_f - LF_f)f(k) + L\theta(k)$$
$$r(k) = WCe(k) + WF_f f(k) - W\theta(k) \qquad (7.5)$$

In the case of $\gamma(k) = 1$,

$$e(k+1) = (A - LC)e(k) + (E_f - LF_f)f(k)$$
$$r(k) = WCe(k) + WF_f f(k) \tag{7.6}$$

It shows that, because of a missing measurement, the residual signal can deviate from zero even if no fault has happened in the process. This is the root of the false alarm rate, which necessitates a residual evaluation stage. The residual dynamics switches between Equation (7.5) and Equation (7.6) depending on the mode (status) of $\gamma(k)$. Thanks to the simultaneous warning system, the possibility of missing measurements is greatly reduced. Thus we can model $\gamma(k)$ as a discrete-time Markov chain with two state space $\Omega = \{1, 2\}$ and stationary transition probability

$$\lambda_{11} = Prob\{\gamma(k+1) = 0 | \gamma(k) = 0\}$$
$$\lambda_{12} = Prob\{\gamma(k+1) = 1 | \gamma(k) = 0\}$$
$$\lambda_{21} = Prob\{\gamma(k+1) = 0 | \gamma(k) = 1\}$$
$$\lambda_{22} = Prob\{\gamma(k+1) = 1 | \gamma(k) = 1\} \tag{7.7}$$
$$\lambda_{11} < \lambda_{21}, \lambda_{11} + \lambda_{12} = 1, \lambda_{21} + \lambda_{22} = 1$$

The residual dynamics is thus characterized by the discrete-time Markovian jump linear system of Equations (7.5)–(7.7).

7.3.2   Residual evaluation

In order to reduce the false alarm rate, we use a threshold selector to evaluate the residual for the modified residual generator Equation (7.3). The decision is made based on the following logic:

$$r_{eval} > J_{th} \Rightarrow \text{A fault is detected.}$$
$$r_{eval} \leq J_{th} \Rightarrow \text{No fault.}$$

where r_{eval} is a function of the residual signal used to measure its size and J_{th} is the threshold.

The issue now is to decide on which norm function is appropriate in our situation. Considering that missing measurements happen with a certain small probability, most of the time the effect of a missing measurement on the residual signal could be seen as driving the (almost) stationary state estimation error (thus, the residual signal) to a nonzero point in the state space. If in the next moment no further measurement is lost, the residual goes again asymptotically to zero. Different from fault, the deviation of the residual signal from zero because of missing measurements appears much more sporadic. Yet, the amplitude of the deviation caused by missing measurements may be comparable to the amplitude of deviation caused by fault.

Based on the foregoing analysis, we propose to use the 2-norm of the residual signal as a residual evaluation function, i.e.

$$r_{eval} = \left(\sum_{j=0}^{\infty} \Re^T(j)r(j) \right)^{1/2} \tag{7.8}$$

The next step is to determine the threshold J_{th}. It is based on the residual dynamics in the fault-free case. So J_{th} depends on δ, the probability of missing measurements $\lambda_{11}, \lambda_{12}, \lambda_{21}, \lambda_{22}$ as well as the parameters L and W of the residual generator, Equation (7.3). To compute $J_{th\lambda}$, we first introduce a lemma concerning the H_{∞} norm of discrete-time Markovian jump linear systems [17, 18].

Lemma 7.1   *Given a scalar $\alpha > 0$ and a stochastic system:*

$$x(k+1) = A_i x(k) + B_i w(k)$$
$$z(k+1) = C_i x(k) + D_i w(k) \tag{7.9}$$
$$A_i = A(\gamma(k))|_{\gamma(k)=i}, \; B_i = B(\gamma(k))|_{\gamma(k)=i},$$
$$C_i = C(\gamma(k))|_{\gamma(k)=i}, \; D_i = D(\gamma(k))|_{\gamma(k)=i},$$

where $\{\gamma(k)\}$ is a discrete-time Markov chain taking values in a finite set $\{1, \ldots, N\}$ with transition probabilities $\lambda_{ij} = Prob\{\gamma(k+1) = j|\gamma(k) = i\}$, the following conditions are equivalent:

1. *The system of Equation (7.9) is internally mean square stable and its H_{∞} norm, defined by*

$$\sup_{w \in l_2, w \neq 0, x(0)=0} \frac{E[\|z\|_2]}{\|w\|_2},$$

   where $E[\bullet]$ stands for mathematical expectation, is less than α.
2. *There exist matrices $P_i, i = 1, \ldots, N$, satisfying the following linear matrix inequalities (LMIs):*

$$\begin{bmatrix} P_i & A_i^T \bar{P}_i & 0 & C_i^T \\ \bar{P}_i A_i & \bar{P}_i & \bar{P}_i B_i & 0 \\ 0 & B_i^T \bar{P}_i & \alpha I & D_i^T \\ C_i & 0 & D_i & \alpha I \end{bmatrix} > 0 \tag{7.10}$$

   where

$$\bar{P}_i = \sum_{j=1}^{N} \lambda_{ij} P_j, \quad i = 1, \ldots, N.$$

3. *There exist matrices P_i and G_i, $i = 1, \ldots, N$, satisfying the following LMIs:*

$$\begin{bmatrix} P_i & A_i^T G_i^T & 0 & C_i^T \\ G_i A_i & G_i + G_i^T - \bar{P}_i & G_i B_i & 0 \\ 0 & B_i^T G_i^T & \alpha I & D_i^T \\ C_i & 0 & D_i & \alpha I \end{bmatrix} > 0 \tag{7.11}$$

$$G_i > 0 \tag{7.12}$$

where

$$\bar{P}_i = \sum_{j=1}^{N} \lambda_{ij} P_j, \quad i = 1, \ldots, \lambda, N.$$

Remark 7.1 *Condition 3 in Lemma 7.1 is presented in a slightly different form from that presented by de Souza [17], as it also requires the LMI Equation (7.12). The proof is similar to the proof of de Souza's Theorem 3.1 by noting that the feasibility of Equation (7.10) implies $P_i > 0$ and thus $\bar{P}_i = \sum_{j=1}^{N} \lambda_{ij} P_j > 0$. Applying Lemma 7.1, we obtain the following theorem which provides a residual evaluation scheme aiming to reduce the false alarm rate.*

Theorem 7.1 *Given the process in Equation (7.1), the residual generator in Equations (7.3)–(7.4) and the probability of measurement missing (see Equation (7.7)), under the residual evaluation function in Equation (7.8), the threshold can be set as*

$$J_{th} = \alpha_{\min} \delta \tag{7.13}$$

where $\alpha_{\min}$ is the optimum of the constrained optimization problem:

$$\min \alpha \tag{7.14}$$

with the following LMIs admitting a solution for P_1 and P_2:

$$\begin{bmatrix} P_1 & \Phi_{1,12} & 0 & C^T W^T \\ \Phi_{1,12}^T & \Phi_{1,22}^T & \Phi_{1,23}^T & 0 \\ 0 & \Phi_{1,23}^T & \alpha I & -W^T \\ WC & 0 & -W & \alpha I \end{bmatrix} > 0 \tag{7.15}$$

$$\begin{bmatrix} P_2 & \Phi_{2,12} & C^T W^T \\ \Phi_{2,12}^T & \Phi_{2,22}^T & 0 \\ WC & 0 & \alpha I \end{bmatrix} > 0 \tag{7.16}$$

where

$$\Phi_{2,12}^T = \lambda_{11}(A^T - C^T L^T)P_1 + \lambda_{12}(A^T - C^T L^T)P_2$$

$$\Phi_{2,22}^T = \lambda_{11} P_1 + \lambda_{12} P_2$$

$$\Phi_{1,23}^T = \lambda_{11} P_1 L + \lambda_{12} P_2 L$$

$$\Psi_{2,12}^T = \lambda_{21}(A^T - C^T L^T)P_1 + \lambda_{22}(A^T - C^T L^T)P_2$$

$$\Phi_{2,22}^T = \lambda_{21} P_1 + \lambda_{22} P_2$$

Proof. The threshold is determined based on the residual dynamics of Equations (7.5)–(7.7) in the fault-free case ($f = 0$). According to Lemma 7.1 and noting

$$A_1 = A - LC, \, B_1 = L, \, C_1 = WC, \, D_1 = -W$$

$$A_2 = A - LC, \, B_2 = 0, \, C_2 = WC, \, D_2 = 0 \qquad (7.17)$$

the feasibility of LMI constraints Equation (7.15) and Equation (7.16) is equivalent to

$$\frac{E[\|r\|_2]}{\|\theta\|_2} < \alpha$$

Thus the threshold can be computed based on a recursive procedure of checking the feasibility of Equation (7.15) and Equation (7.16) till the minimal α is found. ∎ ■

Remark 7.2 *The results of Theorem 7.1 are derived by means of Equation (7.10). Indeed, Equation (7.11) and Equation (7.12) can be used to this aim as well. The derivation and results are similar and thus omitted here. Equation (7.10) involves fewer variables and is suitable for analysis, while Equations (7.11)–(7.12) provide more design freedom and are suitable for synthesis, as shown in the next section.*

We use the following algorithm to select the threshold for the modified residual generator in Equations (7.3)–(7.4):

1. Set an initial value of α.
2. Check the feasibility of the LMI constraints in Equations (7.15)–(7.16).
3. If Equations (7.15)–(7.16) are feasible, reduce the value of α; otherwise, increase α.
4. Repeat Step 2 and Step 3, till α_{min} is found with the desired precision.
5. Compute the threshold J_{th} according to Equation (7.13).

7.3.3 Co-design of residual generator and evaluation

In the previous section, a procedure was given for evaluating the residual generated by the modified residual generator Equation (7.3)–(7.4), where the parameters L and W are designed, as before, under the assumption of perfect communication. In this way, the false alarm rate is greatly reduced.

Recall that, the introduction of a threshold may make small faults undetectable and thus cause missed detections. From the viewpoint of reducing the number of faults missed, a low threshold is desirable. At the same time, the residual dynamics should possess some desired properties, even if some measurements are missing. These call for more active contribution of the residual generator.

In the following discussion, a time-variant residual generator is proposed. The basic idea is to use parameters L_1, W_1 in the case of missing measurements and L_2, W_2 in the case of perfect communication. The values of these parameters are selected to reduce the threshold and simultaneously to ensure satisfactory residual dynamics. Based on this consideration, the residual generator is constructed as

$$\hat{x}(k+1) = A\hat{x}(k) + Bu(k) + L^a(y^a(k) - \hat{y}(k))$$
$$r(k) = W^a(y^a(k) - \hat{y}(k)), \quad \hat{y}(k) = C\hat{x}(k) \tag{7.18}$$

where $y^a(k)$ is defined as Equation (7.4) and

$$\begin{cases} L^a = L_1, W^a = W_1, & \text{if } \gamma(k) = 0 \\ L^a = L_2, W^a = W_2, & \text{if } \gamma(k) = 1 \end{cases} \tag{7.19}$$

As a result, the residual dynamics is now governed by

$$e(k+1) = (A - L_1 C)e(k) + (E_f - L_1 F_f)f(k) + L_1\theta(k)$$
$$r(k) = W_1 Ce(k) + W_1 F_f f(k) - W_1\theta(k) \tag{7.20}$$

in case of $\gamma(k) = 0$ and

$$e(k+1) = (A - L_2 C)e(k) + (E_f - L_2 F_f)f(k)$$
$$r(k) = W_2 Ce(k) + W_2 F_f f(k) \tag{7.21}$$

in case of $\gamma(k) = 1$.

Theorem 7.2 gives a way to select the parameters of the time-variant residual generator of Equations (7.18)–(7.19). As an example, we also show how the pole placement constraint imposed on the residual dynamics

$$\max_i |\rho_i(A - L_1 C)| < \beta_1$$
$$\max_i |\rho_i(A - L_2 C)| < \beta_2 \tag{7.22}$$

is incorporated in the design. However, the procedure below cannot be used to optimally select W_1 and W_2, as $W_1 = W_2 = 0$ is a trivial solution. Hence, we assume that W_1, W_2 are pre-defined and non-zero.

Theorem 7.2 *Assume the process of Equation (7.1), the probability of a missing measurement, Equation (7.7), pre-defined weighting matrices W_1, W_2, and scalars $0 < \beta_1, \beta_2 < 1$. Under the residual evaluation function of Equation (7.8) and the pole placement constraint of*

Equation (7.22), the parameters L_1, L_2 of the residual generator in Equations (7.18)–(7.19) and the threshold $J\beta_{th}$ can be set as

$$L_1 = G_1^{-1} Y_1, \quad L_2 = G_2^{-1} Y_2 \tag{7.23}$$

$$J_{th} = \alpha_{min}\delta \tag{7.24}$$

where α_{min} is the optimum of the constrained optimization problem:

$$\min \alpha \tag{7.25}$$

with the following LMIs admitting a solution for P_1, P_2, G_1, G_2, Y_1, Y_2:

$$G_1 > 0 \tag{7.26}$$

$$G_2 > 0 \tag{7.27}$$

$$\begin{bmatrix} P_1 & \Psi_{1,12} & 0 & C^T W_1^T \\ \Psi_{1,12}^T & \Psi_{1,22} & Y_1 & 0 \\ 0 & Y_1^T & \alpha I & -W_1^T \\ W_1 C & 0 & -W_1 & \alpha I \end{bmatrix} > 0 \tag{7.28}$$

$$\begin{bmatrix} P_2 & \Psi_{2,12} & C^T W_2^T \\ \Psi_{2,12}^T & \Psi_{2,22} & 0 \\ W_2 C & 0 & \alpha I \end{bmatrix} > 0 \tag{7.29}$$

$$\begin{bmatrix} \beta_1^2 G_1 & \Psi_{1,12} \\ \Psi_{1,12}^T & G_1 \end{bmatrix} > 0 \tag{7.30}$$

$$\begin{bmatrix} \beta_2^2 G_2 & \Psi_{2,12} \\ \Psi_{2,12}^T & G_2 \end{bmatrix} > 0 \tag{7.31}$$

where

$$\Psi_{1,12} = A^T G_1^T - C^T Y_1^T$$
$$\Psi_{1,22} = G_1 + G_1^T - \lambda_{11} P_1 - \lambda_{12} P_2$$
$$\Psi_{2,12} = A^T G_2^T - C^T Y_2^T$$
$$\Psi_{2,22} = G_2 + G_2^T - \lambda_{21} P_1 - \lambda_{22} P_2$$

Proof. Define $G_1 L_1 = Y_1$ and $G_2 L_2 = Y_2$. According to Lemma 7.1 and noting

$$A_1 = A - L_1 C, B_1 = L_1, C_1 = W_1 C, D_1 = -W_1$$
$$A_2 = A - L_2 C, B_2 = 0, C_2 = W_2 C, D_2 = 0 \tag{7.32}$$

it is straightforward to check that the feasibility of Equations (7.26)–(7.29) means the the following inequality holds:

$$E[\|r\|_2] < \alpha\|\theta\|_2$$

Using the Schur complement lemma, Equations (7.26), (7.27), and (7.30)–(7.31) can be equivalently re-written as

$$G_1 > 0, \beta_1^2 G_1 - A_1^T G_1 A_1 > 0 \tag{7.33}$$

$$G_2 > 0, \beta_2^2 G_2 - A_2^T G_2 A_2 > 0 \tag{7.34}$$

From classic control theory, the feasibility of Equations (7.33)–(7.34) means that the eigenvalues of $A_1 = A - L_1 C$ and $A_2 = A - L_2 C$ lie, respectively, inside the circles of radius β_1 and β_2. ∎

Theorem 7.2 shows that the residual generator and threshold can be designed by checking the feasibility of LMI constraints of Equations (7.26)–(7.31) recursively. The algorithm can be readily implemented with the aid of the Matlab LMI toolbox.

7.4 Quantized Fault-Tolerant Control

This section addresses the design problem of a fault-tolerant H_∞ controller for linear systems with state quantization. By combining the linear matrix inequality technique and the indirect adaptive method, a new method is proposed for designing a fault-tolerant controller against actuator faults via quantized state feedback. The controller gains update according to the online estimation of eventual faults, which are dependent on the quantized state signals. Meanwhile, the design conditions with variable gains that we propose can be proved to be less conservative than those of the traditional controller with fixed gains.

7.4.1 Introduction

With the widespread use of digital communication channels in modern control systems, where the interface between the controller and the process features some additional information-processing devices, the effects of signal quantization on feedback control systems have been paid more attention [19]. There are two main approaches to quantization: static quantizer design and dynamic quantizer design. Static quantizers [20]–[23] are memoryless feedback quantizers, which presume that data quantization at time k is dependent on the data at time k only, and lead to relatively simple structures for coding–decoding schemes. Dynamic quantizers [24]–[26] are time-varying and their quantization levels are dynamic in order to increase the region of attraction and to attenuate the steady state limit cycle. However, most of the results on quantized feedback with dynamic quantizers are confined to the stabilization problem and control performance issues are not addressed [22, 23]. Recently, the problem of feedback control with static quantizers has been investigated by many researchers. Elia and Mitter [21] investigated the quadratic stabilization for SISO systems by quantized state feedback and indicated that the coarsest, or least dense, quantizer is logarithmic. Fu and Xie [22] extended the method of Elia and Mitter to MIMO systems; the performance of control systems have also been dealt with in the unified framework.

On the other hand, actuator failures may cause severe system performance deterioration that should be avoided in many critical systems, such as aircraft, spacecraft and nuclear power plants [27]. A control system designed to tolerate faults in sensors or actuators while maintaining an acceptable level of stability and performance in the closed-loop system is called a fault-tolerant control (FTC) system. Fault-tolerant design approaches can be broadly classified into passive [28]–[31] and active [27, 32, 33].

Yang and Ye [34] developed a combined adaptive method with active fault-tolerant approach to yield new fault-tolerant H_∞ controllers. The resultant conditions are proved to be more relaxed than those for the traditional reliable controller with fixed gains and can improve the H_∞ performance of closed-loop systems. However, it should be stressed that work on fault-tolerant control often ignores the issue of signal quantization, which may make these methods difficult to apply to actual control systems. If a fault-tolerant controller is applied to an actual control system without considering the signal quantization phenomenon, it may lead to abrupt performance degradation and even instability. Thus, how to design fault-tolerant controllers with quantized feedback becomes a significant research topic. However, there is relatively little work that actually considers the effects of both actuator faults and signal quantization.

In the following discussion, the main result of Yang and Ye [34] has been extended to the fault-tolerant H_∞ control systems design for linear systems against actuator faults via quantized state feedback. Here, the states are quantized before they are passed to the controller, which is similar to that in the work of Fu and Xie [22]. By introducing the adaptive mechanism into the LMI framework, a new fault-tolerant controller with variable gains is designed to guarantee the asymptotic stability and H_∞ performance of a closed-loop system, even in the event of actuator failure. New adaptive laws, which are dependent on the quantized state signals, are also proposed to adjust the controller gains. We now prove both theoretically and through numerical examples that the newly obtained adaptive, fault-tolerant controller design is less conservative than that of the traditional controller with fixed gains.

7.4.2 Problem statement

Consider a linear time-invariant model described by

$$\dot{x}(t) = Ax(t) + B_1\omega(t) + Bu(t)$$
$$z(t) = Cx(t) + Du(t) \tag{7.35}$$

where $x(t) \in \Re^n$ is the state, $u(t) \in \Re^m$ is the control input, $z(t) \in \Re^q$ is the regulated output, and $\omega(t) \in \Re^s$ is an exogenous disturbance in $L_2[0, \infty]$. A, B_1, B, C and D are known constant matrices of appropriate dimensions.

To formulate the fault-tolerant control problem, the following actuator fault model [34] is adopted:

$$u_{ij}^F(t) = \left(1 - \rho_i^j\right)u_i(t), 0 \le \underline{\rho_i^j} \le \rho_i^j \le \bar{\rho}_i^j, \quad i = 1, \ldots, m, \quad j = 1, \ldots, L \tag{7.36}$$

where ρ_i^j is an unknown constant. Here, the index j denotes the jth fault mode and L is the total number of fault modes. Let $u_{ij}^F(t)$ represent the signal from the ith actuator that has failed in the jth fault mode. For every fault mode, $\underline{\rho}_{ij}$ and $\bar{\rho}_{ij}$ represent the lower and upper bounds of ρ_i^j, respectively. Note that, when $\underline{\rho}_i^j = \bar{\rho}_i^j = 0$, there is no fault for the ith actuator u_i in the jth fault mode. When $\underline{\rho}_i^j = \bar{\rho}_i^j = 1$, the ith actuator u_i is an outage in the jth fault mode. When $0 < \underline{\rho}_i^j \le \bar{\rho}_i^j < 1$, in the jth fault mode, the type of actuator fault is "loss of effectiveness".

Denote

$$u_j^F(t) = [u_{1j}^F(t), u_{2j}^F(t), \ldots, u_{mj}^F(t)]^T = (I - \rho^j)u(t)$$

where $\rho^j = diag[\rho_1^j, \rho_2^j, \ldots, \rho_m^j]$, $j = 1, \ldots, L$. Considering the lower and upper bounds $(\underline{\rho}_{ij}, \bar{\rho}_{ij})$, the following set can be defined

$$N_{\rho^j} = \{\rho^j | \rho^j = diag\{\rho_1^j, \rho_2^j, \ldots, \rho_m^j\}, \rho_i^j = \underline{\rho}_i^j \text{ or } \rho_i^j = \bar{\rho}_i^j\}$$

Thus, the set N_{ρ^j} contains a maximum of 2^m elements.

For convenience, in the following sections, for all possible fault modes L, we use a uniform actuator fault model

$$u^F(t) = (I - \rho)u(t), \rho \in \{\rho^1, \ldots, \rho^L\} \tag{7.37}$$

and ρ can be described by $\rho = diag\{\rho_1, \rho_2, \ldots, \rho_m\}$.

The state variable $x(t)$ is assumed to be measurable, quantized via a quantizer $f(\cdot)$, and then transmitted to the controller. The controller structure is chosen as

$$u(t) = K(\hat{\rho}(t))f(x(t)) = (K_0 + K_a(\hat{\rho}(t)) + K_b(\hat{\rho}(t))f(x(t)) \tag{7.38}$$

where $\hat{\rho}(t)$ is the estimate of ρ, $K_a(\hat{\rho}(t)) = \sum_{i=1}^m K_{ai}\hat{\rho}_i(t)$, and $K_b(\hat{\rho}(t)) = \sum_{i=1}^m K_{bi}\hat{\rho}_i(t)$.

Remark 7.3 *It should be noticed that, when $K_{ai} = 0$ and $K_{bi} = 0$ $(i = 1, 2, \ldots, m)$, $K(\hat{\rho}(t))$ reduces into K_0, which is the traditional fixed controller gain. Moreover, although $K_a(\hat{\rho}(t))$ and $K_b(\hat{\rho}(t))$ have the same forms, we deal with them in different ways to give more freedom and less conservativeness to derive our main results.*

On the other hand, the quantizer $f(\cdot) = [f_1(\cdot), f_2(\cdot), \ldots, f_n(\cdot)]$ is assumed to be symmetric, that is, $f_q(-v) = -f_q(-v)$ $(q = 1, 2, \ldots, n)$. The quantizer considered in this section is logarithmic, static, and time-invariant [22].

Following Fu and Xie [22], for each $f_q(\cdot)$, the set of quantized levels is described by

$$\Pi_q = \{\pm s_i^{(q)} | s_i^{(q)} = \xi_q^i s_0^{(q)}, i = \pm 1, \pm 2, \ldots, \} \cup \{\pm s_0^{(q)}\} \cup \{0\}, 0 < \xi_q < 1, s_0^{(q)} > 0 \tag{7.39}$$

Each quantization level s_i^q corresponds to a segment such that the quantizer maps the whole segment to this quantization level. In addition, these segments form a partition of R; that is, they are disjoint and their union is equal to R.

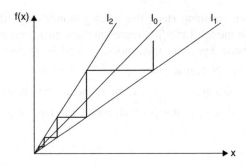

Figure 7.2   Logarithmic quantizer: the functions of l_0, l_1, and l_2 are $f(x) = x$, $f(x) = (1 - \sigma)x$, and $f(x) = (1 + \sigma)x$ respectively.

For the logarithmic quantizer, the associated quantizer $f_q(\cdot)$ are defined as

$$
f_q(v) = \begin{cases} s_i^{(q)}, & \text{if} \quad \frac{1}{1+\sigma_q} s_i^{(q)} < v \le \frac{1}{1+\sigma_q} s_i^{(q)} \\ 0, & \text{if} \quad v = 0 \\ -f_q(-v) & \text{if} \quad v < 0 \end{cases} \tag{7.40}
$$

where $\sigma_q = \frac{1-\xi_q}{1+\xi_q}$.

Following Fu and Xie [22], ξ_q is called the quantization density of quantizer $f_q(\cdot)$. The logarithmic quantizer is illustrated in Figure 7.2, which is very common in the works about quantization (see [22, 23]). We can see from Figure 7.2 that the sector bound is described by a single parameter σ, which is related to the quantization density by $\sigma = \frac{1-\xi}{1+\xi}$.

Denote

$$
\Lambda = \{\delta = [\delta_1, \ldots, \delta_N] \mid \delta_i \in \{\underline{\delta}_i, \bar{\delta}_i\}\}
$$

where $\delta_i (i = 1, \ldots, N)$ is an unknown constant.

The following lemma is given to derive our main results.

Lemma 7.2   *[34]: If there exists a symmetric matrix Θ with*

$$
\Theta = \begin{bmatrix} \Theta_{11} & \Theta_{12} \\ \Theta_{12}^T & \Theta_{22} \end{bmatrix}
$$

and $\Theta_{11}, \Theta_{22} \in \Re^{Nn \times Nn}$ such that the following inequalities hold:

$$
\Theta_{ii} \le 0, i = 1, \ldots, N
$$

For $\delta \in \Lambda$,

$$
\Theta_{11} + \Delta(\delta)\Theta_{12} + (\Delta(\delta)\Theta_{12})^T + \Delta(\delta)\Theta_{22}\Delta(\delta) \ge 0
$$

$$
\begin{bmatrix} Q & E \\ E^T & F \end{bmatrix} + U^T U + G^T \Theta G < 0 \tag{7.41}
$$

then inequality

$$W(\delta) = Q + \sum_{i=1}^{N} \delta_i E_i + \left(\sum_{i=1}^{N} \delta_i E_i \right)^T + \sum_{i=1}^{N} \sum_{j=1}^{N} \delta_i \delta_j F_{ij}$$

$$+ \left(U_0 + \sum_{i=1}^{N} \delta_i U_i \right)^T \left(U_0 + \sum_{i=1}^{N} \delta_i U_i \right) < 0 \qquad (7.42)$$

holds for all $\delta_i \in [\underline{\delta_i}\ \bar{\delta_i}]$, *where*

$$Q = Q^T \in \Re^{n \times n}, \quad F_{ij} = F_{ji}^T \in \Re^{n \times n}, E_i \in \Re^{n \times n}$$

$$\Delta(\delta) = diag[\delta_1 I_{n \times n} \ \dots, \ \delta_N I_{n \times n}]$$

$$E = [E_1 \ E_2 \ \dots, \ E_N], U = [U_0 \ U_1 \ \dots, \ U_N]$$

$$F = \begin{bmatrix} F_{11} & F_{12} & \dots, & F_{1N} \\ F_{21} & F_{22} & \dots, & F_{2N} \\ \vdots & \vdots & \dots, & \vdots \\ F_{N1} & F_{N2} & \dots, & F_{NN} \end{bmatrix}$$

$$G = \begin{bmatrix} \begin{bmatrix} I_{n \times n} \\ \vdots \\ I_{n \times n} \end{bmatrix} & 0 \\ 0 & I_{Nn \times Nn} \end{bmatrix}$$

Denote

$$\Delta = diag\{\sigma_1, \sigma_2, \dots, \sigma_n\}, \Delta_0 = I - \Delta, \Delta_1 = I + \Delta.$$

Lemma 7.3 *[35]: The logarithmic static quantizers $f_q(\cdot)$ $(q = 1, 2, \dots, n)$ are of quantization densities ξ_q, then for any diagonal matrix $S > 0$, the following inequality is true*

$$(f(x(t)) - \Delta_0 x(t))^T S(f(x(t)) - \Delta_1 x(t)) \leq 0 \qquad (7.43)$$

Remark 7.4 *In fact, the logarithmic quantizer is sector bounded and the condition in Lemma 7.3 is just one of the sector conditions to deal with the logarithmic quantizer, which may be more effective and simple for designing controllers for systems with state quantization [35].*

7.4.3 Quantized control design

A new fault-tolerant H_∞ controller design method via quantized state feedback is now proposed based on the adaptive method and the LMI technique. The dynamics with actuator faults,

Equation (7.37), is described by

$$\dot{x}(t) = Ax(t) + B(I - \rho)u(t) + B_1\omega(t)$$
$$Z(t) = Cx(t) + D(I - \rho)u(t) \tag{7.44}$$

The following equality can be easily obtained

$$(I - \rho)u(t) = (I - \rho)(K_0 + K_a(\hat{\rho}(t)) + K_b(\hat{\rho}(t)))f(x(t))$$
$$= [(I - \rho)K_0 + K_a(\rho) - \rho K_a(\hat{\rho})]f(x) + (I - \hat{\rho}(t))K_b(\hat{\rho}(t))f(x(t))$$
$$+ [K_a(\tilde{\rho}) + \tilde{\rho}K_b(\hat{\rho}(t))]f(x(t)) \tag{7.45}$$

where $\tilde{\rho}(t) = \hat{\rho}(t) - \rho$.
   Denote

$$N_{0n} = \begin{bmatrix} \Omega_1 - 2\Delta_0\Delta_1\bar{S} + \frac{1}{\gamma_n^2}B_1B_1^T & \Omega_2 + 2\bar{S} \\ \bullet & -2\bar{S} \end{bmatrix}$$

$$N_{0f} = \begin{bmatrix} \Omega_1 - 2\Delta_0\Delta_1\bar{S} + \frac{1}{\gamma_f^2}B_1B_1^T & \Omega_2 + 2\bar{S} \\ \bullet & -2\bar{S} \end{bmatrix}$$

$$\Omega_1 = AX + XA^T, \quad \Omega_2 = B(I - \rho)Y_0 + B\sum_{i=1}^{m}\rho_iY_{ai}$$

$$G = \begin{bmatrix} \begin{bmatrix} I_{n\times n} \\ \vdots \\ I_{n\times n} \end{bmatrix} & 0 \\ 0 & I_{mn\times mn} \end{bmatrix}, \quad \Psi_i = \begin{bmatrix} 0 & -B\rho Y_{ai} + BY_{bi} \\ 0 & 0 \end{bmatrix},$$

$$\Gamma_i = \begin{bmatrix} 0 & -B^iY_{bj} - (B^jY_{bi})^T \\ 0 & 0 \end{bmatrix}$$

$$\Psi = \begin{bmatrix} \Psi_1 & \Psi_2 & \dots, & \Psi_m \end{bmatrix}, \Gamma = [\Gamma_{ij}], i, j = 1, 2, \dots, m$$
$$U = \begin{bmatrix} U_0 & U_1 & \dots, & U_m \end{bmatrix}, U_0 = \begin{bmatrix} C_1X & D(I - \rho)Y_0 \end{bmatrix},$$
$$U_i = \begin{bmatrix} 0 & D(I - \rho)(Y_{ai} + Y_{bi}) \end{bmatrix},$$

$$\Delta(\hat{\rho}) = [\hat{\rho}_1I_{n\times n} \dots, \hat{\rho}_mI_{n\times n}], \Delta_{\hat{\rho}} = \left\{\hat{\rho} = (\hat{\rho}_1 \dots, \hat{\rho}_m) : \hat{\rho}_i \in \left\{\min_j\{\underline{\rho}_i^j\}, \max_j\{\bar{\rho}_i^j\}\right\}\right\}$$

Theorem 7.3   Let $\gamma_f > \gamma_n > 0$ be given constants if there exist matrices $X > 0, \bar{S} > 0, Y_0, Y_{ai}, Y_{bi}, i = 1, \dots, m$ and a symmetric matrix with

$$\Theta = \begin{bmatrix} \Theta_{11} & \Theta_{12} \\ * & \Theta_{22} \end{bmatrix}$$

and $\Theta_{11}\Theta_{22} \in \Re^{mn \times mn}$ such that the following inequalities hold:

$$\Theta_{22ii} \leq 0, i = 1, \ldots, m$$

$$\Theta_{11} + \Delta(\hat{\rho})\Theta_{12} + (\Delta(\hat{\rho})\Theta_{12})^T + \Delta(\hat{\rho})\Theta_{22}\Delta(\hat{\rho}) \geq 0, \text{ for } \hat{\rho} \in \Delta_{\hat{\rho}}$$

For $\rho = 0$,

$$\begin{bmatrix} N_{on} & \Psi \\ * & \Gamma \end{bmatrix} + U^T U + G^T \Theta G < 0$$

For $\rho \in \{\rho^1, \ldots, \rho^L\}, \rho^j \in N_{\rho^j}$,

$$\begin{bmatrix} N_{of} & \Psi \\ * & \Gamma \end{bmatrix} + U^T U + G^T \Theta G < 0 \tag{7.46}$$

Also, $\hat{\rho}_i(t)$ is determined according to the adaptive law

$$\dot{\hat{\rho}}_i = Proj_{\left[\min_j\{\underline{\rho}_i^j\}, \max_j\{\bar{\rho}_i^j\}\right]}\{L_i\}$$

$$= \begin{cases} 0, & if \quad \begin{aligned} &\hat{\rho}_i = \min_j\{\underline{\rho}_i^j\} \text{ and } L_i \leq 0 \\ &or \; \hat{\rho}_i = \max_j\{\bar{\rho}_i^j\} \text{ and } L_i \geq 0; \end{aligned} \\ L_i, & otherwise \end{cases} \tag{7.47}$$

where

$$L_i = -l_i x^T(t)[PB^i K_b(\hat{\rho}) + PBK_{ai}]f(x(t)) \tag{7.48}$$

with $P = X^{-1}$, $K_{ai} = Y_{ai}X^{-1}$, $K_{bi} = Y_{bi}X^{-1}$, and $l_i > 0 \; (i = 1, \ldots, m)$ is the adaptive law gain to be chosen according to practical applications. Proj$\{\cdot\}$ denotes the projection operator [36], which projects the estimates $\hat{\rho}_i(t)$ to the interval $\left[\min_j\{\underline{\rho}_i^j\}, \max_j\{\bar{\rho}_i^j\}\right]$.

Then, the closed-loop system of Equation (7.44) is asymptotically stable and in normal operation, that is, $\rho = 0$, for $x(0) = 0$, satisfies

$$\int_0^\infty Z^T(t)Z(t)dt \leq \gamma_n^2 \int_0^\infty \omega^T(t)\omega(t)dt + \sum_{i=1}^m \frac{\tilde{\rho}_i^2(0)}{l_i} \tag{7.49}$$

and in the case of actuator failure, that is, $\rho \in \{\rho^1, \ldots, \rho^L\}$, for $x(0) = 0$, satisfies

$$\int_0^\infty Z^T(t)Z(t)dt \leq \gamma_f^2 \int_0^\infty \omega^T(t)\omega(t)dt + \sum_{i=1}^m \frac{\tilde{\rho}_i^2(0)}{l_i} \tag{7.50}$$

where $\tilde{\rho}(t) = diag\{\tilde{\rho}_1(t) \ldots, \tilde{\rho}_m(t)\}$, $\tilde{\rho}_i(t) = \hat{\rho}_i(t) - \rho_i$. *The controller gain is given by*

$$K(\hat{\rho}) = Y_0 X^{-1} + \sum_{i=1}^{m} \hat{\rho}_i Y_{ai} X^{-1} + \sum_{i=1}^{m} \hat{\rho}_i Y_{bi} X^{-1} \tag{7.51}$$

Proof. First, we consider the normal case and choose the following Lyapunov function

$$V = x^T(t) P x(t) + \sum_{i=1}^{m} \frac{\tilde{\rho}_i^2(t)}{l_i} \tag{7.52}$$

then from the derivative of V along the closed-loop system and Lemma 7.3, we can get

$$\dot{V}(t) + Z^T(t)z(t) - \gamma_n^2 \omega^T(t)\omega(t)$$

$$\leq \dot{V}(t) + Z^T(t)z(t) - \gamma_n^2 \omega^T(t)\omega(t) - 2[f(x(t)) - \Delta_0 x(t)]^T S[f(x(t)) - \Delta_1 x(t)]$$

$$\leq x^T(t)\left\{PA + A^T P + \frac{1}{\gamma_n^2}PB_1 B_1^T P - 2\Delta_0\Delta_1 S\right\} x(t)$$

$$+ 2x^T PB[(I - \rho)K_0 + K_a(\rho) - \rho K_a(\hat{\rho})]f(x)$$

$$+ 2x^T PB(I - \hat{\rho})K_b(\hat{\rho})f(x) + 4x^T Sf(x) - 2f^T(x)Sf(x)$$

$$+ (Cx + D(I - \rho)K(\hat{\rho})fx(t))^T (Cx + D(I - \rho)K(\hat{\rho})f(x))$$

$$+ 2x^T PB[(I - \rho)K_a(\tilde{\rho}) + \tilde{\rho}K_b(\tilde{\rho})]f(x) + 2\sum_{i=a}^{m} \frac{\tilde{\rho}_i \dot{\tilde{\rho}}_i}{l_i}. \tag{7.53}$$

Let $B = [b^1, \ldots, b^m]$ and $B^i = [0, \ldots, b^i, \ldots, 0]$, then

$$PB\tilde{\rho}K_b(\hat{\rho}) = \sum_{i=1}^{m} \tilde{\rho}_i PB^i K_b(\hat{\rho}) \tag{7.54}$$

$$PBK_a(\tilde{\rho}) = \sum_{i=1}^{m} \tilde{\rho}_i PBK_{ai}. \tag{7.55}$$

Choose the adaptive law as Equation (7.47), then it is sufficient to show

$$\dot{V}(t) + z'(t)z(t) - \gamma_N^2 \omega'(t)\omega(t) \leq [X'(t) \quad f^T(x(t))]\Pi \begin{bmatrix} x(t) \\ f(x(t)) \end{bmatrix} < 0 \tag{7.56}$$

where

$$\Pi = \begin{bmatrix} \Upsilon_1 & \Upsilon_2 + 2S \\ \bullet & -2S \end{bmatrix} + \begin{bmatrix} C & D(I - \rho)K(\hat{\rho}) \end{bmatrix}^T \begin{bmatrix} C & D(I - \rho)K(\hat{\rho}) \end{bmatrix}$$

$$\Upsilon_1 = PA + A^T P + \frac{1}{\gamma_n^2}PB_1 B_1^T P - 2\Delta_0\Delta_1 S$$

$$\Upsilon_2 = PB[(I - \rho)K_0 + K_a(\rho) - \rho K_a(\hat{\rho}) + (I - \hat{\rho})K_b(\hat{\rho})]$$

Denote $X = P^{-1}$, $Y_0 = K_0 X$, $\bar{S} = XSX$, $Y_{ai} = K_{ai}X$, $Y_{bi} = K_{bi}X$, $i = 1, \ldots, m$, then premultiplying and post-multiplying diag $\{P^{-1}, P^{-1}\}$ and its transpose on both sides of $\Pi < 0$, it follows for any $\rho \in \{\rho^1 \ldots, \rho^L\}$, $\rho^j \in N_{\rho^j}$

$$N_{0n} + N_1(\hat{\rho}_i) + N_2(\hat{\rho}_i) +$$
$$\left\{ \begin{bmatrix} CX & D(I - \rho)Y_0 \end{bmatrix}^T + N_3^T(\hat{\rho}_i) \right\} \begin{bmatrix} CX & D(I - \rho)Y_0 \end{bmatrix} +$$
$$N_3(\hat{\rho}_i) < 0 \qquad (7.57)$$

where

$$N_1(\hat{\rho}_i) = \sum_{i=1}^{m} \hat{\rho}_i \Psi_j, \ N_2(\hat{\rho}_i) = \sum_{i=1}^{m} \sum_{j=1}^{m} \hat{\rho}_i \hat{\rho}_j \Gamma_{ij}, \ (\hat{\rho}_i) = \sum_{i=1}^{m} \hat{\rho}_i U_i$$

It is easy to see that if Equation (7.57) holds, then Equation (7.56) is satisfied for any vector $x \in \mathfrak{R}^n$.

By Lemma 7.2 and Equation (7.46), it follows that Equation (7.57) holds for any $\rho \in \{\rho^1, \ldots, \rho^L\}$, $\rho^j \in N_{\rho^j}$ and $\hat{\rho}$ satisfying Equation (7.47). So, Equation (7.56) holds for any $x \neq 0$, which further implies that $\dot{V}(t) < 0$ for any $x \neq 0$. Thus, the closed-loop system of Equation (7.44) is asymptotically stable for the normal case. Furthermore,

$$\dot{V}(t) + z^T(t)z(t) - \gamma_n^2 \omega^T(t)\omega(t) \leq 0.$$

Integrating the aforementioned inequalities from 0 to 1 on both sides, it follows that

$$V(\infty) - V(0) + \int_0^{\infty} z^T(t)z(t)dt \leq \gamma_n^2 \int_0^{\infty} \omega^T(t)\omega(t)dt$$

Then,

$$\int_0^{\infty} z^T(t)z(t)dt \leq \gamma_n^2 \int_0^{\infty} \omega^T(t)\omega(t)dt + x^T(0)Px(0) + \sum_{i=1}^{m} \frac{\tilde{\rho}_{i^2}(0)}{l_i} \qquad (7.58)$$

which implies that Equation (7.50) holds. The proofs for the system of Equation (7.44) in the fault case are similar and omitted here. ∎

Remark 7.5 *By Equation (7.36) and Equation (7.47), it follows that $\tilde{\rho}_i(0) \leq \max_j\{\bar{\rho}_i^j\} - \min_j\{\underline{\rho}_i^j\}$. Then, one can choose l_i sufficiently large so that $\sum_{i=1}^{m} \frac{\tilde{\rho}_{i^2}(0)}{l_i}$ is sufficiently small.*

Remark 7.6 *In fact, the existence of state quantization has brought some difficulty to the design of fault-tolerant controllers using both an adaptive method and LMI technique. The main difficulty lies in how to design the adaptive law and deal with the cross terms brought by the quantized signal to guarantee the asymptotically stable and optimized H_∞ performance of closed-loop systems in both normal and fault conditions. It can be seen from Equation*

(7.47) that the newly proposed adaptive laws include the quantized state term $f(x(t))$, which indicates how the quantized state signal affects the adaptive law.

Remark 7.7   *Theorem 7.3 gives sufficient conditions for the existence of an adaptive fault-tolerant H_∞ controller via quantized state feedback. If set $Y_{ai} = 0$, $Y_{bi} = 0$, $i = 1, \ldots, m$, then the conditions of Theorem 7.3 reduce to those of fixed gain Y_0. This shows that the design condition for adaptive fault-tolerant H_∞ controllers given in Theorem 7.3 is more relaxed than those for traditional fault-tolerant H_∞ controllers with fixed gains [37].*

Denote γ_n and γ_f as the H_∞ performance indexed in normal and faulty cases, respectively. Then the fault-tolerant H_∞ control problem for linear systems with quantized state-feedback controller, Equation (7.38), can be transformed by solving the following optimization problem:

$$\min \alpha \eta_n + \beta \eta_f \ s.t. \ Equation(7.46) \tag{7.59}$$

where $\eta_n = \gamma_n^2$, $\eta_f = \gamma_f^2$, and α and β are weighting coefficients. Because systems operate in normal conditions most of the time, we can choose $\alpha > \beta$ in Equation (7.59).

7.4.4   Simulation

Consider the system of Equation (7.35) with

$$A = \begin{bmatrix} -1 & 0.9 \\ -8 & -5 \end{bmatrix}, B = \begin{bmatrix} 0.3 & 0 \\ 9 & 3 \end{bmatrix}, B_1 = \begin{bmatrix} 1 \\ 4 \end{bmatrix}$$

The parameters for the quantizer $f(\cdot)$ are $\xi_1 = 0.9$, $\xi_2 = 0.8$.
   In this example, the regulated output $Z(t)$ is chosen as

$$Z(t) = \begin{bmatrix} 0 & 5 \\ 0 & 0 \\ 0 & 0 \end{bmatrix} x(t) + \begin{bmatrix} 0 & 0 \\ 2 & 0 \\ 0 & 1 \end{bmatrix} u(t)$$

to improve the performance of the second state $x_2(t)$. We consider the following three possible fault modes:

- Both of the actuators are normal, that is,

$$\rho_1^1 = \rho_2^1 = 0$$

- The first actuator is in outage and the second actuator may be normal or may have loss of effectiveness, that is,

$$\rho_1^2 = 1, \quad 0 \le \rho_2^2 \le 0.9$$

which denotes that the maximum loss of effectiveness for the second actuator is 0.9.

Table 7.1 Comparison of H_∞ performance index

	Normal case	Fault case
Fixed gain controller	1.8162	2.6166
Adaptive controller	0.3311	1.5632

- The second actuator is in outage and the first actuator may be normal or may have loss of effectiveness, that is,

$$\rho_2^3 = 1, \quad 0 \le \rho_1^3 \le 0.8$$

which denotes that the maximum loss of effectiveness for the first actuator is 0.8.

Using Equation (7.59) with $\alpha = 10$, $\beta = 1$ and Remark 7.7, the H_∞ performances of the fixed-gain controller and the adaptive controller in normal and fault conditions are obtained (see Table 7.1, which further indicates the superiority of our adaptive method).

In the simulation, we use the disturbance $\omega(t) = [\omega_1(t) \quad \omega_2(t)]^T$:

$$\omega_1(t) = \omega_2(t) = \begin{cases} 1, & 2 \le t \le 3(s) \\ 0, & \text{otherwise} \end{cases}$$

The fault cases under consideration are as follows:

1. At 4.5 s, the second actuator is in outage.
2. At 3 s, the loss of effectiveness of the first actuator is 0.7.

To verify the effectiveness and superiority of our adaptive quantized controller compared with the quantized controller with fixed gain, we simulate the response curves of $x_2(t)$ in the normal case, in fault case 1 and in fault case 2 in Figures 7.3–7.5, respectively. It can be seen that both controllers can tolerate the considered faults. Moreover, although state signal quantization exists, our adaptive method has more restraint disturbance ability than the fixed gain one in either normal or fault cases, just as the theory has proved. Figure 7.6 shows the response curves of the controller output with these two controllers in the normal case, in fault case 1 and in fault case 2.

The estimated values $\hat{\rho}_1$, $\hat{\rho}_2$ are given in Figure 7.7. From [38], it follows in general that, when the exogenous input signal is persistently exciting (PE), the estimated values $\hat{\rho}_i$ can converge to true values ρ_i. When the exogenous input signal is not persistently exciting, the estimated values can converge but may not converge to their true values. In this example, $\omega(t)$ is not PE and it takes less than 5 s for the estimated values to converge. It should be noted that in our adaptive quantized controller design, it is not necessary to make the estimated values converge to their true values.

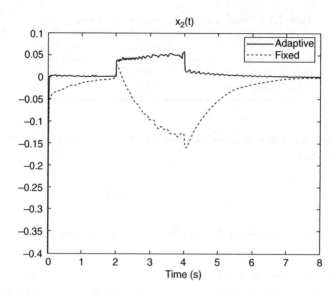

Figure 7.3   Response curves in normal operation with quantized controllers

7.5   Sliding-Mode Observer

7.5.1   Introduction

In recent years, network technology has developed dramatically. Data communication networks have been used in control systems to join together systems with their equipment, devices and components. They are generally referred to as "networked control systems" (NCS). In

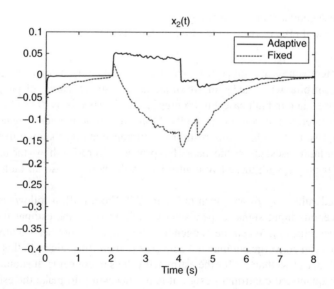

Figure 7.4   Response curves in fault case 1 with quantized controllers

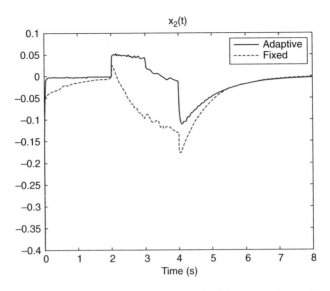

Figure 7.5   Response curves in fault case 2 with quantized controllers

industrial plants, cars, aircraft, and spacecraft, serial communication networks replace data and control signals among locative distributed system components, such as controllers, computers, supervisory devices, and smart sensors (that is, actuators and intelligent I/O). The major features of NCS are simple installation and maintenance, the reduced cost of system wiring, and increased system agility [39]. The two-level control system of a distributed network (DNCS) has been a main trend in commercial and new industrial fields [40]. Because of the complications of networks, this type of automatically controlled system is more susceptible to faults.

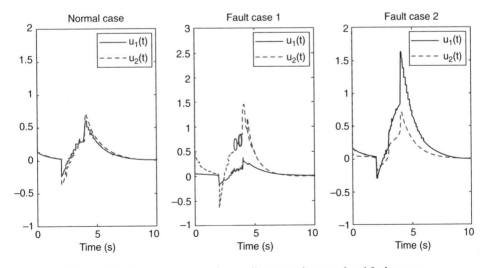

Figure 7.6   Response curves of controller output in normal and faulty cases

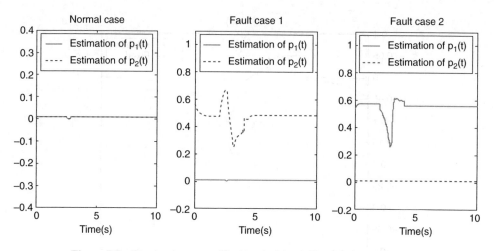

Figure 7.7   Response curves of fault estimation in normal and faulty cases

Therefore, research on this area is theoretical. We must consider fault detection for a two-level DNCS structure, with data exchange between subsystems and a central unit in the network.

7.5.2   Dynamic model

A two-level DNCS is shown in Figure 7.8, where the central unit with subsystems includes controllers, sensors, and actuators.

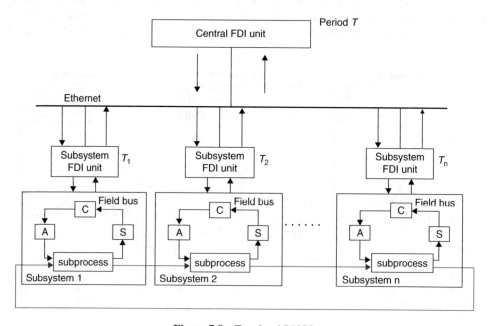

Figure 7.8   Two-level DNCS

A model of the central unit with subsystems is described by:

$$x(k + 1) = Ax(k) + \sum_{j=1}^{\tau_{st}} A_{\tau,j} x(k - j) + Bu(k) + E_d d(k) + E_f f(k)$$

$$y(k) = Cx(k) \tag{7.60}$$

where $x(k) \in \Re^n$ is the state vector, $d(k) \in \Re^r$ is the vector including uncertain items, $y(k) \in \Re^p$ is the output vector, $f(k) \in \Re^q$ is the fault vector, and $u(k) \in \Re^m$ is the control input.

To cope with the faulty system of Equation (7.60), an observer of the form:

$$\hat{x}(k + 1) = A\hat{x}(k) + \sum_{j=1}^{\tau_{st}} A_{\tau,j} \hat{x}(k - j) + Bu(k) + L(y(k) - \hat{y}(k)) + v(k) \tag{7.61}$$

$$\hat{y} = C\hat{x}(k) \tag{7.62}$$

is designed where $\hat{x} \in \Re^n$ is the estimate of the state vector, $\hat{y} \in \Re^p$ is the estimate of output, $L \in \Re^{n \times p}$ is the gain observer, and $v(k) \in \Re^n$ is the input observer. Proceeding further, we can derive the system error model as:

$$e_x(k + 1) = (A - LC)e_x(k) + \sum_{j=1}^{\tau_{st}} A_{\tau,j} e_x(k - j) + Bu(k) + E_d d(k) + E_f f(k) - v(k)$$

$$e_y(k) = Ce_x(k) \tag{7.63}$$

Adopting the sliding-mode theory [41], the sliding-mode manifold $s(k)$ is defined as

$$s(k) = r(k) = e_y(k) \tag{7.64}$$

where the residual $r(k) = 0$ in the absence of fault. To manage the false alarm rate, we select the residual evaluation function J_T and threshold J_{th} as:

$$J_T = \sqrt{\sum_{k=1}^{N} r(k)^T r(k)}$$

$$J_{th} = \sup_{f(k)=0} J_T \tag{7.65}$$

such that

$$\begin{cases} J_T > J_{th} \implies Faults \rightarrow Alarm \\ J_T \le J_{th} \implies \quad No\ Faults \end{cases} \tag{7.66}$$

The formulation of the fault detection design problem reduces to a sliding-mode observer design problem for which it is desired to compute the gain matrix L and the observer's nonlinear input $v(k)$ such that $s(k + 1) = s(k) = 0$ ensures that the error system of Equation (7.63) is asymptotically stable and the sliding-mode manifold satisfies $||s(k + 1)|| < ||s(k)||$.

In the case of full state measurements, the main stability result is summarized by the following theorem.

Theorem 7.4   *Consider the error system of Equation (7.63) subject to the following assumptions*

$$v(k) = v_1(k) + v_2(k), \quad \|d(k)\| \le d_0$$

where

$$v_1(k) = hI_{yx} \frac{y(k) - \hat{y}}{\|y(k) - \hat{y}\|}, \quad [y(k) - \hat{y}] \ne 0$$

$$v_2(k) = \sum_{j=1}^{\tau_{st}} A_{\tau,j} e_x(k - j)$$

$$h = -\|E_d\| d_0$$

If there exist matrices $L \in \Re^{n \times p}$, $P \in \Re^{n \times n}$ satisfying the LMI

$$\begin{bmatrix} -P & (A - LC)^T P \\ \bullet & -P \end{bmatrix} < 0 \tag{7.67}$$

then the error system is asymptotically stable.

Proof. Consider the Lyapunov function:

$$V(k) = e_x(k)^T P e_x(k) \tag{7.68}$$

then we have:

$$\begin{aligned}
\Delta V(k) &= V(k + 1) - V(k) \\
&= e_x(k + 1)^T P e_x(k + 1) - e_x(k)^T P e_x(k) \\
&= e_x(k)^T [(A - LC)^T P(A - LC) - P] e_x(k) \\
&\quad + \sum_{j=1}^{\tau_{st}} A_{\tau,j}^T P \sum_{j=1}^{\tau_{st}} A_{\tau,j} \\
&\quad + 2e_x(k)^T (A - LC)^T P \sum_{j=1}^{\tau_{st}} A_{\tau,j} \\
&\quad + 2e_x(k)^T (A - LC)^T P[E_d d(k) - v(k)] \\
&\quad + 2 \sum_{j=1}^{\tau_{st}} e_x(k - j)^T A_{\tau,j}^T P[E_d d(k) - v(k)] \\
&\quad + [E_d d(k) - v(k)]^T P[E_d d(k) - v(k)]
\end{aligned} \tag{7.69}$$

Letting $v(k) = v_1(k) + v_2(k)$ and manipulating using v_1 and $h = -||E_d||d_0$, we obtain:

$$2e_x(k)^T(A - LC)^T P[E_d d(k) - v(k)] + 2\sum_{j=1}^{\tau_{st}} e_x(k - j)^T A_{\tau,j}^T P[E_d d(k) - v(k)]$$

$$+[E_d d(k) - v(k)]^T P[E_d d(k) - v(k)]$$

$$\leq 2||e_x(k)^T||||(A - LC)^T P||[||E_d d(k)|| + ||v_1(k)||]$$

$$+2||\sum_{j=1}^{\tau_{st}} e_x(k - j)^T A_{\tau,j}^T P||[||E_d d(k)|| + ||v_1(k)||]$$

$$+||[||E_d d(k)|| + ||v_1(k)||]^T P||[||E_d d(k)|| + ||v_1(k)||]$$

$$-2\sum_{j=1}^{\tau_{st}} e_x(k - j)^T A_{\tau,j}^T P v_2(k) - 2e_x(k)^T(A - LC)^T P v_2(k) + v_2(k)^T P v_2(k)$$

$$\leq ||e_x(k)^T||||(A - LC)^T P||[||E_d d(k)||d_0 + h] + 2||\sum_{j=1}^{\tau_{st}} e_x(k - j)^T A_{\tau,j}^T P||$$

$$\times[||E_d d(k)||d0 + h] + [||E_d d(k)||d0 + h][||E_d d(k)||d0 + h]$$

$$-2\sum_{j=1}^{\tau_{st}} e_x(k - j)^T A_{\tau,j}^T P v_2(k) - 2e_x(k)^T(A - LC)^T P v_2(k)$$

$$+v_2(k)^T P v_2(k)$$

$$\leq -2\sum_{j=1}^{\tau_{st}} e_x(k - j)^T A_{\tau,j}^T P v_2(k)$$

$$-2e_x(k)^T(A - LC)^T P v_2(k) + v_2(k)^T P v_2(k). \tag{7.70}$$

This yields

$$\Delta V(k) = V(k + 1) - V(k)$$

$$\leq e_x(k)^T[(A - LC)^T P(A - LC) - P]e_x(k)$$

$$+\sum_{j=1}^{\tau_{st}} e_x(k - j)^T A_{\tau,j}^T P \sum_{j=1}^{\tau_{st}} e_x(k - j)^T A_{\tau,j}$$

$$+2e_x(k)^T(A - LC)^T P v_2(k) + v_2(k)^T P v_2(k)$$

$$-2\sum_{j=1}^{\tau_{st}} e_x(k - j)^T A_{\tau,j}^T P v_2(k) \tag{7.71}$$

Substituting $v_2(k) = \sum_{j=1}^{\tau_{st}} A_{\tau,j} e_x(k - j)$ into Equation (7.71) and manipulating, we arrive at

$$\Delta V(k) = V(k + 1) - V(k)$$

$$\leq e_x(k)^T[(A - LC)^T P(A - LC) - P]e_x(k) \tag{7.72}$$

By Schur complements on Equation (7.67), the asymptotic stability of the error system of Equation (7.63) follows from Equation (7.72) and the proof is completed. ∎

The associated reachability result is summarized by the following theorem.

Theorem 7.5 *Consider the error system of Equation (7.63) subject to the following assumptions*

$$v(k) = v_1(k) + v_2(k), \quad ||d(k)|| \leq d_0$$

where

$$v_1(k) = hI_{yx} \frac{y(k) - \hat{y}}{||y(k) - \hat{y}||}, \quad [y(k) - \hat{y}] \neq 0$$

$$v_2(k) = \sum_{j=1}^{\tau_{st}} A_{\tau,j} e_x(k - j)$$

$$h = -||E_d||d_0.$$

If there exists a matrix $L \in \Re^{n \times p}$ and a scalar $\gamma^2 < 1$ satisfying the LMI:

$$\begin{bmatrix} -\gamma^2 C^T C & (A - LC)^T C^T \\ \bullet & -I \end{bmatrix} < 0 \tag{7.73}$$

then the error system will get into the sliding surface in finite time.

Proof. Starting from $e_y(k + 1) = C e_x(k + 1)$, it follows that

$$e_y(k + 1)^T e_y(k + 1) - \gamma^2 e_y(k)^T e_y(k) = e_x(k + 1)^T C^T C e_x(k + 1)$$
$$- \gamma^2 e_x(k)^T C^T C e_x(k) \tag{7.74}$$

where $\gamma^2 < 1$. Substituting Equation (7.63) into Equation (7.74) and manipulating, it yields:

$$e_y(k + 1)^T e_y(k + 1) - \gamma^2 e_y(k)^T e_y(k)$$
$$= e_x(k)^T [(A - LC)^T C^T C(A - LC) - \gamma^2 C^T C] e_x(k)$$
$$+ [2e_x(k)^T [(A - LC)^T + \sum_{j=1}^{\tau_{st}} A_{\tau,j}^T e_x(k - j)^T] C^T C \sum_{j=1}^{\tau_{st}} A_{\tau,j} e_x(k - j)$$
$$+ 2e_x(k)^T [(A - LC)^T C^T C[E_d d(k) - v(k)]$$
$$+ 2 \sum_{j=1}^{\tau_{st}} A_{\tau,j} e_x(k - j) C^T C[E_d d(k) - v(k)]$$
$$+ [E_d d(k) - v(k)]^T C^T C[E_d d(k) - v(k)] \tag{7.75}$$

Letting $v(k) = v_1(k) + v_2(k)$ and manipulating v_1 and $h = -||E_d||d_0$, we reach:

$$2e_x(k)^T[(A - LC)^T C^T C[E_d d(k) - v(k)]$$

$$+2\sum_{j=1}^{\tau_{st}} A_{\tau,j} e_x(k - j) C^T C[E_d d(k) - v(k)]$$

$$+[E_d d(k) - v(k)]^T C^T C[E_d d(k) - v(k)]$$

$$\leq v_2(k)^T C^T C v_2(k) - 2e_x(k)^T[(A - LC)^T C^T C v_2(k)$$

$$-2\sum_{j=1}^{\tau_{st}} A_{\tau,j} e_x(k - j) C^T C v_2(k) \tag{7.76}$$

Further manipulations with $v_2(k) = \sum_{j=1}^{\tau_{st}} A_{\tau,j} e_x(k - j)$ yield:

$$e_y(k + 1)^T e_y(k + 1) - \gamma^2 e_y(k)^T e_y(k)$$

$$\leq e_x(k)^T[(A - LC)^T C^T C(A - LC) - \gamma^2 C^T C]e_x(k)$$

$$+2e_x(k)^T(A - LC)^T C^T C + \sum_{j=1}^{\tau_{st}} A_{\tau,j} e_x(k - j)$$

$$+\sum_{j=1}^{\tau_{st}} e_x^T(k - j)A_{\tau,j}^T C^T C \sum_{j=1}^{\tau_{st}} A_{\tau,j} e_x(k - j)$$

$$-2e_x(k)^T[(A - LC)^T E_d d(k) - v(k)]$$

$$-2\sum_{j=1}^{\tau_{st}} e_x^T(k - j)A_{\tau,j}^T C^T C v_2(k)$$

$$+v_2(k)^T C^T C v_2(k)$$

$$\leq e_x(k)^T[(A - LC)^T C^T C(A - LC) - \gamma^2 C^T C]e_x(k) \tag{7.77}$$

By Schur complements on Equation (7.73), it follows from Equation (7.77) that

$$e_y(k + 1)^T e_y(k + 1) - \gamma^2 e_y(k)^T e_y(k) < 0$$

which implies that

$$||s(k + 1)|| < ||s(k)||$$

and the proof is completed.   ∎

7.5.3 Limited state measurements

When some of the states are not measurable, we have to use a different sliding-mode observer. It has the form:

$$\begin{bmatrix} x_1(k+1) \\ x_2(k+1) \\ x_3(k+1) \end{bmatrix} = \begin{bmatrix} A_1 \\ A_2 \\ A_3 \end{bmatrix} x(k) + \sum_{j=1}^{\tau_{st}} \begin{bmatrix} A_{\tau,j,1} \\ A_{\tau,j,2} \\ A_{\tau,j,3} \end{bmatrix} x(k-j) + \begin{bmatrix} B_1 \\ B_2 \\ B_3 \end{bmatrix} u(k)$$

$$+ \begin{bmatrix} E_{d,1} \\ E_{d,2} \\ E_{d,3} \end{bmatrix} d(k) + \begin{bmatrix} E_{f,1} \\ E_{f,2} \\ E_{f,3} \end{bmatrix} f(k)$$

$$y(k) = \begin{bmatrix} 0 & C_2 & 0 \\ 0 & 0 & C_3 \end{bmatrix} x(k) \tag{7.78}$$

where $x_1(k) \in \Re^{n-r}$, $x_2(k) \in \Re^{r-s}$, $x_3 \in \Re^s$. Note that $x_2(k)$, $x_3(k)$ can be obtained from $y(k)$ and thus $x_1(k)$ is unknown. To proceed further, we recall the following assumption:

Assumption 7.1

$$rank \begin{bmatrix} C \begin{bmatrix} E_{f,1} \\ E_{f,2} \\ E_{f,3} \end{bmatrix} \end{bmatrix} = s \tag{7.79}$$

and $E_{f,3}$ is a non-singular matrix.

Using the coordinate transition

$$Q = \begin{bmatrix} I & 0 & E_{f,1}E_{f,3}^{-}1 \\ 0 & I & E_{f,2}E_{f,3}^{-}1 \\ 0 & 0 & I \end{bmatrix} \tag{7.80}$$

with Equation (7.78) and manipulating, it yields:

$$x_1(k+1) = A_{11}x_1(k) + A_{12}C^{-}1_2 y_2(k)$$

$$+ A_{13}C^{-}1_3 y_3(k) + \sum_{j=1}^{\tau_{st}} A_{\tau,j,11}x_1(k-j))$$

$$+ \sum_{j=1}^{\tau_{st}} A_{\tau,j,12}C^{-}1 y_2(k-j))$$

$$+ \sum_{j=1}^{\tau_{st}} (A_{\tau,j,13}C^{-}1 y_3(k-j))$$

$$+ B_1 u(k) + E_1 d(k) + E_{f,1}E_{f,3}^{-}1C^{-}1_3 y_3(k+1) \tag{7.81}$$

$$x_2(k+1) = A_{21}x_1(k) + A_{22}x_2(k)$$

$$+ A_{23}C^-1_3y_3(k) + \sum_{j=1}^{\tau_{st}} (A_{\tau,j,21}x_1(k-j))$$

$$+ \sum_{j=1}^{\tau_{st}} (A_{\tau,j,22}C^-1_2y_2(k-j))$$

$$+ \sum_{j=1}^{\tau_{st}} (A_{\tau,j,23}C^-1_3y_3(k-j))$$

$$+ B_2u(k) + E_2d(k) + E_{f,2}E_{f,3}^-1C^-1_3y_3(k+1) \qquad (7.82)$$

$$x_3(k+1) = A_{31}x_1(k) + A_{32}C^-1_2y_2(k)$$

$$+ A_{33}x_3(k) + \sum_{j=1}^{\tau_{st}} (A_{\tau,j,31}x_1(k-j))$$

$$+ \sum_{j=1}^{\tau_{st}} (A_{\tau,j,32}C^-1_2y_2(k-j))$$

$$+ \sum_{j=1}^{\tau_{st}} (A_{\tau,j,33}x_3(k-j)) + B_3u(k)$$

$$+ E_3d(k) + E_{f,3}f(k) \qquad (7.83)$$

$$y_2(k) = C_2x_2(k), \quad y_3(k) = C_3x_3(k) \qquad (7.84)$$

where

$$A_1 - E_{f,1}E_{f,3}^-1A_3 = [A_{1,1}A_{1,2}A_{1,3}]$$
$$A_{\tau,j,1} - E_{f,1}E_{f,3}^-1A_{\tau,j,3} = [A_{\tau,1,1}A_{\tau,1,2}A_{\tau,1,3}]$$
$$A_2 - E_{f,2}E_{f,3}^-1A_3 = [A_{2,1}A_{2,2}A_{2,3}]$$
$$A_{\tau,j,2} - E_{f,2}E_{f,3}^-1A_{\tau,j,3} = [A_{\tau,2,1}A_{\tau,2,2}A_{\tau,2,3}]$$
$$B_1 = (B_1 - E_{f,1}E_{f,3}^-1B_3), \quad B_2 = (B_2 - E_{f,2}E_{f,3}^-1B_3)$$
$$E_1 = (E_{d,1} - E_{f,1}E_{f,3}^-1E_{d,3})$$
$$E_2 = (E_{d,2} - E_{f,2}E_{f,3}^-1E_{d,3})$$
$$A_3 = [A_{3,1}A_{3,2}A_{3,3}]$$
$$A_{\tau,j,3} = [A_{\tau,3,1}A_{\tau,3,2}A_{\tau,3,3}]$$

Introduce

$$\bar{u}(k) = A_{12}C^{-}1_2 y_2(k) + A_{13}C^{-}1_3 y_3(k) + \sum_{j=1}^{\tau_{st}}(A_{\tau,j,12}C^{-}1 y_2(k-j))$$

$$+ \sum_{j}^{\tau}(A_{\tau,13}C^{-}1 y_3(k-j))$$

$$+ B_1 u(k) + E_1 d(k) + E_{f,1}E_{f,3}^{-}1C^{-}1_3 y_3(k+1)$$

$$w(k) = C^{-}1_2 y_2(k) - A_{22}C^{-}1_2 y_2(k)$$

$$- A_{23}C^{-}1_3 y_3(k) - \sum_{j=1}^{\tau_{st}}(A_{\tau,j,22}C^{-}1_2 y_2(k-j))$$

$$- \sum_{j=1}^{\tau_{st}}(A_{\tau,j,23}C^{-}1_3 y_3(k-j))B_2 u(k) - E_2 d(k)$$

$$+ E_{f,2}E_{f,3}^{-}1C^{-}1_3 y_3(k+1) \tag{7.85}$$

The design of observer for the state components $x_1(k)$ and $x_3(k)$ is therefore given by:

$$\hat{x}_1(k+1) = A_{11}\hat{x}_1(k) + \sum_{j=1}^{\tau_{st}}(A_{\tau,j,11}\hat{x}_1(k-j)) + \bar{u}(k)$$

$$+ L_1[w(k) - A_{21}\hat{x}_1(k) + \sum_{j=1}^{\tau_{st}}(A_{\tau,j,21}\hat{x}_1(k-j))] \tag{7.86}$$

$$\hat{x}_3(k+1) = A_{31}\hat{x}_1(k) + A_{32}C^{-}1_2 y_2(k) + A_{33}\hat{x}_3(k) + \sum_{j=1}^{\tau_{st}}(A_{\tau,j,31}\hat{x}_1(k-j))$$

$$+ \sum_{j=1}^{\tau_{st}}(A_{\tau,j,32}C^{-}1_2 y_2(k-j)) + \sum_{j=1}^{\tau_{st}}(A_{\tau,j,33}\hat{x}_3(k-j))$$

$$+ B_3 u(k) + L_2(y_3(k) - \hat{y}_3(k)) - v(k) \tag{7.87}$$

$$\hat{y}_3(k) = C_3\hat{x}_3(k) \tag{7.88}$$

where $\hat{x}_1(k)$ and $\hat{x}_3(k)$ the estimated value of $x_1(k)$ and $x_3(k)$, $\hat{y}(k)$ the estimate of the output, and $L_1 \in \mathfrak{R}^{(n-r)\times(r-s)}$ and $L_2 \in \mathfrak{R}^{(r)\times p}$ are unknown gains. Algebraic manipulations lead to the error system as:

$$e_1(k+1) = (A_{11} - L_1 A_2 1)e_1(k)$$

$$+ \sum_{j=1}^{\tau_{st}}(A_{\tau,j,11} - L_1 A_{\tau,21})e_1(k-j) + (E_1 - L_1 E_2)d(k) \tag{7.89}$$

$$e_y(k+1) = C_3 A_{31} e_1(k) + (C_3 A_{33} C_3^- 1 - C_3 L_2) e_y(k)$$

$$+ \sum_{j=1}^{\tau_{st}} C_3 (A_{\tau,j,31} e_1(k-j)$$

$$+ C_3 \sum_{j=1}^{\tau_{st}} C_3^- 1 (A_{\tau,j,33} e_y(k-j) + C_3 v$$

$$+ E_{d,3} d(k) + E_{f,3} f(k) \tag{7.90}$$

$$y_3(k) = C_3 \hat{x}_3(k) \tag{7.91}$$

In terms of the sliding-mode manifold $s(k) = e_y(k)$, the objective is to find matrices $L_1 \in \Re^{(n-r)\times(r-s)}$ and $L_2 \in \Re^{(r)\times p}$ and signal $v(k)$ such that the error system of Equation (7.89) is asymptotically stable when $s(k+1) = s(k) = 0$ and error systems of Equation (7.89) and Equation (7.90) satisfies

$$||s(k+1)|| < ||s(k)||$$

The error system stability and reachability are summarized by the following theorems.

Theorem 7.6 *Consider the error system of Equation (7.89). If there exist matrices $L_1 \in \Re^{(n-r)\times(r-s)}$ and $P \in \Re^{(n-r)\times(n-r)}$, $Q \in \Re^{(n-r)\times(r-r)}$ satisfying the LMI*

$$
\begin{bmatrix}
-P + \tau_{st} Q & 0 & 0 & 0 & 0 & (A_{11} - L_1 A_{21})^T P \\
\bullet & -Q & 0 & \cdots & 0 & (A_{\tau,1,11} - L_1 A_{\tau,1,21})^T P \\
\bullet & \bullet & -Q & \cdots & 0 & (A_{\tau,2,11} - L_1 A_{\tau,2,21})^T P \\
\bullet & \bullet & \bullet & \ddots & \vdots & \vdots \\
\bullet & \bullet & \bullet & \bullet & -Q & (A_{\tau,\tau_{st},11} - L_1 A_{\tau,\tau_{st},21})^T P \\
\bullet & \bullet & \bullet & \bullet & \bullet & -P
\end{bmatrix} < 0 \tag{7.92}
$$

then the error system of Equation (7.89) is asymptotically stable.

Theorem 7.7 *Consider the error system of Equation (7.90) subject to the following assumptions*

$$v(k) = v_1(k) + v_2(k), \quad ||d(k)|| \le d_0$$

where

$$v_1(k) = h I_{yx} \frac{y(k) - \hat{y}}{||y(k) - \hat{y}||}, \quad [y(k) - \hat{y}] \ne 0$$

$$v_2(k) = \sum_{j=1}^{\tau_{st}} A_{\tau,j} e_x(k-j)$$

$$h = -||\Psi|| d_0 - ||C_3 E_{d,3}|| d_0, \quad \Psi = [C_3 A_{31} C_3 A_{\tau,1,31} C_3 A_{\tau,\tau_{st},31}]$$

If there exists a matrix $L_2 \in \Re^{n \times p}$ and a scalar $\gamma^2 < 1$ satisfying the LMI:

$$\begin{bmatrix} -\gamma^2 I & (C_3 A_{33} C_3^{-1} - C_3 L_2)^T \\ \bullet & -I \end{bmatrix} < 0 \qquad (7.93)$$

then the error system will get into the sliding surface in finite time.

The proofs of Theorems 7.6 and 7.7 follow by parallel development to Theorems 7.4 and 7.5; they are left to the reader as an exercise.

7.5.4   Simulation results: full state measurements

Consider a discrete-time system of the type of Equation (7.60) with matrices:

$$A = \begin{bmatrix} 0 & 0 & 0.897 & 0 & 0.007 & 0 \\ 0 & 0 & 0 & 0.928 & 0 & 0.05 \\ 0 & 0 & 0.951 & 0.003 & 0.010 & 0 \\ 0 & 0 & 0 & 0.97 & 0 & 0.020 \\ 0 & 0 & 0.011 & 0.021 & 0.951 & 0.002 \\ 0 & 0 & 0 & 0.011 & 0 & 0.951 \end{bmatrix}$$

$$B = \begin{bmatrix} 0.056 & 0 & 0 \\ 0.0523 & 0 & 0 \\ 0 & 0.061 & 0 \\ 0 & 0.054 & 0 \\ 0 & -0.013 & 0 \\ 0 & -0.004 & 0 \end{bmatrix}, \quad E_d = \begin{bmatrix} 0.013 & 0 & 0 \\ 0.061 & 0 & 0 \\ 0 & 0.012 & 0 \\ 0 & 0.061 & -0.001 \\ 0 & -0.002 & 0.031 \\ 0 & 0 & 0.061 \end{bmatrix}$$

$$E_f = \begin{bmatrix} 0.039 & 0 & 0 \\ 0.011 & 0 & 0 \\ 0 & 0.041 & 0 \\ 0 & 0.011 & -0.002 \\ 0 & -0.005 & 0.005 \\ 0 & -0.001 & 0.011 \end{bmatrix}$$

$$C = \begin{bmatrix} 0 & 0 & 0.928 & 0.488 & 0.005 & 0.005 \\ 0 & 0 & 0.951 & 0.495 & 0.012 & 0.01 \\ 0 & 0 & 0.061 & 0.0158 & 0.588 & 0.396 \end{bmatrix}$$

$$u(k) = \begin{bmatrix} u_1(kT + 0.05) \\ u_1(kT + 0.1) \\ u_2(kT + 0.1) \end{bmatrix}, \quad f(k) = \begin{bmatrix} f_1(kT + 0.05) \\ f_1(kT + 0.1) \\ f_2(kT + 0.1) \end{bmatrix}$$

$$d(k) = \begin{bmatrix} d_1(kT + 0.05) \\ d_1(kT + 0.1) \\ d_2(kT + 0.1) \end{bmatrix}, \quad y(k) = \begin{bmatrix} y_1(kT + 0.05) \\ y_1(kT + 0.1) \\ y_2(kT + 0.1) \end{bmatrix}$$

$$x(k) = \begin{bmatrix} x_1(kT + 0.05) \\ x_1(kT + 0.1) \\ x_2(kT + 0.1) \end{bmatrix}$$

When the time-delay is $\tau_{st} = 2$, then

$$A_{\tau,1} = \begin{bmatrix} 0 & 0 & 0.251 & 0 & -14.375 & 7.188 \\ 0 & 0 & 0 & 0.011 & 0 & 0 \\ 0 & 0 & 0.063 & 0 & -3.594 & 1.797 \\ 0 & 0 & 0 & 0.001 & 0 & 0 \\ 0 & 0 & 0 & 0 & 0 & 0 \\ 0 & 0 & 0 & 0 & 0 & 0 \end{bmatrix}$$

$$A_{\tau,2} = \begin{bmatrix} 0 & 0 & 0.03 & 0 & -1.725 & 0.863 \\ 0 & 0 & 0 & 0.18 & 0 & 0 \\ 0 & 0 & 0.001 & 0 & -0.052 & 0.026 \\ 0 & 0 & 0 & 0.041 & 0 & 0 \\ 0 & 0 & 0 & 0 & 0 & 0 \\ 0 & 0 & 0 & 0 & 0 & 0 \end{bmatrix}$$

Application of Theorems 7.4 and 7.5 yields the observer gains as:

$$L = \begin{bmatrix} 43.028 & -38.572 & 0.384 \\ 73.791 & 74.553 & -0.775 \\ -32.354 & -29.502 & 0.324 \\ -60.765 & 61.704 & -0.604 \\ 5.624 & -5.545 & 0.792 \\ -8.785 & 9.765 & 0.623 \end{bmatrix}$$

$$r = 0.822, \quad h = -0.007$$

Figures 7.9 and 7.10 give the simulation results of the system residual signal without fault and with fault, respectively. Figure 7.11 illustrates the residual evaluation.

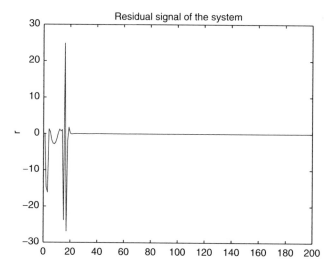

Figure 7.9 System residual signal without fault

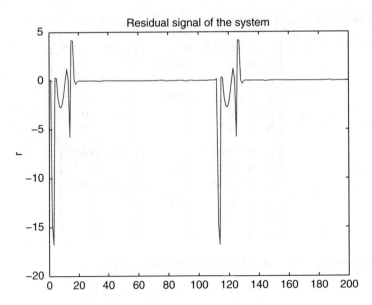

Figure 7.10   System residual signal with fault

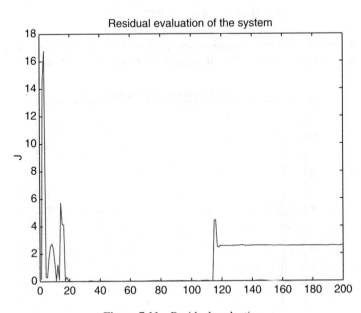

Figure 7.11   Residual evaluation

7.5.5 Simulation results: partial state measurements

In this case, the central unit system is modified as:

$$
\begin{bmatrix} x_1(k+1) \\ x_2(k+1) \\ x_3(k+1) \end{bmatrix} = \begin{bmatrix} A_1 \\ A_2 \\ A_3 \end{bmatrix} x(k) + \sum_{j=1}^{\tau_{st}} \begin{bmatrix} A_{\tau,j,1} \\ A_{\tau,j,2} \\ A_{\tau,j,3} \end{bmatrix} x(k-j) + \begin{bmatrix} B_1 \\ B_2 \\ B_3 \end{bmatrix} u(k)
$$
$$
+ \begin{bmatrix} E_{d,1} \\ E_{d,2} \\ E_{d,3} \end{bmatrix} d(k) + \begin{bmatrix} E_{f,1} \\ E_{f,2} \\ E_{f,3} \end{bmatrix} f(k)
$$
$$
y(k) = \begin{bmatrix} 0 & C_2 & 0 \\ 0 & 0 & C_3 \end{bmatrix} x(k)
$$

Application of Theorems 7.6 and 7.7 yields the observer gains as

$$
L_1 = \begin{bmatrix} 0.011 \\ 0.276 \end{bmatrix}, \quad L_2 = \begin{bmatrix} 7.056 & 5.721 & 11.895 \\ 8.34 & 4.789 & 6.522 \\ -10.756 & 12.552 & -0.078 \end{bmatrix}, \quad h = -0.008 \quad (7.94)
$$

The simulation result in Figure 7.12 illustrates the system residual signal without fault and Figure 7.13 shows the system residual signal with fault. Figure 7.14 illustrates the residual evaluation.

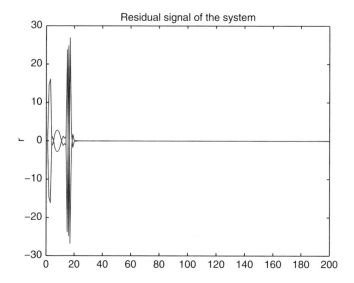

Figure 7.12 System residual signal without fault

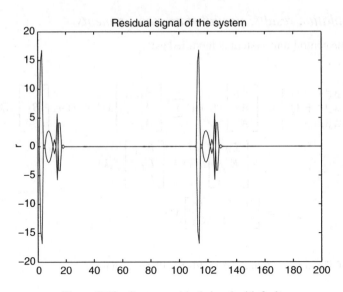

Figure 7.13   System residual signal with fault

7.6   Control of Linear Switched Systems

This section is concerned with fault-tolerant controller methodology for a class of discrete-time switched systems with time-varying delay. The mathematical development incorporates an improved Lyapunov functional and the average dwell time scheme to construct a mode-dependent (delay-dependent) controller which ensures exponential stability of the closed-loop system. Furthermore, an optimization problem with LMI constraints is formulated to estimate the domain of attraction.

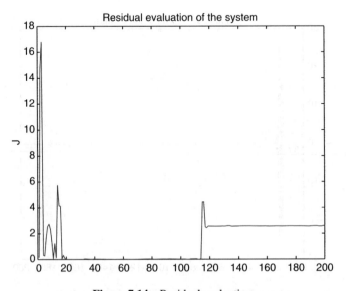

Figure 7.14   Residual evaluation

7.6.1   Introduction

Switched systems consist of multiple systems with their own parameterizations and a switching law. It is agreed that the stabilization of these type of systems is composed of three basic problem [42]:

1. Find the conditions for stability under arbitrary switching.
2. Identify the limiting class of switching signals.
3. Construct a stabilizing switching signal.

It appears that Problem 2 has been explored in relatively fewer studies than Problems 1 and 3. These works reveal that the average dwell time technique is a powerful method for solving this problem; however, these studies have only considered stability and filtering problems; the feedback stabilization problem has not been fully addressed. This section considers this issue and develops some procedures to stabilize these types of system via an average dwell time scheme. Focus is also given to actuator saturation and unexpected faults as these occur in most practical problems. A feedback controller is designed for each of the systems to ensure exponential stability of the closed-loop system. Based on the dwell-time approach and free weighting matrix method, some delay dependent sufficient conditions for the existence of such a controller are derived in the formulation of LMI. Then an optimization problem with LMI constraints is presented to estimate the domain of attraction of the origin of the underlying systems.

7.6.2   Problem formulation

The mathematical description for discrete-time switched systems with time-varying delay is given as

$$S_{\sigma(k)} : x(k+1) = A_{\sigma(k)}x(k) + A_{d\sigma(k)}x(k - d(k)) + B_{\sigma(k)}\vartheta(u(k))$$

$$x(\varphi) = \phi(\varphi), \qquad \varphi = -d_2, -d_2 + 1, \cdots, 0, \tag{7.95}$$

where $x(k) \in \Re^n$ is the state, $u(k) \in \Re^m$ is the control input and $\phi(\varphi)$ is a vector-valued initial function. The time-varying delay $d(k)$ is assumed to satisfy $0 < d_m \le d(k) \le d_M$. $\sigma(k) : Z \to M = \{1, 2, \ldots, N\}$ is the switching signal, N is the finite integer and Z is the set of positive integers. Meanwhile, for switching time sequence $k_0 < k_1 < k_2 < \ldots$ of the switching signal $\sigma(k)$, the holding time between $[k_l, k_{l+1}]$ is called the "dwell time" of the currently engaged subsystem, where l is a nonnegative integer. A_i, A_{di} and $B_i, i = 1, 2, \ldots, N$ are constant matrices with appropriate dimensions. The function $\vartheta(.) : \Re^m \to \Re^m$ is the standard saturation function.

Actuator failure is taken as the percentage loss in control signal effectiveness. When an actuator experiences failure, we use $u_F(k)$ to describe the control signal sent from the actuator. In the following discussion, the actuator failure model with failure parameter F is considered as

$$u_F(k) = Fu(k),$$

where

$$0 \le \underline{F} = diag\{\underline{f}_1, \ldots, \underline{f}_m\} \le F = diag\{\overline{f}_1, \ldots, \overline{f}_m\} \le I$$

and variables

$$f_r (r = 1, 2, \ldots, m)$$

quantify the actuator failures. Proceeding further, we introduce

$$F_0 = diag\{f_{01}, \ldots, f_{0m}\} := \{\underline{F} + \overline{F}\}/2$$
$$= diag\{\{\underline{f}_1 + \overline{f}_1\}/2, \ldots, \{\underline{f}_m + \overline{f}_m/2\}\} \tag{7.96}$$

$$\tilde{F} = diag\{\tilde{f}_1, \ldots, \tilde{f}_m\} := \{\overline{F} - \underline{F}\}/2 \tag{7.97}$$

$$= diag\{\{\overline{f}_1 - \underline{f}_1\}/2, \ldots, \{\overline{f}_m - \underline{f}_m/2\}\}. \tag{7.98}$$

Let the controller be $u(k) = K_i x(k)$ $(i = 1, \ldots, N)$, where K_i are the controller gain matrices to be determined. Then, the closed-loop, discrete-time, switched time-delay system with actuator saturation becomes

$$x(k + 1) = (A_i x(k) + B_i F \vartheta (K_i x(k))) + A_{di} x(k - d(k)) \tag{7.99}$$

In the following discussion, we denote Ξ the set of $m \times m$ diagonal matrices for the ith subsystem with diagonal elements 1 or 0. Each element of Ξ is labeled E_{ij}, $j = 1, 2, \ldots, 2^m$ and we denote $E_{ij}^- = I - E_{ij}$.

7.6.3   Stability of a closed-loop system

The set of sufficient conditions for the stability of the closed-loop system of Equation (7.99) is summarized without proof by the following theorem:

Theorem 7.8   *The closed-loop system of Equation (7.99) is exponentially stable under any switching signal with average dwell time T_a satisfying Equation (7.105) if there exist some matrices $P_i > 0$, $Q_{1i} > 0$, Q_{2i}, $Z_{1i} > 0$, $Z_{2i} > 0$,*

$$X_i = \begin{bmatrix} X_{11i} & X_{12i} \\ * & X_{22i} \end{bmatrix} > 0, \quad Y_i = \begin{bmatrix} Y_{11i} & Y_{12i} \\ * & Y_{22i} \end{bmatrix} > 0,$$

$$M_i = \begin{bmatrix} M_{1i} \\ M_{2i} \end{bmatrix}, \quad N_i = \begin{bmatrix} N_{1i} \\ N_{2i} \end{bmatrix}, \quad S_i = \begin{bmatrix} S_{1i} \\ S_{2i} \end{bmatrix},$$

with appropriate dimensions such that the following inequalities hold:

$$
\Phi = \begin{bmatrix} \Phi_1 & d_1\Phi_2^T & d_{12}\Phi_2^T & \Phi_3^T P_i \\ \bullet & -d_1 Z_{1i}^{-1} & 0 & 0 \\ \bullet & \bullet & -(d_M - d_m)Z_{2i}^{-1} & 0 \\ \bullet & \bullet & \bullet & -P_i \end{bmatrix} < 0, \tag{7.100}
$$

$$
\psi_1 = \begin{bmatrix} X_i & M_i \\ * & Z_{1i} \end{bmatrix} \geq 0, \tag{7.101}
$$

$$
\psi_2 = \begin{bmatrix} Y_i & N_i \\ * & Z_{2i} \end{bmatrix} \geq 0, \tag{7.102}
$$

$$
\psi_3 = \begin{bmatrix} Y_i & S_i \\ * & Z_{2i} \end{bmatrix} \geq 0, \tag{7.103}
$$

$$
P_i \leq \mu P_\upsilon, \qquad Q_{1i} \leq \mu Q_{1\upsilon}, \qquad Q_{2i} \leq \mu Q_{2\upsilon}
$$
$$
Z_{1i} \leq \mu Z_{1\upsilon}, \qquad Z_{2i} \leq \mu Z_{2\upsilon}, \qquad \forall i, \upsilon \in M, \tag{7.104}
$$

$$
T_a > T_a^* = -\frac{ln(\mu)}{ln(1-\alpha)} \tag{7.105}
$$

for all $i = 1, 2, \ldots, N$ and $j = 1, 2, \ldots, 2^m$ and

$$
\Phi_1 = \begin{bmatrix} \Phi_{11} & \Phi_{12} & \Phi_{13} & \Phi_{14} \\ \bullet & \Phi_{22} & \Phi_{23} & \Phi_{24} \\ \bullet & \bullet & \Phi_{33} & 0 \\ \bullet & \bullet & \bullet & \Phi_{44} \end{bmatrix}
$$

$$
\Phi_{11} = Q_{1i} + Q_{2i} + (1-\alpha)^{d_m}(M_{1i} + M_{1i}^T + d_m X_{11i})
$$
$$
+ (1-\alpha)^{d_M}(d_M - d_m)Y_{11i} - (1-\alpha)P_i
$$
$$
\Phi_{12} = (1-\alpha)^{d_m}(d_m X_{12i} + M_{2i}^T) + (1-\alpha)^{d_M}(N_{1i} - S_{1i} + (d_M - d_m)Y_{12i})
$$
$$
\Phi_{13} = -(1-\alpha)^{d_m}M_{1i} + (1-\alpha)^{d_M}S_{1i}, \quad \Phi_{14} = -(1-\alpha)^{d_M}N_{1i}
$$
$$
\Phi_{22} = d_m(1-\alpha)^{d_m}X_{22i} + (1-\alpha)^{d_M}(N_{2i} + N_{2i}^T - S_{2i} - S_{2i}^T + (d_M - d_m)Y_{22i})
$$
$$
\Phi_{23} = (1-\alpha)^{d_M}S_{2i} - (1-\alpha)^{d_m}M_{2i}, \quad \Phi_{24} = -(1-\alpha)^{d_M}N_{2i}
$$
$$
\Phi_{33} = -(1-\alpha)^{d_m}Q_{1i}, \quad \Phi_{44} = -(1-\alpha)^{d_M}Q_{2i}
$$
$$
\Phi_2 = [A_i - I + B_i F(E_{ij}K_i + E_{ij}^- H_i) \quad A_{di} \quad 0 \quad 0]
$$
$$
\Phi_3 = [A_i + B_i F(E_{ij}K_i + E_{ij}^- H_i) \quad A_{di} \quad 0 \quad 0]
$$

The estimate of domain of attraction for the system of Equation (7.99) is given by $\Gamma \leq 1$, where

$$
\Gamma = \delta^2 [\max_{i \in M} \lambda_{max}(P_i) + d_m \max_{i \in M} \lambda_{max}(Q_{1i}) + d_M \max_{i \in M} \lambda_{max}(Q_{2i})
$$
$$
+ 2d_m(d_m + 1)\max_{i \in M} \lambda_{max}(Z_{1i}) + 2d_M(d_M - d_m + 1)\max_{i \in M} \lambda_{max}(Z_{2i})] \tag{7.106}
$$

Remark 7.8 *By choosing $\mu > 1$ and $\alpha \to 0$ in Equation (7.105), we have $T_a^* \to \infty$ which corresponds to the non-switching case. On the other hand, by choosing $\mu = 1$, we have $T_a^* = 0$, which implies that the switching signal can be arbitrary. The proof of Theorem 7.8 is quite standard using Lyapunov–Krasoviskii theorem and can be found in the work of Zhang and Yu [43].*

The unknown gain can be determined by the following theorems.

Theorem 7.9 *The closed-loop system of Equation (7.99) is exponentially stable under any switching signal with the average dwell time T_a satisfying Equation (7.105) if there exist matrices $W_1 > 0$, $\overline{Q}_{1i} > 0$, $\overline{Q}_{2i} > 0$, $\overline{Z}_{1i} > 0$, $\overline{Z}_{2i} > 0$,*

$$\overline{X}_i = \begin{bmatrix} \overline{X}_{11i} & \overline{X}_{12i} \\ \bullet & \overline{X}_{22i} \end{bmatrix} > 0, \quad \overline{Y}_i = \begin{bmatrix} \overline{Y}_{11i} & \overline{Y}_{12i} \\ * & \overline{Y}_{22i} \end{bmatrix} > 0,$$

$$\overline{M}_i = \begin{bmatrix} \overline{M}_{1i} \\ \overline{M}_{2i} \end{bmatrix}, \quad \overline{N}_i = \begin{bmatrix} \overline{N}_{1i} \\ \overline{N}_{2i} \end{bmatrix}, \quad \overline{S}_i = \begin{bmatrix} \overline{S}_{1i} \\ \overline{S}_{2i} \end{bmatrix}$$

Matrices R_i, T_i and $\overline{K}_i$ have appropriate dimensions such that the following inequalities hold:

$$\overline{\Phi} = \begin{bmatrix} \overline{\Phi}_1 & d_m\overline{\Phi}_2^T & (d_M - d_m)\overline{\Phi}_2^T & \overline{\Phi}_3^T \\ \bullet & -d_m\overline{Z}_{1i}^{-1} & 0 & 0 \\ \bullet & \bullet & -(d_M - d_m)\overline{Z}_{2i}^{-1} & 0 \\ \bullet & \bullet & \bullet & -R_i - R_i^T + W_i \end{bmatrix} < 0 \qquad (7.107)$$

$$\tilde{\psi}_1 = \begin{bmatrix} \tilde{X}_i & \tilde{M}_i \\ \bullet & R_i^T + R_i - \tilde{Z}_{1i} \end{bmatrix} \geq 0, \qquad (7.108)$$

$$\tilde{\psi}_2 = \begin{bmatrix} \tilde{Y}_i & \tilde{N}_i \\ \bullet & R_i^T + R_i - \tilde{Z}_{2i} \end{bmatrix} \geq 0, \qquad (7.109)$$

$$\tilde{\psi}_3 = \begin{bmatrix} \tilde{Y}_i & \tilde{S}_i \\ \bullet & R_i^T + R_i - \tilde{Z}_{2i} \end{bmatrix} \geq 0, \qquad (7.110)$$

$$W_i \leq \mu W_\upsilon, \qquad \overline{Q}_{1i} \leq \mu \overline{Q}_{1\upsilon}, \qquad \overline{Q}_{2i} \leq \mu \overline{Q}_{2\upsilon}$$

$$\overline{Z}_{1i} \leq \mu \overline{Z}_{1\upsilon}, \qquad \overline{Z}_{2i} \leq \mu \overline{Z}_{2\upsilon}, \qquad \forall i, \upsilon \in M, \qquad (7.111)$$

$$\overline{W}_i = \begin{bmatrix} -W_i & t_{ir}^T \\ \bullet & -1 \end{bmatrix} \leq 0, \qquad (7.112)$$

for all $i = 1, 2, \ldots, N$ and $j = 1, 2, \ldots, 2^m$, where t_{ir} is the rth row of T_i and

$$\overline{\Phi}_1 = \begin{bmatrix} \overline{\Phi}_{11} & \overline{\Phi}_{12} & \overline{\Phi}_{13} & \overline{\Phi}_{14} \\ \bullet & \overline{\Phi}_{22} & \overline{\Phi}_{23} & \overline{\Phi}_{24} \\ \bullet & \bullet & \overline{\Phi}_{33} & 0 \\ \bullet & \bullet & \bullet & \overline{\Phi}_{44} \end{bmatrix},$$

$$\overline{\Phi}_{11} = \overline{Q}_{1i} + \overline{Q}_{2i} + (1 - \alpha)^{d_m}(\overline{M}_{1i} + \overline{M}_{1i}^T + d_m\overline{X}_{11i})$$
$$+ d_{12}(1 - \alpha)^{d_M}\overline{Y}_{11i} - (1 - \alpha)\overline{W}_i,$$

$$\overline{\Phi}_{12} = (1 - \alpha)^{d_m}(d_m\overline{X}_{12i} + \overline{M}_{2i}) + (1 - \alpha)^{d_M}(\overline{N}_{1i} - \overline{S}_{1i} + (d_M - d_m)\overline{Y}_{12i}),$$

$$\overline{\Phi}_{13} = -(1-\alpha)^{d_m}\overline{M}_{1i} + (1-\alpha)^{d_M}\overline{S}_{1i},$$

$$\overline{\Phi}_{22} = d_1(1-\alpha)^{d_m}\overline{X}_{22i} + (1-\alpha)^{d_M}(\overline{N}_{2i} + \overline{N}_{2i}^T - \overline{S}_{2i} - \overline{S}_{2i}^T + \overline{Y}_{22i})$$

$$\overline{\Phi}_{23} = (1-\alpha)^{d_M}\overline{S}_{2i} - (1-\alpha)^{d_m}\overline{M}_{2i},$$

$$\overline{\Phi}_{24} = -(1-\alpha)^{d_M}\overline{N}_{2i},$$

$$\overline{\Phi}_{33} = -(1-\alpha)^{d_m}\overline{Q}_{1i},$$

$$\overline{\Phi}_{44} = -(1-\alpha)^{d_M}\overline{Q}_{2i}$$

The controller gain is given as $K_i = \overline{K}_i R_i^{-1}$.

The proof follows by manipulating Theorem 7.8 using Schur complements.

Theorem 7.10   *The closed-loop system of Equation (7.99) is exponentially stable under any switching signal with the average dwell time T_a satisfying Equation (7.105) if there exist matrices $W_1 > 0$, $\overline{Q}_{1i} > 0$, $\overline{Q}_{2i} > 0$, $\overline{Z}_{1i} > 0$, $\overline{Z}_{2i} > 0$ and a diagonal matrix $U > 0$,*

$$\overline{X}_i = \begin{bmatrix} \overline{X}_{11i} & \overline{X}_{12i} \\ \bullet & \overline{X}_{22i} \end{bmatrix} > 0, \tag{7.113}$$

$$\overline{Y}_i = \begin{bmatrix} \overline{Y}_{11i} & \overline{Y}_{12i} \\ \bullet & \overline{Y}_{22i} \end{bmatrix} > 0, \tag{7.114}$$

$$\overline{M}_i = \begin{bmatrix} \overline{M}_{1i} \\ \overline{M}_{2i} \end{bmatrix}, \quad \overline{N}_i = \begin{bmatrix} \overline{N}_{1i} \\ \overline{N}_{2i} \end{bmatrix}, \quad \overline{S}_i = \begin{bmatrix} \overline{S}_{1i} \\ \overline{S}_{2i} \end{bmatrix}, \tag{7.115}$$

with matrices R_i, T_i and K_i of appropriate dimensions such that the inequalities of Equation (7.105), Equations (7.107)–(7.112), and the following inequality hold

$$\hat{\Phi} = \begin{bmatrix} \overline{\Phi}_1 & d_1\hat{\Phi}_2^T & d_{12}\hat{\Phi}_2^T & \hat{\Phi}_3^{\ T} & 0 & \overline{K}_i^T E_{ij}^T \tilde{F} + T_i^T E_{ij}^{-T} \tilde{F} \\ \bullet & -d_1\overline{Z}_{1i}^{-1} & 0 & 0 & d_1 B_i U & 0 \\ \bullet & \bullet & -d_{12}\overline{Z}_{2i}^{-1} & 0 & d_{12} B_i U & 0 \\ \bullet & \bullet & \bullet & -R_i - R_i^T + W_i & B_i U & 0 \\ \bullet & \bullet & \bullet & \bullet & -U & 0 \\ \bullet & \bullet & \bullet & \bullet & \bullet & -U \end{bmatrix}$$

$$< 0 \tag{7.116}$$

for all $i = 1, 2, \ldots, N$ *and* $j = 1, 2, \ldots, 2^m$, *where*

$$\hat{\Phi}_2 = [(A_i - I)R_i + B_i F_0(E_{ij}\overline{K}_i + E_{ij}^- T_i) \quad A_{di} \quad R_i \quad 0 \quad 0]$$

$$\hat{\Phi}_3 = [A_i R_i + B_i F_0(E_{ij}\overline{K}_i + E_{ij}^- T_i) \quad A_{di} \quad R_i \quad 0 \quad 0]$$

With all the feasible solutions satisfying the conditions of Equation (7.105) and Equations (7.107)–(7.116) we are encouraged to proceed to finding a large estimate of the domain of attraction.

Based on the foregoing results, an optimization problem with LMI constraints can be formulated as follows:

$$\min \quad \gamma = [\rho_0 + d_m \rho_1 + d_M \rho_2 + 2d_m(d_m + 1)\rho_3$$
$$+ 2(d_m + d_M)(d_m + d_M + 1)\rho_4]$$

subject to Equation (7.105), Equation (7.107)–(7.116),

$$\begin{bmatrix} \rho_0 I & I \\ I & R_i + R_i^T - W_i \end{bmatrix} \geq 0,$$

$$\begin{bmatrix} \rho_1 I & I \\ I & R_i + R_i^T - \overline{Q}_{1i} \end{bmatrix} \geq 0, \tag{7.117}$$

$$\begin{bmatrix} \rho_2 I & I \\ I & R_i + R_i^T - \overline{Q}_{2i} \end{bmatrix} \geq 0,$$

$$\begin{bmatrix} \rho_3 I & I \\ I & Z_{1i} \end{bmatrix} \geq 0,$$

$$\begin{bmatrix} \rho_4 I & I \\ I & Z_{2i} \end{bmatrix} \geq 0,$$

7.6.4 Simulation

A simulation example is now presented to illustrate the foregoing methodology. Consider a system with two modes corresponding to two subsystems as follows:

System 1:

$$A_1 = \begin{bmatrix} 0.3 & 0.2 \\ 0 & 0.3 \end{bmatrix}, \quad A_{d1} = \begin{bmatrix} -0.2 & 0 \\ -0.1 & -0.2 \end{bmatrix}, \quad B_1 = \begin{bmatrix} 0 \\ 1 \end{bmatrix}$$

System 2:

$$A_2 = \begin{bmatrix} 0.6 & 0.8 \\ 0.2 & 0.9 \end{bmatrix}, \quad A_{d2} = \begin{bmatrix} 0.1 & 0.3 \\ 0.1 & -0.1 \end{bmatrix}, \quad B_2 = \begin{bmatrix} 1 \\ 0 \end{bmatrix},$$

$$\underline{F} = 0.7, \quad \overline{F} = 0.9, \quad \text{and} \quad 1 \leq d(k) \leq 2$$

We assume that the above system has an average dwell time of $T_a = 2$. Selecting $\alpha = 0.2$ and $\mu = 1.2$ yields $T_a^* = -\frac{ln(\mu)}{ln(1-\alpha)} = 0.8171$. Thus the condition from Equation (7.105) is satisfied. Moreover, we get $\gamma_m = 29.5001$ by solving the optimization problem of Equation (7.117) along with the controller gains $K_1 = [-0.0785 \quad 0.5875]$ and $K_2 = [0.0724 \quad -1.6533]$. The corresponding domain of attraction estimate is $\delta_m = 1/\sqrt{\gamma} = 0.1841$ and the decay rate is $R = (1 - \alpha)\mu^{1/T_a} = 0.8764 < 1$. This assures the exponential stability of the closed-loop system.

Remark 7.9 *The explanation for matrix E_{ij} can be traced to the work of Cao and Lin [44]. In this regard, Ξ is defined as a set of $m \times m$ diagonal matrices with either 1 or 0 as diagonal elements. In our case, $m = 1$; we get*

$$System\ 1:$$

$$\Xi_1 = \{ \underbrace{[1]}_{E_{11}} , \underbrace{[0]}_{E_{12}} \}$$

$$System\ 2:$$

$$\Xi_2 = \{ \underbrace{[1]}_{E_{21}} , \underbrace{[0]}_{E_{22}} \}$$

Furthermore, LMI of Equation (7.116) can be expanded as follows:

$$
\begin{bmatrix}
\overline{\Phi}_{11} & \overline{\Phi}_{12} & \overline{\Phi}_{13} & \overline{\Phi}_{14} & \beta^T & \Delta^T & \Lambda^T & 0 & \rho \\
\bullet & \overline{\Phi}_{22} & \overline{\Phi}_{23} & \overline{\Phi}_{24} & d_1 A_{di}^T & d_{12} A_{di}^T & A_{di}^T & 0 & 0 \\
\bullet & \bullet & \overline{\Phi}_{33} & \overline{\Phi}_{34} & d_m R_i^T & (d_m - d_m) R_i^T & R_i^T & 0 & 0 \\
\bullet & \bullet & \bullet & \overline{\Phi}_{44} & 0 & 0 & 0 & 0 & 0 \\
\bullet & \bullet & \bullet & \bullet & -d_1 \overline{Z}_{1i} & 0 & 0 & d_1 B_i U & 0 \\
\bullet & \bullet & \bullet & \bullet & \bullet & -d_{12} \overline{Z}_{2i} & 0 & d_{12} B_i U & 0 \\
\bullet & \bullet & \bullet & \bullet & \bullet & \bullet & -\Gamma & B_i U & 0 \\
\bullet & \bullet & \bullet & \bullet & \bullet & \bullet & \bullet & -U & 0 \\
\bullet & \bullet & \bullet & \bullet & \bullet & \bullet & \bullet & \bullet & -U
\end{bmatrix},
$$

where

$$\beta = d_1((A_i - I)R_i + B_i F_0(E_{ij}\overline{K}_i + E_{ij}^-T_i)),$$

$$\Delta = d_{12}((A_i - I)R_i + B_i F_0(E_{ij}\overline{K}_i + E_{ij}^-T_i)),$$

$$\Lambda = A_i R_i + B_i F_0(E_{ij}\overline{K}_i + E_{ij}^-T_i),$$

$$\Gamma = R_i + R_i^T - W_i,$$

$$\rho = \overline{K}_i^T E_{ij}^{T}\tilde{(F)} + T_i^T E_{ij}^- \tilde{T}(F)$$

In the simulation setup, the sequence of activation of the subsystems is assumed to be $S_1, S_1, S_2, S_2, S_1, S_1, \ldots$. The initial states are taken as $x_1(-2, -1, 0) = 1$ and $x_2(-2, -1, 0) = -1$. The time-varying delay and actuator faults are taken as

$$d(k) = \begin{cases} 1 & 0 \le k \le 10 \\ 2 & 10 \le k \le 20 \end{cases}$$

$$F = \begin{cases} 0.7 & 0 \le k \le 10 \\ 0.9 & 10 \le k \le 20 \end{cases}$$

The state trajectories are given in Figures 7.15 and 7.16 and the control input is shown in Figure 7.17. It can be seen that the states are regulated within six seconds, which proves the effectiveness of the methodology.

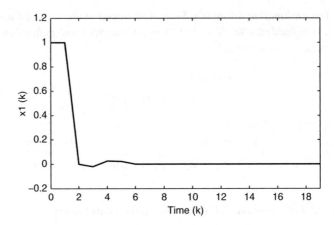

Figure 7.15 State trajectory of x_1 for the closed-loop system of Equation (7.105)

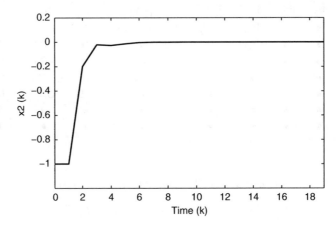

Figure 7.16 State trajectory of x_2 for the closed-loop system of Equation (7.105)

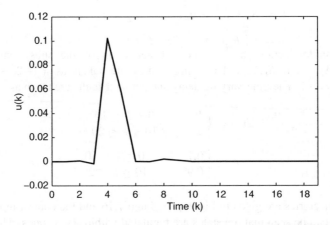

Figure 7.17 Control input u for the closed-loop system of Equation (7.105)

7.7   Notes

On the basis of the adaptive method and LMI technique, the problem of designing adaptive fault-tolerant H_∞ controllers for linear systems with quantized state feedback was studied. An adjustable control law is designed to automatically compensate the effect of actuator fault and quantized state signal on the system. The resultant design conditions are proved to be more relaxed than those of a traditional fault-tolerant controller with fixed gain.

The fault-detection problem of networked control systems with missing measurements was then studied. In order to cope with missing measurements, the structure of the standard model-based residual generator is modified and dynamic network resource allocation is suggested. The dynamics of the residual generator is shown to be characterized by a discrete-time Markovian jump linear system. Then a residual evaluation scheme is developed, aiming to reduce the false alarm rate caused by missing measurements. Further, we have proposed a co-design approach of a time-variant residual generator and threshold to improve the dynamics and the sensitivity of the FD system to the faults.

Next, we discussed how to apply sliding-mode observer on two levels of DNCS with time delay and also with fault detection. In addition, we presented the sliding-mode observer when the state system is measured as well as when the state system is not measured; in this case we used the transformation matrix to separate the measurable states and the unmeasurable states and a special sliding-mode observer was designed for this case.

Finally, a methodology [43] was presented which gives fault-tolerant control of the discrete-time switched systems under actuator saturation. The results obtained by implementing the proposed methodology on a discrete-time system with two subsystems show that the scheme is effective.

References

[1] Frank, P. M., and Ding, X. (1944) "Frequency domain approach to optimally robust residual generation and evaluation for model-based fault diagnosis", *Automatica*, **30**:789–904.

[2] Gertler, J. J. (1998) *Fault Detection and Diagnosis in Engineering Systems*, New York: CRC Press.

[3] Patton, R. J., Frank, P. M., and Clark, R. N. (Eds) (2000) *Issues of Fault Diagnosis for Dynamic Systems*, London: Springer-Verlag.

[4] Hu, S., and Zhu, Q. (2003) "Stochastic optimal control and analysis of stability of networked control systems with long delay", *Automatica*, **39**:1877–1884.

[5] Lian, F., Moyne, J., and Tilbury, D. (2003) "Modelling and optimal controller design of networked control systems with time delays", *Int. J. Control*, **76**(6):591–606.

[6] Matveev, A. S., and Savkin, A. V. (2003) "The problem of state estimation via asynchronous communication channels with irregular transmission times", *IEEE Trans. Automatic Control*, **48**(4):670–676.

[7] Savkin, A. V., and Petersen, I. R. (2003) "Set-valued state estimation via a limited capacity communication channel", *IEEE Trans. Automatic Control*, **48**(4):676–680.

[8] Walsh, G. C., and Hong, Y. (2001) "Scheduling of networked control systems", *IEEE Control Systems Magazine*, **11**(2):57–65.

[9] Walsh, G. C., Beldiman, O., and Bushnell, L. G. (2002) "Error encoding algorithms for networked control systems", *Automatica*, **38**:261–267.

[10] Zhang, W., Branicky, S. M., and Philips, S. M. (2001) "Stability of networked control systems", *IEEE Control Systems Magazine*, **21**(1):84–99.

[11] Zhivoglyadov, P. V., and Middleton, R. H. (2003) "Networked control design for linear systems", *Automatica*, **39**:743–750.

[12] Wang, Z., Ho, D. W. C., and Liu, X. (2003) "Variance constrained filtering for uncertain stochastic systems with missing measurements", *IEEE Trans. Automatic Control*, **48**:1254–1258.

[13] Emami-Naeini, A., Akhter, M. M., and Rock, S. M. (1988) "Effect of model uncertainty on failure detection: The threshold selector", *IEEE Trans. Automatic Control*, **33**:1106–1115.

[14] Stoustrup, J., Niemann, H., and la Cour-Harbo, A. 2003 "Optimal threshold functions for fault detection and isolation", in *Proc. American Control Conference*, Denver, Colorado USA, 1782–1787.

[15] Ding, S. X., Zhang, P., Frank, P. M., and Ding, E. L. (2003) "Threshold calculation using LMI-technique and its integration in the design of fault detection systems", in *Proc. 42nd IEEE Conference on Decision and Control*, Maui, Hawaii USA, 469–474.

[16] Ding, S. X., Zhang, P., Frank, P. M., and Ding, E. L. (2003) "Application of probabilistic robustness technique to the fault detection system design", in *Proc. 42nd IEEE Conference on Decision and Control*, Maui, Hawaii USA, 972–977.

[17] de Souza, C. E. (2003) "A mode-independent H_∞ filter design for discrete-time Markovian jump linear systems", in *Proc. 42nd IEEE Conference on Decision and Control*, Maui, Hawaii USA, 2811–2816.

[18] Seiler, P., and Sengupta, R. (2003) "A bounded real lemma for jump systems", *IEEE Trans. Automatic Control*, **48**(9):1651–1654.

[19] Liberzon, D., and Hespanha, J. P. (2005) "Stabilization of nonlinear systems with limited information feedback", *IEEE Trans. Automatic Control*, **50**(6):910–915.

[20] Delchamps, D. F. 1990 "Stabilizing a linear system with quantized state feedback", *IEEE Trans. Automatic Control*, **35**(8):916–924.

[21] Elia, N., and Mitter, S. K. 2001 "Stabilization of linear systems with limited information", *IEEE Trans. Automatic Control*, **46**(9):1384–1400.

[22] Fu, M., Xie, L. (2005) "The sector bound approach to quantized feedback control", *IEEE Trans. Automatic Control*, **50**(11):1698–1711.

[23] Gao, H., Chen, T., and Lam, J. (2008) "A new delay system approach to network based control", *Automatica*, **44**(1):39–52.

[24] Brockett, R. W., and Liberzon, D. (2000) "Quantised feedback stabilization of linear systems", *IEEE Trans. Automatic Control*, **45**(7):1279–1289.

[25] Ling, Q., and Lemmon, M. D. (2005) "Stability of quantized control systems under dynamic bit assignment", *IEEE Trans. Automatic Control*, **50**(5):734–740.

[26] Nair, G. N., and Evans, R. J. 2003 "Exponential stabilisability of finite-dimensional linear systems with limited data rates", *Automatica*, **39**(4):585–593.

[27] Boskovic, J. D., Mehra, R. K. (2002) "An adaptive retrofit reconfigurable flight controller", in *Proc. 2002 Conference on Decision and Control*, Las Vegas, Nevada, USA, 1257–1262.

[28] Geromel, J. C., Bernussou, J., and de Oliveira, C. M. (1993) "H_2 norm optimization with constrained dynamic output feedback controllers: decentralized and reliable control", *IEEE Trans. Automatic Control*, **44**(7):1449–1454.

[29] Veillette, R. J. (1995) Reliable linear-quadratic state-feedback control, *Automatica*, **31**(1):137–143.

[30] Hsieh, C. S. (2002) "Performance gain margins of the two-stage LQ reliable control", *Automatica*, **38**:1985–1990.

[31] Wu, H. N. (2005) "Reliable mixed $L_2 - H_\infty$ fuzzy static output feedback control for nonlinear systems with sensor faults", *Automatica*, **41**:1925–1932.

[32] Zhang, Y. M. and Jiang, J. (2001) "Integrated active fault-tolerant control using IMM approach", *IEEE Trans. Aerospace and Electronic Systems*, **37**(4):1221–1235.

[33] Tao, G., Joshi, S. M., and Ma, X. L. (2001) "Adaptive state feedback and tracking control of systems with actuator failures", *IEEE Trans. Automatic Control*, **46**(1):78–95.

[34] Yang, G. H., and Ye, D. (2010) "Reliable H_∞ control of linear systems with adaptive mechanism", *IEEE Trans. Automatic Control*, **55**(1):242–247.

[35] Zhu, X. L., and Yang, G. H. (2009) "New controller design method for networked control-systems with quantized state feedback", *Proc. American Control Conference*, St Louis, MO, USA, 5103–5108.

[36] Ioannou, P. A., and Sun, J. (1996) *Robust Adaptive Control*, Englewood Cliffs, NJ: Prentice Hall.

[37] Boyd, S., Ghaoui, L. E., Fern, E., Balakrishnan, V. (1994) *Linear Matrix Inequalities in System and Control Theory*, Philadelphia, PA: Society for Industrial and Applied Mathematics.

[38] Stengel, R. F. (1991) "Intelligent failure-tolerant control", *IEEE Control Systems Magazine*, **11**(4):14–23.

[39] Rauch, H. E. (1995) "Autonomous control reconfiguration", *IEEE Control Systems Magazine*, **15**(6):37–48.

[40] Dash, S., and Venkatasubramanian, V. (2000) "Challenges in the industrial applications of fault diagnostic systems", *Computers and Chemical Engineering*, **24**(2):785–791.

[41] Edwards, C., Spurgeon, S. K., Patton, R. J. (2000) "Sliding mode observers for fault detection and isolation", *Automatica*, **36**(4):541–553.

[42] Mahmoud, M. S. 2009 *Switched Time-Delay Systems*, USA: Springer-Verlag.

[43] Zhang, D., and Yu, L. (2011) "Fault-tolerant control for discrete-time switched linear systems with time varying delay and actuator saturation", *J. of Optimal Theory Applications*, **153**(1):157–176.

[44] Cao, Y. Y., and Lin, Z. (2003) "Stability analysis of discrete-time systems with actuator saturation by a saturation dependent Lyapunov function", *Automatica*, **39**:1235–1241.

8

Industrial Fault-Tolerant Architectures

In distributed industrial control systems, control and supervision elements are based on a microcontroller, in the simplest and in the most complex form (that is, hard real-time systems). Microcontrollers enable low-cost systems and a high degree of integration and, in most cases, eliminate the need for additional peripherals. Besides, modern microcontrollers offer high computational power, in comparison with previous ones.

In systems that may involve a risk for the users or the process under control, it is necessary to study and improve dependability. When a system is able to continue its work in the presence of faults, it is called a "fault-tolerant system". Further, a system is safe when, in the event of a non-recoverable failure, it stops in a known state that does not cause a risk to the process. These aspects justify the incorporation of different mechanisms in order to detect and tolerate the majority of faults that may appear in these systems.

The development of a distributed system for industrial applications with a high degree of dependability is presented in this chapter.

8.1 Introduction

Distributed systems for industrial applications with a high degree of dependability have basic principles of operation in common. There is little difference in the communication channel or links between elements and their purpose. The following common aspects can be mentioned:

- There is a microcontroller devoted to communication tasks.
- Another microcontroller is devoted to the execution of the application.
- The interface between the microcontrollers is based on a dual-port RAM or FIFO memory.

The fault-tolerant distributed system addressed below is based on a set of nodes connected by two local area networks based on the controller area network (CAN) protocol. Over CAN, we use the CAN Application Layer (CAL) proposed by CAN in Automation (CiA).

Analysis and Synthesis of Fault-Tolerant Control Systems, First Edition. Magdi S. Mahmoud and Yuanqing Xia.
© 2014 John Wiley & Sons, Ltd. Published 2014 by John Wiley & Sons, Ltd.

In order to determine domains where the distributed system can be applied, it is necessary to analyze its performance and dependability. Thus, the first stage in the project is focused on the design of tools for the performance and reliability measurement of distributed systems. The main goals of this stage are:

- to analyze fault-tolerant architectures for the nodes of the distributed system;
- to evaluate the use of checkpoints in order to determine their effects on the reliability of the whole system;
- to determine the performance of a CAN network used as the backbone of an industrial distributed system, considering different architectures of the nodes and recovery point techniques with different CAN controllers;
- to develop hardware and software tools to monitor the performance of the system; and
- to develop hardware tools to inject faults into the system, in order to obtain the fault coverage and latency of the fault-tolerance mechanisms included in the system.

8.2   System Architecture

The basic idea of the architecture of the distributed system is presented in Figure 8.1. We can see that the structure of a node is based on:

- a 16-bit microcontroller (C167CR) (chosen because it is a powerful microcontroller with a built-in CAN controller and many input–output ports) that is in charge of controlling and supervising the industrial process;
- an 8-bit microcontroller (80C251) devoted to communication;
- a dual-port memory that connects the microcontrollers;
- two communication networks that can be accessed by both microcontrollers.

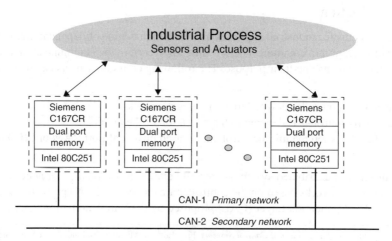

Figure 8.1   System architecture

8.3 Architecture of a Fault-Tolerant Node

Markov-chain models have been used to develop the structure of a fault-tolerant node. Each model represents the structure of a single node. In this section, we calculate the reliability and safety of each structure, and compare the results of each of three modeled and analyzed architectures.

8.3.1 Basic architecture

The main idea of a distributed architecture based on small nodes is that the system nodes should be cheap, simple and flexible. The basic architecture used in our studies is shown in Figure 8.2. There are two communication networks, two microcontrollers (an 8-bit microcontroller for communication tasks and a 16-bit microcontroller for the execution of control algorithms), and one dual-port RAM memory. The nodes store the current state of the system in their dual-port memory. In case of a non-recovered error of the main processor (transient or permanent), the communication processor can send information about the current state to other nodes in the system. In such a case, other nodes will continue the processes of the failed node.

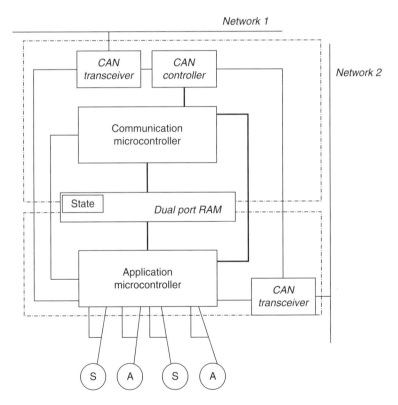

Figure 8.2 Basic architecture (Architecture I)

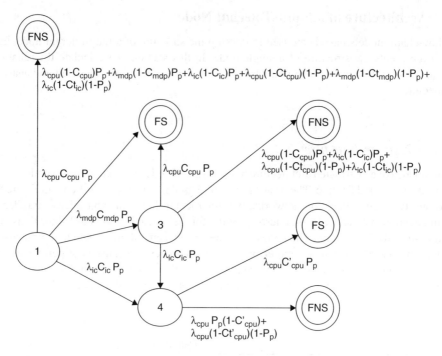

Figure 8.3   Markov model of the basic architecture

The Markov model [1] used to obtain the reliability and security of the basic architecture is shown in Figure 8.3.

To the basic architecture, we should be able to add fault-tolerance mechanisms to increase the reliability and safety of the nodes at a low cost. As a first approach, we propose that the communication microcontroller should also be the watchdog processor of the application microcontroller. This approach will be able to recover from errors caused by a transient fault of the application microcontroller.

8.3.2   Architecture with improved reliability

The second approach, shown in Figure 8.4, is intended to improve reliability. The main idea is to replicate the most complex component of the node, because it has the highest failure probability. In our case, this component is the main processor. This approach aims to tolerate both transient and permanent failures of the application processor.

8.3.3   Symmetric node architecture

As a third approach, a symmetric node architecture can be considered in which there are two identical 16-bit microcontrollers, one in charge of the communications and one executing the application code. In the case of failure, the fault-free microcontroller will perform all

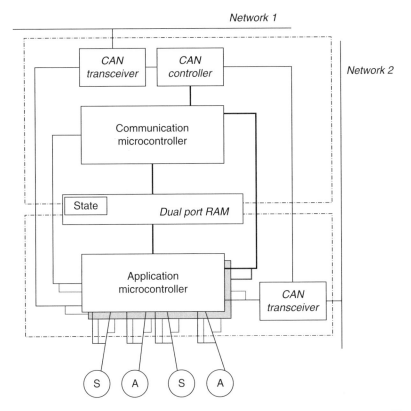

Figure 8.4   Basic architecture with replicated application processor (Architecture II)

the functions of the system. The cost that must be paid is that system performance will be degraded. As can be seen in Figure 8.5, both microcontrollers receive signals from the sensors and send signals to the actuators; they are also connected to both local area networks.

8.3.4   Results

We have obtained reliability and safety measures for the three architectures by solving the Markov models with Surf-2 [2], using the parameters shown in Table 8.1. Analyzing the results (see Table 8.2), Architecture II is the most dependable [1] of the three architectures presented but it is the most expensive.

Architecture I is least expensive and simplest, but its fault tolerance mechanisms are the poorest. It is an appropriate architecture when some minimal dependability, at low cost, is required. When it is used, it increases the dependability of the system by a factor of 30–40 compared to a similar system without fault-tolerance mechanisms.

Architecture III presents characteristics similar to Architecture I if the coverage of transient faults is low. However when fault coverage is higher its reliability is almost the same as the reliability of Architecture II.

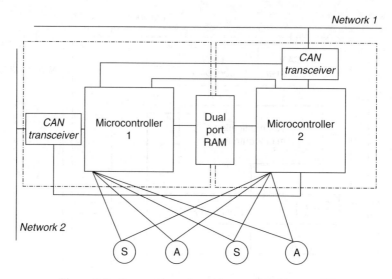

Figure 8.5   Symmetric node architecture (Architecture III)

The graph in Figure 8.6 shows that, for low coverage values, Architecture II is best. For high levels of coverage, Architecture III shows similar behavior to Architecture II. Architecture I shows the worst results.

Figure 8.7 shows the results for node safety (the larger the better). In this case, Architectures I and II show similar levels of safety. Architecture III shows a low level of safety when the coverage factor is low. Only with high coverage is its safety close to the safety obtained with the others.

8.4   Recovery Points

The GSTF group is deploying the analysis of the recovery techniques of a system similar to that shown in Figure 8.1. The impact of recovery points on the performance of a distributed

Table 8.1   Parameters for the resolution of the models

Application failure rate (AFR)	$1 \times 10^{-5} - 1 \times 10^{-6}$ (f/h)
Dual-port memory failure rate	1/10 of AFR
Permanent fault coverage (PFC)	0.99
PFC under fault	0.5
PFC of AFR	0.8
Transient fault coverage (TFC)	0.8
TFC for dual-port memory	0.999
TFC for AFR	0.75, 0.80, 0.85, 0.90, 0.99, 0.999
Permanent fault probability	0.2
Time	100 000 h

Table 8.2   Node reliability

	Architecture III	Architecture I	Architecture II
Ct = 0.75	0.6727868	0.6573024	0.7715912
Ct = 0.80	0.7187097	0.6839038	0.8028328
Ct = 0.85	0.7680003	0.7115879	0.8353461
Ct = 0.9	0.8215457	0.7403987	0.8691828
Ct = 0.99	0.9282881	0.7952455	0.9335975
Ct = 0.999	0.9397809	0.8009500	0.9402971

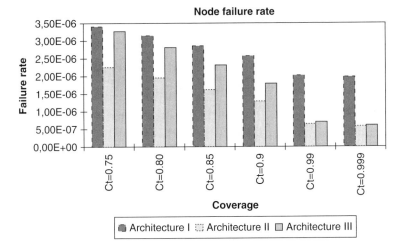

Figure 8.6   Failure rate of the node versus fault coverage

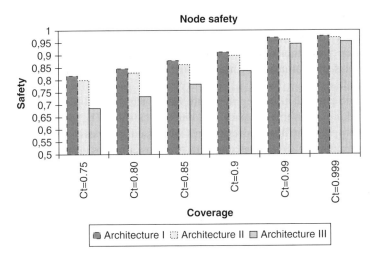

Figure 8.7   Node safety versus coverage

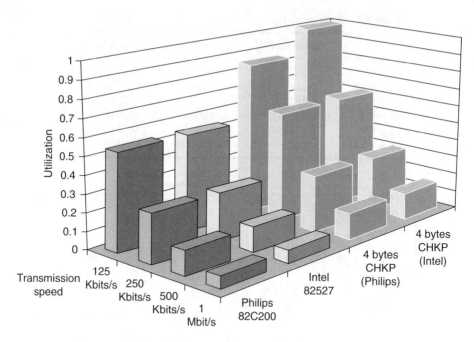

Figure 8.8   Analysis of CAN controllers

system is presented in the work of [3]. Several simulations have been run to assess the network performance as a function of different CAN controllers.

The recovery-point mechanisms are developed at two levels. At the first level, the communication processor is in charge of storing the local checkpoint, which enables the system to tolerate transient faults inside the node. Since the checkpoint is local to the node, the network traffic is not affected. At a second level, in order to tolerate permanent faults, the state is sent to other nodes in the system, as a function of several requirements (for instance, timing constraints, priority assignment, and security constraints). These mechanisms work in such a way that if there is an error inside the node, another node in the network may continue the task of the failed node.

The obtained results can be seen in Figure 8.8, where the influence of the use of CAN controllers working at different transmission rates is analyzed. The tests were made with and without recovery points. When recovery points were used, the message length was 4 bytes.

As can be seen the bus is overloaded for low transmission rates (128 and 256 Kbps), thus it is not suitable for real-time applications. However, for high transmission rates (500 Kbps to 1 Mbps), it can be applied in soft real-time systems.

8.5   Networks

We use two different networks and evaluate three different application layers: the CAN application layer (CAL), a smart distributed system (SDS) and a manufacturing message specification (MMS). The goal is to analyze these application layers in the presence of faults, to detect their weaknesses in order to fix them or, if necessary, develop a new fault-tolerant application layer.

8.6   System Fault Injection and Monitoring

The actual performance of a system is obtained when the application is being executed. Therefore, it is necessary to develop hardware tools capable of monitoring the system during program execution. We have planned the development of several types of monitor.

As a first approach, we develop a centralized monitor in order to measure the system utilization, its productivity, and the percentage of messages that are lost in the network. To verify the monitor it is necessary for a traffic generator to create a characteristic workload by injecting messages into the network. The main goal is to determine its performance, reliability and the performance when recovery points are inserted.

As a second approach, we develop local monitors that will be used to study the nodes (process time, communication time, etc.). We believe that knowledge of these parameters will be of interest when implementing the fault-tolerant mechanisms inside the nodes. A new high-speed physical fault injector has also been developed [4]. The idea of this approach is that the monitors and the physical fault injector work together to:

- validate the fault-tolerant mechanisms included in the node;
- inject faults during the critical states of the system (observed by the monitor);
- obtain the actual fault latency times and the fault coverage; these factors are of great value in validating the analytical models of the system to decide if the system can be used in real-time applications.

In addition, we have developed a software tool for fault injection in very high-speed integrated circuit (VHSIC) hardware description language (VHDL) models.

8.6.1   Monitoring systems

Monitoring systems can be classified as software, hardware or mixed. As a rule of thumb, their complexity increases as the performance increases or the degree of intrusion into the system decreases. Software monitors are the most simple and flexible but they are the most intrusive. Hardware monitors are the most complex and least flexible but also the least intrusive.

In the case of real-time systems, any intrusion made during the monitoring process can alter the timing characteristics of the system. We have to make a great effort to ensure that the overhead suffered by the system is minimal but that, at the same time, it is constant and measurable in order to maintain the timing characteristics of the system.

The need for synchronization between the monitor and the fault-injection system, and the fact that we need to interact with the system as soon as a predefined state is detected, makes it impossible to use the hybrid monitor. This is why we are developing a hardware monitor.

We have already designed a distributed monitoring system as depicted in Figure 8.9. In this system, the nodes are monitored using non-intrusive monitors. Local monitors gather information from the nodes and send it to the central monitor through a special network. The central monitor is also in charge of the timing synchronization of the local monitors. This information is used to order the events of the system.

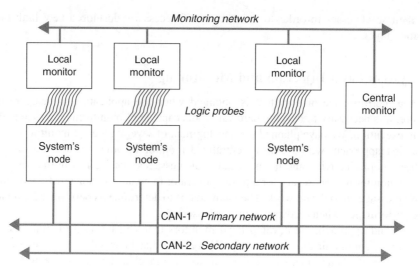

Figure 8.9   Bus utilization

8.6.2   *Design methodology*

We are interested in using a design methodology based on co-design. The complexity involved in the design of a distributed system motivates the use of advanced techniques that cover all the stages and parts of the system (from hardware to software). A high-level methodology facilitates the design process (see Figure 8.10).

The goal is to use a workbench where we can study (at a simulation level) the characteristics of industrial distributed systems. This workbench is a framework with libraries of nodes, processors and the underlying protocols of a network. The user can connect these elements and, based on these connections, the performance of the system is obtained.

The framework is based on Ptolemy [5], which provides a research laboratory to test and explore design methodologies that support multiple design styles and implementation technologies. In short, Ptolemy is a flexible foundation upon which to build prototyping environment.

the design process starts with the specification of the application; in our case, this means the typeS and characteristics of the nodes (actuators, sensors and supervisors), the network protocol and the time necessary to send a bit across the network (the "bit time"). The topology of the network can be seen in Figure 8.11.

The main characteristics of the nodes are the message length (in bytes), the priority and, if necessary, the rate at which the node sends the data. Based on this information and using the framework a simulation model is built. This model is created using a schematic editor with three types of node:

- Sensors collect information about the industrial processes, such as the temperature and level of the water in a tank.
- Actuators generate changes in industrial processes. For instance, an actuator may open a valve to fill the tank or control a smelting furnace.

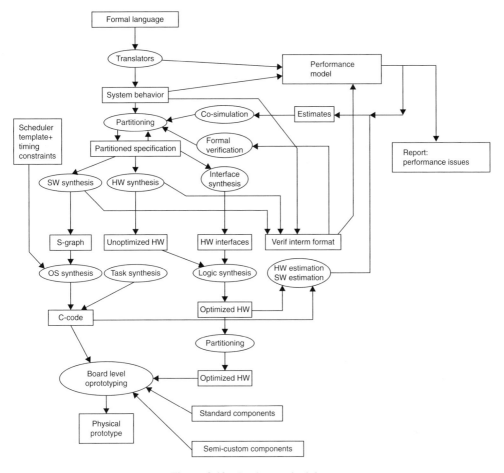

Figure 8.10   Design methodology

- Supervisors collect data (such as network utilization, message latency time etc.) that can be used by engineers in charge of the industrial process. They are also used to send data to the actuators or to get data from the sensors as requested by the users.

Once the simulation model is finished, it is necessary to verify that messages can be sent and received by the nodes and to verify the times involved.

In the literature, if a sensor is capable of collecting and transmitting data by itself it is called a "smart sensor". To be consistent with the literature, we consider Processor 1's communication control sensor to be a smart sensor and Processor 1's communication control actuator to be a smart actuator. In our particular case the communication controller is a CAN controller.

The workload of the network is composed of the set of messages that will be transmitted. We can ensure that the messages can be scheduled. In other words, taking into account the message lengths, the bit time and the network protocol, it is possible to send and receive data across the network in a predictable way. This study is called the "schedulability" test. If such

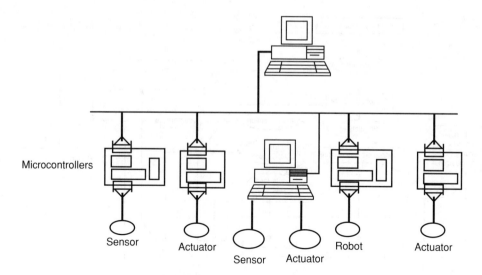

Figure 8.11   Industrial local area network (ILAN)

a test fails (because a given message does not fit into the schedule), it is necessary to reduce the bit time, change the network protocol or use more than one network. This is the main advantage of the methodology. We start with the workload of the network and most of the design decisions are based on the schedulability test, which guarantees that the data is sent and received by the nodes.

Once we have verified that messages can be scheduled, the next job is to ensure that the data is sent and received by the appropriate nodes. This is made by exercising the simulation model; each of the components in the framework has a VHDL model that is obtained from the schematic which represents the industrial process. The model of each component writes information that allows the others to determine message latencies and network utilization.

The design process continues with the refinement and specification of the implementation characteristics of each node. We use co-design because it allows us to explore different options in the implementation of the nodes, for example, if they are made with hardware or if they are made using stored programs and processors. Once the architecture of the node is chosen, it is be implemented using custom or semi-custom components.

8.7   Notes

Modern manufacturing systems need improvements in their dependability. In this chapter, we have studied fault-tolerant techniques applicable to distributed industrial control systems. Several fault tolerance mechanisms are presented at different levels in order to improve the reliability and safety of these systems. We presented several approaches to spatial redundancy at node level. At system level, we have presented the use of checkpointing techniques at two levels: local recovery points for tolerance of transient faults without increasing the network traffic and global recovery points in order for tolerance of permanent faults.

To improve the dependability it is necessary to develop fault-tolerant application layers. We have studied the common application layers used in CAN networks to add the necessary mechanisms to develop a fault-tolerant node. It is also important for the set of tools to analyze the system behavior. We use both physical and VHDL fault injectors in order to achieve fault-tolerance measures, such as fault coverage and error latencies. We have also presented several ways to study the network behavior through the use of monitors. Finally, we noted the need for advanced tools (such as hardware–software co-design) during the design process to cope with the complexity of systems devoted to industrial applications.

References

[1] Campelo, J. C., Rodriguez, F., Serrano, J. J., and Gil, P. (1997) "Dependability evaluation of fault-tolerant architectures in distributed industrial control systems", in *Proc. Second IEEE Int. Workshop on Factory Communication Systems (WFCS'97)*, Barcelona, 1–3 October, 1244–1250.

[2] Beounes, C., Aguera, M., Arlat, J., Bachman, S., *et al.* (1993) "SURF-2: a program for dependability evaluation of complex hardware and software systems", in *Proc. 23rd IEEE Int. Symposium Fault-Tolerant Computing*, Toulouse, France, 668–673.

[3] Rubio, A., Campelo, J. C., Ors, R., and Serrano, J. J. (1997) "Checkpointing in a CAN based distributed computer control system", *Preprints of the 14th Workshop on Distributed Computer Control Systems (DCCS'97)*, Seoul, Korea, 283–289.

[4] Gil, P., Baraza, C., Gil, D., and Serrano, J. J. (1997) "High speed fault injector for safety validation of industry machinery", *EWDC8, European Workshop on Dependable Computing: Experimental Validation of Dependable Systems*, Göteborg, 291–298.

[5] Albaladejo, J., and Lemus, L. (1997) *Herramientas de Codiseño*, Internal Report GSTF-160997, Technical University of Valencia, Department of Computer Engineering.

References

9

Fault Estimation for Stochastic Systems

9.1 Introduction

Modern control systems are highly complex and a substantial amount of research is in progress to make the system under consideration fault tolerant [1]. Various types of fault can occur in a system, ranging from sensor faults to faults in the plant itself, and they tend to decrease the performance of the system considerably or even make the system unstable. As we know, the objective of fault-tolerant control (FTC) architecture is to mitigate the effects of faults to make the system behave in an acceptable way or, at least, to define post-fault objectives for the system to avoid catastrophic failures [1].

It has been pointed out [1] that there are two possible ways to deal with system faults:

- passively: the control is made robust with respect to possible system faults;
- actively: fault detection and isolation and accommodation techniques are used.

This chapter presents an active technique which not only estimates the fault but also accommodates it. The type of fault considered is a loss in actuator effectiveness. The proposed methodology is unique in three ways:

- It rapidly and accurately estimates the faults.
- The control law to counter faults is easy to implement.
- Process and sensor noise are taken into account, which makes the methodology practical.

In the following discussion, we deal with discrete-time fault estimation of a helicopter model. The system is divided into two subsystems: one is decoupled from faults and therefore a Kalman filter can be designed under some standard assumptions; the other is affected by faults but its states are available for measurement. The fault estimation process is online and it utilizes the state estimates from the subsystem unaffected by faults. Output of the nominal controller is then modified to accommodate for the faults.

Analysis and Synthesis of Fault-Tolerant Control Systems, First Edition. Magdi S. Mahmoud and Yuanqing Xia.
© 2014 John Wiley & Sons, Ltd. Published 2014 by John Wiley & Sons, Ltd.

9.2   Actuator Fault Diagnosis Design

Consider the following linear stochastic system with actuator fault:

$$x(k+1) = Ax(k) + Bu(k) + Ef(k) + \omega(k) \tag{9.1}$$

$$y(k) = Cx(k) + \upsilon(k)$$

$$= [0 \quad I]x(k) + \upsilon(k) \tag{9.2}$$

where $x(k) \in \mathfrak{R}^n$ is the state vector, $u(k) \in \mathfrak{R}^m$ is the control input vector, $y(k) \in \mathfrak{R}^r$ is the measured output vector, $f(k) \in \mathfrak{R}^q$ is the vector function to model the actuator fault, the process noise $\omega(k)$ and sensor noise $\upsilon(k)$ are zero mean random sequences with covariances matrices $S = S^T > 0$ and $Q = Q^T > 0$, respectively. A, B, C, and E are real constant matrices of appropriate dimensions. The pair (A, B) is controllable and (A, C) is observable. Note the special form of the C matrix. If C is of full row rank, $C = [0 \quad C_1]$ where C_1 is an $r \times r$ non-singular matrix, then there exists a similarity transformation

$$x = \begin{bmatrix} I_{n-r} & 0 \\ 0 & C_1 \end{bmatrix} \tag{9.3}$$

that can bring Equation (9.2) into the desired form.

For the purpose of estimating the fault, the system of Equations (9.1)–(9.2) can be written as

$$\begin{bmatrix} x_1(k+1) \\ x_2(k+1) \\ x_3(k+1) \end{bmatrix} = \begin{bmatrix} A_1 \\ A_2 \\ A_3 \end{bmatrix} x(k) + \begin{bmatrix} B_1 \\ B_2 \\ B_3 \end{bmatrix} u(k) + \begin{bmatrix} E_1 \\ E_2 \\ E_3 \end{bmatrix} f(k) + \begin{bmatrix} \omega_1(k) \\ \omega_2(k) \\ \omega_3(k) \end{bmatrix} \tag{9.4}$$

$$y(k) = \begin{bmatrix} y_1(k) \\ y_2(k) \end{bmatrix} = \begin{bmatrix} 0 & I_{r-q} & 0 \\ 0 & 0 & I_q \end{bmatrix} x(k) + \begin{bmatrix} \upsilon_1 \\ \upsilon_2 \end{bmatrix} \tag{9.5}$$

where $x_1(k) \in \mathfrak{R}^{n-r}$, $x_2(k) \in \mathfrak{R}^{r-q}$, $x_3(k) \in \mathfrak{R}^q$. Note that $\mathbf{E}(x_2(k)) = \mathbf{E}(y_1(k))$ and $\mathbf{E}(x_3(k)) = \mathbf{E}(y_2(k))$, where $\mathbf{E}(.)$ stands for the expected value. Hence only $x_1(k)$ needs to be estimated. The following assumption is made:

Assumption 9.1   *Rank(CE) = q and E_3 is nonsingular.*

Define

$$\bar{A}_1 := A_1 - E_1 E_3^{-1} A_3, \quad \bar{A}_2 = A_2 - E_2 E_3^{-1} A_3$$

$$\bar{B}_1 := B_1 - E_1 E_3^{-1} B_3, \quad \bar{B}_2 = B_2 - E_2 E_3^{-1} B_3$$

$$\bar{A}_1 := [\bar{A}_{11} \quad \bar{A}_{12} \quad \bar{A}_{13}], \quad \bar{A}_2 = [\bar{A}_{21} \quad \bar{A}_{22} \quad \bar{A}_{23}] \tag{9.6}$$

with

$$\bar{A}_{11} \in \mathfrak{R}^{(n-r)\times(n-r)}, \quad \bar{A}_{12} \in \mathfrak{R}^{(n-r)\times(r-q)}, \quad \bar{A}_{13} \in \mathfrak{R}^{(n-r)\times q}$$

$$\bar{A}_{21} \in \mathfrak{R}^{(r-q)\times(n-r)}, \quad \bar{A}_{22} \in \mathfrak{R}^{(r-q)\times(r-q)}, \quad \bar{A}_{23} \in \mathfrak{R}^{(r-q)\times q}.$$

The methodology to estimate the state x_1 of the system of Equations (9.4)–(9.5) is given as follows: assume that the pair $(\overline{A}_{11}, \overline{A}_{21})$ is observable and Assumption 9.1 holds and that $\mathbf{E}[(x_1(0) - \hat{x}_1(0))(x_1(0) - \hat{x}_1(0))^T] := P(0)$ is given. Then the algorithm to find unbiased minimum variance estimate of $x_1(k)$ is given in steps as follows:

1.

$$T_1 := [I \quad 0 \quad -E_1 E_3^{-1} \quad -\overline{A}_{12} \quad -\overline{A}_{13} - E_1 E_3^{-1}]$$
$$T_2 := [0 \quad I \quad -E_2 E_3^{-1} \quad I + \overline{A}_{22} \quad -\overline{A}_{23} - E_2 E_3^{-1}] \tag{9.7}$$

2.

$$\overline{S} := T_1 \begin{bmatrix} S & 0 \\ 0 & Q \end{bmatrix} T_1^T$$

$$\overline{Q} := T_2 \begin{bmatrix} S & 0 \\ 0 & Q \end{bmatrix} T_2^T \tag{9.8}$$

3. The covariance matrix

$$P(k) := \mathbf{E}[(x_1(k) - \hat{x}_1(k))(x_1(k) - \hat{x}_1(k))^T]$$

is updated by

$$P(k+1) = \overline{A}_{11} P(k)(\overline{A})_{11}^T + \overline{Q} - K(k)[\overline{A}_{21} P(k)\overline{A}_{21}^T + \overline{S}]K^T(k) \tag{9.9}$$

4. The Kalman filter gain is given by

$$K(k) = \overline{A}_{11} P(k)\overline{A}_{21}^T [\overline{A}_{21} P(k)\overline{A}_{21}^T + \overline{S}]^{-1} \tag{9.10}$$

5. The estimate of state $x_1(k)$ is now given as follows

$$\hat{x}_1(k) = \overline{A}_{11}\hat{x}_1(k-1) + \rho(k) + K(k)(\lambda(k) - \overline{A}_{21}\hat{x}_1(k-1)) \tag{9.11}$$

where $\rho(k)$ and $\lambda(k)$ are defined as

$$\rho(k) := \overline{A}_{12} y_1(k-1) + \overline{A}_{13} y_2(k-1) + E_1 E_3^{-1} y_2(k) + \overline{B}_1 u(k-1)$$
$$\lambda(k) := y_1(k) - E_2 E_3^{-1} y_2(k) - \overline{A}_{22} y_1(k-1)$$
$$\quad - \overline{A}_{23} y_2(k-1) - \overline{B}_2 u(k-1) \tag{9.12}$$

The proof of this algorithm is quite standard and the equations given above can be obtained by pre-multiplying

$$
\begin{bmatrix}
I & 0 & -E_1 E_3^{-1} \\
0 & I & -E_2 E_3^{-1} \\
0 & 0 & I
\end{bmatrix}
\tag{9.13}
$$

into Equation (9.4).

The augmented vector for all the states, that is, those estimated and those available on the output is given as

$$
\hat{x}(k) = \begin{bmatrix} \hat{x}_1(k) \\ y(k) \end{bmatrix}
\tag{9.14}
$$

and the actuator fault estimate $\hat{f}(k)$ is given as

$$
\hat{f}(k-1) = (E_3)^{-1}[y_2(k) - A_3 \hat{x}(k-1) - B_3 u(k-1)].
\tag{9.15}
$$

We note that the fault for time instant k can only be estimated when the output for $k + 1$ is available. In other words, there will be a delay of one time instant between the output and the fault estimation. This limitation does not make a significant difference for the practical system due to high sampling rates.

9.3   Fault-Tolerant Controller Design

This section presents fault-tolerant controller design to counter the effects of actuator faults. The loss of actuator effectiveness can be modeled by the matrix $E = -B$ and $q = m$ such that the function $f(k) = R(k)u(k) \in \mathfrak{R}^m$, which gives the actuator failure in a multiplicative form. The matrix $R(k) \in \mathfrak{R}^{m \times m}$ in this case consists of only the diagonal elements $r_i(k), i = 1, 2, \ldots m$ such that $r_i(k) = 0$ represents a no-fault nominal condition and $0 < r_i(k) \leqslant 1$ represents the percentage degradation in each actuator input channel i at time instant k.

To recover the actuator effectiveness, the compensation law $u_R(k)$ is defined as follows

$$
u_R(k) = (I - \hat{R}(k))^{-1} u(k)
\tag{9.16}
$$

where $\hat{R}(k)$ is the estimation of $R(k)$ obtained using Equation (9.15) and $(I - \hat{R}(k))$ is assumed to be non-singular. With this reconfigured law, the closed loop equation can be written as

$$
x(k+1) = Ax(k) + B(I - R(k))(I - \hat{R}(k))^{-1} u(k) + \omega(k)
\tag{9.17}
$$

9.4 Extension to an Unknown Input Case

The results obtained are extended to a case where an input to the system is unknown. The system is thus expressed as

$$x(k + 1) = Ax(k) + Bu(k) + Ef(k) + D\zeta(k) + \omega(k)$$
$$y(k) = Cx(k) + \upsilon(k) = [0 \quad I]x(k) + \upsilon(k) \tag{9.18}$$

where $\zeta(k)$ denotes the unknown input vector of the system. The unknown state of the system and the unknown input are augmented in a single vector and an augmented Kalman filter (AKF) is used to obtain the estimates. If the augmented vector is given as follows

$$\xi(k) := \begin{bmatrix} x_1(k) \\ \zeta(k) \end{bmatrix} \tag{9.19}$$

and

$$\overline{D}_1 := D_1 - E_1 E_3^{-1} D_3$$
$$\overline{D}_2 := D_2 - E_2 E_3^{-1} D_3$$

and it is assumed that the augmented system is observable, then the estimate $\hat{\xi}(k)$ is given by following steps:

1.

$$P(k + 1) = \begin{bmatrix} \overline{A}_{11} & \overline{D}_1 \\ 0 & I \end{bmatrix} P(k) \begin{bmatrix} \overline{A}_{11} & \overline{D}_1 \\ 0 & I \end{bmatrix}^T + \begin{bmatrix} \overline{Q} & 0 \\ 0 & 0 \end{bmatrix}$$
$$- L(k)\{[\overline{A}_{21} \quad \overline{D}_2]P(k)[\overline{A}_{21} \quad \overline{D}_2]^T + \overline{S}\}L^T(k). \tag{9.20}$$

2. The filter gain $L(k)$ is given by

$$L(k) = \begin{bmatrix} \overline{A}_{11} & \overline{D}_1 \\ 0 & I \end{bmatrix} P(k)[\overline{A}_{21} \quad \overline{D}_2]^T$$
$$\times \{[\overline{A}_{21} \quad \overline{D}_2]P(k)[\overline{A}_{21} \quad \overline{D}_2]^T + \overline{S}\}^{-1} \tag{9.21}$$

3.

$$\hat{\xi}(k + 1) = \begin{bmatrix} \overline{A}_{11} & \overline{D}_1 \\ 0 & I \end{bmatrix} \hat{\xi}(k) + \begin{bmatrix} \rho(k + 1) \\ 0 \end{bmatrix}$$
$$+ L(k)[\lambda(k + 1) - [\overline{A}_{21} \quad \overline{D}_2]\hat{\xi}(k)] \tag{9.22}$$

where the definitions of $x(k)$, $\rho(k)$ and $\lambda(k)$ are the same as in Equation (9.12).

The preceding steps are obtained by using the same pre-multiplier as in Equation (9.13) or Equation (9.18). From the model of Equation (9.18), the actuator fault $f(k)$ can be estimated in mean sense as follows

$$\hat{f}(k-1) = (E3)^{-1}[y_2(k) - A_3\hat{x}(k-1) - B_3u(k-1)] - D_3\hat{\zeta}(k-1) \qquad (9.23)$$

9.5 Aircraft Application

This section presents an application of the fault estimation and accommodation technique on a helicopter model in a vertical plane. The discrete-time system is given as follows:

$$x(k+1) = \begin{bmatrix} 0.9996 & 0.0003 & 0.0002 & -0.0037 \\ 0.0005 & 0.9900 & -0.0002 & -0.0406 \\ 0.0010 & 0.0037 & 1.0453 & 10.5644 \\ 0 & 0 & 0.0101 & 1.0524 \end{bmatrix} x(k)$$

$$+ \begin{bmatrix} 0.0044 & 0.0018 \\ 0.0353 & -0.0755 \\ -0.0559 & 0.0454 \\ -0.0003 & 0.0002 \end{bmatrix} u(k) + Ef(k) + \omega(k) \qquad (9.24)$$

$$y(k) = \begin{bmatrix} 1 & 0 & 0 & 0 \\ 0 & -1 & 0 & 0 \\ 0 & 0 & 1 & 0 \end{bmatrix} x(k) + \upsilon(k) \qquad (9.25)$$

where $E = -B$ and the sampling rate T is 0.01 s. The state vector $x(t) \in \Re^4$ is composed of:

- $x_1 = v$ longitudinal velocity;
- $x_2 = \omega$ vertical velocity;
- $x_3 = \omega_y$ rate of pitch;
- $x_4 = \theta$ pitch angle;

and the components of the command vector are:

- u_1 general cyclic command;
- u_2 longitudinal cyclic command.

Remark 9.1 *It is important to observe that the system of Equations (9.24)–(9.25) is not in the standard form as given in Equations (9.4)–(9.5). Firstly, the unmeasurable state $x_4 = \theta$ should be placed as the first element of the state vector, that is, the state vector should be permuted. Then, using the similarity transformation Equation (9.3), the system should be brought into the standard form.*

9.5.1 Transforming the system into standard form

9.5.1.1 Permutation

The vector x is permuted such that $x_4 = \theta$ is placed as the first element. We define a permutation matrix F as

$$F = \begin{bmatrix} 0 & 0 & 0 & 1 \\ 1 & 0 & 0 & 0 \\ 0 & 1 & 0 & 0 \\ 0 & 0 & 1 & 0 \end{bmatrix} \tag{9.26}$$

When multiplied with the state Equations (9.24) and (9.25), this gives the system in the form

$$z(k+1) = \begin{bmatrix} 0 & 0 & 0.0101 & 1.0524 \\ 0.9996 & 0.0003 & 0.0002 & -0.0037 \\ 0.0005 & 0.9900 & -0.0002 & -0.0406 \\ 0.0010 & 0.0037 & 1.0453 & 10.5644 \end{bmatrix} z(k) \tag{9.27}$$

$$+ \begin{bmatrix} -0.0003 & 0.0002 \\ 0.0044 & 0.0018 \\ 0.0353 & -0.0755 \\ -0.0559 & 0.0454 \end{bmatrix} u(k) + \begin{bmatrix} 0.0003 & -0.0002 \\ -0.0044 & -0.0018 \\ -0.0353 & 0.0755 \\ 0.0559 & -0.0454 \end{bmatrix} f(k) + \omega(k)$$

$$y(k) = \begin{bmatrix} 0 & 1 & 0 & 0 \\ 0 & 0 & -1 & 0 \\ 0 & 0 & 0 & 1 \end{bmatrix} z(k) + \upsilon(k) \tag{9.28}$$

where $z(k)$ is the permuted state vector given as

$$z(k) = \begin{bmatrix} \theta \\ \upsilon \\ \omega \\ \omega_y \end{bmatrix}$$

9.5.1.2 Similarity transformation

After permuting the state vector, a similarity transformation of the form Equation (9.3) is applied to the system of Equation (9.27). The transformation matrix M thus formed, following Equation (9.3), is given as

$$M = \begin{bmatrix} 1 & 0 & 0 & 0 \\ 0 & 1 & 0 & 0 \\ 0 & 0 & -1 & 0 \\ 0 & 0 & 0 & 1 \end{bmatrix} \tag{9.29}$$

The new state vector $\bar{x}(k)$ is obtained by $\bar{x}(k) = M^{-1}z(k)$ and the system equation becomes

$$\bar{x}(k+1) = \overbrace{MAM^{-1}}^{\bar{A}}\bar{x}(k) + \overbrace{MB}^{\bar{B}}u(k) + \overbrace{ME}^{\bar{E}}f(k) + \omega(k)$$

$$y(k) = \overbrace{CM^{-1}}^{\bar{C}}\bar{x}(k) + \upsilon(k) \qquad (9.30)$$

where $\bar{A}, \bar{B}, \bar{E}$ and $\bar{C}$ are the new system matrices under the transformation.

The sensor and process noise $\upsilon(k)$ and $\omega(k)$ are taken as white noise sequences with covariance matrices given as $Q = 0.2^2 I_{3\times3}$ and $S = diag\{0.01^2 \quad 0.1^2 \quad 0.1^2 \quad 0.01^2\}$. The loss of actuator effectiveness in u_1 and u_2 are considered. $f(k)$ can be described as follows

$$f(k) = \begin{bmatrix} r_1(k) & 0 \\ 0 & r_2(k) \end{bmatrix} u(k)$$

with $0 \leqslant r_i(k) \leqslant 1(i = 1, 2)$ representing the percentage degradation in the ith control input.

Remark 9.2 *In the following discussion, simple computations show that the correct values of $\bar{A}_{11}$ and $\bar{A}_{21}$ are given by:*

$$\bar{A}_{11} = 6.42 \times 10^{-6}, \quad \bar{A}_{21} = 0.9998$$

In the simulation, both constant and time-varying faults are considered and are given as follows:

$$r_1(k) = \begin{cases} 0, & t < 4(s) \\ 0.4, & 4 \leqslant t \leqslant 10(s) \end{cases}$$

$$r_2(k) = \begin{cases} 0, & t < 2(s) \\ 0.5 + 0.3sin(\pi t), & 2 \leqslant t \leqslant 10(s) \end{cases}$$

Equations (9.7)–(9.12) are implemented in MATLAB.

Remark 9.3 *Observe that the system given by Equation (9.30) is unstable since it has an eigenvalue $10.564 > 1$. For this purpose the system must be stabilized by a feedback controller which ensures a stable closed-loop system before applying a fault-tolerant control scheme.*

To stabilize the system, an LQR feedback controller was designed and implemented in such a way that the control gains were calculated in every iteration; that is, the controller changed in every iteration because of the occurrence of a fault. To calculate the gains, the system given by Equation (9.30) was used with a modified B matrix because of the presence of the fault. This adaptive method guaranteed regulation of the states at the cost of increased computational load.

The system matrix of Equation (9.30) was deliberately changed to give stable eigenvalues and the system was then simulated to reproduce the results.

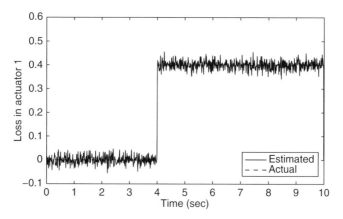

Figure 9.1 Loss in actuator 1

9.5.2 Simulation results

The fault estimation results are presented in Figures 9.1 and 9.2. The state trajectories of the system are plotted in Figures 9.3 and 9.4.

The unknown input case is also simulated where $\zeta(k)$ is taken as a constant, of value 0.1 and the fault is given as

$$r_3(k) = \begin{cases} 0, & t < 2(s) \\ 0.3 + 0.1cos(2\pi t), & 2 \leq t \leq 10(s) \end{cases} \tag{9.31}$$

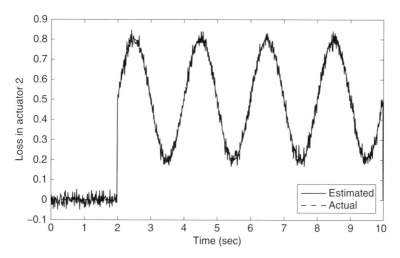

Figure 9.2 Loss in actuator 2

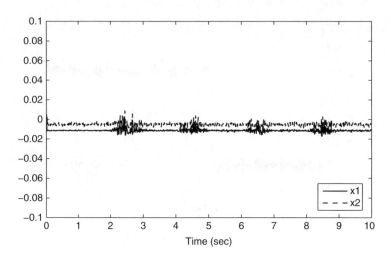

Figure 9.3 Trajectories of states 1 and 2

The results are shown in Figures 9.5 and 9.6 and it can be seen that the unknown input and the fault are estimated well. Furthermore, an estimate of the state $x_1(k)$ is shown in Figure 9.7, which strengthens the results.

9.6 Router Fault Accommodation in Real Time

Recent research focuses attention on achieving fault-tolerant control (FTC) without employing a fault diagnosis module [2, 3]. The complete FTC is composed of two parts: fault detection–identification (FDI) and fault accommodation (FA). These two modules help to

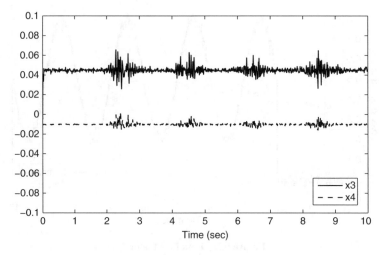

Figure 9.4 Trajectories of states 3 and 4

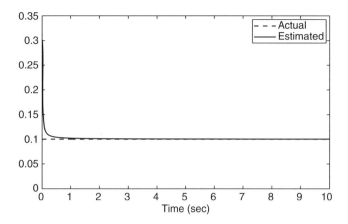

Figure 9.5   Estimation of the unknown input $\zeta(k)$

achieve a successful fault-tolerant system. In fact, a proper exchange of information is required between the two, which makes an integrated system difficult to realize. When a fault occurs, FDI becomes active and FA waits for information about the fault behavior, which induces delay, referred to as "fault detection delay". After receiving information about the fault, the FA module rectifies it and FDI is adapted to the new scenario to get rid of false detection. Though these modules work in parallel, in independent paths or different time axes, they require the exchange of information, such as the fault occurrence time, fault behavior, its severity etc. [4]. Research into the robustness of both modules has been carried out over a good few years. Various techniques have evolved to achieve FTC [5] but, in some cases, knowledge of the fault is pre-assumed [6] and only issues about the impact of fault detection and correction delays on the system behavior are considered. An important requirement for an FDI system used in

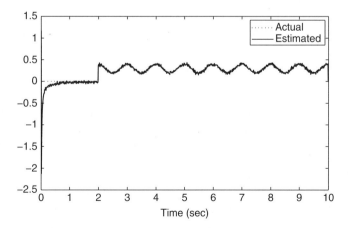

Figure 9.6   Fault estimation in the presence of unknown input

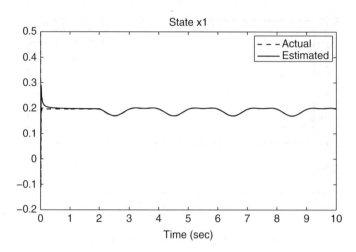

Figure 9.7   State estimation $\hat{x}_1(k)$ in the presence of unknown input

FTC is that the faults should be estimated with accuracy in real time. Second, knowledge of the plant must be available to carry out FDI.

Looking out for these shortcomings, the concept of a behavioral system theory from the perspective of FTC that does not determine the fault behavior is proposed by Yame and Sauter [7]. This section looks at the extension in terms of synthesizing the controllers using the concept of canonical controllers [8] with respect to the desired behavior. Yame and Sauter propose the behavioral approach to FTC in conjunction with the unfalsified control theory [9]. The advantage of employing the behavioral concept is that it works on the trajectories generated by the system in real time. This helps to eliminate the need for an exact model of the plant, and hence the FDI. The main point which attracts attention is the controller synthesis under the unfalsified framework when the effect of the fault is not taken into account beforehand [7]. It is obvious that controller synthesis that achieves the given specification utilizing the dynamics of plant is one of the central issues of control system theory. We handle this problem in a way inspired by the achievable behavior approach followed by Van der Schaft [8] (see also the references in his work). In the work of Willems [10], the central role of synthesizing the controller is played by the set of trajectories along which the dynamics of a system evolves. Moreover, a controller yielding the given specification is regarded as a "canonical controller".

A natural way of coping with the fault accommodation problem is to synthesize the control parameters based upon the system trajectories when the fault occurs. This approach is applied to an internet router for congestion control. On occurrence of a fault, the network characteristics change instantaneously, provoking congestion at different levels of communication. The problem of internet congestion is tackled from the point of view of linear control theory by Hollot et al. [11] and nonlinear control by Ren, Lin and Wei [12]. Recently, some work has been done on internet congestion as an FTC problem [13, 14]. From this background, this section provides a synthesis of a linear canonical controller for a given desired behavior without knowledge of the exact mathematical model of the system.

9.6.1 Canonical controller and achievable behavior

We give a brief overview of achievable (or implementable) behavior using partial interconnection and canonical controllers [8, 15]. For a linear, time-invariant differential system, the full plant behavior, $P_f \in L^{q+k}$ is the full behavior comprising a set of solutions (w, c) such that

$$R\left(\frac{d}{dt}\right)w(t) = M\left(\frac{d}{dt}\right)c(t) \quad w \in \Re^q, c \in \Re^k \tag{9.32}$$

where $R(\xi) \in \Re^{l \times q}[\xi]$ and $M(\xi) \in \Re^{l \times k}[\xi]$. Here c denotes the variables which are accessible to the controller and w denotes the variables whose behavior we intend to restrict. This restriction is imposed by the interconnection of Equation (9.32) with another subsystem having behavior, $C \in L^k$

$$H\left(\frac{d}{dt}\right)c(t) = 0 \tag{9.33}$$

where $H(\xi) \in \Re^{h \times k}[\xi]$ is the controller. The interconnection of the subsystems of Equations (9.32) and (9.33) gives the controlled subsystem which is shared by c variables as

$$P_f \|_c C = \{(w, c) | (w, c) \in P_f \text{ and } c \in C\} \tag{9.34}$$

It is assumed that the manifest (to-be-controlled) variables are observable from the shared auxiliary variables (control variables). Eliminating c from the full controlled behavior using the elimination theorem [10], the restriction $(P_f \|_c C)_w$ on the behavior of the manifest variable w, is defined by

$$(P_f \|_c C)_w = \{w \in W | \exists c \in C \text{ such that } (w, c) \in P_f\} \tag{9.35}$$

Therefore, for given $P_f \in L^{q+k}$ it is required to find $C \in L^k$ that implements $K \in L^q$ through c, where $K = (P_f \|_c C)_w$. This is known as the implementability theorem. It depends only on the projected full plant behavior $(P_f)_w$ and on the behavior consisting of the plant trajectories with the interconnection variables equal to zero. This behavior is denoted by $N_w(P_f)$ and is called the "hidden behavior". It is defined as

$$N_w(P_f) = \{w | (w, 0) \in P_f\} \tag{9.36}$$

Theorem 9.1 *[10] Let $P_f \in L^{q+k}$ be the full plant behavior. Then $K \in L^q$ is implementable by a controller $C \in L^k$ acting on the interconnection variable c if and only if*

$$N_w(P_f) \subseteq K \subseteq (P_f)_w$$

Theorem 9.2   *Theorem 9.1 shows that K can be any behavior that is wedged between the given behaviors $N_w(P_f)$ and $(P_f)_w$. Here we define the stabilizability of the manifest variable as the desired behavior which states that there exists a $C \in L^k$ such that the monic* *characteristic polynomial* $r \in (P_f \| _c C)_w$ *is Hurwitz, assuming* $(P_f)_w$ *is stabilizable. Hence the controller C should yield the controlled behavior* $K = S$. *We call S the "desired controlled behavior". The controllers to be implemented are "stabilizing controllers".*

A controller imposes restrictions only on the control variables and it is propagated through the plant to the variables to be controlled. Designing controllers that satisfy the desired behavior requires the set of differential equations in $\Re^{q \times k}$. We use the idea of a canonical controller that incorporates the concept of the internal model principle (IMP). The construction of the canonical controller requires the use of a plant model that has the same behavior as the plant. The restriction on the control variables can be made by interconnecting the plant model to the desired behavior S using the variables to be controlled. This is how the canonical controller is constructed. We denote the plant model by P_{imp} and the behavior of the canonical controller obtained using this construction as C_{can}^{imp}:

$$C_{can}^{imp} = \{c \in C | \exists v \in W \text{ such that}$$

$$(v, c) \in P_{imp} \text{ and } v \in S\} \tag{9.37}$$

Van der Schaft states [8] that if the plant model P_{imp} in the canonical controller C_{can}^{imp} differs from the actual model P_f then the desired behavior is still achievable, i.e. $S \subset (P_f \| _c C_{can}^{imp})_w$ if $P_{imp} \subset P_f$, while $(P_f \| _c C_{can}^{imp})_w \subset S$ if $P_f \subset P_{imp}$.

Van der Schaft makes no distinction between P_f and P_{imp} [8]; they are considered the same. We seek to analyze the behavior of fault and attempt to shrink or enlarge S such that the interconnection $(P_f \| _c C_{can}^{imp})_w$ approximates the desired behavior as well as possible. Note that the canonical controller is the class of all controllers that follow the desired behavior; here, the desired behavior is the stability of the interconnected system as.

9.6.2 Router modeling and desired behavior

A window-based nonlinear fluid-flow dynamic model for TCP networks is considered in this study. A detailed explanation of the model is presented by Hollot *et al.* [11] and Silva *et al.* [16]. A coupled set of nonlinear differential equations that reflects the dynamics of TCP accurately with the average TCP window size and the average queue length are given as:

$$\dot{w}(t) = \frac{1}{R(t)} - \frac{w(t)}{2} \cdot \frac{w(t - R(t))}{R(t - R(t))} \cdot p(t - R(t)) \tag{9.38a}$$

$$\dot{q}(t) = -C + \frac{N(t)}{R(t)} \cdot w(t) \tag{9.38b}$$

$$R(t) = \frac{q(t)}{C} + T_p \tag{9.38c}$$

where w is the average TCP window size (in packets); q is the instantaneous queue length (in packets); T_p is the propagation delay (in seconds); R is the transmission roundtrip time (RTT); C is the link capacity (in packets per second); N is the number of TCP connections; and p is the packet-dropping probability, which is the control input to decrease the sending rate and maintain the bottleneck queue length. All the above variables are assumed to be non-negative. Equation (9.38a) uses the additive increase and multiplicative decrease (AIMD) congestion control algorithm to evaluate the average window size during the TCP flow while Equation (9.38b) is the dynamics of the queue length accumulated as the transmission rate surpasses the link capacity.

Given the vector of network parameters $\eta = (N, C, d)$, a set of feasible operating points Ω_η is defined by

$$\Omega_\eta = \{(w_0, q_0, p_0) : w_0 \in (0, w_{\max}), q_0 \in (0, q_{\max}),$$
$$p_0 \in (0, 1) \text{ and } \dot{w} = 0; \dot{q} = 0\} \tag{9.39}$$

The network parameters η are feasible if Ω_η is nonempty. A linearized model for TCP congestion control, delays, and queues is expressed by

$$P(s) = \frac{C^2/2N}{(s + 2N/d^2C)(s + 1/d)} \cdot e^{-sd}$$
$$\text{or } P(s) = P_0(s) \cdot e^{-Ls} \tag{9.40}$$

where d is the RTT. Here $P(s)$ is the characterization of the internal model, discussed in the previous section, which depends only on η. The proportional–integral (PI) controller is suggested by Hollot *et al.* [17] for the stability of queue length denoted by $G(s; k_p, k_i)$.

So we have fixed the structure of the canonical controller and now its parameters are to be explored using the internal model and the desired behavior. We use an algorithm ([16], Section 11.5) to define the stability region because to Routh–Hurwitz stability criterion is not applicable to Equation (9.38a)–(9.38c). The region S_R contains the complete set of points (k_p, k_i) for which the desired behavior is implemented for all delays between 0 and d. As stated by Van der Schaft [8], if one controller can achieve the desired behavior then the same behavior can also be implemented by different controllers of the same class. The stability region S_R is expressed as $S_R = S_1 S_L$, where $S_1 = S_0 S_N$ (S_0 is the set of k_p and k_i values that stabilize the delay-free system $P_0(s)$ and S_N is the set of k_p and k_i values such that $G(s; k_p, k_i)P_0(s)$ is an improper transfer function and S_L is the set of (k_p, k_i) values such that $G(s; k_p, k_i)P_0(s)$ has a minimal destabilizing delay that is less than or equal to d). Formally, S_L is

$$S_L = \{(k_p, k_i) \notin S_N : \exists L \in [0, d], \omega \in R$$
$$\text{such that } G(j\omega; k_p, k_i)P_0(s) \cdot e^{-jL\omega} = -1\}.$$

To compute S_R, first define the projection of the stability region S_R on the line $k_p = \hat{k}_p$ as:

$$S_{R,\hat{k}_p} = \{(k_p, k_i) \in S_R : k_p = \hat{k}_p\}$$

so that the stability region can be calculated for each value of the proportional gain $\hat{k}_p$:

$$S_R = \bigcup_{\hat{k}_p} S_{R,\hat{k}_p}.$$

To compute $S_R, \hat{k}_p$, define the projections

$$S_{1,\hat{k}_p} = \{(k_p, k_i) \in S_1 : k_p = \hat{k}_p\},$$

$$S_{N,\hat{k}_p} = \{(k_p, k_i) \in S_N : k_p = \hat{k}_p\},$$

$$S_{L,\hat{k}_p} = \{(k_p, k_i) \in S_L : k_p = \hat{k}_p\}.$$

Then, $S_{R,\hat{k}_p} = S_{1,\hat{k}_p} S_{L,\hat{k}_p}$. The set $S_{L,\hat{k}_p}$ can be further decomposed and computed as:

$$S_{L,\hat{k}_p} = S_{L,\hat{k}_p}^+ \cup S_{L,\hat{k}_p}^-$$

where

$$S_{L,\hat{k}_p}^+ = \left\{ \left(\hat{k}_p, k_i \right) \notin S_{N,\hat{k}_p} : \exists \omega \in \Omega^+, k_i = \sqrt{M(\omega)} \right\},$$

$$S_{L,\hat{k}_p}^- = \left\{ \left(\hat{k}_p, k_i \right) \notin S_{N,\hat{k}_p} : \exists \omega \in \Omega^-, k_i = -\sqrt{M(\omega)} \right\},$$

$$\Omega^+ = \{\omega : \omega > 0, M(\omega) \geq 0,$$

$$\frac{\pi + \angle\left[\left(\sqrt{M(\omega)} + j\hat{k}_p\omega \right) R_0(j\omega) \right]}{\omega} \leq d \right\},$$

$$\Omega^- = \{\omega : \omega > 0, M(\omega) \geq 0,$$

$$\frac{\pi + \angle\left[\left(-\sqrt{M(\omega)} + j\hat{k}_p\omega \right) R_0(j\omega) \right]}{\omega} \leq d \right\},$$

$$M(\omega) = \frac{1}{|R_0(j\omega)|^2} - \hat{k}_p^2\omega^2,$$

$$R_0(s) = \frac{P_0}{s}(s).$$

where $\angle$ [W] represents the phase-angle of W.

Remark 9.4   *Using this algorithm, the sensitivity of S_R to the set η is studied. On varying one parameter while keeping the other two constant, it is found that the region S_R increases as values of N increase and values of d, C decrease.*

9.6.3   *Description of fault behavior*

We know from previous chapters that faults are categorized into three types:

- actuator faults;
- process parameter faults;
- sensor faults.

In the remainder of this section, we study only process parameter faults, where the parameters (given by set η) of the plant changes instantaneously resulting in queue length instability. To maintain stability in the transmission control protocol with active queue management (TCP/AQM) network, this kind of fault should be accommodated in real time. Internet congestion control as an FTC problem has been discussed by Aubrun, Join, and Menighed [13] using a multiple random early detection (multi-RED) controller configuration but their work is lacking in the determination of the parameters of the controller in real time: the fault scenarios and the parameters of the controllers required to maintain the stability of the queue length is known *a priori*. Fliess, Join, and Mounier [14] studied only the diagnosis module, for the possible occurrence of actuator faults and sensor faults. In this section, we achieve fault-tolerant control without using an explicit diagnosis module and the parameters of controllers are determined without *a priori* knowledge of the faults.

In an AQM router, when the number of connections or users increases, it leads to a proportional increase in RTT for fixed capacity [18]. Interestingly, N and RTT oppose each other from the viewpoint of defining the stability region. Now we define the internal model of the plant for which the set of controllers is to be designed. We make one minor assumption, that the router with fixed capacity is always connected to a user and at any point in time the number of users can only increase to a maximum level that the router can support. Referring to Equation (9.40), we see that RTT depends on T_p which is constant and $(q(t), C)$ is always positive.

For example, we define $\eta_{imp} = (3750, 60, 0.1)$ for the internal model shown in Figure 9.8 and η_{cur} for the current set of parameters of the plant. This gives the set of all controllers C_{can}^{imp}. If we see the interconnection $(P_f \|_c C_{can}^{imp})_w$, considering $P_f = P_{imp}$ or $\eta_{cur} = \eta_{imp}$, all sets of controllers implement the desired behavior. Now suppose the parametric fault occurs and η_{cur} changes to $(3750, 70, 0.15)$. The stability region in Figure 9.8 for η_{cur} shows that the

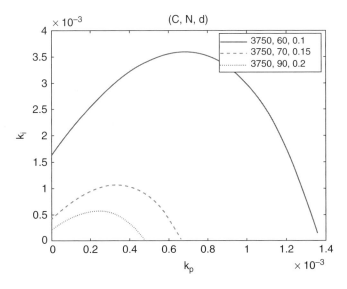

Figure 9.8   Stability region (under the curve) for different sets of η

interconnected system $(P_f \|_c C_{can}^{imp})_w \subset S$. Therefore, the composed system satisfies at least the "specifications" given by S.

9.6.4 A least restrictive controller

In the previous section, the set of all possible controllers was determined by the connection of the internal plant model and the desired behavior. For optimal performance of the TCP/AQM network, there must exist a set of controllers that achieves it. By "optimal performance" we mean the cost performance index J which is to be minimized. Now the problem is formulated as the need to find a set of controllers $C_{res} \subset C_{can}^{imp}$ such that $D \subset (P_f \|_c C_{can}^{imp})_w \subset S$, where D is the optimal behavior and denotes $D = (P_f \|_c C_{res})_w$. The set D is given by $D = \{w \in W : J(w) < \gamma\}$ where γ is a real bound. The cost function J is assumed to capture the control objective and examples of such functions are the integral square error (ISE) and the plant output variance.

Now, consider the fault-tolerant control problem. Here we introduce the theory of switching control, which uses the concept of "fictitious reference". A reference signal is generated by the input–output data collected in real time and by controller inversion. This method has been successfully studied for application to hydraulic processes [7]. One of the problems with this work is that the possible faults are defined beforehand and the controller configuration with respect to each mode is designed. In real time, if any of the faulty modes occurs the correct controller is selected from the controller bank. This switching control can also be seen in the work of Aubrun, Join, and Menighed [13] but no supervisor or reconfiguration mechanism is installed in the process. With the canonical controller, every unknown faulty mode is characterized in the region S_R. In real-time operation, only this region is explored to select the set of restricted controllers. Therefore, we can say S_R is the region of exploration. This means that for any faulty behavior, there exists some set (k_p, k_i) which can accommodate the fault.

Assume the system is working satisfactorily with the interconnection of the least restrictive controller (LRC) and the plant. After the occurrence of the fault, the current controller is not able to achieve the optimal behavior. The set of canonical controllers might achieve the desired behavior but now the problem is to select the LRC that can accommodate the fault successfully and guarantee optimal performance. To achieve this task, input–output data (u_{data}, y_{data}) is collected up to the current time. This data is used to generate the fictitious reference, $\tilde{r}$ for each controller without putting them into the feedback loop. Now based upon the cost function $J(w)$ where $w = (\tilde{r}, u_{data}, y_{data})$, only those controllers are considered to be the least restrictive and satisfy the optimal desired behavior.

9.7 Fault Detection for Markov Jump Systems

In this section, we investigate the fault detection (FD) problem for a class of discrete-time Markov jump linear systems (MJLS) with partially known transition probabilities. The proposed systems are more general, which relaxes the traditional assumption in Markov jump systems that all the transition probabilities must be completely known. A residual generator is constructed and the corresponding FD is formulated as an H_∞ filtering problem by which the error between the residual and the fault are minimized in the H_∞ sense. Sufficient conditions for the existence of FD filter are derived, based on linear matrix inequality. We give a numerical example on a multiplier–accelerator model economic system to illustrate the potential of the developed theoretical results.

9.7.1 Introduction

The past two decades have witnessed valuable research into stochastic hybrid systems, where Markov jump systems have remained a hot topic because of their widely practical applications in manufacturing systems, power systems, aerospace systems, networked control systems (NCS), etc. [19, 20, 21, 22]. For Markov jump systems with completely known transition probabilities, many problems have been addressed [19, 20]. Some recent extensions have considered uncertain transition probabilities, with the aim of utilizing robust methodologies to deal with the norm-bounded or polytopic uncertainties presumed in the transition probabilities [23, 24]. Unfortunately, the structure and "nominal" terms of the uncertain transition probabilities under consideration have to be known *a priori*.

The ideal assumption on the transition probabilities inevitably limits the application of the traditional Markov jump systems theory. In fact, the likelihood of obtaining complete knowledge about the transition probabilities is questionable and the cost is probably high. A typical example is found in NCS, where the time-varying delays and random packet loss induced by communication channels can be modeled as Markov chains and, accordingly, the resulting closed-loop system can be studied by means of Markov jump systems theory [25, 21, 22]. However, the variation of delays and packet dropouts in almost all types of communication network are different in different running periods of the network, and all or part of the elements in the desired transition probabilities matrix are difficult or costly to obtain. The same problems may arise in other practical systems with Markovian jumps. Thus, rather than the large complexity of measuring or estimating all the transition probabilities, it is significant and necessary from a control perspective to study more general Markov jump systems with partially known transition probabilities.

On another research front line, fault detection for many dynamic systems has been much investigated [26]. The key point of FD is to construct the residual generator, determine the residual evaluation function and the threshold, then make a judgement whether a fault alert should be generated by comparing the values of the evaluation function with the prescribed threshold. Usually, the residual generator is realized by formulating FD as a filtering problem with some performance index, such as H_∞ filtering, and therefore the FD filter gains and the optimal H_∞ performance index for the augmented systems can be obtained by some optimization method, such as linear matrix inequality (LMI) [27, 28, 29]. Recently, some attention has been drawn to Markov jump linear systems (MJLSs) [30], with uncertainties [28], with time delays [28, 29] or with applications [31] (more discussion of the H_∞ performance index can be found in the work of de Farias *et al.* [32] and of Seiler and Sengupta [21]). However, such results are still obtained based on the traditional assumption of complete knowledge about transition probabilities. It is more practical and challenging to find an FD filter for underlying systems with partially known transition probabilities, which motivates us in this study.

In this section, we investigate the problem of FD for a class of discrete-time MJLSs with partially known transition probabilities. More precisely, the considered systems relax the assumption that all the transition probabilities have to be completely known and cover traditional MJLS as a special case. The residual generator is constructed and the FD problem is formulated as an H_∞ filtering problem such that the error between residual and fault are minimized in the H_∞ sense. Sufficient conditions for the existence of the FD filter for the underlying systems are derived via LMIs. A numerical example is presented to show the validity and potential of the developed theoretical results.

9.7.2 Problem formulation

Fix the probability space (Ω, F, P) and consider the following class of discrete-time MJLSs:

$$x(k+1) = A(r_k)x(k) + B(r_k)u(k) + E(r_k)d(k) + F(r_k)f(k),$$

$$y(k) = C(r_k)x(k) + D(r_k)d(k) + G(r_k)f(k), \qquad (9.41)$$

where $x(k) \in \mathfrak{R}^n$ is the state vector, $u(k) \in \mathfrak{R}^u$ is the control input, $d(k) \in \mathfrak{R}^d$ is the unknown input, and $f(k) \in \mathfrak{R}^f$ is the fault to be detected. $u(k)$, $d(k)$, and $f(k)$ are assumed to belong to $l_2[0, \infty)$. $y(k) \in \mathfrak{R}^y$ is the output vector. $\{r_k, k \geq 0\}$ is a discrete-time homogeneous Markov chain, which takes values in a finite set $I \triangleq \{1, \ldots, N\}$ with a transition probabilities matrix $\Lambda = \{\pi_{ij}\}$ namely, for $r_k = i$, $r_{k+1} = j$, one has

$$P_r(r_{k+1} = j | r_k = i) = \pi_{ij},$$

where $\pi_{ij} \geq 0$, $\forall i, \in j \in I$, and $\sum_{j=1}^{N} \pi_{ij} = 1$. The set I contains N modes of the system in Equation (9.41) and for $r_k = i \in I$, the system matrices of the ith mode are denoted by A_i, B_i, C_i, D_i, E_i, F_i and G_i, which are considered here to be known real constants with appropriate dimensions.

In addition, the transition probabilities of the jumping process $\{r_k, k \geq 0\}$ are assumed to be partially accessible, i.e. some elements in matrix Λ, are unknown. For instance, for the system of Equation (9.41) with five operation modes, the transition probability matrix may be:

$$\begin{bmatrix} \pi_{11} & * & \pi_{13} & * & \pi_{15} \\ * & * & * & \pi_{24} & \pi_{25} \\ \pi_{31} & \pi_{32} & \pi_{33} & * & * \\ * & * & \pi_{43} & \pi_{44} & * \\ * & \pi_{52} & * & \pi_{54} & * \end{bmatrix}$$

where $*'$ represents the inaccessible elements. For notation clarity, $\forall i \in I$, we denote that

$$I_k^i \triangleq \{j : \pi_{ij} \text{ is known}\}, \qquad I_{UK}^i \triangleq \{j : \pi_{ij} \text{ is unknown}\}. \qquad (9.42)$$

Also, we denote $\pi_K^i \triangleq \sum_{j \in I_k^i} \pi_{ij}$ throughout.

Remark 9.5   *In the literature, the transition probabilities of the Markov chain $\{r_k, k \geq 0\}$ are generally assumed to be completely known ($I_{UK}^i = \emptyset$, $I_K^i = I$) or completely unknown ($I_K^i = \emptyset$, $I_{UK}^i = I$). Therefore, the transition probability matrix considered in this section is more natural to Markov jump systems and covers the previous two cases. Note that transition probabilities with polytopic or norm-bounded uncertainties can still be viewed as completely "known" in the sense of this section.*

We are interested in designing an FD filter for the underlying system. Its desired structure is considered to be

$$x_F(k+1) = A_F(r_k)x_F(k) + B_F(r_k)y(k),$$
$$r_e(k) = C_F(r_k)x_F(k) + D_F(r_k)y(k), \tag{9.43}$$

where $x_F(k) \in \mathfrak{R}^n$, $r_e(k) \in \mathfrak{R}^f$ is the residual, and $A_F(r_k)$, $B_F(r_k)$, $C_F(r_k)$ and $D_F(r_k)$, $\forall r_k \in I$ are the matrices with compatible dimensions to be determined. The FD filter with the above structure is assumed to jump synchronously with the modes in the system of Equation (9.41).

Denoting $\tilde{x}(k) \triangleq [x^T(k)\ x_F^T(k)]^T$, $e(k) \triangleq r_e(k) - f(k)$ and augmenting the model of Equation (9.41) to include Equation (9.43), we can obtain the following augmented system:

$$\tilde{x}_F(k+1) = \tilde{A}(r_k)\tilde{x}(k) + \tilde{B}(r_k)w(k),$$
$$e(k) = \tilde{C}(r_k)\tilde{x}(k) + \tilde{D}(r_k)w(k), \tag{9.44}$$

where $w(k) = [u^T(k)\ d^T(k)\ f^T(k)]^T$ and

$$\tilde{A}(r_k) = \begin{bmatrix} A(r_k) & 0 \\ B_F(r_k)C(r_k) & A_F(r_k) \end{bmatrix},$$

$$\tilde{B}(r_k) = \begin{bmatrix} B(r_k) & E(r_k) & F(r_k) \\ 0 & B_F(r_k)D(r_k) & B_F(r_k)G(r_k) \end{bmatrix},$$

$$\tilde{C}(r_k) = \begin{bmatrix} D_F(r_k)C(r_k) & C_F(r_k) \end{bmatrix},$$

$$\tilde{D}(r_k) = \begin{bmatrix} 0 & D_F(r_k)D(r_k) & D_F(r_k)G(r_k) - I \end{bmatrix}.$$

Obviously, the resulting system of Equation (9.44) is also an MJLS with partially known transition probabilities Equation (9.42). Now, to present the main objective of this section more precisely, we also introduce the following definitions for the system of Equation (9.44), which are essential for later developments.

Definition 9.1   *The system of Equation (9.44) is said to be stochastically stable if for $w(k) \equiv 0$, $k \geq 0$ and every initial condition $\tilde{x}_0 \in \mathfrak{R}^n$ and $r_0 \in I$, the following holds:*

$$E\left\{ \sum_{k=0}^{\infty} \|\tilde{x}(k)\|^2 | \tilde{x}_0, r_0 \right\} < \infty$$

Definition 9.2 *Given a scalar $\gamma > 0$, the system of Equation (9.44) is said to be stochastically stable and has an H_∞ model error performance index γ if it is internally stochastically stable and under zero initial condition, $\|e\|^{E_2} < \gamma\|w\|_2$ holds for all nonzero $w(k) \in l_2[0, \infty)$.*

Therefore, our objective is to determine matrices $\{A_F(r_k), B_F(r_k), C_F(r_k), D_F(r_k)\}$ of the FD filter such that the augmented system of Equation (9.44) is stochastically stable and has a guaranteed H_∞ performance index. Note that the original system of Equation (9.41) is assumed in the following discussion to be stable [28, 29], as an usual precondition in the **FD** problems. Furthermore, as commonly adopted in the literature, the fault $f(k)$ can be detected by the following steps:

1. Select a residual evaluation function $J(r_e(k)) = \sqrt{\sum_{l=k_0}^{k} \Re_e^T(l)r_e(l)}$, where k_0 denotes the initial evaluation time instant.
2. Select a threshold $J_{th} \triangleq \sup_{d \in l_2, f=0, k>0} E[J(r_e(k))]$.
3. Test:

$$J(r_e(k)) > J_{th} \Rightarrow \text{with faults} \Rightarrow \text{alarm}, \tag{9.45}$$

$$J(r_e(k)) < J_{th} \Rightarrow \text{no faults}. \tag{9.46}$$

Remark 9.6 *It is worth noting that a better definition for $J(r_e(k))$, which facilitates the detection of intermittent faults, can be found in the work of Meskin and Khorasani [31].*

Before ending the section, we give the following lemma for system of Equation (9.44), which will be used in the proof of our main results.

Lemma 9.1 *[20]: The system of Equation (9.44) is stochastically stable with a prescribed H_∞ performance index $\gamma > 0$ if and only if there exists a set of symmetric and positive-definite matrices $P_i, i \in I$ satisfying*

$$\Xi_i \triangleq \begin{bmatrix} -\bar{P}_i & 0 & \bar{P}_i\tilde{A}_i & \bar{P}_i\tilde{B}_i \\ * & -I & \tilde{C}_i & \tilde{D}_i \\ * & * & -P_i & 0 \\ * & * & * & -\gamma^2 I \end{bmatrix} < 0, \tag{9.47}$$

where $\bar{P}_i \triangleq \sum_{j \in I} \pi_{ij} P_j$.

In the following discussion, based on Lemma 9.1, we first give two H_∞ bounded real lemmas (BRLs) for the underlying augmented system of Equation (9.44) and then the design of the FD filter for the system of Equation (9.41).

9.7.3 H_∞ bounded real lemmas

Lemma 9.2   *Consider the system of Equation (9.44) with partially known transition probabilities Equation (9.42) and let $\gamma > 0$ be a given constant. If there exist matrices $P_i > 0, \forall i \in I$ such that*

$$\begin{bmatrix} -P_k^i & 0 & P_k^i \tilde{A}_i & P_k^i \tilde{B}_i \\ \bullet & -\pi_k^i I & \pi_k^i \tilde{C}_i & \pi_k^i \tilde{D}_i \\ \bullet & \bullet & -\pi_k^i P_i & 0 \\ \bullet & \bullet & \bullet & -\pi_k^i \gamma^2 I \end{bmatrix} < 0, \qquad (9.48)$$

$$\begin{bmatrix} -P_j & 0 & P_j \tilde{A}_i & P_j \tilde{B}_i \\ \bullet & -I & \tilde{C}_i & \tilde{D}_i \\ \bullet & \bullet & -P_i & 0 \\ \bullet & \bullet & \bullet & -\gamma^2 I \end{bmatrix} < 0, \quad \forall j \in T_{uk}^i, \qquad (9.49)$$

where $P_K^i \triangleq \sum_{j \in I_K^i} \pi_{ij} P_j$, then the augmented system of Equation (9.44) is stochastically stable with an H_∞ performance index γ.

Proof. Note that Equation (9.47) can be rewritten as

$$\Xi_i = \begin{bmatrix} \sum_{j \in I_K^i} \pi_{ij} P_j & 0 & \sum_{j \in I_K^i} \pi_{ij} P_j \tilde{A}_i & \sum_{j \in I_K^i} \pi_{ij} P_j \tilde{B}_i \\ \bullet & -\left(\sum_{j \in I_K^i} \pi_{ij}\right) I & \sum_{j \in I_K^i} \pi_{ij} \tilde{C}_i & \sum_{j \in I_K^i} \pi_{ij} \tilde{D}_i \\ \bullet & \bullet & -\sum_{j \in I_K^i} \pi_{ij} P_j & 0 \\ \bullet & \bullet & \bullet & -\left(\sum_{j \in I_K^i} \pi_{ij}\right) \gamma^2 I \end{bmatrix}$$

$$+ \sum_{j \in I_{uK}^i} \pi_{ij} \begin{bmatrix} -P_j & 0 & P_j \tilde{A}_i & P_j \tilde{B}_i \\ \bullet & -I & \tilde{C}_i & \tilde{D}_i \\ \bullet & \bullet & -P_i & 0 \\ \bullet & \bullet & \bullet & -\gamma^2 I \end{bmatrix}$$

$$= \begin{bmatrix} -P_k^i & 0 & P_k^i \tilde{A}_i & P_k^i \tilde{B}_i \\ \bullet & -\pi_k^i I & \pi_k^i \tilde{C}_i & \pi_i \tilde{D}_i \\ \bullet & \bullet & -\pi_k^i P_i & 0 \\ \bullet & \bullet & \bullet & -\pi_k^i \gamma^2 I \end{bmatrix} + \sum_{j \in I_{uK}^i} \pi_{ij} \begin{bmatrix} -P_j & 0 & P_j \tilde{A}_i & P_j \tilde{B}_i \\ \bullet & -I & \tilde{C}_i & \tilde{D}_i \\ \bullet & \bullet & -P_i & 0 \\ \bullet & \bullet & \bullet & -\gamma^2 I \end{bmatrix}$$

Therefore, the inequalities of Equation (9.48) and Equation (9.49) guarantee $\Xi_i < 0$, i.e., $J < 0$ which means that $\|e\|_{E2} < \gamma \|w\|_2$ (obviously, no knowledge of π_{ij}, $\forall j \in I_{UK}^i$ is required in Equations (9.48) and (9.49)). This completes the proof.   ■

Remark 9.7   *Although Lemma 9.2 gives a BRL for the MJLS with partially known transition probabilities, it is hard to apply it to obtain the desired reduced-order model here because of the cross coupling of matrix product terms among different system operation modes, as shown* *in Equations (9.48) and (9.49). To overcome this difficulty, the technique using a slack matrix developed by Zhang et al. [33] can be adopted to obtain an improved criterion for the system of Equation (9.44).*

Lemma 9.3   *Consider system of Equation (9.44) with the partially known transition probabilities of Equation (9.42) and let $\gamma > 0$ be a given constant. If there exist matrices $P_i > 0$, and R_i, $\forall i \in I$ such that*

$$
\begin{bmatrix}
\Upsilon_j - R_i - \Re_i^T & 0 & R_i \tilde{A}_i & R_i \tilde{B}_i \\
\bullet & -I & \tilde{C}_i & \tilde{D}_i \\
\bullet & \bullet & -P_i & 0 \\
\bullet & \bullet & \bullet & \gamma^2 I
\end{bmatrix} < 0,
\tag{9.50}
$$

where if $\pi_K^i = 0$, $\Upsilon_j \triangleq P_j$, $j \in I_{UK}^i$ otherwise

$$
\begin{cases}
\Upsilon_j \triangleq \frac{1}{\pi_k^i} P_k^i, \\
\Upsilon_j \triangleq P_j, & \forall j \in I_{UK}^i
\end{cases}
\tag{9.51}
$$

and $P_K^i = \sum_{j \in I_k^i} \pi_{ij} P_j$, then the augmented system of Equation (9.44) is stochastically stable with an H_∞ performance index γ.

Proof. By Lemma 9.2, we conclude that the system of Equation (9.44) is stochastically stable with an H_∞ performance index γ if the inequalities of Equations (9.48) and (9.49) hold. Notice that if $\pi_K^i = 0$, the conditions of Equations (9.48)–(9.49) are reduced to Equation (9.49). Then, for $\pi_K^i \neq 0$, Equation (9.48) can be rewritten as

$$
\begin{bmatrix}
-\frac{1}{\pi_k^i} P_k^i & 0 & \frac{1}{\pi_k^i} P_k^i \tilde{A}_i & \frac{1}{\pi_k^i} P_k^i \tilde{B}_i \\
\bullet & -I & \tilde{C}_i & \tilde{D}_i \\
\bullet & \bullet & -P_i & 0 \\
\bullet & \bullet & \bullet & \gamma^2 I
\end{bmatrix} < 0.
\tag{9.52}
$$

On the other hand, for an arbitrary matrix R_i, $\forall i \in I$, we have the following facts:

$$
\left(\frac{1}{\pi_k^i} P_k^i - R_i \right)^T \left(\frac{1}{\pi_k^i} P_k^i \right)^{-1} \left(\frac{1}{\pi_k^i} P_k^i - R_i \right) \geq 0,
$$

$$
(P_j - R_i)^T P_j^{-1} (P_j - R_i) \geq 0.
$$

Using Equation (9.51), one has

$$
\Upsilon_j - R_i - \Re_i^T \geq -\Re_i^T \Upsilon_j^{-1} R_i.
$$

Furthermore, from Equation (9.50), we can obtain that

$$
\begin{bmatrix}
-\mathfrak{R}_i^T \Upsilon_j^{-1} - R_i & 0 & R_i \tilde{A}_i & R_i \tilde{B}_i \\
\bullet & -I & \tilde{C}_i & \tilde{D}_i \\
\bullet & \bullet & -P_i & 0 \\
\bullet & \bullet & \bullet & \gamma^2 I
\end{bmatrix} < 0.
$$

Performing a congruence transformation using $diag\{R-1_i \Upsilon_j, I, I, I\}$ yields Equations (9.52) and (9.49) for $j \in I_K^i$ and $j \in I_{UK}^i$, respectively (note that R_i is invertible if it satisfies Equation (9.50)). This completes the proof. ∎

9.7.4 H_∞ FD filter design

As an application of Lemma 9.3, the following theorem presents sufficient conditions for the existence of an admissible H_∞ FD filter with the form of Equation (9.43).

Theorem 9.3 *Consider the system of Equation (9.41) with partially known transition probabilities Equation (9.42) and let $\gamma > 0$ be a given constant. If there exist matrices $P_{1i} > 0$, $Y_i > 0$ and $P_{3i} > 0, \forall i \in I$, and matrices $P_{2i}, X_i, Z_i, A_{fi}, B_{fi}, C_{fi}, D_{fi}, \forall i \in I$, with $\pi_K^i = 0$, $\Upsilon_{1j} \triangleq P_{1j}, \Upsilon_{2j} \triangleq P_{2j}, \Upsilon_{3j} \triangleq P_{1j}, j \in I_{UK}^i$, otherwise,*

$$
\begin{cases}
\Upsilon_{1j} \triangleq \frac{i}{\pi_k^i} P_k^{1i} \triangleq \frac{i}{\pi_k^i} \sum_{j \in I_K^i} \pi_{1j} P_{1j}, \\[2mm]
\Upsilon_{2j} \triangleq \frac{i}{\pi_k^i} P_k^{2i} = \frac{i}{\pi_k^i} \sum_{j \in I_K^i} \pi_{1j} P_{2j}, \\[2mm]
\Upsilon_{3j} \triangleq \frac{i}{\pi_k^i} P_k^{3i} = \frac{i}{\pi_k^i} \sum_{j \in I_K^i} \pi_{1j} P_{3j},
\end{cases}
\tag{9.53}
$$

$$
\begin{cases}
\Upsilon_{1j} \triangleq P_{1j}, \\
\Upsilon_{2j} \triangleq P_{2j}, \quad \forall j \in I_{UK}^i, \\
\Upsilon_{3j} \triangleq P_{3j},
\end{cases}
\tag{9.54}
$$

then there exists an FD filter such that the resulting model error system, Equation (9.44), is stochastically stable with an H_∞ performance index γ. Moreover, if a feasible solution exists, the gains of an admissible FD filter in the form of Equation (9.43) are given by

$$
A_{Fi} = Y_i^{-1} A_{fi}, \quad B_{Fi} = Y_i^{-1} B_{fi},
$$

$$
C_{Fi} = C_{fi}, \quad D_{Fi} = D_{fi}, \quad i \in I.
\tag{9.55}
$$

Proof. Consider the system of Equation (9.44) and assume the matrices P_i and R_i in Lemma 9.3 have the following forms:

$$
P_i \triangleq \begin{bmatrix} P_{1i} & P_{2i} \\ \bullet & P_{3i} \end{bmatrix}, \quad R_i \triangleq \begin{bmatrix} X_i & Y_i \\ Z_i & Y_i \end{bmatrix},
$$

then we have

$$\pi_k^i \triangleq \sum_{j \in I_K^i} \pi_{ij} P_j = \sum_{j \in I_K^i} \pi_{1j} \begin{bmatrix} P_{1i} & P_{2i} \\ \bullet & P_{3i} \end{bmatrix}, \quad R_i \triangleq \begin{bmatrix} P_k^{1i} & P_k^{2i} \\ \bullet & P_k^{3i} \end{bmatrix}.$$

Further define matrix variables

$$A_{fi} = Y_i A_{Fi}, \quad B_{fi} = Y_i B_{Fi}, \quad C_{fi} = C_{Fi}, \quad D_{fi} = D_{Fi}$$

$$\triangleq \begin{bmatrix} \Upsilon_{1j} & \Upsilon_{2j} \\ \bullet & \Upsilon_{3j} \end{bmatrix}.$$

where Υ_{1j}, Υ_{2j} and Υ_{3j} are defined in Equations (9.53) and (9.54) for $j \in \Upsilon I_K^i$ and $j \in I_{UK}^i$, respectively. One can readily obtain

$$\begin{bmatrix} \Phi_{1i} & \Phi_{2i} & 0 & X_i A_i + B_{fi} C_i & A_{fi} & \Theta_{1i} \\ \bullet & \Phi_{3i} & 0 & Z_i A_i + B_{fi} C_i & A_{fi} & \Theta_{2i} \\ \bullet & \bullet & -I & D_{fi} C_i & C_{fi} & \Theta_{3i} \\ \bullet & \bullet & \bullet & -P_{1i} & -P_{2i} & 0 \\ \bullet & \bullet & \bullet & \bullet & -P_{3i} & 0 \\ \bullet & \bullet & \bullet & \bullet & \bullet & -\gamma^2 I \end{bmatrix} < 0$$

$$\Phi_{1i} = \Upsilon_{1j} - X_i - X_i^T$$
$$\Phi_{2i} = \Upsilon_{2j} - Y_i - Z_i^T$$
$$\Phi_{3i} = \Upsilon_{3j} - Y_i - Y_i^T$$
$$\Theta_{1i} = \begin{bmatrix} X_i B_i & X_i E_i + B_{fi} D_i & X_i F_i + B_{fi} G_i \end{bmatrix}$$
$$\Theta_{2i} = \begin{bmatrix} Z_i B_i & Z_i E_i + B_{fi} D_i & Z_i F_i + B_{fi} G_i \end{bmatrix}$$
$$\Theta_{3i} \leq \begin{bmatrix} 0 & D_{fi} D_i & D_{fi} G_i - I \end{bmatrix} \tag{9.56}$$

by replacing $\tilde{A}_i$, Bi, $\tilde{C}_i$, $\tilde{D}_i$, Υ_j, P_i and R_i into Equation (9.50); if Equation (9.56) holds, the augmented system of Equation (9.44) will be stochastically stable with an H_∞ performance. Meanwhile, if a solution of Equation (9.56) exists, the parameters of the admissible filter are given by Equation (9.55). This completes the proof. ∎

Remark 9.8   *By setting $\delta = \gamma^2$ and minimizing δ, subject to Equation (9.56), we can obtain the optimal H_∞ performance index γ^* (by $\gamma = \sqrt{\delta}$) and the corresponding filter gains. Also, it can be deduced from Equation (9.56) that, given different degrees of unknown elements in the transition probabilities matrix, the optimal γ^* achieved for the system of Equation (9.44) and the corresponding filter should be different, which we illustrate with a numerical example in the next section.*

9.7.5 Simulation

In this section, we revisit an economic system to show the usefulness of the results developed above. Consider the economic system based on a multiplier–accelerator model proposed by Samuelson [34], which aims to relate government expenditure (control input) and national income (system state). The interaction between the multiplier and the accelerator in the model appears in the difference equation form. For more details on the model, we refer readers to the work of Blair and Sworder [35] and Costa *et al.* [36] and the references therein.

$$C_t = cY_{t-1},$$
$$I_t = w(Y_{t-1} - Y_{t-2}),$$
$$Y_t = C_t + I_t + G_i^E, \tag{9.57}$$

where

C is the consumption expenditure,

Y is national income,

I is induced private investment,

G^E is government expenditure,

s is marginal propensity to save,

$c = (1 - s)$ is marginal propensity to consume,

$\frac{1}{s}$ is the multiplier,

w is the accelerator coefficient,

t is time, $t = kT = k, \ (T = 1)$.

By denoting $x_{k+1} = [x_1(k + 1) x_2(k + 1)]^T$ and considering

$$x_1(k + 1) = x_2(k),$$
$$x_2(k + 1) = Y_k,$$

one can derive from Equation (9.57) that

$$x_{k+1} = Ax_k + BG_k^E, \tag{9.58}$$

where

$$A = \begin{bmatrix} 0 & 1 \\ -w & 1 - s + w \end{bmatrix}, \ B = \begin{bmatrix} 0 \\ 1 \end{bmatrix}.$$

Therefore, the dynamic behaviour between national income and government expenditure is described in Equation (9.58). The variation of the group of s and w corresponds to three different economic situations ("norm", "boom" and "slump"), which change from one mode to another according to a Markov chain [35]. The parameters for the three modes of the systems are computed [35] as

$$\text{Norm: } A_1 = \begin{bmatrix} 0 & 1 \\ -2.5 & 3.2 \end{bmatrix},$$

$$\text{Boom: } A_2 = \begin{bmatrix} 0 & 1 \\ -43.7 & 45.4 \end{bmatrix}, \tag{9.59}$$

$$\text{Slump: } A_3 = \begin{bmatrix} 0 & 1 \\ 5.3 & -5.2 \end{bmatrix}.$$

Some control issues, such as the regulator problem and feedback control [35] and the constrained control problem ([36]), have been investigated for the system. Here, we are concerned with the problem of detecting faults possibly encountered in the system. The fault signal could be some emergent economic event or other factors which will intervene in the above equations.

First of all, for the three unstable modes in Equation (9.59), we would like to choose the stabilizing controller gains to meet the prerequisite in FD problems that the original system should be stable (note that this is not the point of this section – the gains can be determined by a certain design procedure such that the system mode is stable):

$$K_1 = \begin{bmatrix} 2 & -3 \end{bmatrix}, \quad K_2 = \begin{bmatrix} 43 & -45 \end{bmatrix}, \quad K_3 = \begin{bmatrix} -5 & 5 \end{bmatrix},$$

by which the system matrices become (taking the mode corresponding to the Norm scenario as an example)

$$\text{Norm: } \bar{A}_1 = A_1 + BK_1$$

$$= \begin{bmatrix} 0 & 1 \\ -2.5 & 3.2 \end{bmatrix} + \begin{bmatrix} 0 \\ 1 \end{bmatrix} \begin{bmatrix} 2 & -3 \end{bmatrix}$$

$$= \begin{bmatrix} 0 & 1 \\ -0.5 & 0.2 \end{bmatrix}.$$

Then, reconsider the system of Equation (9.41) with the following data

$$A_1 = \begin{bmatrix} 0 & 1 \\ -0.5 & 0.2 \end{bmatrix}, \quad A_2 = \begin{bmatrix} 0 & 1 \\ -0.7 & 0.4 \end{bmatrix},$$

$$A_3 = \begin{bmatrix} 0 & 1 \\ 0.3 & -0.2 \end{bmatrix},$$

$$B_1 = B_2 = B_3 = \begin{bmatrix} 0 & 0 \end{bmatrix}^T, C_1 = C_2 = C_3 = \begin{bmatrix} 0 & 1 \end{bmatrix},$$

$$D_1 = D_2 = D_3 = 0.4, E_1 = E_2 = E_3 = \begin{bmatrix} 0 & 0.1 \end{bmatrix}^T,$$

$$F_1 = F_2 = F_3 = \begin{bmatrix} 0 & -0.5 \end{bmatrix}^T, G_1 = G_2 = G_3 = 0.9.$$

Table 9.1   Transition probability matrices A

	Completely known		
	1	2	3
1	0.67	0.17	0.16
2	0.30	0.47	0.23
3	0.26	0.10	0.64

Table 9.2   Transition probability matrices B

	Completely known	Partially known	(case I)
	1	2	3
1	0.67	0.17	0.16
2	0.30	*	*
3	0.26	0.10	0.64

Table 9.3   Transition probability matrices C

	Completely known	Partially known	(case II)
	1	2	3
1	0.67	0.17	0.16
2	0.30	*	*
3	*	*	0.64

The problem here is to validate the FD filter design results for the system with the above parameters. The four cases for the transition probability matrix considered in this example are shown in Tables 9.1–9.4 (note that "*" means "unknown number").

For $k = 0, 1, \ldots, 300$, the unknown input $d(k)$ is simulated by a white noise signal with amplitude less than 0.5 (given in Figure 9.9). The fault signal is set up as

$$f(k) = \begin{cases} 0.8, & \text{for } k = 100, 101, \ldots, 200, \\ 0, & \text{others.} \end{cases}$$

Table 9.4   Transition probability matrices D

	Completely known	Partially known	(case I)
	1	2	3
1	*	*	*
2	*	*	*
3	*	*	*

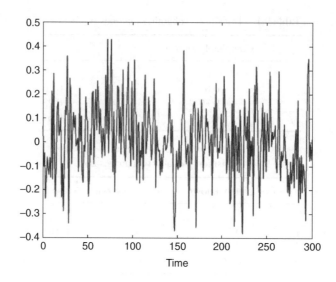

Figure 9.9   Unknown input $d(k)$

By solving the convex problem in Equation (9.56), the optimal H_∞ performance indices and the corresponding FD filter are obtained for the four transition probability cases. The corresponding results are listed in Table 9.5. We omit the filter gains for simplicity.

Now, consider the transition probability matrix with completely known elements as the practical one for the other three cases in Table 9.1; we can generate a possible evolution of system modes as shown in Figure 9.10. Accordingly, Figure 9.11 shows the generated residual signals $r_e(k)$ and Figure 9.12 presents the evolution of $J(r_e(k)) = \sqrt{\sum_{l=0}^{k} \Re_e^T(l)r_e(l)}$ for both the faulty case and the fault-free case for the four transition probability matrices in Tables 9.1–9.4. Then, based on the path in Figure 9.10 and the selected threshold $J_{th} = \sup_{d \in l_2, f=0} = E[\sqrt{\sum_{k_0}^{300} \Re_e^T(k)r_e(k)}]$, the time steps N_d for the FD by the evaluation function $J(r_e(k))$ and test Equations (9.45)–(9.46) can be calculated (see Table 9.5). It can be seen that the more transition probability knowledge we have, the better H_∞ performance index the augmented system can achieve and the less time is needed for the FD. Therefore, a tradeoff can be built, in practice, between the complexity of obtaining transition probabilities and the performance benefits and efficiency of detection by means of our ideas and approaches.

Table 9.5   Minimum γ^*, detection time steps N_d for the different cases

Transition probabilities	γ^*	N_d
Completely known	0.8203	6
Partially known (Case I)	0.8796	7
Partially known (Case II)	0.9258	7
Completely unknown	0.9817	8

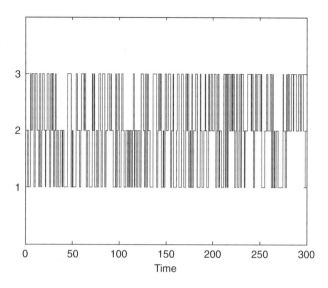

Figure 9.10   Mode evolution

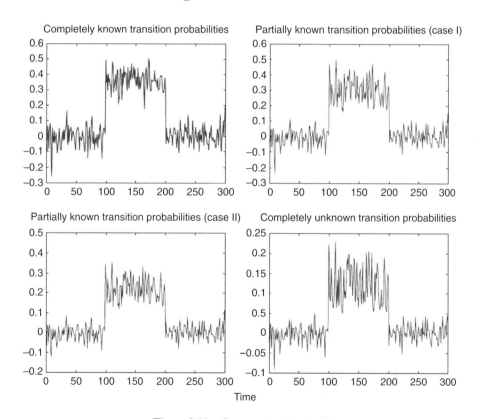

Figure 9.11   Generated residual $r_e(k)$

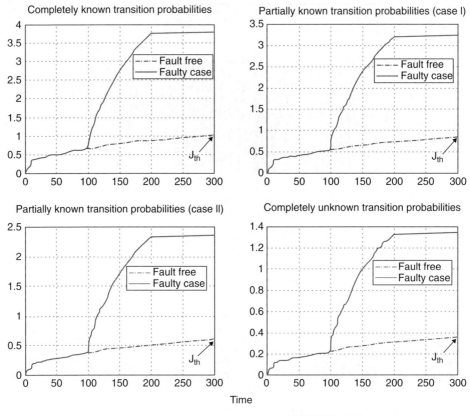

Figure 9.12   Evolution of $J(r_e) = \sqrt{\sum_{l=0}^{k} \Re_e^T(l) r_e(l)}$

9.8   Notes

For a linear, MIMO, discrete-time system with noise, a fault-tolerant control (FTC) problem has been addressed in this chapter by first estimating the fault and then using a reconfigured control law to counter the effects of fault. The unmeasurable states are estimated and the results are subsequently extended to a case where an unknown input, which can be regarded as a disturbance, is estimated. The methodology was successfully applied to a helicopter model.

Following this, fault-tolerant control is achieved based upon the trajectories of variables to be controlled and control variables in the behavioral framework. The problem of internet congestion control was discussed from the viewpoint of fault tolerance theory. Moreover, the other types of faults mentioned in Section 9.6 needs to be explored to achieve complete FTC. As TCP/AQM is a SISO system after linearization, the modeling of loss in actuator efficiency would be the point of interest. The synthesis of the controller in real time is the major issue in model-free FTC approaches; it is handled successfully for the application to the internet router.

Finally, the FD problem for discrete-time MJLS with partially known transition probabilities was investigated. The underlying systems are more general than the traditional MJLS, where all the transition probabilities are assumed to be completely known. The LMI-based sufficient conditions of the FD filter are obtained and a tradeoff can be observed between the complexity of obtaining all the transition probabilities and the time steps to detect the fault. Note that the threshold we set, as commonly selected in the literature, will not detect the removal of a fault.

References

[1] Jiang, B., and Chowdhury, F. N. (2005) "Fault estimation and accommodation for linear MIMO discrete-time systems", *IEEE Trans. Control Systems Technology*, **13**(3):493–499.

[2] Yame, J. J., and Kinnaert, M. (2003) "Performance based switching for fault tolerant control", *Preprints the 5th IFAC SAFEPREOCESS*, 555–560.

[3] Ingimundarson, A., Pena, R. S. S. (2008) "Using the unfalsified control concept to achieve fault tolerance", in *Proc. IFAC World Congress*, 1236–1242.

[4] Jiang, J. (2005) "Fault-tolerant control systems: An introductory overview", *Automatica SINCA*, **31**(1):161–174.

[5] Noura, H., Sauter, D., Hamelin, F., and Theilliol, D. (2000) "Fault-tolerant control in dynamic systems: application to a winding machine", *IEEE Control System Magazine*, 33–49.

[6] Theilliol, D., Join, C., and Zhang, Y. (2008) "Actuator fault tolerant control design based on a reconfigurable reference input", *Int. J. Appl. Math. Comput. Sci.*, **18**(4):2008, 553–560.

[7] Yame, J. J., Sauter, D. (2008) "A real-time model-free reconfiguration mechanism for fault-tolerance: application to a hydraulic process", in *Proc. 10th Intl. Conf. on Control, Automation, Robotics and Vision*, 91–96.

[8] Van der Schaft, A. J. (2003) "Achievable behavior of general systems", *Systems and Control Letters*, **49**(2):141–149.

[9] Safonov, M. G., and Tsao, T. C. (1997) "The unfalsified control concept and learning", *IEEE Trans. Automatic Control*, **42**(6):843–847.

[10] Willems, J. C. (1997) "On interconnection, control and feedback", *IEEE Trans. Automatic Control*, **42**(3):326–339.

[11] Hollot, C. V., Misra, V., Towsley, D., and Gong, W. (2000) *A control theoretic analysis of RED*, Tech. Report UM-CS-2000-041.

[12] Ren, F., Lin, C., and Wei, B. (2005) "A nonlinear control theoretic analysis to TCPRED system", *Computer Networks*, **49**:580–592.

[13] Aubrun, C., Join, C., and Menighed, M. (2006) "Multi-RED controller for router fault accommodation, *IFAC SAFEPROCESS*, 319–324.

[14] Fliess, F., Join, C., and Mounier, H. (2004) "An introduction to nonlinear fault diagnosis with an application to a congested internet router", *Advances in Comm. Control Networks-LNCIS*, 327–343.

[15] Julius, A. A., Willems, J. C., Bellur, M. N., and Trentelman, H. L. (2005) "The canonical controllers and regular interconnection", *Systems and Control Letters*, **54**(8):787–797.

[16] Silva, G. J., Datta, A., Bhattacharyya, S. P. (2005) *PID Controller for Time- Delay Systems*, Boston, Mass, USA: Birkhäuser.

[17] Hollot, C. V., Misra, V., Towsley, D., and Gong, W. (2002) "Analysis and design of controllers for AQM routers supporting TCP flows", *IEEE Trans. Automatic Control*, **47**(6):945–959.

[18] Kellet, C. M., Shorten, R. N., and Leith, D. J. (2006) "Sizing internet router buffers, active queue management, and the Lure problem", in *Proc. Decision and Control Conference*, 650–654.

[19] Boukas, E. K. (2005) *Stochastic Switching Systems: Analysis and Design*, Basel, Berlin: Birkhäuser.

[20] Costa, O. L. V., Fragoso, M. D., and Marques, R. P. (2005) *Discrete-time Markovian Jump Linear Systems*, London: Springer-Verlag.

[21] Seiler, P., and Sengupta, R. (2005) "An H_∞ approach to networked control", *IEEE Trans. Automatic Control*, **50**:356–364.

[22] Zhang, L., Shi, Y., Chen, T., and Huang, B. (2005) "A new method for stabilization of networked control systems with random delays", *IEEE Trans. Automatic Control*, **50**:1177–1181.

[23] Karan, M., Shi, P., and Kaya, C. Y. (2006) "Transition probability bounds for the stochastic stability robustness of continuous- and discrete-time Markovian jump linear systems", *Automatica*, **42**:2159–2168.

[24] Xiong, J. L., Lam, J., Gao, H., and Daniel, W. C. (2005), "On robust stabilization of Markovian jump systems with uncertain switching probabilities", *Automatica*, **41**:897–903.

[25] Krtolica, R., Ozguner, U., Chan, H., Goktas, H., et al. (1994) "Stability of linear feedback systems with random communication delays", *Int. J. Control*, **59**:925–953.

[26] Patton, R. J., Frank, P. M., and Clark, R. N. (Eds) (2000) *Issues of Fault Diagnosis for Dynamic Systems*, London: Springer-Verlag.

[27] Zhong, M. Y., Ding, S., Lam, J., and Wang, H. (2003) "An LMI approach to design robust fault detection filter for uncertain LTI systems", *Automatica*, **39**(3):543–550.

[28] Zhong, M., Ye, H., Shi, P., and Wang, G. (2005), "Fault detection for Markovian jump systems", *IET Control Theory and Applications*, **152**:397–402.

[29] Wang, H., Wang, C., Gao, H., and Wu, L. (2006), "An LMI approach to fault detection and isolation filter design for Markovian jump systems with mode-dependent time-delays", in *Proc. American Control Conference*, Minneapolis, Minnesota, 5686–5691.

[30] Kaoutar, A., and Boukas, E. K. (2005) "H_∞ based fault detection and isolation for Markovian jump systems", in *Proc. 16th IFAC Triennial World Congress*, Prague, Czechoslovakia, 43–47.

[31] Meskin, N., and Khorasani, K. (2009) "Fault detection and isolation of discrete-time Markovian jump systems with application to a network of multi-agent systems having imperfect communication channels", *Automatica*, **45**:2032–2040.

[32] de Farias, D. P., Geromel, J. C., do Val, J. B. R., and Costa O. L. V. (2000) "Output feedback control of Markov jump linear systems in continuous-time", *IEEE Trans. Automatic Control*, **45**:944–949.

[33] Zhang, L., Shi, P., Wang, C., and Gao, H. (2006) "Robust H_∞ filtering for switched linear discrete-time systems with polytopic uncertainties", *Int. J. Adaptive Control and Signal Processing*, **20**:291–304.

[34] Samuelson P. A. (1939) *The Review of Economic Statistics*, **152**, Cambridge, Mass., USA: Harvard University Press, 75–76.

[35] Blair, W. P., and Sworder, D. D. (1975) "Feedback control of a class of linear discrete systems with jump parameters and quadratic cost criteria", *Int. J. Control*. **21**:833–844.

[36] Costa, O. L. V., Filho, E. O. A., Boukas, E. K., and Marques, R. (1999) "Constrained quadratic state feedback control of discrete-time Markovian jump linear systems", *Automatica*, **35**:617–626.

10

Applications

This chapter is devoted to applications that use fault detection algorithms:

- abrupt changes in electrocardiograms (ECG), based on the maximum likelihood ratio;
- a multiple-input–multiple-output (MIMO) linear system based on frequency domain data;
- an electromechanical positioning system;
- fermentation processes;
- flexible-joint robots.

10.1   Detection of Abrupt Changes in an Electrocardiogram

This section is devoted to the detection of abrupt changes in an electrocardiogram (ECG). A linear time-variant model with Gaussian white noise is used to describe the real ECG signal, based on the estimated system parameters and tuned covariances of noise, and off-line and online generalized likelihood ratio (GLR) tests for ECG signal are developed for change detection.

10.1.1   Introduction

Change detection is receiving more and more attention, and many studies have shown that model-based change detection methods are efficient and easy to compute [1, 2]. Some of them focus on the solutions of the off-line and online generalized likelihood ratio (GLR) tests in biomedical signal processing by means of Kalman filter technology. The GLR test allows the detection of changes in signal behavior. Such change detection is of importance in automatic signal analysis. Changes may refer to external artefacts, that is, they may be caused by electrode movement during recording of a ventricular fibrillation (VF) ECG or refer to changes in the proper signal itself [3]. An example of the latter would be the sudden appearance of ventricular fibrillation in sinus rhythm ECG [4, 5, 6] or, conversely, the spontaneous termination of VF.

Change detection also has the role of locating the fault occurrence in time and giving a quick alarm [7]. After the alert, isolation is often needed to locate the faulty component. Furthermore, change detection is important in automated application of algorithms for signal analysis, which can be distorted by artefacts. The location of artefacts in time allows exclusion of certain corrupted parts of the signal from the analysis.

Analysis and Synthesis of Fault-Tolerant Control Systems, First Edition. Magdi S. Mahmoud and Yuanqing Xia.
© 2014 John Wiley & Sons, Ltd. Published 2014 by John Wiley & Sons, Ltd.

In a Kalman state space description, the detection of change can be recast into the detection of change in parameters or states. The GLR test proposed by Willsky and Jones [8] utilizes this approach. GLR's general applicability has contributed to it now being a standard tool in change detection. As summarized by Kerr [9], GLR has an appealing analytic framework, is widely understood by many researchers, and is readily applicable to systems already utilizing a Kalman filter. The Kalman filter is used for online implementation easily. The main advantage of the Kalman filter over the recursive least squares (RLS) algorithm and least mean square (LMS) algorithm is that it can be assigned different tracking speeds for the slip slope and slip offset [10].

Another advantage with GLR is that it partially solves the isolation problem in fault detection, that is, locating the physical cause of the change. Kerr points out a number drawbacks with GLR [9]. Among them, we mention problems with choosing decision thresholds and, for some applications, an untenable computational burden.

The use of likelihood ratios in hypothesis testing is motivated by the Neyman–Pearson Lemma; more and more research attention has recently been paid to testing likelihood ratios. For example, Shanableh, Assaleh, and Al-Rousan [11] propose a variety of feature extraction methods based on likelihood ratios test for ArSL recognition. In the application they consider, the likelihood ratio is the optimal test statistic when the change magnitude is known and just one change time is considered. This is not the case here, but a sub-optimal extension is immediate: the test is computed for each possible change time or a restriction to a sliding window; if several tests indicate a change, the most significant is taken as the estimated change time. In GLR, the actual change in the state of a linear system is estimated from data and then used in the likelihood ratio.

In this section, off-line and online GLR test methods are presented in the context of biomedical signal processing; more efficiently, a windowed online GLR test scheme is proposed.

10.1.2   Modeling ECG signals with an AR model

In this chapter, an autoregressive (AR) model is used to describe the ECG signal:

$$z_t = a_1(t)z_{t-1} + a_2(t)z_{t-2} + \ldots, +a_n(t)z_{t-n} + \omega_t \tag{10.1}$$

where time-varying parameters $a_k(t)$, $k = 1, 2, \ldots, n$ are to be identified, $\omega_t \in N(0, \Omega)$ is white noise and $z_t, t = 1, 2, \ldots$, are measurements of the ECG signal. In order to estimate the parameters of the AR model in Equation (10.1), the RLS algorithm is used to identify these parameters online. The RLS algorithm updates the parameter estimate by the recursion:

$$\hat{\theta}_t^{AR} = \hat{\theta}_{t-1}^{AR} + K_t^{AR}(z_t - \phi_t^T \hat{\theta}_{t-1}^{AR}) \tag{10.2}$$

$$K_t^{AR} = \frac{P_{t-1}^{AR}\phi_t}{\lambda + \phi_t^T P_{t-1}^{AR}\phi_t} \tag{10.3}$$

$$P_t^{AR} = \frac{1}{\lambda}\left(P_{t-1}^{AR} - \frac{P_{t-1}^{AR}\phi_t\phi_t^T P_{t-1}^{AR}}{\lambda + \phi_t^T P_{t-1}^{AR}\phi_t}\right) \tag{10.4}$$

where $\hat{\theta}_t^{AR}, \phi_t^T, K_t^{AR}$ and P_t^{AR} are the parameter estimate vector, historical measured data, filter gain and estimation-error covariance, respectively. $\hat{\theta}_t^{AR}$ and ϕ_t^T are defined as

$$\hat{\theta}_t^{AR} = [\hat{a}_1(t)\,\hat{a}_2(t)\,\ldots,\,\hat{a}_n(t)]^T \tag{10.5}$$

$$\phi_t^T = [z_{t-1}\,z_{t-2}\,\ldots,\,z_{t-n}] \tag{10.6}$$

and the tuning forgetting factor λ should be set less than 1. Furthermore, we can write the AR model of Equation (10.1) with states and system matrices as follows:

$$x_{t+1} = A_t x_t + B_t u_t + v_t \tag{10.7}$$

$$y_t = C_t x_t + D_t u_t + e_t, \tag{10.8}$$

where

$$x_t = \begin{bmatrix} z_t \\ z_{t-1} \\ \vdots \\ z_{t-n+1} \end{bmatrix}, A_t = \begin{bmatrix} (\hat{\theta}_t^{AR})^T \\ - \quad - \quad - \\ I_{n-1} \quad | \quad 0_{n-1} \end{bmatrix}$$

$$v_t^T = [\omega_t\,0\,\ldots,\,0]^T, Q_t = diag_n\{\Omega,\,0,\,\ldots,\,0\}, e_t = 0$$

$$B_t = 0, D_t = 0, C_t = [1\,0\,\ldots,\,0]$$

The parameters of the AR model are estimated by implementation of the RLS algorithm. Other algorithms also can be used to identify the parameters of ECG signal, for example, the Levinson algorithm or LMS algorithm. The estimate of parameters does not differ largely between algorithms, see Figure 10.1.

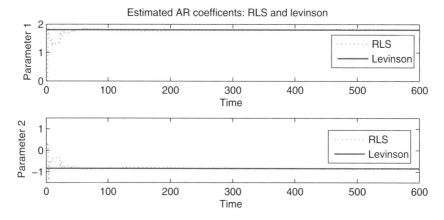

Figure 10.1   Comparison of RLS and Levinson estimates of model parameters

10.1.3 Linear models with additive abrupt changes

In this section, the detection of additive abrupt changes in linear state space models is introduced. Sensor and actuator faults as a sudden offset or drift can all be modeled as additive changes. In addition, disturbances are traditionally modeled as additive state changes. The likelihood ratio formulation provides a general framework for detecting such changes and isolating the fault or disturbance. Consider the state space model of Equation (10.7) with additive abrupt changes described as follows:

$$x_{t+1} = A_t x_t + B_t u_t + v_t + \sigma_{t-k} v \tag{10.9}$$

$$y_t = C_t x_t + D_t u_t + e_t, \tag{10.10}$$

where $x_t \in \Re^n$, $y_t \in \Re^m$ and $u_t \in \Re^l$ are states, measurements, and inputs, respectively. A_t, B_t, C_t and D_t are parameter matrices, which are known with proper dimensions. Operator σ_{t-k} denotes the step function, and the additive change (fault) $v \in \Re^n$ enters at time k as a step. Here v_t, e_t and x_0 are assumed to be independent Gaussian variables:

$$
\begin{aligned}
v_t &\in & N(0, Q_t) \\
e_t &\in & N(0, R_t) \\
x_0 &\in & N(0, X_0).
\end{aligned}
$$

The GLR algorithm can be implemented as follows [12].

Algorithm 10.1 *(GLR test)*
Given the signal model of Equation (10.9):

1. *Calculate the innovations from the Kalman filter assuming no change:*

$$\varepsilon_t = y_t - C_t \hat{x}_{t|t-1} - D_{u,t} u_t \tag{10.11}$$

   where $\hat{x}_{t|t-1}$ is the estimate of state x_t at time instant $t-1$ and the covariance of measurement y_t is

$$S_t = C_t^T P_{t|t-1} C_t + \tilde{R} \tag{10.12}$$

   where $P_{t|t-1} = A_t P_{t-1} A_t^T + \tilde{Q}$. Note that the real covariances Q_t and R_t are unknown; tuning parameters $\tilde{Q}$ and $\tilde{R}$ are used in Kalman filtering, and the convergence of the estimation-error P_t is given in Remark 10.1.
2. *Compute the regressors $\varphi_t(k)$:*

$$\varphi_{t+1}^T(k) = C_{t+1} \left(\prod_{i=k}^{t} A_i - A_t \mu_t(k) \right) \tag{10.13}$$

$$\mu_{t+1}(k) = A_t \mu_t(k) + K_{t+1} \varphi_{t+1}^T(k) \tag{10.14}$$

initialized by zero at time $t = k$.

3. *Compute the linear regression quantities:*

$$R_t(k) = \sum_{i=1}^{t} \varphi_i(k) S_i^{-1} \varphi_i^T(k) \qquad (10.15)$$

$$f_t(k) = \sum_{i=1}^{t} \varphi_i(k) S_i^{-1} \varepsilon_i \qquad (10.16)$$

where ε_t is the Kalman filter innovations and, for each k, $1 \le k \le t$.
4. *At time $t = N$, the test statistic is given by*

$$L_N(k, \hat{v}(k)) = f_N^T(k) \mathfrak{R}_N^{-1}(k) f_N(k) \qquad (10.17)$$

$$\hat{v}(k) = \mathfrak{R}_N^{-1}(k) f_N(k). \qquad (10.18)$$

5. *A change candidate is given by $\hat{k} = \arg\max L_N(k, v(k))$.*

The change candidate is accepted if $L_N(\hat{k}, \hat{v}(k))$ is greater than some threshold h (otherwise $\hat{k} = N$) and the corresponding estimate of the change magnitude is given by $\hat{v}_N(\hat{k}) = R_N^{-1}(\hat{k}) f_N(\hat{k})$.

Remark 10.1 *A second-order, state-space growth, dynamic linear model was proposed by Harrison and Stevens to represent the heart-rate trend [13]. In their model, the state transition matrix $A_t \in \mathfrak{R}^{2 \times 2}$ is assumed to be constant and known; one of the states is defined as the real signal level; the other state is defined as the real signal's incremental change; and both states are corrupted by independent Gaussian white noise. Furthermore, the covariance of the state noise Q_t and covariance of the observation noise R_t are assumed to be time-variant and unknown. Based on this model, Yang, Dumont and Ansermino developed an adaptive change detection method in heart-rate-trend monitoring [14]. Their main contribution is to estimate the covariance matrices, Q_t and R_t, and to detect change points using adaptive cumulative sum (CUSUM) testing.*

The state transition matrix A_t of the model in Equation (10.7), used in this section, is assumed to be time-variant and unknown, which is obtained by the RLS algorithm. The covariance of state noise Q_t and the covariance of observation noise R_t are assumed to be unknown as well and a pair of tuning parameters $(\tilde{Q}, \tilde{R})$, instead of real covariances (Q_t, R_t), is used in Kalman filtering. The key step of the proposed method is to estimate the states based on a given pair $(\tilde{Q}, \tilde{R})$. For the same set of historical observation data y_t, $t = 0, 1, \ldots, N$, different estimation-error sequences P_t, $t = 0, 1, \ldots, N$ are obtained with different given pairs $(\tilde{Q}, \tilde{R})$.

Now we give the analysis of the convergence of estimation-error P_t. According to the Kalman filter algorithm, we have

$$P_t = (I - K_t C_t) A_t P_{t-1} A_t^T (I - K_t C_t)^T + (I - K_t C_t) \tilde{Q} (I - K_t C_t)^T + K_t \tilde{R} K_t^T \qquad (10.19)$$

Note that K_t does not need to be a Kalman gain. Define $P_{1,t}$ as the estimation-error covariance when $Q_t = 0$, and $P_{2,t}$ as the estimation-error covariance when $R_t = 0$. If the following equation holds

$$P_0 = P_{1,0} + P_{2,0} \tag{10.20}$$

then we have

$$P_t = P_{1,t} + P_{2,t}, \forall t = 1, 2, \ldots \tag{10.21}$$

It is assumed that

$$Q_t = (1 + \alpha_t)\tilde{Q}, \ R_t = (1 + \beta_t)\tilde{R} \tag{10.22}$$

where α_t and β_t are unknown scalars. According to Simon [15], the true estimation-error covariance P_t^{true} can be described by

$$P_t^{true} = P_t + \alpha_t P_{1,t} + \beta_t P_{2,t} \tag{10.23}$$

If a suitable pair of tuning parameters $(\tilde{Q}, \tilde{R})$ is chosen, $E\{\alpha_t\} \approx 0$, $E\{\beta_t\} \approx 0$, and

$$E\{P_t^{true}\} \approx E\{P_t\} \tag{10.24}$$

If the pair $(\tilde{Q}, \tilde{R})$ is suitable, the convergence of estimation-error covariance is close to that of the standard Kalman filter.

Remark 10.2   *We can use RLS to estimate the change v and the equation $x_t \approx \hat{x}_{t|t-1} + \mu_t v$ can be used to solve the compensation problem after detection of change and estimation (isolation) of v.*

Remark 10.3   *The formulation in Algorithm 10.1 is off-line. Since the test statistic involves a matrix inversion of R_N, a more efficient online method is as follows:*

$$L_t(k, \hat{v}(k)) = f_t^T(k)\hat{v}_t(k) \tag{10.25}$$

where t is used as time index instead of N. The RLS scheme can now be used to update $\hat{v}_t(k)$ recursively, eliminating the matrix inversion of $R_t(k)$. Thus, the best implementation requires t parallel RLS schemes and one Kalman filter.

In the following discussion, detailed simulation studies illustrate the effectiveness of the developed algorithms.

10.1.4   Off-line detection of abrupt changes in ECG

In a biomedical signal, abrupt changes of introduced noise (such as artificial CPR noise or an electronic pulse) can be considered as the observation noise and noise that comes from humans or animals (such as breathing or the body moving) can be considered as the process noise. We can construct an AR model for the medical signal with a suitable order and identify the model parameters with methods such as RLS, Levinson, Kalman filters, etc.

In the GLR test algorithm, we can use different methods to estimate the AR model parameters; here we use RLS and Levinson [16]. From Figure 10.1, we can see that the RLS method and Levinson method can obtain the same AR parameters (the same as the Kalman filter model parameters), if the signal is long enough.

Figure 10.2 shows an off-line GLR test on a 15 000 ECG signal series (using real data obtained in Innsbruck Medical University). It uses a constant Kalman filter model via the Levinson method and all the signal data. From the figure, we can see that the GLR test statistic $L_N(k, v(k))$ has a pulsed high value; once we can set an efficient threshold h, we can easily get the abrupt change points, for at these points the signal frequency changes obviously. There will always be some points that cannot be detected despite abrupt changes happening there; because of the threshold h being assigned for all the time, the maximum value of the test statistics cannot exceed the threshold at these points.

Figures 10.3–10.7 show the off-line GLR test on an ECG signal from an ECG database created in Innsbruck Medical University. The signal is mixed with VF and CPR (noise signal)

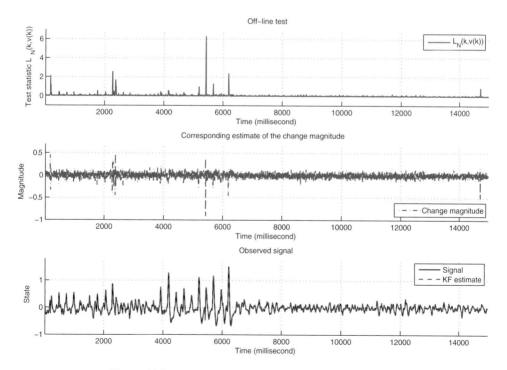

Figure 10.2   Off-line GLR test statistic for change detection

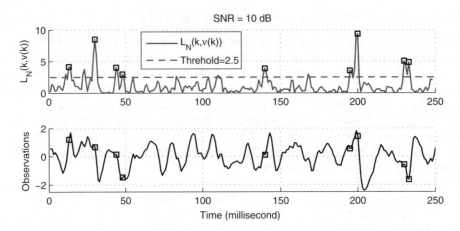

Figure 10.3   Levinson estimated constant filter model: off-line GLR test on 10 dB

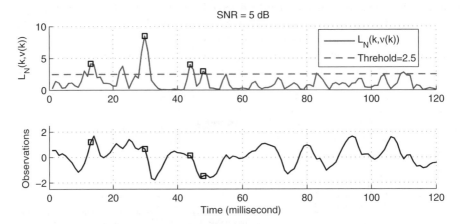

Figure 10.4   Levinson estimated constant filter model: off-line GLR test on 5 dB

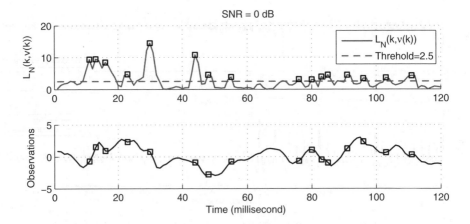

Figure 10.5   Levinson estimated constant filter model: off-line GLR test on 0 dB

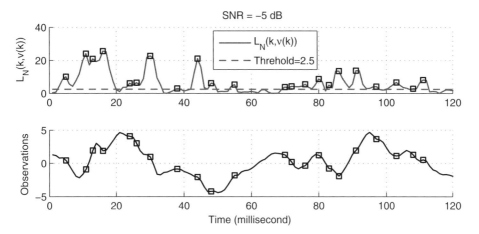

Figure 10.6   Levinson estimated constant filter model: off-line GLR test on −5 dB

artificially with different signal-to-noise ratios (SNR). We set the same threshold $h = 2.5$ for each test and compared the test statistic $L_N(k, v(k))$ and the original observation signal to obtain the abrupt change points according the setting threshold h.

From Figures 10.3–10.7, we can conclude that the abrupt changes obtained from test statistics with the same threshold is hard to detect efficiently. The main difficulty is the appropriate choice of threshold h.

10.1.5   Online detection of abrupt changes in ECG

In Remark 10.3, we presented an efficient way of obtaining an online GLR test method, by applying the algorithm. Figure 10.8 shows a simulation result from the online GLR test. In

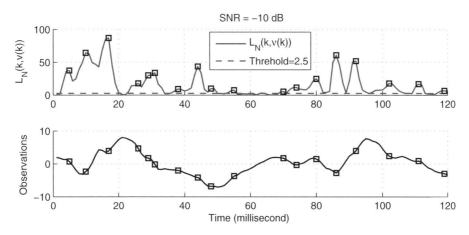

Figure 10.7   Livenson estimated constant filter model: off-line GLR test on −10 dB

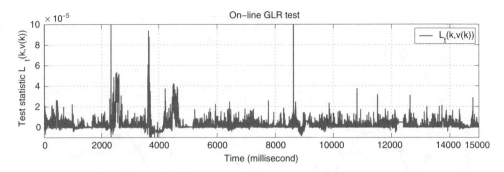

Figure 10.8   Online test statistic $L_t(k, v(k))$ with RLS estimation

this simulation, a third-order AR model with white noise is used to describe the measured signal; the parameters of this model are estimated using an RLS scheme with a forgetting factor $\lambda = 0.55$. In real applications, since the observed signal data size increases, the memory size required to restore the computed data increases by the square and computing time for each step increases too, so it is important to set a suitable data structure.

We propose a trade-off of a windowed online GLR method:

$$L_t(t - W, \hat{v}(t - W)) = f_t^T(t - W)\hat{v}(t - W) \tag{10.26}$$

where W is the length of data window. Better performance for the estimate can be obtained if a larger W is chosen, however, a larger W means that more time is needed to detect an abrupt change after it occurs. Computing $\hat{v}(t - W)$ recursively needs data from steps $t - W$ to t, so that the computing time of every step is fixed. Figure 10.9 shows a simulation result of the windowed online GLR test. In this simulation, a third-order AR model with white noise is used to describe the measured signal, and the parameters of this model are estimated using an RLS scheme with a forgetting factor $\lambda = 0.55$. The window length of this simulation is $W = 100$, so that the memory used is much less than with online GLR tests and the computing speed increases notably. From Figure 10.9, we can see that the performance of the proposed

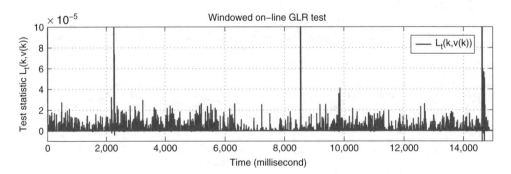

Figure 10.9   Windowed online test statistic $L_t(k, v(k))$ with RLS estimation

windowed online GLR test method does not differ from the online method at the abrupt change points, so the proposed windowed online GLR method is definitely suitable for application in real-time tests.

10.2   Detection of Abrupt Changes in the Frequency Domain

This section is devoted to the detection of abrupt changes for multiple-input–multiple-output (MIMO) linear systems based on frequency domain data. The real discrete-time Fourier transform is used to map the measured inputs and outputs from the time domain to the frequency domain. Under the hypothesis that the state change occurrence time is k, the system is split into two systems at time instant k: one of them describes the frequency dynamics before the state change occurs and the other describes the frequency dynamics after the occurrence; thus the latent state change is modeled as an initial state disturbance to be estimated based on frequency domain samples. Furthermore, the occurrence time is estimated by maximizing a likelihood ratio function.

10.2.1   Introduction

Dynamic systems sometimes incur abnormal changes. In control systems, these changes are mostly undesirable. If a change is not strong enough to cause the control system to collapse, we call it a fault, otherwise it is a failure. Fault detection in the automation field is of the utmost importance for safety and it is receiving more and more attention. The most popular method of fault detection, namely, model-based fault detection, makes decisions by comparing the virtual outputs of the mathematical model with the actual outputs of the system considered. Many studies have shown that model-based fault detection methods are efficient and easy to compute [17]–[21]. However, the changes in dynamic systems are not always undesirable since they may contain much useful information. For example, Xia *et al.* [21] proposed an algorithm for detecting abnormal state changes, which contain biological movements and can be modeled as abrupt state changes in biological signals. Ukil *et al.* [22] gave an analysis of abrupt changes occurring in power delivery processes. Some researchers pay attention detecting changes in communications [23] and others to analyzing the stability when changes occur [3].

By virtue of the state-space description of dynamic systems, most abrupt changes in signal can be modeled as state changes, i.e, the state jumps at some unknown time instant with known or unknown magnitudes. Hence, the change detection problem becomes one of estimating the occurrence times and magnitudes of changes. Maximum likelihood (ML) estimation is a powerful tool for estimating the unknown parameters by maximizing the likelihood of given measurements. However, it is more popular to use the ratio of likelihoods under different hypotheses to estimate the unknown parameters. For estimating the occurrence time and magnitude of an unknown abrupt state change for linear systems, Willsky and Jones [8] proposed the generalized likelihood ratio test algorithm in the time domain. In their method, a Kalman filter is applied to estimate the states under the hypothesis that there is no state change; at the same time the innovations of the Kalman filter are refiltered under the hypothesis that a change occurs at a hypothetical time instant with unknown magnitude to obtain the estimate of magnitude; then the estimate of occurrence time is solved by maximizing the likelihood ratio.

In practical engineering, frequency domain data is widely used and many methods in the frequency domain have been proposed in much of the literature over the past several decades. Pintelon and Schoukens [24] extended system identification methods to the frequency domain and McKelvey [25] proposed a finite discrete-time Fourier-transform-based identification approach for SISO linear systems. Agüro *et al.* [26] compared maximum likelihood estimation in the time domain and in the frequency domain, then proved that maximum likelihood estimation in the frequency domain is equal to maximum likelihood estimation in the time domain. The frequency domain approach has special features that, in some applications, can favor methods using frequency domain data. For example, it is sufficient to find a model that accurately describes the true system in a limited frequency band [25]. A low-order model could thus be sufficient rather than trying to fit a more complex model to all frequencies. Additionally, if data is obtained by different experiments, all frequency data can be merged into one data set. Continuous-time models that are valid for large frequency ranges can be estimated from data sets obtained from several experiments, each using a different sampling frequency. Thirdly, the field of digital signal processing relies heavily on operations in the frequency domain (i.e. on the Fourier transform). For example, several lossy image and sound compression methods employ the discrete Fourier transform: the signal is cut into short segments, each is transformed, and then the Fourier coefficients of high frequencies, which assumed to be unnoticeable, are discarded. The decompressor computes the inverse transform based on this reduced number of Fourier coefficients [27, 28].

In this section, we consider MIMO, time-invariant, linear systems with a latent additive state change and only frequency samples available for observation. First, we apply the finite discrete-time Fourier transform (DFT) to show that the system representations in the time and frequency domains are equal to each other. We then define two hypotheses; one assumes that no change happens and the other assumes a state change occurs at a hypothetical time instant with an unknown magnitude. We then split the system into two systems at the hypothetical occurrence time. One of the systems describes the frequency dynamics after the hypothetical time instant. Based on this consideration, the state change is modeled as an initial state disturbance of the system that describes the dynamics after the hypothetical occurrence time. Finally, the magnitudes at every hypothetical time instant are estimated and the maximum likelihood ratio estimate of occurrence time is solved by maximizing the likelihood ratio of two hypotheses with estimated magnitudes.

10.2.2   Problem formulation

Consider a general MIMO, time-invariant, discrete-time state-space model of the form:

$$x_{t+1} = Ax_t + Bu_t + \omega_t + f_t; \qquad (10.27)$$

$$y_t = Cx_t + Du_t + \upsilon_t \qquad (10.28)$$

where $x_t \in \mathfrak{R}^n$, $u_t \in \mathfrak{R}^l$, and $y_t \in \mathfrak{R}^m$ are the state, input and output vectors of the system, respectively, and $A \in \mathfrak{R}^{n \times n}$, $B \in \mathfrak{R}^{n \times l}$, $C \in \mathfrak{R}^{m \times n}$ and $D \in \mathfrak{R}^{m \times l}$ are the system matrices. The state abrupt change $f_t = \delta_{t-k} \upsilon$ with unknown magnitude υ occurs at the unknown time instant

k and δ_j is the pulse function ($\delta_j = 1$ if $j = 0$ and zero otherwise). Furthermore, we define that

$$n_t = \begin{bmatrix} \omega_t \\ \upsilon_t \end{bmatrix} \tag{10.29}$$

is a discrete-time Gaussian white noise sequence with covariance structure:

$$E\{n_l n_k^T\} = P\delta[l - k] \tag{10.30}$$

where $E\{\cdot\}$ denotes the mathematical expectation and

$$P = \begin{bmatrix} Q & S \\ S^T & R \end{bmatrix} \tag{10.31}$$

The initial value of state x_0 is also assumed to be an independent Gaussian variable: $x_0 \sim \mathcal{N}(\mu_{x_0}; D_{x_0})$, where μ_{x_0} and D_{x_0} are known.

Note that the aim of this chapter is to detect both the magnitude and the occurrence time of the abrupt state change based on the frequency domain data. The system represented by Equation (10.27)–(10.28) is presented to introduce the basic assumption of the system and the state change considered. In the next section, we present the relationship between the state change and the system representation in the frequency domain. In hypothesis testing, the likelihood ratios rather than the likelihoods are usually used [29]. The likelihood radio (LR) test is a multiple hypothesis test, where the change hypotheses are paired. Now, we define two hypotheses:

$$H_0 : \text{no change} \tag{10.32}$$

$$H_1(k, v) : \text{a change of magnitude } v \text{ at time instant } k \tag{10.33}$$

and introduce the log likelihood ratio for the hypotheses as the test statistic:

$$g_t(k, v) \triangleq \log \frac{p(y^t | H_1(k, v))}{p(y^t | H_0)} \tag{10.34}$$

where y^t denotes the measurement set $\{y_i | i = 0, \ldots, t\}$. The notation g_t denotes the distance measure between two hypotheses. It is assumed that $H_1(k, v) = H_0$ if $k = t$. Estimating v and k is coupled: if v is known, the LR estimate is described by

$$\hat{k}^{LR} = \arg_k \max g_t(k, v) \tag{10.35}$$

However, v is an unknown parameter. By virtue of the generalized likelihood ratio (GLR) test proposed by Willsky and Jones [8], it is possible to eliminate the unknown nuisance parameter v as follows:

$$\hat{v}(k) = \arg_v \max \log \frac{p(y^t | k, v)}{p(y^t | k = t)} \tag{10.36}$$

$$\hat{k} = \arg_k \max \log \frac{p(y^t | k, \hat{v}(k))}{p(y^t | t)} \tag{10.37}$$

then the two-variable maximum problem becomes two single-variable maximum problems, which are easy to solve in the time domain. In the rest of this chapter, an equivalent form of GLR test is proposed in frequency domain and an algorithm is developed in terms of the real discrete-time Fourier transform.

10.2.3   Frequency domain ML ratio estimation

In this section, we transfer the MLR estimation problem to the frequency domain from the time domain. First, we use the finite DFT given by:

$$S_k = \frac{1}{\sqrt{N}} \sum_{t=0}^{N-1} s_t z_k^{-t}; \quad k = 0, 1, \ldots, N-1 \tag{10.38}$$

where $S_k \in C$ is the frequency domain signal, $s_t \in R$ is the time domain signal, $z_k = e^{j\omega_k}$ and $\omega_k = 2\pi k/N$. Let $\Re(S_k)$ and $\Im(S_k)$ denote the real and imaginary parts of S_k, respectively, then we have

$$S_k = \Re(S_k) + j\Im(S_k) \tag{10.39}$$

The finite DFT samples defined in Equation (10.38) can be expressed in matrix form as

$$\vec{S} = \mathcal{F}_N \vec{s} \tag{10.40}$$

where $\vec{S} = [S_0 \ldots S_{N-1}]^T$, $\vec{s} = [s_0 \ldots s_{N-1}]^T$ and matrix $\mathcal{F}_N$ is the $N \times N$ finite DFT matrix given by

$$\mathcal{F}_N = \frac{1}{\sqrt{N}} \begin{bmatrix} 1 & 1 & 1 & \cdots & 1 \\ 1 & z_1 & z_1^2 & \cdots & z_1^{N-1} \\ 1 & z_2 & z_2^2 & \cdots & z_2^{N-1} \\ \vdots & \vdots & \vdots & \ddots & \vdots \\ 1 & z_{N-1} & z_{N-1}^2 & \cdots & z_{N-1}^{N-1} \end{bmatrix} \tag{10.41}$$

Likewise, the inverse transformation, often abbreviated as IDFT, can be expressed in matrix form as

$$\vec{s} = \mathcal{F}_N^{-1} \vec{S} \tag{10.42}$$

where $\mathcal{F}_N^{-1}$ is the $N \times N$ matrix given by

$$\mathcal{F}_N^{-1} = \mathcal{F}_N^H \tag{10.43}$$

where the superscript H denotes the complex conjugate transpose.

In the more general case, the signal $s_t \in \mathfrak{R}^n$ is a real vector, i.e.,

$$s_t = [s_t^1 \, s_t^2 \, \ldots \, s_t^n]^T \tag{10.44}$$

where $s_t^i \in R$ denotes the ith element of vector s_t. Let $S_k \in C^n$ denote the finite DFT of signal s_t and we can rewrite Equation (10.38) as follows

$$S_k^i = \frac{1}{\sqrt{N}} \sum_{t=0}^{N-1} s_t^i z_k^{-t}; \quad k = 0, 1, \ldots, N-1 \quad i = 0, 1, \ldots, n \tag{10.45}$$

where $S_k^i \in C$ is the ith element of the complex vector S_k. To develop the vector finite DFT in matrix form, we also define $\vec{S} = [S_0^T \ldots S_{N-1}^T]^T$ and $\vec{s} = [s_0^T \ldots s_{N-1}^T]^T$. Based on Equation (10.45), we arrive at

$$\vec{S} = \mathcal{F}_N^n \vec{s} \tag{10.46}$$

where

$$\mathcal{F}_N^n = \mathcal{F}_N \otimes I_n, \tag{10.47}$$

in which $\otimes$ denotes the Kronecker products. Since $\mathcal{F}_N^{-1} = \mathcal{F}_N^H$, it is easy to verify that

$$(\mathcal{F}_N^n)^{-1} = (\mathcal{F}_N \otimes I_n)^{-1} = \mathcal{F}_N^{-1} \otimes I_n^{-1} = \mathcal{F}_N^H \otimes I_n^H = (\mathcal{F}_N \otimes I_n)^H = (\mathcal{F}_N^n)^H \tag{10.48}$$

Now, applying the finite DFT Equation (10.38) to the time domain linear system of Equations (10.27)–(10.28) gives

$$z_k X_k = A X_k + B U_k + \Omega_k + \alpha \frac{z_k}{\sqrt{N}} + F_k; \tag{10.49}$$

$$Y_k = C X_k + D U_k + \Upsilon_k \tag{10.50}$$

where $\alpha = x_0 - x_N$ and $X_k, Y_k, U_k, \Omega_k, \Upsilon_k$ and F_k are the frequency domain signals transferred from the time domain signals $x_t, y_t, u_t, \omega_t, \upsilon_t$ and f_t, respectively. We rewrite Equations (10.49)–(10.50) as

$$Y_k = \Lambda_k \alpha + G_k U_k + H_k W_k + \Xi_k F_k \tag{10.51}$$

where $W_k = [\Omega_k^T \quad \Upsilon_k^T]^T$ and

$$\Lambda_k = C(z_k I - A)^{-1} \frac{z_k}{\sqrt{N}} \tag{10.52}$$

$$G_k = C(z_k I - A)^{-1} B + D \tag{10.53}$$

$$H_k = [C(z_k I - A)^{-1} \mid I_{m \times m}] \tag{10.54}$$

$$\Xi_k = C(z_k I - A)^{-1} \tag{10.55}$$

Furthermore, we have

$$\vec{Y} = \Lambda_D \alpha + G_D \vec{U} + H_D \vec{W} + \Xi_D \vec{F} \tag{10.56}$$

where $\vec{S} = [S_0^T \ \ldots \ S_{N-1}^T]^T$, $S = Y, X, U, W$ and

$$\Lambda_D = [\Lambda_0 \ldots \Lambda_{N-1}]^T \tag{10.57}$$

$$G_D = diag\{G_0 \ldots G_{N-1}\} \tag{10.58}$$

$$H_D = diag\{H_0 \ldots H_{N-1}\} \tag{10.59}$$

$$\Xi_D = diag\{\Xi_0 \ldots \Xi_{N-1}\} \tag{10.60}$$

Since there is a deterministic relationship between $\vec{y}$ and $\vec{Y}$, and $\bar{Y}_l = Y_{N-l}$, where $\bar{Y}_l$ denotes the complex conjugate of Y_l, the probability distribution function (PDF) of the sequence $\{Y_k\}$, $k = 0, \ldots, N-1$ is singular. This means that $L = \lfloor \frac{N}{2} \rfloor$ components of the sequence $\{Y_k\}$, $k = 0, \ldots, N-1$ are necessary, where $\lfloor a \rfloor$ denotes the smallest integer greater than or equal to a (see Appendix A2.1 for more detail). Let

$$l_{F.N}(k, \nu) \triangleq \log p(Y_0, \ldots, Y_{N-1}|k, \nu) \tag{10.61}$$

then we have

$$l_{F.N}(k, \nu) = \log p(Y_0, \ldots, Y_L|k, \nu) \tag{10.62}$$

The real discrete Fourier transform (RDFT) [30] is a map from real sequence $\{y_k\}$, $k = 0, \ldots, N-1$ to real sequence $\{Y_{R.k}\}$, $k = 0, \ldots, N-1$. The following lemma presents this map formally.

Lemma 10.1 *[26] Given the time domain vector $\vec{y} \in \Re^N$, there is a real unitary matrix transformation $\mathcal{R}_N \in C^{N \times N}$ that gives a real-valued frequency domain representation $\vec{Y}_R$:*

$$\vec{Y}_R = \mathcal{R}_N \vec{y} \tag{10.63}$$

where

$$\text{if } N \text{ is even,} \quad \vec{Y}_R = [Y_0, \sqrt{2}\Re(Y_1), \sqrt{2}\Im(Y_1), \ldots, Y_L]^T \tag{10.64}$$

$$\text{if } N \text{ is odd,} \quad \vec{Y}_R = [Y_0, \sqrt{2}\Re(Y_1), \sqrt{2}\Im(Y_1), \ldots, \sqrt{2}\Re(Y_L), \sqrt{2}\Im(Y_L)]^T \tag{10.65}$$

Here we introduce a matrix $\mathcal{T}_N \in C^{N \times N}$:

$$\mathcal{T}_N = \begin{bmatrix} 1 & 0 & 0 & \ldots & 0 & 0 \\ 0 & a & 0 & \ldots & 0 & a \\ 0 & b & 0 & \ldots & 0 & -b \\ \vdots & \vdots & \vdots & \ddots & \vdots & \vdots \end{bmatrix} \tag{10.66}$$

where $a = \frac{\sqrt{2}}{2}$ and $b = \frac{\sqrt{2}}{2j}$. If N is even, then the $(L + 1)$th element of the last row of $\mathcal{T}_N$ is 1. If N is odd, the Lth element and the $(L + 1)$th element are b and $-b$, respectively, where $L = [N/2]$. In fact, the matrix $\mathcal{T}_N$ is non-singular and unitary, thus the relationship between real vector $\vec{y}$ and real vector $\vec{Y}_R$ can be expressed as

$$\vec{Y}_R = \mathcal{T}_N \vec{Y} = \mathcal{T}_N \mathcal{F}_N \vec{y} = \mathcal{R}_N \vec{y}$$

Remark 10.4 *If y_t are vectors with n-dimension and we define a vector as $\vec{y} = [y_0^T \ \cdots \ y_{N-1}^T]^T$, then developing the real discrete Fourier transform to the vector case gives*

$$\vec{Y}_R = \mathcal{T}_N^n \vec{Y} = \mathcal{T}_N^n \mathcal{F}_N^n \vec{y} = \mathcal{R}_N^n \vec{y} \tag{10.67}$$

where $\mathcal{T}_N^n = \mathcal{T}_N \otimes I_n$ and $\vec{Y}_R = [Y_0^T, \sqrt{2}\Re(Y_1)^T, \sqrt{2}\Im(Y_1)^T, \ldots]^T$.

In the rest of this chapter, we prefer to use symbols $\mathcal{F}$, $\mathcal{T}$ and $\mathcal{R}$ instead of $\mathcal{F}_N^n$, $\mathcal{T}_N^n$ and $\mathcal{R}_N^n$ (or $\mathcal{F}_N$, $\mathcal{T}_N$ and $\mathcal{R}_N$) when the proper dimensions of these matrices are clear according to the context.

Based on the relationship mentioned above, the following lemma clarifies the equivalence between the time and frequency domain maximum likelihood.

Lemma 10.2 *[26] Given a dynamic system model and a set of output measurements, the estimates obtained by maximizing the likelihood function are the same irrespective of the representation of the data, either in the time or the frequency domain. That is*

$$\arg_\beta \max p_{\vec{y}}(\vec{y}|\beta) = \arg_\beta \max p_{\vec{Y}_R}(\vec{Y}_R|\beta) \tag{10.68}$$

where β is a vector that contains the parameters to be estimated and equality holds with probability 1.

According to Lemma 10.2, the GLR test method of Equations (10.36)–(10.37) can be expressed as

$$\hat{v}(k) = \arg_v \max \log \frac{p(\vec{Y}_R|k, v)}{p(\vec{Y}_R|k = N)} \tag{10.69}$$

$$\hat{k} = \arg_k \max \log \frac{p(\vec{Y}_R|k, \hat{v}(k))}{p(\vec{Y}_R|N)} \tag{10.70}$$

which mean the same estimates are obtained when maximizing the corresponding likelihood ratios irrespective of whether a time domain or a frequency domain representation is used for the data.

In the following discussion, we present the likelihood ratio detection of abrupt changes for linear systems from measured outputs and inputs in the frequency domain.

10.2.4   Likelihood of the hypothesis of no abrupt change

According to linear systems with a state jump at time instant k (described by Equations (10.27)–(10.28)), we have

$$x_t = A^t x_0 + \sum_{i=0}^{t-1} A^{t-i-1}(Bu_i + \omega_i) + \sigma_{t-k-1} A^{t-k-1} v \tag{10.71}$$

where σ_t denotes the step function, i.e, $\sigma_t = 1$ when $t \geq 0$, and $\sigma_t = 0$ when $t < 0$. Since the initial value of the state and the noise are variables with Gaussian distribution, the state x_t is also variable with Gaussian distribution:

$$x_t \sim \mathcal{N}\left(\mu_{x_t}; D_{x_t}\right) \tag{10.72}$$

where

$$\mu_{x_t} = A^t \mu_{x_0} + \sum_{i=0}^{t-1} A^{t-i-1} Bu_i + \sigma_{t-k-1} A^{t-k-1} v,$$

and

$$D_{x_t} = A^t D_{x_0}(A^t)^T + \sum_{i=0}^{t-1} A^{t-i-1} Q(\sum_{i=0}^{t-1} A^{t-i-1})^T$$

Based on hypothesis H_0, i.e., the abrupt change does not occur during the time interval from instant 0 to instant $N - 1$, the difference between the initial value and the end value is

$$\alpha = (I - A^N)x_0 - \sum_{i=0}^{N-1} A^{N-i-1}(Bu_i + \omega_i) \tag{10.73}$$

We present the logarithmic likelihood computation for the system of Equations (10.27)–(10.28) from the measured outputs and inputs in the frequency domain as follows.

Lemma 10.3   *Consider the linear system with latent state change described by Equations (10.27)–(10.28). If the real discrete Fourier transform (defined by Equation (10.40) and Equation (10.63)) of the measured outputs and inputs are $\vec{Y}_R$ and $\vec{U}_R$, respectively, then $\vec{Y}_R$ with Gaussian distribution under hypothesis H_0 (defined by Equation (10.32)) is:*

$$\vec{Y}_R \sim \mathcal{N}(\mu_{\vec{Y}_R}, D_{\vec{Y}_R}) \tag{10.74}$$

where

$$\mu_{\vec{Y}_R} = T \Lambda_D (I - A^N) \mu_{x_0} - T \Lambda_D \sum_{i=0}^{N-1} A^{N-i-1} B u_i + T G_D \vec{U} \tag{10.75}$$

$$D_{\vec{Y}_R} = T \Lambda_D D_\alpha \Lambda_D^H T^H + T H_D \mathcal{F} \vec{P} \mathcal{F}^H H_D^H T^H \tag{10.76}$$

$$D_\alpha = (I - A^N) D_{x_0} (I - A^N)^T + (\sum_{i=0}^{N-1} A^{N-i-1}) Q (\sum_{i=0}^{N-1} A^{N-i-1})^T \tag{10.77}$$

and $\vec{P} = diag\{P, \ldots, P\} \in \Re^{(n+m)N \times (n+m)N}$. Furthermore, the logarithmic likelihood under hypothesis H_0 is given by

$$L(N, v) = -\log p(\vec{Y}_R | H_0) \tag{10.78}$$

$$= \frac{1}{2} \left(N \log(2\pi) + \log \det D_{\vec{Y}_R} + (\vec{Y}_R - \mu_{\vec{Y}_R})^T D_{\vec{Y}_R}^{-1} (\vec{Y}_R - \mu_{\vec{Y}_R}) \right) \tag{10.79}$$

Proof. Based on hypothesis H_0, no abrupt change occurs, thus we can rewrite Equation (10.56) as

$$\vec{Y} = \Lambda_D \alpha + G_D \vec{U} + H_D \vec{W}$$

where α is given by Equation (10.73). Pre-multiplying T gives

$$\vec{Y}_R = T \Lambda_D \alpha + T G_D \vec{U} + T H_D \vec{W}$$

The mathematical expectation and covariance of α in case H_0 are

$$\mu_\alpha = (I - A^N) \mu_{x_0} - \sum_{i=0}^{N-1} A^{N-i-1} B u_i \tag{10.80}$$

$$D_\alpha = (I - A^N) D_{x_0} (I - A^N)^T + (\sum_{i=0}^{N-1} A^{N-i-1}) Q (\sum_{i=0}^{N-1} A^{N-i-1})^T \tag{10.81}$$

thus the mathematical expectation and covariance of $\vec{Y}_R$ are

$$\mu_{\vec{Y}_R} = T \Lambda_D \mu_\alpha + T G_D \vec{U} \tag{10.82}$$

$$D_{\vec{Y}_R} = T \Lambda_D D_\alpha \Lambda_D^H T^H + T H_D \mathcal{F} \vec{P} \mathcal{F}^H H_D^H T^H \tag{10.83}$$

where $\vec{P} \triangleq E\{\vec{n}\vec{n}^T\} = diag\{P, \dots, P\} \in \Re^{(n+m)N \times (n+m)N}$. According to Equation (10.62), we have

$$L(N, v) = -\log p(\vec{Y}_R | H_0) \tag{10.84}$$

$$= \frac{1}{2}\left(N\log(2\pi) + \log\det D_{\vec{Y}_R} + (\vec{Y}_R - \mu_{\vec{Y}_R})^T D_{\vec{Y}_R}^{-1}(\vec{Y}_R - \mu_{\vec{Y}_R})\right) \tag{10.85}$$

∎

The assumption of this chapter is that only the frequency samples are available for measurement; samples u_i, where $i = 0, \dots, N-1$, can be obtained from the frequency samples using inverse Fourier transform.

10.2.5 Effect of an abrupt change

In this subsection, we present a maximum likelihood estimation of the change magnitude v under hypothesis H_1, i.e., an abrupt change of magnitude v occurs at time instant k. Under this hypothesis, we have

$$\vec{Y}_R = T\Lambda_D\alpha + TG_D\vec{U} + TH_D\vec{W} + T\Xi_D\vec{F}$$

where two input signals, the extra term α and the state change signal $\vec{F}$, are relative to the abrupt state change (k, v).

In the rest of this section, we propose a transform to remove the input signal $\vec{F}$. First, we define that $N_1 \triangleq N - k - 1$ and $N_2 \triangleq k + 1$, so that $N = N_1 + N_2$. The data set $\{s_t\}$, $s = u, y, t = 0, 1, \dots, N-1$ is also separated as two vectors $\vec{s}^{k-} = [s_0^T \dots s_k^T]^T$ and $\vec{s}^{k+} = [s_{k+1}^T \dots s_{N-1}^T]^T$. Hence, from time instant $k+1$ to time instant $N-1$ we have N_1 dependent measurements $\{y_t\}, t = k+1, k+2, \dots, N-1$. We transform the data of Equations (10.27)–(10.28) from the time to the frequency domain according to the following finite DFT:

$$\vec{S}^{k+} = \mathcal{F}_{N_1}^n \vec{s}^{k+}; \tag{10.86}$$

$$s = y, u, x;$$

$$S = Y, U, X$$

The above transform means that only the last N_1 measured outputs and inputs are under consideration. If we define

$$\mathcal{M}_{N_1} \triangleq [0_{N_1 \times (N-N_1)} | I_{N_1 \times N_1}] \tag{10.87}$$

$$\mathcal{M}_{N_1}^n \triangleq \mathcal{M}_{N_1} \otimes I_{n \times n} \tag{10.88}$$

where $0_{N_1 \times (N-N_1)}$ denotes an $N_1 \times (N - N_1)$ dimensional zero matrix, then we have a direct transform from frequency domain data $\vec{S}$ to frequency domain data $\vec{S}^{k+}$:

$$\vec{S}^{k+} = \mathcal{F}_{N_1}^n \mathcal{M}_{N_1}^n (\mathcal{F}_{N_1}^n)^H \vec{S} \tag{10.89}$$

Applying the transform operation of Equation (10.89) to the frequency system of Equations (10.49) and (10.50) gives

$$z_t X_t^{k+} = A X_t^{k+} + B U_t^{k+} + \Omega_t^{k+} + \alpha_k \frac{z_t}{\sqrt{N_1}}; \tag{10.90}$$

$$Y_t^{k+} = C X_t^{k+} + D U_t^{k+} + \Upsilon_t^{k+} \tag{10.91}$$

where the extra term is defined as

$$\alpha_k \triangleq x_{k+1} - x_N$$

$$= (I - A^{N_1})x_{k+1} + \sum_{i=k+1}^{N-1} A^{N-i-1}(Bu_i + \omega_i) \tag{10.92}$$

$$= (A^{k+1} - A^N)x_0 + \sum_{i=0}^{k} A^{k-i}(Bu_i + \omega_i)$$

$$- \sum_{i=0}^{N-1} A^{N-i-1}(Bu_i + \omega_i) + (I - A^{N_1})v \tag{10.93}$$

Note that the term α_k is with Gaussian PDF; we write $\alpha_k \sim \mathcal{N}(\mu_{\alpha,k}; D_{\alpha,k})$, where

$$\mu_{\alpha,k} \triangleq E\{\alpha_k\} \tag{10.94}$$

$$= (A^{k+1} - A^N)\mu_{x_0} + \sum_{i=0}^{k} A^{k-i}Bu_i - \sum_{i=0}^{N-1} A^{N-i-1}Bu_i + (I - A^{N_1})v$$

$$D_{\alpha,k} \triangleq E\{[\alpha_k - \mu_{\alpha,k}][\alpha_k - \mu_{\alpha,k}]^T\} \tag{10.95}$$

$$= (A^{k+1} - A^N)D_{x_0}(A^{k+1} - A^N)^T + (\sum_{i=0}^{k} A^{k-i})Q(\sum_{i=0}^{k} A^{k-i})^T$$

$$+ (\sum_{i=0}^{N-1} A^{N-i-1})Q(\sum_{i=0}^{N-1} A^{N-i-1})^T$$

Remark 10.5 *Under the hypothesis that a state change with unknown magnitude occurs at time instant k, the frequency domain system of Equations (10.90) and (10.91), whose initial time instant is $k + 1$, is without the state change signal inputs F_t^{k+}. This means that the state change is expressed not as an initial state value disturbance after the occurrence but as an input signal. Hence, according to Equation (10.93), the extra term α_k contains the initial state value disturbance $(I - A^{N_1})v$.*

The finite DFT representations of Equations (10.90) and (10.91) can also be written in terms of transfer functions as

$$\vec{Y}^{k+} = \Lambda_D^{k+}\alpha_k + G_D^{k+}\vec{U}^{k+} + H_D^{k+}\vec{W}^{k+} \tag{10.96}$$

where

$$\Lambda_D^{k+} = [\Lambda_0^H \ldots \Lambda_{N_1-1}^H]^H \tag{10.97}$$

$$G_D^{k+} = diag\{G_0 \ldots G_{N_1-1}\} \tag{10.98}$$

$$H_D^{k+} = diag\{H_0 \ldots H_{N_1-1}\} \tag{10.99}$$

$$\Xi_D^{k+} = diag\{\Xi_0 \ldots \Xi_{N_1-1}\} \tag{10.100}$$

Pre-multiplying $\mathcal{T}$ on both sides of Equation (10.96) gives

$$\vec{Y}_R^{k+} = \mathcal{T}\Lambda_D^{k+}\alpha_k + \mathcal{T}G_D^{k+}\vec{U}^{k+} + \mathcal{T}H_D^{k+}\vec{W}^{k+} \tag{10.101}$$

Before proposing the maximum likelihood estimation of the change magnitude ν under hypothesis $H_1(k, \nu)$, we introduce a useful lemma (see Lemma A.6) for solving the weighted least square (WLS) problem. The key problem in this chapter is to estimate the change magnitude ν at a hypothetical time instant based on the frequency domain data. The following lemma represents that this problem can be transformed to a WLS problem.

Lemma 10.4 *Consider the linear system with unknown initial state disturbance ν described by Equations (10.90) and (10.91), whose initial time instant is $k + 1$. If the real discrete Fourier transform (defined by Equation (10.86) and Equation (10.63)) of the measured outputs and inputs are $\vec{Y}_{\Re}^{k+}$ and $\vec{U}_{\Re}^{k+}$, respectively, then the maximum likelihood estimate of the initial state disturbance magnitude*

$$\hat{\nu}(k) = X\vec{Y}_S^{k+} \tag{10.102}$$

is given by solving the following equations:

$$A^{k+}XA^{k+} = A^{k+} \tag{10.103}$$

$$(D_{\vec{Y}_R^{k+}}^{-1}A^{k+}X)^H = D_{\vec{Y}_R^{k+}}^{-1}A^{k+}X \tag{10.104}$$

where

$$D_{\vec{Y}_R^{k+}} = \mathcal{T}\Lambda_D^{k+}D_{\alpha_k}(\mathcal{T}\Lambda_D^{k+})^H + \mathcal{T}H_D^{k+}\mathcal{F}\vec{P}^{k+}(\mathcal{T}H_D^{k+}\mathcal{F})^H \tag{10.105}$$

$$\vec{Y}_S^{k+} = \vec{Y}_R^{k+} - \mathcal{T}G_D^{k+}\vec{U}^{k+} \tag{10.106}$$

$$-\mathcal{T}\Lambda_D^{k+}\left[(A^{k+1} - A^N)\mu_{x_0} + \sum_{i=0}^{k}A^{k-i}Bu_i - \sum_{i=0}^{N-1}A^{N-i-1}Bu_i\right] \tag{10.107}$$

$$A^{k+} = \mathcal{T}\Lambda_D^{k+}(I - A^{N_1}) \tag{10.108}$$

and the definitions of transfer functions Λ_D^{k+}, H_D^{k+}, G_D^{k+} can be found in Equation (10.101).

Proof. Rewrite Equation (10.101) as

$$\vec{Y}_R^{k+} = \mathcal{T}\Lambda_D^{k+}\alpha_k + \mathcal{T}G_D^{k+}\vec{U}^{k+} + \mathcal{T}H_D^{k+}\mathcal{F}\vec{n}^{k+} \tag{10.109}$$

The real vector $\vec{Y}_R^{k+}$ is with Gaussian distribution since the vectors α_k and $\vec{n}^{k+}$ are with Gaussian distributions, and $\vec{U}^{k+}$ is a deterministic vector, i.e.,

$$\vec{Y}_R^{k+} \sim \mathcal{N}\left(\mu_{\vec{Y}_R^{k+}}, D_{\vec{Y}_R^{k+}}\right) \tag{10.110}$$

where

$$\mu_{\vec{Y}_R^{k+}} = \mathcal{T}\Lambda_D^{k+}\mu_{\alpha_k} + \mathcal{T}G_D^{k+}\vec{U}^{k+} \tag{10.111}$$

$$D_{\vec{Y}_R^{k+}} = \mathcal{T}\Lambda_D^{k+}D_{\alpha_k}(\mathcal{T}\Lambda_D^{k+})^H + \mathcal{T}H_D^{k+}\mathcal{F}\vec{P}^{k+}(\mathcal{T}H_D^{k+}\mathcal{F})^H \tag{10.112}$$

According to Equation (10.62), we have

$$L(k, v) = -\log p(\vec{Y}_R^{k+}|\vec{Y}_R^{k-}, H_1) \tag{10.113}$$

$$= -\log p(\vec{Y}_R^{k+}|\vec{Y}_R^{k-}, k, v) \tag{10.114}$$

$$= \frac{1}{2}\left(N_1\log(2\pi) + \log\det D_{\vec{Y}_R^{k+}}\right. \tag{10.115}$$

$$\left. +(\vec{Y}_R^{k+} - \mu_{\vec{Y}_R^{k+}})^T D_{\vec{Y}_R^{k+}}^{-1}(\vec{Y}_R^{k+} - \mu_{\vec{Y}_R^{k+}})\right) \tag{10.116}$$

Hence, the maximum likelihood estimation of the disturbance magnitude v is proposed as follows:

$$\hat{v}(k) = \arg_v \min L(k, v) \tag{10.117}$$

$$= \arg_v \min(\vec{Y}_R^{k+} - \mu_{\vec{Y}_R^{k+}})^T D_{\vec{Y}_R^{k+}}^{-1}(\vec{Y}_R^{k+} - \mu_{\vec{Y}_R^{k+}}) \tag{10.118}$$

$$= \arg_v \min(\vec{Y}_S^{k+} - A^{k+}v)^T D_{\vec{Y}_R^{k+}}^{-1}(\vec{Y}_S^{k+} - A^{k+}v) \tag{10.119}$$

where

$$\vec{Y}_S^{k+} = \vec{Y}_R^{k+} - \mathcal{T}G_D^{k+}\vec{U}^{k+}$$

$$-\mathcal{T}\Lambda_D^{k+}\left[(A^{k+1} - A^N)\mu_{x_0} + \sum_{i=0}^{k}A^{k-i}Bu_i - \sum_{i=0}^{N-1}A^{N-i-1}Bu_i\right] \tag{10.120}$$

$$A^{k+} = \mathcal{T}\Lambda_D^{k+}(I - A^{N_1}) \tag{10.121}$$

According to Lemma A.6, the maximum likelihood test of v satisfies Equations (10.103)–(10.104). ∎

In practical engineering, the computational burden of solving Equation (10.103) and Equation (10.104) is very heavy if N_1 is large. We now derive a recursive implementation for estimating $v(k)$ under the hypothesis that the state change occurs at time instant k.

Algorithm 10.2   *The recursive form of estimating $v(k)$ under the hypothesis that the state change occurs at time instant k is given as follows:*

$$\hat{v}(k)_i = \hat{v}(k)_{i-1} + P(k)_{i-1}\varphi_i^H[\varphi_i P(k)_{i-1}\varphi_i^H + D_i]^{-1}(z_i - \varphi_i\hat{v}(k)_{i-1}) \quad (10.122)$$

$$P(k)_i = P(k)_{i-1} + P(k)_{i-1}\varphi_i^H[\varphi_i P(k)_{i-1}\varphi_i^H + D_i]^{-1}\varphi_i P(k)_{i-1} \quad (10.123)$$

where

$$z_i = \vec{Y}_i^{k+} - G_i\vec{U}_i^{k+} - \Lambda_i\left[(A^{k+1} - A^N)\mu_{x_0} + \sum_{i=0}^{k} A^{k-i}Bu_i - \sum_{i=0}^{N-1} A^{N-i-1}Bu_i\right]$$

$$\varphi_i = \Lambda_i(I - A^{N_1})$$

and the matrix $P(k)_i$ denotes the covariance of the estimation error.

Proof. According to Lemma A.7(i), we have $\mathcal{F}\vec{P}^{k+}\mathcal{F}^H = \vec{P}^{k+}$, hence

$$D_{\vec{Y}_R^{k+}} = T\Lambda_D^{k+}D_{\alpha_k}(T\Lambda_D^{k+})^H + TH_D^{k+}\vec{P}^{k+}(TH_D^{k+})^H \quad (10.124)$$

$$= TD_{\vec{y}^{k+}}T^H \quad (10.125)$$

where

$$D_{\vec{y}^{k+}} = diag\{D_0, D_1, \ldots, D_{N_1-1}\}, \quad D_i = \Lambda_i\alpha_k\Lambda_i^H + H_iPH_i^H, \quad k = 0, \ldots, N_1 - 1$$

The inverse of $D_{\vec{y}^{k+}}$ is obtained as follows:

$$D_{\vec{Y}_R^{k+}}^{-1} = (TD_{\vec{y}^{k+}}T^H)^{-1} = TD_{\vec{y}^{k+}}^{-1}T^H \quad (10.126)$$

where

$$D_{\vec{y}^{k+}}^{-1} = diag\{D_0^{-1}, D_1^{-1}, \ldots, D_{N_1-1}^{-1}\}$$

According to Lemma A.7(iii), Equation (10.126) is not always diagonal. However, we can rewrite the result of Lemma 10.4 as

$$\hat{v}(k) = \arg_v \min(\vec{Y}_S^{k+} - A^{k+}v)^T D_{\vec{Y}_R^{k+}}^{-1}(\vec{Y}_S^{k+} - A^{k+}v)$$

$$= \arg_v \min(T^H\vec{Y}_S^{k+} - T^HA^{k+}v)^T D_{\vec{y}^{k+}}^{-1}(T^H\vec{Y}_S^{k+} - T^HA^{k+}v)$$

$$= \arg_v \min \sum_{i=0}^{N_1-1}(z_i - \varphi_i v)^H D_i^{-1}(z_i - \varphi_i v) \quad (10.127)$$

where

$$z_i = \vec{Y}_i^{k+} - G_i \vec{U}_i^{k+} - \Lambda_i \left[(A^{k+1} - A^N)\mu_{x_0} + \sum_{i=0}^{k} A^{k-i} B u_i - \sum_{i=0}^{N-1} A^{N-i-1} B u_i \right]$$

$$\varphi_i = \Lambda_i (I - A^{N_1})$$

Now, let $P(k)_i$ denote the covariance matrix of the estimation error. Then we have the following recursive form by minimizing the differentiation of Equation (10.127):

$$\hat{v}(k)_i = \hat{v}(k)_{i-1} + P(k)_{i-1}\varphi_i^H [\varphi_i P(k)_{i-1}\varphi_i^H + D_i]^{-1}(z_i - \varphi_i \hat{v}(k)_{i-1})$$

$$P(k)_i = P(k)_{i-1} + P(k)_{i-1}\varphi_i^H [\varphi_i P(k)_{i-1}\varphi_i^H + D_i]^{-1}\varphi_i P(k)_{i-1}$$

∎

Remark 10.6 *Algorithm 10.2 gives a recursive implementation of estimating $v(k)$. In this algorithm, $\hat{v}(k)_i$ denotes the estimate of $v(k)$ based on the sampled frequency points from 0 to i. If the measured outputs and inputs have considerable small values at high-frequency sampled points, a reduced set of data is used to obtain a satisfactory estimate without a heavy computational burden.*

Remark 10.7 *Here, we give a geometric explanation for Lemma 10.4. Let*

$$\vec{Y}_P^{k+} \triangleq \mathcal{T}\Lambda_D^{k+} \left[(A^{k+1} - A^N)\mu_{x_0} + \sum_{i=0}^{k} A^{k-i} B u_i - \sum_{i=0}^{N-1} A^{N-i-1} B u_i \right] + \mathcal{T} G_D^{k+} \vec{U}^{k+}$$

then $\vec{Y}_P^{k+}$ denotes the real Fourier transform of the predictive outputs from $k+1$ to $N-1$, and $\vec{Y}_S^{k+} = \vec{Y}_R^{k+} - \vec{Y}_P^{k+}$, defined in Equation (10.120), denotes the real Fourier transform of the prediction errors, or innovations. According to the linear projection theory, the minimization problem of Equation (10.118) is to find an estimated quantity $\hat{v}(k)$ on a plane, which is spanned by the innovations $\vec{Y}_S^{k+}$.

Once we have the estimates of magnitude $v(k)$ under the hypotheses $H_1(k, v)$, $k = 0, \ldots, N-1$, the next step is to estimate the occurrence time instant k by maximizing the likelihood under the hypothesis $H_1(k, \hat{v}(k))$.

Lemma 10.5 *Consider the MIMO time-invariant linear system with latent state change described by Equations (10.90) and (10.91). If the real discrete Fourier transform (defined by Equations (10.86) and (10.63)) of the measured outputs and inputs are $\vec{Y}_{\Re}^{k+}$ and $\vec{U}_{\Re}^{k+}$, respectively, and the state change magnitude is v, then the maximum likelihood estimate of the change occurrence time instant k is given by:*

$$\hat{k} = \arg_k \min \left(-k \log(2\pi) + \log \det D_{\vec{Y}_R^{k+}} \right.$$

$$\left. + (\vec{Y}_R^{k+} - \mu_{\vec{Y}_R^{k+}})^T D_{\vec{Y}_R^{k+}}^{-1} (\vec{Y}_R^{k+} - \mu_{\vec{Y}_R^{k+}}) \right)$$

where

$$\mu_{\vec{Y}_R^{k+}} = \mathcal{T} \Lambda_D^{k+} \mu_{\alpha_k} + \mathcal{T} G_D^{k+} \vec{U}^{k+}$$

$$D_{\vec{Y}_R^{k+}} = \mathcal{T} \Lambda_D^{k+} D_{\alpha_k} (\mathcal{T} \Lambda_D^{k+})^H + \mathcal{T} H_D^{k+} \mathcal{F} \vec{P}^{k+} (\mathcal{T} H_D^{k+} \mathcal{F})^H$$

and the definitions of transfer functions Λ_D^{k+}, H_D^{k+}, G_D^{k+}, μ_{α_k} *and* D_{α_k} *can be found in Equation (10.101), Equation (10.94) and Equation (10.95).*

Proof. Rewrite Equation (10.101) as

$$\vec{Y}_R^{k+} = \mathcal{T} \Lambda_D^{k+} \alpha_k + \mathcal{T} G_D^{k+} \vec{U}^{k+} + \mathcal{T} H_D^{k+} \mathcal{F} \vec{n}^{k+} \tag{10.128}$$

The real vector $\vec{Y}_R^{k+}$ is with Gaussian distribution since the vectors α_k and $\vec{n}^{k+}$ are with Gaussian distributions, and $\vec{U}^{k+}$ is a deterministic vector, i.e.,

$$\vec{Y}_R^{k+} \sim \mathcal{N}\left(\mu_{\vec{Y}_R^{k+}}, D_{\vec{Y}_R^{k+}}\right) \tag{10.129}$$

where

$$\mu_{\vec{Y}_R^{k+}} = \mathcal{T} \Lambda_D^{k+} \mu_{\alpha_k} + \mathcal{T} G_D^{k+} \vec{U}^{k+}$$

$$D_{\vec{Y}_R^{k+}} = \mathcal{T} \Lambda_D^{k+} D_{\alpha_k} (\mathcal{T} \Lambda_D^{k+})^H + \mathcal{T} H_D^{k+} \mathcal{F} \vec{P}^{k+} (\mathcal{T} H_D^{k+} \mathcal{F})^H$$

According to Equation (10.62), we have

$$\begin{aligned}
L(k, v) &= -\log p(\vec{Y}_R^{k+} | \vec{Y}_R^{k-}, H_1) \\
&= -\log p(\vec{Y}_R^{k+} | \vec{Y}_R^{k-}, k, v) \\
&= \frac{1}{2}\left(N_1 \log(2\pi) + \log \det D_{\vec{Y}_R^{k+}}\right. \\
&\quad \left. + (\vec{Y}_R^{k+} - \mu_{\vec{Y}_R^{k+}})^T D_{\vec{Y}_R^{k+}}^{-1} (\vec{Y}_R^{k+} - \mu_{\vec{Y}_R^{k+}})\right)
\end{aligned}$$

Hence, the maximum likelihood estimation of change occurrence time instant k based on the hypothesis that the change magnitude is known is proposed as follows

$$\begin{aligned}
\hat{k} &= \arg_k \min L(k, v) \\
&= \arg_k \min \frac{1}{2}\left(N_1 \log(2\pi) + \log \det D_{\vec{Y}_R^{k+}}\right. \\
&\quad \left. + (\vec{Y}_R^{k+} - \mu_{\vec{Y}_R^{k+}})^T D_{\vec{Y}_R^{k+}}^{-1} (\vec{Y}_R^{k+} - \mu_{\vec{Y}_R^{k+}})\right) \\
&= \arg_k \min \left(-k \log(2\pi) + \log \det D_{\vec{Y}_R^{k+}}\right. \\
&\quad \left. + (\vec{Y}_R^{k+} - \mu_{\vec{Y}_R^{k+}})^T D_{\vec{Y}_R^{k+}}^{-1} (\vec{Y}_R^{k+} - \mu_{\vec{Y}_R^{k+}})\right)
\end{aligned}$$

Based on the analysis and the lemmas mentioned above, we formally propose the maximum likelihood ratio estimation of state change for MIMO linear systems based on the frequency domain data.

Theorem 10.1   *Consider the time-invariant MIMO linear system with latent state change described by Equations (10.27)–(10.28), where the latent state change occurs at time instant k with magnitude v. If the real discrete Fourier transform (defined by Equation (10.40) and Equation (10.63)) of the measured outputs and inputs from time instant 0 to time instant $N - 1$ are, respectively, $\vec{Y}_R$ and $\vec{U}_R$, then the frequency maximum likelihood ratio estimation of pair (k, v), denoted by $(\hat{k}, v(\hat{k}))$, is given by*

$$A^{k+} X A^{k+} = A^{k+} \tag{10.130}$$

$$(D_{\vec{Y}_R^{k+}}^{-1} A^{k+} X)^H = D_{\vec{Y}_R^{k+}}^{-1} A^{k+} X \tag{10.131}$$

$$\hat{v}(k) = X \vec{Y}_S^{k+} \tag{10.132}$$

$$\hat{k} = \arg_k \min \left(-k \log(2\pi) + \log \det D_{\vec{Y}_R^{k+}} \right.$$
$$\left. + (\vec{Y}_R^{k+} - \mu_{\vec{Y}^{k+}})^T D_{\vec{Y}_R^{k+}}^{-1} (\vec{Y}_R^{k+} - \mu_{\vec{Y}^{k+}}) \right) \tag{10.133}$$

where

$$\mu_{\vec{Y}_R^{k+}} = \mathcal{T} \Lambda_D^{k+} \mu_{\alpha_k} + \mathcal{T} G_D^{k+} \vec{U}^{k+}$$

$$D_{\vec{Y}_R^{k+}} = \mathcal{T} \Lambda_D^{k+} D_{\alpha_k} (\mathcal{T} \Lambda_D^{k+})^H + \mathcal{T} H_D^{k+} \mathcal{F} \vec{P}^{k+} (\mathcal{T} H_D^{k+} \mathcal{F})^H$$

$$\mu_{\alpha_k} = (A^{k+1} - A^N) \mu_{x_0} + \sum_{i=0}^{k} A^{k-i} B u_i - \sum_{i=0}^{N-1} A^{N-i-1} B u_i + (I - A^{N_1}) \hat{v}(k)$$

and the definitions of transfer functions Λ_D^{k+}, H_D^{k+}, G_D^{k+}, D_{α_k} can be found in Equation (10.101) and Equation (10.95).

Proof. According to the conditional probability theory, we have

$$p(\vec{y}|\beta) = p(\vec{y}^{k-}|\beta) p(\vec{y}^{k+}|\vec{y}^{k-}, \beta) \tag{10.134}$$

Thus, rewrite Equation (10.69) as follows:

$$\hat{v}(k) = \arg_v \max \left(\log p(\vec{Y}_R|k, v) - \log p(\vec{Y}_R|k = N) \right) \tag{10.135}$$

$$= \arg_v \max \left(\log p(\vec{Y}_R^{k-}|k, v) + \log p(\vec{Y}_R^{k+}|\vec{Y}_R^{k-}, k, v) \right.$$
$$\left. - \log p(\vec{Y}_{\Re}^{k-}|k = N) - \log p(\vec{Y}_{\Re}^{k+}|\vec{Y}_R^{k-}, k = N) \right)$$

Note that $p(\vec{Y}_R^{k-}|k, v) = p(\vec{Y}_{\Re}^{k-}|k = N)$ and $p(\vec{Y}_{\Re}^{k+}|\vec{Y}_R^{k-}, k = N)$ is not a function of v, and thus Equation (10.135) can be rewritten as

$$\hat{v}(k) = \arg_v \max \log p(\vec{Y}_R^{k+}|\vec{Y}_R^{k-}, k, v)$$

According to Lemma 10.4, the above problem is equivalent to estimating an unknown initial state disturbance, and the estimates can be found by solving Equations (10.130) and (10.131). The next step is to estimate the occurrence time instant k. From Equation (10.70) we have

$$\hat{k} = \arg_k \max \log p(\vec{Y}_{\Re}^{k+}|\vec{Y}_R^{k-}, k, \hat{v}(k))$$

and substituting $\hat{v}(k)$ for v in Lemma 10.5 gives Equation (10.133).　　　　　■

10.2.6　Simulation results

In this section we present a numerical example to highlight the estimates obtained by the proposed method. We consider a simple model expressed in state-space form described by Equations (10.27) and (10.28):

$$A = \begin{bmatrix} 0.95 & 0 \\ 0.01 & 0.85 \end{bmatrix}, \quad B = \begin{bmatrix} 1 & 1 \\ 1 & 1 \end{bmatrix}, \quad C = \begin{bmatrix} 1 & 0 \\ 0 & 1 \end{bmatrix}, \quad D = \begin{bmatrix} 0 & 0 \\ 0 & 0 \end{bmatrix},$$

and

$$Q = \begin{bmatrix} 0.1 & 0 \\ 0 & 0.1 \end{bmatrix}, \quad R = \begin{bmatrix} 0.2 & 0 \\ 0 & 0.2 \end{bmatrix}, \quad S = \begin{bmatrix} 0 & 0 \\ 0 & 0 \end{bmatrix},$$

where the input $u_t \in \Re^2$ is a zero-mean Gaussian noise sequence with variance

$$D_u = \begin{bmatrix} 0.3 & 0 \\ 0 & 0.3 \end{bmatrix}.$$

The system was simulated over N data points, where $N = 31$. The abrupt state change occurred at $k = 16$ with magnitude $v = [-8 \; 13]^T$. The states are initialized as follows

$$x_0 \sim \mathcal{N}\left(\begin{bmatrix} 7.4 \\ -9.3 \end{bmatrix}, \begin{bmatrix} 0.1 & 0 \\ 0 & 0.1 \end{bmatrix} \right)$$

Figure 10.10 shows the RDFT of measured inputs and outputs, i.e. $\vec{U}_R$ and $\vec{Y}_R$, respectively. The least square estimation (given by Equation (10.102)) in this simulation is solved by applying the *lscov* command of MATLAB®. Figure 10.11 shows the Monte Carlo simulation

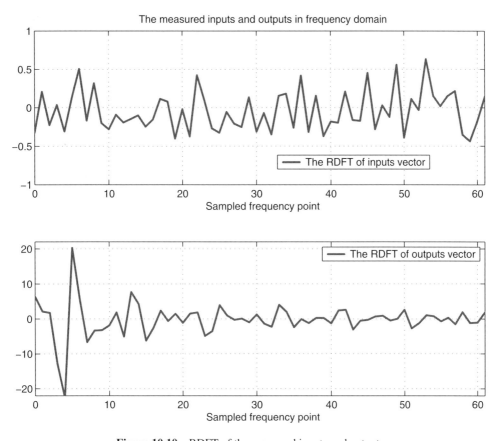

Figure 10.10 RDFT of the measured inputs and outputs

results with 1000 repeats, where $\hat{k} = 16$ and $\hat{v}(\hat{k}) = [-9.1378 \ 12.6482]^T$. Table 10.1 shows the estimation errors of the Monte Carlo simulations, where $[e_1 \ e_2]^T \triangleq v^T - \hat{v}^T(\hat{k})$.

10.3 Electromechanical Positioning System

In this section, we examine an observer-based fault detection identification and accommodation (FDIA) system for an electromechanical actuator. The setup consists of a pilot-scale plant equipped with three sensors and three separate linear observers to detect and identify faults and provide a reconstructed measure of the output signal.

10.3.1 Introduction

Engineers strive to produce high-reliability systems which require little or no attention over their design life. However, the adoption of a fault detection isolation and accommodation

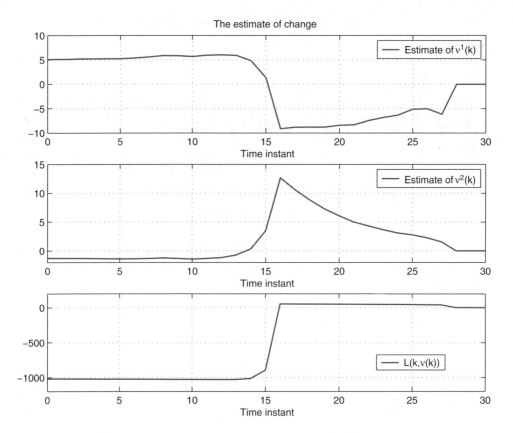

Figure 10.11 Estimates of the change magnitude at the hypothetical time instants and likelihood of occurrence time

(FDIA) policy is usually considered a necessary safeguard against the vagaries of machining quality, material integrity, the environment, etc. This is particularly true in aerospace environments, where safety is of paramount importance. Improved aircraft maintainability and availability are two of the benefits promised by smart electromechanical actuators. However, the question of the algorithms to be incorporated in smart actuators has not yet been fully answered and in many cases there remains some doubt as to their effectiveness and reliability. Generally, in a system without FDIA, a sensor fault would cause undesirable operation and

Table 10.1 Monte Carlo simulation results

Repeat Number	10	50	100	500	1000	5000	10 000
$|e_1|$	2.3845	1.9137	1.0702	1.3579	1.1378	0.6839	0.4745
$|e_2|$	1.5730	1.2079	1.0189	0.3776	0.3518	0.1510	0.0110
$|e_1|+|e_2|$	3.9575	3.1216	2.0891	1.7355	1.4896	0.8349	0.4855

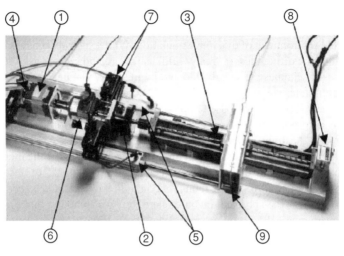

1. Brushless DC motor   6. Torque transducer
2. Gearbox   7. Pneumatic brakes
3. Ballscrew & carriage   8. Position sensor
4. Encoder (Velocity)   9. Load cell
5. Pneumatic actuators

Figure 10.12   Electromechanical test bed

could ultimately result in either a shut-down or catastrophic failure of the actuator (and possibly of the system in which the actuator is but one part). Hence, the FDIA design goal for this unit is to provide a means by which the actuator can continue to operate (ideally, as close to its specification as possible) even after a sensor fault or failure.

On a typical electromechanical system, fault detection is often carried out by means of monitoring measurable output signals (e.g., current) against pre-determined limit values. Accommodation of the detected fault can then be provided by a redundant sensor, model-based (analytical) redundancy or by reconfiguration of the control scheme to run without that sensor. Such an approach to fault detection has been applied to the test-bed system of Figure 10.12 and was found to yield reasonable results for detection and accommodation of sensor faults. In that case, alternative synthesized sensor outputs were provided by a nonlinear model of the system running in real time. However, the use of more advanced algorithms gives the potential for detecting incipient faults earlier, for better identification of the root cause, and for provision of improved sensor fault accommodation.

10.3.2   Problem formulation

For this application, it is desired that the FDIA functionality be provided based on the use of analytical rather than physical redundancy. This essentially means making use of mathematical models of the system (or parts of it). The application of such models in parameter estimation, parity equations or fault detection schemes based on a state estimator (an observer) has been widely discussed in the literature (see the work of Blanke [31] Frank & Ding [32];

Isermann [33] Isermann & Balle [34] and the references therein). However, there still appears to be a shortage of practical applications, with many researchers favouring the easier and more controllable environment of dynamic simulation. Whilst such approaches have produced impressive results, the difficulties of "real-world" hardware and signals need to be addressed before model based schemes are widely adopted by industry. The desire to develop a unified design that encapsulates detection, isolation and accommodation of sensor faults suggests that a prediction error or residual-based approach is most appropriate. In its simplest form, this implies use of a model running in parallel with the plant and driven by the same input signal as the plant. A slight extension of this is the observer-based approach - involving feedback of actual measured outputs - which improves the robustness of the model predictions to measurement and modeling uncertainties. Indeed the use of a "bank" of observers not only leads to the unique isolation of sensor faults, but also directly facilitates accommodation of the faulty sensor. This is the approach favoured here. It is worth noting that a variety of more advanced and mathematically complex approaches have been proposed to further improve robustness to unknown inputs and sensitivity to faults.

10.3.3   Test bed

The prototype actuator under development consists of a brushless DC motor (with two separate three-phase windings) driving into a ballscrew via a gearbox. The experimental test-bed system shown in Figure 10.12 was commissioned as a research vehicle. All the results described were obtained on this system. The system is fully instrumented and linked to a dSPACE digital signal processor (DSP) interfaced to a PC that provides a facility for test waveform generation, data collection, data transmission to Matlab/Simulink, control loop closure, fault detection, and condition monitoring implementation.

   The position control algorithm applied to the system is of the form shown in Figure 10.13 and comprises both inner and outer loops. The inner loop is a conventional P+I current control-loop. The outer (position control) loop is of a proportional–integral-plus (PIP) design. Here, the discrete-time PIP controller was based on the discrete linear model of the test-bed with the inner loop closed. The directly measurable states are the sampled values of the position,

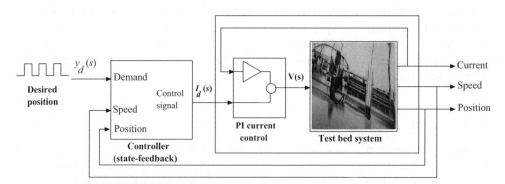

Figure 10.13   Electromechanical control system

velocity and integral of the position error. Naturally, both control loops are implemented with anti-windup provision, in order to obviate the potential problem of integrator wind-up when the system reaches its physical saturation limits.

10.4   Application to Fermentation Processes

It is becoming clear that a fault takes place in a dynamic system when the difference between the behaviors of the system and its nominal model exceeds an acceptable limit. As indicated before, one way to describe the fault is to define it as a deviation of the system parameter vector from its nominal value [34, 35]. Therefore, the task of fault isolation is to find the faulty parameter in the system parameter vector. Fault detection and isolation (FDI) is usually used in a fault-tolerant control (FTC) system [31, 36], in which estimation of the faulty parameter value is very important for controller reconfiguration. The task of this estimation is known as "fault identification".

In the following discussion, only fault isolation and identification are considered; it is assumed that the fault detection is fast enough (relative to the process of fault isolation). Once a fault occurs, it is detected immediately and isolation is triggered. The time of fault occurrence and of the beginning of isolation are considered to be the same, denoted by t_f.

The residual variable is usually used in the FDI field. The residual is defined as a variable based on a deviation between measurements and the model output of the system under consideration [34]. Generation of the residual usually uses an observer.

Thanks to the available powerful analytical tools of linear systems theory, many results for FDI in for linear dynamic systems have been achieved [37]–[40]. The FDI problem for nonlinear dynamic systems is more difficult. Achievements in recent years mainly depend on three kinds of method: those based on nonlinear geometric theory [41]–[43]; those based on parity space theory [44]–[46]; and those based on adaptive observers [35]. The application of the first is limited because a suitable decoupling framework with respect to the possible faults does not always exist. The method based on parity space theory has been used only for linear, bilinear, state affine, and algebraic systems. Methods based on adaptive observers fit many kinds of nonlinear system but its isolation speed is not ideal because of its classical parameter identification approach.

Existing FDI methods pay little attention to fault identification. The methods based on nonlinear geometric theory and on parity space theory do not provide estimation of faulty parameters. The method based on adaptive observers can provide estimation of faulty parameter but its speed is not ideal. In order to fit more general nonlinear dynamic systems and get ideal speed, this section puts forward a new method of fault isolation and identification. The practical domain of the value of each system parameter is divided into a certain number of intervals. After verifying all the intervals, the faulty parameter value is found and the fault is therefore isolated and identified. The principle of verification is based on the local monotonous characteristic of the observer prediction error, i.e. the residual. The only condition for the application of this method is that the system dynamic is a monotonous function with respect to the considered parameter. Therefore, it fits many kinds of nonlinear systems. We proves that this method has ideal speed.

Concerning the use of intervals, we would mention the methodology of interval arithmetic [47] and its related technique of set membership approach [48], although our method is not based on this theory. In this methodology the uncertainties of the system parameters are expressed by an interval. According to the interval and the initial condition of the system, the envelope is calculated. This envelope represents the trajectories across time of the possible maximum and minimum values of the system output variable. The envelope, hence, includes a whole family of temporal curves. The different temporal curves are indiscernible, so that the qualitative features may not be captured, particularly as the envelope bounding curve often does not correspond to an unique temporal trajectory. Moreover, an interval arithmetic does not always provide an exact envelope, that is to say the envelope may be overbounded or underbounded. Because of this, existing interval arithmetic is not suitable for our method – we need an exact boundary trajectory with respect to a certain boundary value of the parameter. In the work of Valdes-Gonzalez *et al.* [49] and Raissi *et al.* [50], interval arithmetic is used to estimate system states. Examples of the use of this methodology for FDI can be found in the work of Boukhris *et al.* [51], Sainz *et al.* [52], and Fagarasan *et al.* [48]. In these examples, the purpose of using interval arithmetic is to enhance the robustness of the FDI method under the consideration of system uncertainties. But essentially the basic mechanism of fault isolation has no relationship with interval arithmetic.

There are other examples of the use of interval observers that are based on the conception of cooperative systems [53, 54]. The method concerns the use of interval observers for dynamic estimation of bounds on unmeasurable variables (or parameters) of uncertain dynamical systems. This method has a special requirement for a Jacobian matrix of the observer prediction error dynamic. This requirement limits the method application.

10.4.1 Nonlinear faulty dynamic system

The considered nonlinear dynamic system is as follows:

$$\dot{x} = f(x, \theta, u),$$
$$y = cx, \tag{10.136}$$

where $x \in \Re^n$ is the system state vector, $\theta \in \Re^p$ the system parameter vector, $u \in \Re^m$ the system measurable input vector, and $y \in \Re^l$ the system measurable output vector. $c \in \Re^{l \times n}$ is a known system output matrix. $f(x, \theta, u)$ is a known nonlinear vector function that characterizes the dynamic of the system. $f(x, \theta, u)$ and its first partial derivatives on x and θ are continuous, bounded, and Lipschitz in x and θ. The nominal value of the system parameter vector θ is denoted by θ^0 and is known.

The hth component of the system output vector is noted as

$$y_h = h^T x, \tag{10.137}$$

where $h^T \in \Re^{1 \times n}$ is the hth horizontal vector of the matrix c. The scalar output y_h will be used for fault analysis, therefore, the choice of h should guarantee that the observable state subspace based on y_h can reflect the considered faults.

Definition 10.1   *There is a fault in the dynamic system of Equation (10.136), if the dynamic difference*

$$\Delta f(x, \theta, \theta^0, u) = f(x, \theta, u) - f(x, \theta^0, u) \qquad (10.138)$$

between the system of Equation (10.136) and its nominal model $\dot{x} = f(x, \theta^0, u)$, caused by the difference of parameter vectors $\Delta\theta = \theta - \theta^0$, is large.

A parameter vector difference $\Delta\theta$ which causes a fault should be large because $f(x, \theta, u)$ is Lipschitz in θ. A fault-free system has a parameter vector close to θ^0 and is regarded as θ^0 for convenience in this paper. While a system with fault is called a "post-fault" system, its parameter vector is denoted by θ^f. θ represents θ^0 or θ^f according to the context.

Remark 10.8   *In Definition 10.1, the sense of the word "large" is relative to the normal dynamic difference $\Delta f(x, \theta, \theta^0, u)$ of the fault-free system caused by parameter uncertainties or modeling error.*

The mathematical description of a real dynamic system always possesses parameter uncertainties and modeling error, but their effects i.e. $\Delta f(x, \theta, \theta^0, u)$, should be limited to a normal range, related to the practical engineering problem. If the system dynamic difference greatly exceeds this normal range, we say that there is a fault.

10.4.2   Residual characteristics

The considered observer is given by

$$\dot{\hat{x}} = f(\hat{x}, \theta^{ob}, u) + k(y - \hat{y}),$$
$$\hat{y} = c\hat{x},$$
$$\varepsilon = y_h - \hat{y}_h. \qquad (10.139)$$

Its internal form is

$$\dot{\hat{x}} = f(\hat{x}, \theta^{ob}, u) + kc(x - \hat{x}),$$
$$e = x - \hat{x},$$
$$\varepsilon = he, \qquad (10.140)$$

where x is the system state vector in Equation (10.136), $\hat{x} \in \mathfrak{R}^n$ is the observer state vector, $e \in \mathfrak{R}^n$ is the state prediction error. $\varepsilon \in R$ is the residual – it is a scalar and calculable. $\hat{y}_h = h\hat{x}.\theta^{ob} \in \mathfrak{R}^p$ is the observer parameter vector. $\theta^{ob} = \theta^0$ for $t < t_f$. $k \in \mathfrak{R}^{n \times l}$ is the observer gain matrix. The expression Equation (10.140) is used for theoretical analysis.

As the proposed method is based on observers, we have to assume that the considered system is observable with their working excitations and that their availability for the considered system and the observers' states can converge to the system state when the observer parameter vectors equate to the system parameter vector. We also assume that the faults under consideration can be detected by each observer using corresponding residual ε. In this section, we do not deal with the problem of nonlinear system observability. Moreover, the problems of nonlinear observer design and observer design with respect to fault detection are not the objective of this section. For these problems, the interested reader is referred to [54]–[58]. For the problem of adaptive observer design for nonlinear systems, the reader is referred to [59, 60, 61]. In the proposed method, an adaptive observer only estimates one parameter; this condition facilitates the design problem [35, 62]. Moreover, the adaptive observers in our method work in small parameter intervals – the needed parameter convergence characteristic is almost local.

In view of the preceding discussions, we can state the following assumption.

Assumption 10.1   *The model of Equation (10.139) is an observer of the system of Equation (10.136). It is well designed therefore has good stability and convergence characteristics, and fault detection ability. Before the occurrence of the fault, $\theta^{ob} = \theta^0$; the observer state $\hat{x}$ has converged to the system state x, so*

$$e(t_f) = 0, \quad \varepsilon(t_f) = 0. \tag{10.141}$$

After the occurrence of the fault, θ^{ob} is a preselected parameter value. At the point x of the system state, if $\Delta f(x, \theta^f, \theta^{ob}, u) = f(x, \theta^f, u) - f(x, \theta^{ob}, u) = 0$, then $e(t) = 0$, $\varepsilon(t) = 0 \forall t \geq t_f$. If $\Delta f(x, \theta^f, \theta^{ob}, u) \neq 0$, then $e(t) \neq 0$ and $\varepsilon(t) \neq 0$. If $\Delta f(x, \theta^f, \theta^{ob}, u)$ is large, then $e(t)$ and $\varepsilon(t)$ are large.

As mentioned earlier, t_f represents the time of fault occurrence as well as the time that the fault is detected. It means that the fault detection is very quick. If this condition is not satisfied, it can be proven that the method is also valid after a small modification. In this case, we should add a short waiting time after a fault is detected before the isolation operation is completed and before the decision of isolation is made. t_f is determined by the fault detection procedure. At t_f, the sth system parameter changes because of the fault occurrence and the jth observer parameter is switched to a preselected value by the isolation procedure immediately in order to isolate the fault:

$$\begin{cases} \theta_s^f = \theta_s^0 + \Delta^f, \\ \theta_l^f = \theta_l^0, l \neq s, \end{cases} t \geq t_f, \qquad \begin{cases} \theta_j^{ob} = \theta_j^0 + \Delta^{ob}, \\ \theta_l^{ob} = \theta_l^0, l \neq j, \end{cases} t \geq t_f,$$

where Δ^f, Δ^{ob} are real numbers. Δ^f is the change value of the faulty parameter caused by the fault and $\theta_j^0 + \Delta^{ob}$ is the preselected value of the observer; it is a bound of a parameter interval of the jth system parameter. One defines

$$\delta\theta = (\theta^{ob} - \theta^f)|_{t \geq t_f}.$$

In order to facilitate the following discussions, two isolation observers are defined corresponding to two bounds of the parameter interval $[\theta_j^b, \theta_j^a]$ of the jth parameter. They are two particular cases of the isolation observer, Equation (10.140), on the interval bounds:

$$\dot{\hat{x}}^a = f(\hat{x}^a, \theta^{oba}, u) + kc(x - \hat{x}^a),$$

$$e^a = x - \hat{x}^a,$$

$$\varepsilon^a = he^a, \tag{10.142}$$

$$\dot{\hat{x}}^b = f(\hat{x}^b, \theta^{obb}, u) + kc(x - \hat{x}^b),$$

$$e^b = x - \hat{x}^b,$$

$$\varepsilon^b = he^b, \tag{10.143}$$

where

$$\theta_j^{oba} = \begin{cases} \theta_j^0, t < t_f, \\ \theta_j^a, t \geq t_f, \end{cases} \theta_l^{oba} = \theta_l^0 \forall t, l \neq j,$$

$$\theta_j^{obb} = \begin{cases} \theta_j^0, t < t_f, \\ \theta_j^b, t \geq t_f, \end{cases} \theta_l^{obb} = \theta_l^0 \forall t, l \neq j.$$

Figure 10.14 shows that the interval bounded by the two observers does not contain the faulty parameter value. In this example the system parameter vector has two parameters. The nominal value of the parameter vector is $\theta^0 = (\theta_1^0, \theta_2^0) = (1.0, 3.1)$. The fault takes place at parameter θ_1. The faulty parameter value is $\theta_1^f = 6.7$. After t_f, θ_2^{obb} is switched to 6.5 and θ_2^{oba} is switched to 8.0, while θ_1^{obb} and θ_1^{oba} do not change. The interval [6.5, 8.0] of the parameter θ_2 does not contain the faulty parameter value $\theta_1^f = 6.7$. In this case, the difference $\delta\theta^a = (\theta^{oba} - \theta^f)|_{t \geq t_f}$ and the difference $\delta\theta^b = (\theta^{obb} - \theta^f)|_{t \geq t_f}$ between the observer parameter vectors and the post-fault system parameter vector are large.

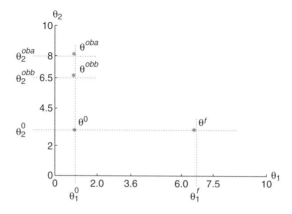

Figure 10.14   The interval does not contain a faulty parameter value

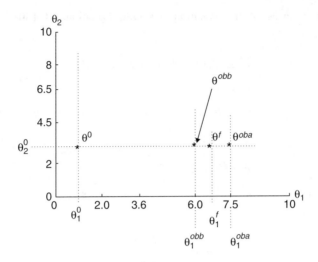

Figure 10.15   The interval contains a faulty parameter value

An example in which the interval bounded by the two observers contains the faulty parameter value is shown by Figure 10.15. In this example, the interval $[6.0, 7.5]$ of parameter θ_1 contains the faulty parameter value $\theta_1^f = 6.7$. Though the difference $\delta\theta = \theta^f - \theta^0$ between the post-fault system parameter vector θ^f and the system nominal parameter vector θ^0 is large, the differences $\delta\theta^a = (\theta^{oba} - \theta^f)|_{t \geq t_f}$ and $\delta\theta^b = (\theta^{obb} - \theta^f)|_{t \geq t_f}$ between the observer parameter vectors and the post-fault system parameter vector are small because the interval is small.

It should be remembered that $\delta\theta(\delta\theta^a, \delta\theta^b)$ represents the parameter vector difference between the observer and the post-fault system, while $\Delta\theta = \theta^f - \theta^0$ is the difference between the post-fault system and the nominal system. $\Delta\theta$ is always large for a post-fault system but $\delta\theta$ will be small if the faulty parameter value is contained in a small interval.

Assumption 10.2   *At each point x in the state space with the same control variable u, the function $f(x, \theta, u)$ in Equation (10.136) (also in Equations (10.139) and (10.140)) satisfies the following conditions:*

- *Any component $f_i(x, \theta, u), i \in \{1, \ldots, n\}$ which is an explicit function of the considered parameter θ_j is a monotonous function of parameter θ_j.*
- *$\theta_f(x, \theta, u)$ is a monotonous function of the considered parameter θ_j.*

Theorem 10.2   *It is assumed that the parameter changes of the system and of the observers are at the same parameter, that is to say $s = j$. Two bounds of a considered parameter interval are denoted by θ_j^b and θ_j^a. The interval is smaller than a certain size which will be mentioned* *later. The residuals* $\varepsilon^a(t)$ *and* $\varepsilon^b(t)$ *correspond to the two interval bounds while* $\varepsilon(t)$ *represents the residual in the general case.*

- *If* $\theta_j^f \in [\theta_j^b, \theta_j^a]$ *then* $sgn(\varepsilon^a(t)) = sgn(\varepsilon^b(t))$, $\forall t$ *and* $\varepsilon(t)$ *is a monotonous function of the parameter difference* $\delta\theta_j$ *when* $\theta_j^{ob} \in [\theta_j^b, \theta_j^a]$. *Specifically,* $\lim_{\delta\theta_j \to 0} \varepsilon(t) = 0$, $\forall t \geq t_f$.
- *If* $\theta_j^f \notin [\theta_j^b, \theta_j^a]$ *then, at the period beginning after the fault occurrence, the equality* $sgn(\varepsilon^a(t)) = sgn(\varepsilon^b(t))$ *is satisfied.*

Proof. Equation (10.136) This case corresponds to the example of Figure 10.15 in which the interval contains the faulty parameter value. First of all, we consider the situation at the point $(x, \hat{x}, \theta^f, \delta\theta_j) = (x, x, \theta^f, 0)$.In an infinitely small neighborhood of the point $(x, \hat{x}, \theta^f, \delta\theta_j) = (x, x, \theta^f, 0)$, by making a linearization of the observer prediction error equation:

$$\dot{e}(t) = f(x, \theta^f, u) - f(x, \theta^{ob}, u) - kc(x - \hat{x})$$

At this point and by neglecting high-order terms, the following can be obtained:

$$\dot{e} = \left(\left. \frac{\partial f(x, \theta, u)}{\partial x} \right|_{\theta^f, x} - kc \right) e - \left. \frac{\partial f(x, \theta, u)}{\partial \theta} \right|_{\theta^f, x} \delta\theta.$$

In the above equation, the matrices $(\partial f(x, \theta, u)/\partial x|_{\theta^f, x} - kc)$ and $\partial f(x, \theta, u)/\partial\theta|_{\theta^f, x}$ are not related by $\delta\theta = \theta^{ob} - \theta^f$ but by θ^f, x and u. They are considered as time-varying matrices due to x and u, which are variables of the time. They are re-written as

$$A(t) = \left(\left. \frac{\partial f(x, \theta, u)}{\partial x} \right|_{\theta^f, x} - kc \right), \quad B(t) = \left. \frac{\partial f(x, \theta, u)}{\partial \theta} \right|_{\theta^f, x}.$$

So, the observer prediction error equation is

$$\dot{e} = A(t) - B(t)\delta\theta.$$

The solution of prediction error is

$$e(t) = \Phi(t, t_0)e(t_0) - \int_{t_0}^{t} \Phi(t, \tau)B(\tau)\delta\theta d\tau,$$

where t_0 is the initial time of calculation. $\Phi(\cdot, \cdot)$ is the state transfer matrix which is only a function of the matrix $A(t)$. Because $e(t_f) = 0$, we have

$$e(t) = - \int_{t_f}^{t} \Phi(t, \tau)B(\tau)\delta\theta d\tau.$$

When $s = j$, $\delta\theta = V_j\delta\theta_j$. And $V_j = [0, \ldots, 1, \ldots, 0]^\top$, its jth component is 1 while other components are zero. So at the infinitely small neighborhood of the point $(x, \hat{x}, \theta^f, \delta\theta_j) = (x, x, \theta^f, 0)$ 0 we have

$$e(t) = -\left(\int_{t_f}^{t} \Phi(t, \tau)B(\tau)V_j d\tau\right)\delta\theta_j,$$

$$\varepsilon(t) = -h\left(\int_{t_f}^{t} \Phi(t, \tau)B(\tau)V_j d\tau\right)\delta\theta_j. \tag{10.144}$$

The function $\int_{t_f}^{t} \Phi(t, \tau)B(\tau)V_j d\tau$ is independent of $\delta\theta_j$.

At the point $(x, \hat{x}, \theta^f, \delta\theta_j) = (x, x, \theta^f, 0)$, the derivative

$$\frac{d\varepsilon(t)}{d(\delta\theta_j)} = -h\left(\int_{t_f}^{t} \Phi(t, \tau)B(\tau)V_j d\tau\right),$$

is independent of $\delta\theta_j$, where

$$B(t)V_j = \left.\frac{\partial f(x, \theta, u)}{\partial\theta_j}\right|_{\theta_f, x}.$$

Therefore, in an infinitely small neighborhood of the point $(x, \hat{x}, \theta^f, \delta\theta_j) = (x, x, \theta^f, 0)$, $\varepsilon(t)$ is a monotonous function of $\delta\theta_j$.

Now we consider the situation in a domain $D \subseteq \Re^{2n+p+1}$ which contains the point $(x, x, \theta^f, 0)$. Because $f(x, \theta, u)$ and its first partial derivatives on x and θ are continuous in x and θ, so the function $\varepsilon(t) = \varepsilon(x, \hat{x}, \theta^f, \delta\theta_j)$ and its first partial derivatives on $x, \hat{x}, \theta^f, \delta\theta_j$ and its first derivative on $\delta\theta_j$ are also continuous in $x, \hat{x}, \theta^f$ and $\delta\theta_j$.

Because of the continuity of $d\varepsilon(t)/d(\delta\theta_j)$, there is a compact domain $D \subseteq \Re^{2n+p+1}$ which contains the point $(x, x, \theta^f, 0)$. For any point $(x, \hat{x}, \theta^f, \delta\theta_j) \in D$, the sign of $d\varepsilon(t)/d(\delta\theta_j)$ is the same as the sign of $-h(\int_{t_f}^{t} \Phi(t, \tau)B(\tau)V_j d\tau)$.

Because $\hat{x}$ is a continuous function of $\delta\theta_j$, when $\delta\theta_j \to 0$, in other words when $\theta_j^{ob} \to \theta_j^f$, we have $\hat{x} \to x$ and $(x, \hat{x}, \theta^f, \delta\theta_j) \to (x, x, \theta^f, 0)$. So in the space of θ_j^{ob} there is a compact domain $Q \subseteq R$ which contains the point $\theta_j^{ob} = \theta_j^f$, when $\theta_j^{ob} \in Q$, and the point $(x, \hat{x}, \theta^f, \delta\theta_j) \in D$.

On the domain Q, $\varepsilon(t)$ is a monotonous function of single parameter difference $\delta\theta_j$ due to the sign of $d\varepsilon(t)/d(\delta\theta_j)$ which is independent of $\delta\theta_j$.

It is obvious that domain Q will not contain the point $\theta_j^{ob} = \theta_j^*$ which satisfies

$$\left.\frac{\partial f(x, \theta, u)}{\partial\theta_j}\right|_{\theta_j^*} = 0 \tag{10.145}$$

If domain Q contains the point θ_j^*, there will be a point θ_j^d in the left neighborhood of the point θ_j^* and another point θ_j^u in the right neighborhood of the point θ_j^* such that

$$f(x,\theta,u)\bigg|_{\theta_j=\theta_j^d} = f(x,\theta,u)|_{\theta_j=\theta_j^u} \neq f(x,\theta,u)|_{\theta_j=\theta_j^*}.$$

Then we will obtain

$$\varepsilon(t)|_{\theta_j^{ob}=\theta_j^d} = \varepsilon(t)|_{\theta_j^{ob}=\theta_j^u} \neq \varepsilon(t)|_{\theta_j^{ob}=\theta_j^*}.$$

Thus $\varepsilon(t)$ will not be a monotonous function with respect to θ_j. Therefore, there must exist two points θ_j^a, θ_j^b in the interval $[\theta_j^d, \theta_j^u]$ such that the sign of $d\varepsilon(t)/d(\delta\theta_j)|_{\theta_j^{ob}=\theta_j^a}$ and the sign of $d\varepsilon(t)/d(\delta\theta_j)|_{\theta_j^{ob}=\theta_j^b}$ are different. This result contradicts with the definition of domain Q.

For different excitations of the system of Equation (10.136), Q is not the same. For all considered excitations, we note the intersection of all possible Q as Q_{min}. So, for a θ_j^f value, $\exists Q_{min}$, we re-note it as $Q_{min}(\theta_j^f)$, when $\theta_j^{ob} \in Q_{min}(\theta_j^f)$, $\varepsilon(t)$ is a monotonous function of $\delta\theta_j$.

Now we consider the interval $[\theta_j^b, \theta_j^a]$. For all the points of θ_j^f in the interval, the intersection of all the corresponding $Q_{min}(\theta_j^f)$ of the points is noted as $Q_{min}(a,b)$. If the interval is smaller than a certain size, $Q_{min}(a,b)$ is not an empty set, and we can also ensure that $[\theta_j^b, \theta_j^a] \subseteq Q_{min}(a,b)$.

In this case, for any values $\theta_j^f, \theta_j^{ob} \in [\theta_j^b, \theta_j^a]$, we have:

- $\varepsilon(t)$ is a monotonous function of $\delta\theta_j$.
- $\varepsilon(t)|_{\delta\theta_j=0} = 0, \forall t \geq t_f$.
- Because $\theta_j^f \in [\theta_j^b, \theta_j^a]$, so $sgn(\delta\theta_j^b) = -sgn(\delta\theta_j^a)$, and because θ_j^b, θ_j^a are in the subset $Q_{min}(a,b)$ and, $\varepsilon(t)|_{\delta\theta_j=0} = 0$, so we have

$$sgn(\varepsilon^a(t)) = -sgn(\varepsilon^b(t)) \quad \forall t.$$

If Assumption 10.2 is not satisfied, there exists maybe a point θ_j^* satisfying the condition of Equation (10.145). This point will be in one of the intervals. We assume that this point is not at the bounds of the interval, therefore, the interval will not belong to a compact domain Q and the monotonous property will not be satisfied in this interval. If Assumption 10.2 is satisfied, this case can be avoided.

Now, the first part of Theorem 10.2 is proven.

Equation (10.137) In this case, though the parameter changes of the system and of the observer are at the same parameter, the interval does not contain the faulty parameter value. According to Equation (10.141), at t_f, we have $x(t_f) = \hat{x}^a(t_f) = \hat{x}^b(t_f)$, so it will be

$$\dot{\varepsilon}^a(t_f) = hf(x(t_f), \theta^f, u) - hf(x(t_f), \theta^{oba}, u),$$
$$\dot{\varepsilon}^b(t_f) = hf(x(t_f), \theta^f, u) - hf(x(t_f), \theta^{obb}, u).$$

Because $\theta_j^f \notin [\theta_j^b, \theta_j^a]$, it implies that $sgn(\theta_j^a - \theta_j^f) = sgn(\theta_j^b - \theta_j^f)$ and $sgn(\theta_j^{oba} - \theta_j^f) = sgn(\theta_j^{obb} - \theta_j^f)$. Because $hf(x, \theta, u)$ is a monotonous function of θ and $(\theta_l^{oba} - \theta_l^f)|_{l \neq j} = 0$, $(\theta_l^{obb} - \theta_l^f)|_{l \neq j} = 0$, we can obtain

$$sgn(\dot{\varepsilon}^a(t_f)) = sgn(\dot{\varepsilon}^b(t_f)).$$

In the neighborhood of t_f, the variations of $\dot{\varepsilon}^a(t)$ and $\dot{\varepsilon}^b(t)$ are small, then

$$\varepsilon^a(t_f + \Delta t) = \varepsilon^a(t_f) + \dot{\varepsilon}^a(t_f)\Delta t = \dot{\varepsilon}^a(t_f)\Delta t,$$
$$\varepsilon^b(t_f + \Delta t) = \varepsilon^b(t_f) + \dot{\varepsilon}^b(t_f)\Delta t = \dot{\varepsilon}^b(t_f)\Delta t,$$

therefore

$$sgn(\varepsilon^a(t_f + \Delta t)) = sgn(\varepsilon^b(t_f + \Delta t)).$$

The second part of Theorem 10.2 is thus proven.   ■

Remark 10.9   *The proof of Theorem 10.2 provides us a local monotonous characteristic of $\varepsilon(t)$ with respect to $\delta\theta_j$. For some kinds of nonlinear dynamic system, we can prove Theorem 10.2 more easily using the concept of "cooperative systems" [53, 54] and the obtained monotonous characteristic is more global, but more limiting for the system type. Since our method is based on small intervals (small sizes of the intervals are necessary to achieve a fast speed of the method and precise estimation of the faulty parameter value), the local monotonous characteristic given by the proof in Theorem 10.2 is sufficient, while the manner of this proof can be used for more general nonlinear systems. Thanks to the progress of computer technology, we can use the consideration that the parameter value domain can be partitioned into the desired small intervals. This idea enables us to solve the problems caused by the nonlinearities.*

In short, cooperative systems are dynamical systems for which the non-diagonal terms of the Jacobian matrix are non-negative. In the observer prediction error equation, the Jacobian matrix is $((\partial f(x, \theta, u)/\partial x)|_{\theta^f, x} - kc)$. This condition of a cooperative system is strong; it limits the application. For the observer prediction error dynamic, if this condition is satisfied, after the fault occurrence, for the initial value $e(t_0)|_{t_0 \geq t_f} \geq 0$ (for the notations $e(t) \geq 0$ and $e(t) \leq 0$, see the work of [63], the prediction error will satisfy that $e(t) \geq 0, \forall t \geq t_0$. It is similar for the initial value $e(t_0)|_{t_0 \geq t_f} \leq 0$. Our objective is to get a relation that for two deviations of parameter $\delta\theta_j^a \neq 0$ and $\delta\theta_j^b \neq 0$, if $sgn(\delta\theta_j^a)sgn(\delta\theta_j^b) = -1$, it should be $e^a(t) \geq 0, e^b(t) \leq 0, \forall t \geq t_f$, or $e^a(t) \leq 0, e^b(t) \geq 0, \forall t \geq t_f$. So we need a relation between $e(t_0)$ and $\delta\theta_j$ that if $sgn(\delta\theta_j^a)sgn(\delta\theta_j^b) = -1$, then $e^a(t_0) \geq 0, e^b(t_0) \leq 0$ or $e^a(t_0) \leq 0, e^b(t_0) \geq 0$. But the condition of the cooperative system does not give any information about this relation; it can be provided by Assumption 10.2 at least in the neighborhood of t_f. So even if we limit the application in a cooperative system, Assumption 10.2 is also needed.

Assumption 10.3 *After the fault occurrence, if the parameter changes of the system and of the observer are not at the same parameter, that is to say $s \neq j$, no matter what the value of the change of the isolation observer parameter, the dynamic difference between the isolation observer and the post-fault system at point $\hat{x} = x$ is large. That is to say*

$$\Delta f(x, \theta^f, \theta^{ob}, u) = f(x, \theta^f, u) - f(x, \theta^{ob}, u)$$

is large.

Assumption 10.3 is reasonable because the difference between the faulty parameter of the system and the corresponding parameter of the observer (this parameter is maintained at its nominal value, see Figure 10.14) is large. This large parameter difference should cause a large dynamic difference.

Lemma 10.6 *After the fault occurrence, if the parameter changes of the system and of the observer are not at the same parameter, that is to say $s \neq j$, no matter what the value of the change of the isolation observer parameter, the amplitude of the residual $\varepsilon(t)$ of the isolation observer is large.*

Proof. According to Assumptions 10.3 and 10.1 this is true. ∎

Lemma 10.7 *After the fault occurrence, whether $s = j$ or not, and whatever the value of the system parameter change, the differences of the prediction errors and of the residuals given by*

$$E^{ab}(t) = e^a(t) - e^b(t), \ \varepsilon^{ab}(t) = \varepsilon^a(t) - \varepsilon^b(t)$$

are monotonous functions of the parameter difference between the two interval bounds which is noted by $\nabla\theta_j^{ab} = \theta_j^a - \theta_j^b$, and

$$\lim_{\nabla\theta_j^{ab} \to 0} E^{ab}(t) = 0, \ \forall t \geq t_f, \quad \lim_{\nabla\theta_j^{ab} \to 0} \varepsilon^{ab}(t) = 0, \ \forall t \geq t_f.$$

Proof.

$$\dot{E}^{ab}(t) = f(\hat{x}^b, \theta^{obb}, u) - f(\hat{x}^a, \theta^{oba}, u) - kcE^{ab}(t). \tag{10.146}$$

Because the difference $\theta_j^{oba} - \theta_j^{obb} = \theta_j^a - \theta_j^b$ is small, and $(\theta_l^{oba} - \theta_l^{obb})|_{l \neq j} = 0$, it can be proven that $E^{ab}(t) = \hat{x}^b(t) - \hat{x}^a(t)$ is not large. By doing a linearization of the function $f(\hat{x}^a, \theta^{oba}, u)$ at point $(\hat{x}^b, \theta^{obb})$ and by neglecting high-order terms, we can obtain

$$\dot{E}^{ab}(t) = A_1(t)E^{ab}(t) + B_1(t)(\theta^{oba} - \theta^{obb}),$$

where

$$A_1(t) = ((\partial f(x, \theta, u)/\partial x)|_{\hat{x}^b, \theta^{obb}} - kc), \quad B_1(t) = -(\partial f(x, \theta, u)/\partial \theta)|_{\hat{x}^b, \theta^{obb}}.$$

The solution of $E^{ab}(t)$ is

$$E^{ab}(t) = \int_{t_f}^{t} \Phi^{ab}(t, \tau)(\theta^{oba} - \theta^{obb})d\tau,$$

where $\Phi^{ab}(t, \tau)$ is a state transfer matrix, the only function of the matrix $A_1(t)$. $\theta^{oba} - \theta^{obb} = V_j \nabla \theta_j^{ab}$. $V_j = [0, \ldots, 1, \ldots, 0]^{\top}$, i.e. its jth component is one while other components are zero. So at the neighborhood of the point $(\hat{x}^b, \hat{x}^a, \theta^{obb}, \nabla \theta_j^{ab}) = (\hat{x}^b, \hat{x}^b, \theta^{obb}, 0)$ we have

$$E^{ab}(t) = \left(\int_{t_f}^{t} \Phi^{ab}(t, \tau)B_1(\tau)V_j d\tau \right) \nabla \theta_j^{ab},$$

$$\varepsilon^{ab}(t) = h\left(\int_{t_f}^{t} \Phi^{ab}(t, \tau)B_1(\tau)V_j d\tau \right) \nabla \theta_j^{ab}.$$

The term $\int_{t_f}^{t} \Phi^{ab}(t, \tau), B_1(\tau)V_j d\tau$ is independent of $\nabla \theta_j^{ab}$. It is evident that

$$\lim_{\nabla \theta_j^{ab} \to 0} E^{ab}(t) = 0, \ \forall t \geq t_f, \quad \lim_{\nabla \theta_j^{ab} \to 0} \varepsilon^{ab}(t) = 0, \ \forall t \geq t_f.$$

Using the manner of the proof of Theorem 10.2, it can be proven that $\varepsilon^{ab}(t)$ is a monotonous function of $\nabla \theta_j^{ab}$ when the parameter interval is small. ∎

Theorem 10.3   *It is assumed that, after the fault occurrence, the parameter changes of the system and of the observers are not at the same parameter, that is to say $s \neq j$. If the parameter interval $[\theta_j^b, \theta_j^a]$ is small enough, then the time t_e exists that*

$$sgn(\varepsilon^a(t_e)) = sgn(\varepsilon^b(t_e)).$$

Proof. According to Lemma 10.6, after the fault occurrence, the amplitudes of the residuals $\varepsilon^a(t)$ and $\varepsilon^b(t)$ of the isolation observers are large; two curves $\varepsilon^a(t)$ and $\varepsilon^b(t)$ will be far from the axis of abscissa at any time. On the other hand, when $\nabla \theta_j^{ab} = \theta_j^a - \theta_j^b$ is small enough, according to Lemma 10.7, two curves $\varepsilon^a(t)$ and $\varepsilon^b(t)$ are very close. So, there must exist a time t_e that $sgn(\varepsilon^a(t_e)) = sgn(\varepsilon^b(t_e))$. ∎

According to Theorems 10.2 and 10.3, the conclusion below can be made.

Conclusion For a parameter interval $[\theta_j^b, \theta_j^a]$ which is small enough, if it does not contain the faulty parameter value of the post-fault system, the isolation index:

$$v(t) = sgn(\varepsilon^a(t))sgn(\varepsilon^b(t)) \qquad (10.147)$$

will be 1 some time after the fault occurrence. If the parameter interval contains the faulty parameter value, the isolation index $v(t)$ will be maintained as -1 all the time. As soon as the isolation index $v(t)$ becomes 1, it can be determined that the interval does not contain the faulty parameter value of the post-fault system, in spite of $v(t)$ being -1 or 1 afterwards.

10.4.3   The parameter filter

One assumes that p parameters $\theta_1, \theta_2, \ldots, , \theta_j, \ldots, \theta_p$ of the system parameters vector θ have probability of fault occurrence. The practical domain of each parameter is partitioned into a certain number of intervals. For example, parameter θ_j is partitioned into q intervals, their bounds denoted by $\theta_j^{(0)}, \theta_j^{(1)}, \ldots, \theta_j^{(i)}, \ldots, \theta_j^{(q)}$. The bounds of the ith interval are $\theta_j^{(i-1)}$ and $\theta_j^{(i)}$, also denoted as $\theta_j^{b(ij)}(t)$ and $\theta_j^{a(ij)}(t)$.

The size of each interval should satisfy the condition that the interval is in the corresponding domain $Q_{\min(a,b)}$ which is mentioned in the proof of Theorem 10.2. We can check it by simulation test because the dynamic model of the system is known.

To verify if an interval contains the faulty parameter value of the post-fault system, a parameter filter is built for this interval. To facilitate the explanation, only the parameter filter of the ith interval of the jth parameter is discussed.

Each bound of the intervals is used as a parameter to build an isolation observer. For N intervals in series, there are $(N + 1)$ bounds, therefore $(N + 1)$ isolation observers are built. A parameter filter consists of two isolation observers which correspond to two interval bounds. On the other hand, each isolation observer serves two neighboring intervals. An interval which contains a parameter nominal value is unable to contain the faulty parameter value, so a parameter filter is not built for it.

The parameter filter for the ith interval of the jth parameter is given below. The isolation observers are

$$\dot{\hat{x}}^{a(ij)} = f(\hat{x}^{a(ij)}, \theta^{oba(ij)}(t), u) + k(y - \hat{y}^{a(ij)}),$$

$$\hat{y}^{a(ij)} = c\hat{x}^{a(ij)},$$

$$\varepsilon^{a(ij)} = y_h - h\hat{x}^{a(ij)}, \qquad (10.148)$$

$$\dot{\hat{x}}^{b(ij)} = f(\hat{x}^{b(ij)}, \theta^{obb(ij)}(t), u) + k(y - \hat{y}^{b(ij)}),$$

$$\hat{y}^{b(ij)} = c\hat{x}^{b(ij)},$$

$$\varepsilon^{b(ij)} = y_h - h\hat{x}^{b(ij)}. \qquad (10.149)$$

Their internal forms are given by

$$\dot{\hat{x}}^{a(ij)} = f(\hat{x}^{a(ij)}, \theta^{oba(ij)}(t), u) + kc(x - \hat{x}^{a(ij)}),$$

$$e^{a(ij)} = x - \hat{x}^{a(ij)},$$

$$\varepsilon^{a(ij)} = he^{a(ij)}, \tag{10.150}$$

$$\dot{\hat{x}}^{b(ij)} = f(\hat{x}^{b(ij)}, \theta^{obb(ij)}(t), u) + kc(x - \hat{x}^{b(ij)}),$$

$$e^{b(ij)} = x - \hat{x}^{b(ij)},$$

$$\varepsilon^{b(ij)} = he^{b(ij)}, \tag{10.151}$$

where

$$\theta_j^{oba(ij)}(t) = \begin{cases} \theta_j^0, t < t_f, \\ \theta_j^{(i)}, t \ge t_f, \end{cases} \quad \theta_l^{oba(ij)}(t) = \theta_l^0 \quad \forall t, l \neq j,$$

$$\theta_j^{obb(ij)}(t) = \begin{cases} \theta_j^0, t < t_f, \\ \theta_j^{(i-1)}, t \ge t_f, \end{cases} \quad \theta_l^{obb(ij)}(t) = \theta_l^0 \quad \forall t, l \neq j.$$

We would emphasize that, in general cases, a single observer plays the roles of the two observers of $\hat{x}^{a(ij)}$ and $\hat{x}^{b(i+1,j)}$.

The isolation index of this parameter filter is calculated by

$$v(ij)(t) = sgn(\varepsilon^{a(ij)}(t))sgn(\varepsilon^{b(ij)}(t)). \tag{10.152}$$

As soon as $v^{(ij)}(t) = 1$, the parameter filter sends the "non-containing" signal to indicate that this interval does not contain the faulty parameter value.

In the proof of Lemma 10.7, the derivation of Equation (10.146) is based on an implicit assumption that the observer of Equation (10.142) and the observer of Equation (10.143) have the same gain k. Because each observer works for two neighboring intervals, therefore, all the observers in series have the same gain. It is sometimes not suitable for a nonlinear system because it may cause a problem with convergence for large domains of the system parameters. Therefore, sometimes we need different gains for two neighboring intervals. If we want to choose different gains for different observers in two neighboring intervals, we should build two observers with different gains between the two neighboring intervals rather than only one.

10.4.4 Fault filter

The set of all parameter filters of a parameter is called the fault filter of this parameter. To facilitate the explanation, we discuss only the fault filter of the jth parameter. The discussion consists of two cases:

- *Case 1:* The fault is not on the jth parameter.

  After the fault occurrence, all the parameter filters of the jth parameter send non-containing signals, some time after t_f. When all these non-containing signals have been sent, a "fault exclusion" signal for the jth parameter is sent to indicate that the fault is not on the jth parameter.

- *Case 2:* The fault occurs on the jth parameter.

  After the fault occurrence, for any value of the faulty parameter, the fault is in one of the parameter intervals of the jth parameter. According to the Conclusion in the previous section, the parameter filter whose interval contains the faulty parameter value cannot send the non-containing signal all the time, therefore, the fault exclusion signal for jth parameter cannot be sent all the time.

10.4.5   Fault isolation and identification

After a fault is detected, all the fault filters are triggered by the isolation procedure. After a time period, if $p - 1$ fault filters except one have sent fault exclusion signals, the fault is isolated. The fault filter which did not send the fault exclusion signal is related to the faulty parameter and the parameter interval whose parameter filter did not send the non-containing signal contains the faulty parameter value. Assuming that this interval is the ith interval of the jth parameter, then the estimation of the faulty parameter value can be calculated by

$$\hat{\theta}_j^f(t) = \frac{1}{2}(\theta_j^{a(ij)}(t) + \theta_j^{b(ij)}(t)) \tag{10.153}$$

and the two bounds $\theta_j^{a(ij)}(t)$ and $\theta_j^{b(ij)}(t)$ of the interval can be used as the bounds of this parameter.

This estimation of the faulty parameter value and the obtained parameter bounds do not rely on classic parameter identification methods but on the proposed fault isolation method. As soon as the fault is isolated, they are obtained. We prove in the next section that the fault isolation of the proposed method is fast – generally classic parameter identification needs a longer time, so we can get this estimation more quickly than by using classic parameter identification methods.

10.4.6   Isolation speed

After the fault occurrence, in each parameter filter whose interval does not contain the faulty parameter value, the residuals of the two isolation observers will leave zero and increase immediately with the initial speeds

$$\dot{\varepsilon}^a(t_f) = h(f(x(t_f), \theta^f, u) - f(x(t_f), \theta^{oba}, u)),$$
$$\dot{\varepsilon}^b(t_f) = h(f(x(t_f), \theta^f, u) - hf(x(t_f), \theta^{obb}, u)).$$

According to Assumptions 10.1 and 10.3, for a parameter filter which does not correspond to the faulty parameter, $\varepsilon^a(t)$ and $\varepsilon^b(t)$ are large, so $\dot{\varepsilon}^a(t_f) = \lim_{\Delta t \to 0}(1/\Delta t)\varepsilon^a(t_f + \Delta t)$ and $\dot{\varepsilon}^b(t_f) = \lim_{\Delta t \to 0}(1/\Delta t)\varepsilon^b(t_f + \Delta t)$ are large. On the other hand, according to Lemma 10.7, the residual difference $\varepsilon^{ab}(t) = \varepsilon^a(t) - \varepsilon^b(t)$ is a monotonous function of the parameter difference $\nabla \theta_j^{ab} = \theta_j^a - \theta_j^b$. So, from the theoretical point of view, if the parameter interval

$[\theta_j^b, \theta_j^a]$ is small enough, two residual curves $\varepsilon^a(t)$ and $\varepsilon^b(t)$ will be close enough to each other and the case in which these two residuals have the same sign will occur early enough. That is to say, fault isolation will be fast enough.

10.4.7 Parameter partition

According to the preceding discussion, in order to satisfy the condition that the interval is in the corresponding domain $Q_{min}(a, b)$ (which is mentioned in the proof of Theorem 10.2) and in order to get good speed of fault isolation, the parameter partition should be fine. On the other hand, the finer the parameter partition, the more online calculation is needed. So, a trade-off should be made between the number of parameter intervals and the speed of fault isolation. The optimal partition depends on the system dynamic model. We can find a better trade-off by means of simulation tests because the system dynamic model is known.

10.4.8 Adaptive intervals

For the parameter filter in the case $j \neq s$, large amplitudes of $\varepsilon^a(t)$, $\varepsilon^b(t)$, and a small difference between $\varepsilon^a(t)$ and $\varepsilon^b(t)$ are favorable for isolation. If Assumption 10.3 is not very well satisfied, the amplitudes of $\varepsilon^a(t)$ and $\varepsilon^b(t)$ will not be ideally large. In this case, smaller interval sizes and more intervals are needed and it will cost more online calculation time. To confront this case an improved method using adaptive intervals is used. Another purpose of using adaptive intervals is to improve the precision of fault identification.

After the fault occurrence, in every parameter filter, the parameters of the two isolation observers are continually modified online by the parameter identification procedures which identify the corresponding system parameter. Consequently, the two interval bounds are modified. In this case, the isolation observers are called "adaptive isolation observers". For the parameter filter of the ith interval of the jth parameter, the parameter identification procedures are

$$\frac{d}{dt}\theta_j^{oba(ij)}(t) = k_\theta^{a(ij)}h(x - \hat{x}^{a(ij)}),$$

$$\frac{d}{dt}\theta_l^{oba(ij)}(t) = 0, \quad l \neq j, \tag{10.154}$$

$$\frac{d}{dt}\theta_j^{obb(ij)}(t) = k_\theta^{b(ij)}h(x - \hat{x}^{b(ij)}),$$

$$\frac{d}{dt}\theta_l^{obb(ij)}(t) = 0, \quad l \neq j \tag{10.155}$$

and

$$\theta_j^{a(ij)}(t) = \theta_j^{oba(ij)}(t), \theta_j^{b(ij)}(t) = \theta_j^{obb(ij)}(t),$$

where $k_\theta^{a(ij)}$ and $k_\theta^{b(ij)} \in R$ are identification gains and $hx = y_h$ is the hth component of the system output vector.

Initially, we discussed the adjustment of the parameter difference between the two adaptive isolation observers. Now we consider the case of $k_\theta^{a(ij)} = k_\theta^{b(ij)}$. From Equations (10.150)–(10.155), we obtain

$$\frac{d}{dt}(\theta_j^{oba(ij)}(t) - \theta_j^{obb(ij)}(t)) = k_\theta^{a(ij)} h E^{ab(ij)}(t), \tag{10.156}$$

$$\dot{E}^{ab(ij)}(t) = f(\hat{x}^{b(ij)}, \theta_j^{obb(ij)}(t), u) - f(\hat{x}^{a(ij)}, \theta_j^{oba(ij)}(t), u) - kc E^{ab(ij)}(t)$$

$$E^{ab(ij)}(t) = \hat{x}^{b(ij)}(t) - \hat{x}^{a(ij)}(t) \tag{10.157}$$

According to Lemma 10.7, $E^{ab(ij)}(t)$ and $\varepsilon^{ab(ij)}(t) = h E^{ab(ij)}(t)$ are monotonous functions of the parameter difference $\nabla\theta_j^{ab(ij)}(t) = \theta_j^{oba(ij)}(t) - \theta_j^{obb(ij)}(t)$ and

$$\lim_{\nabla\theta_j^{ab(ij)}(t)\to 0} E^{ab(ij)}(t) = 0, \qquad \lim_{\nabla\theta_j^{ab(ij)}(t)\to 0} \varepsilon^{ab(ij)}(t) = 0, \forall t \geq t_f$$

So, by using a gradient algorithm as indicated in Equation (10.156), the adjustment of $\nabla\theta_j^{ab(ij)}(t)$ implies that $\nabla\theta_j^{ab(ij)}(t)$, $E^{ab(ij)}(t)$ and $\varepsilon^{ab(ij)}(t)$ can converge to zero.

For the case in which $k_\theta^{a(ij)} \neq k_\theta^{b(ij)}$, the adjustment law of the parameter difference is given by

$$\frac{d}{dt}(\theta_j^{oba(ij)}(t) - \theta_j^{obb(ij)}(t)) = k_\theta^{a(ij)} h E^{ab(ij)}(t)$$

$$+ (k_\theta^{a(ij)} - k_\theta^{b(ij)})h(x(t) - \hat{x}^{b(ij)}(t)). \tag{10.158}$$

If $s = j$, the prediction error $e^{b(ij)} = x(t) - \hat{x}^{b(ij)}(t)$ will converge to zero according to the adjustment law of Equation (10.155) when $\theta_j^{obb(ij)}$ converges to $\theta_s^f|_{s=j}$. So, $E^{ab(ij)}(t)$ and $\varepsilon^{ab(ij)}(t)$ will converge to zero according to the adjustment law of Equation (10.158) when $\theta_j^{oba(ij)}(t) - \theta_j^{obb(ij)}(t)$ converges to zero.

If $s \neq j$, the identification procedure of Equation (10.155) cannot correctly operate. The term $(k_\theta^{a(ij)} - k_\theta^{b(ij)})h(x(t) - \hat{x}^{b(ij)}(t))$ in Equation (10.158) cannot converge to zero, so $E^{ab(ij)}(t)$, $\varepsilon^{ab(ij)}(t)$ and $\theta_j^{oba(ij)}(t) - \theta_j^{obb(ij)}(t)$ cannot converge to zero according to Equation (10.158).

We are now in a position to discuss the fault isolation and identification under the adaptive manner. There are two cases:

- *Case 1:* The interval under consideration contains the faulty parameter value of the post-fault system.

  Because $s = j$, $\theta_j^{oba(ij)}(t)$, $\theta_j^{obb(ij)}(t)$ and $\theta_s^f|_{s=j}$ are on the same parameter, so the parameter identification procedures of Equations (10.154) and (10.155) can identify correctly the faulty parameter θ_s^f. It is assumed that the gains $k_\theta^{a(ij)}$, $k_\theta^{b(ij)}$ guarantee the parameter identifications of Equations (10.154) and (10.157) without overshoot. So along with the operation of the parameter identifications, $\nabla\theta_j^{ab(ij)}(t) = \theta_j^{oba(ij)}(t) - \theta_j^{obb(ij)}(t)$ becomes smaller and smaller but the faulty parameter θ_j^f of the post-fault system is still between the parameters $\theta_j^{oba(ij)}(t)$ and $\theta_j^{obb(ij)}(t)$ all the time. According to Theorem 10.2, the condition: $v^{(ij)}(t) = -1$ is satisfied all the time.

• *Case 2:* The interval under consideration does not contain the faulty parameter value of the post-fault system.

There are two sub-cases:

– *Sub-case 1: s = j*

In the period beginning after the occurrence of the fault, the parameter variations of $\theta_j^{oba(ij)}(t)$ and $\theta_j^{obb(ij)}(t)$ caused by the parameter identification procedures are small.

According to Theorem 10.2, the condition $v^{(ij)}(t) = -1$ is not satisfied, we can therefore decide that the interval does not contain the faulty parameter value of the post-fault system.

– *Sub-case 2: s ≠ j*

The adjusted parameters $\theta_j^{oba(ij)}(t)$ and $\theta_j^{obb(ij)}(t)$ of the two adaptive isolation observers are not on the same parameter as the faulty parameter θ_s^f. So, the parameter identification procedures of Equations (10.154) and (10.155) cannot correctly identify the faulty parameter θ_s^f.

According to Lemma 10.6, the residuals $\varepsilon^{a(ij)}(t)$ and $\varepsilon^{b(ij)}(t)$ will maintain large amplitudes. On the other hand, as discussed above, if $k_\theta^{a(ij)} \neq k_\theta^{b(ij)}$, along with the operation of parameter identification, the difference $\varepsilon^{ab(ij)}(t) = \varepsilon^{a(ij)}(t) - \varepsilon^{b(ij)}(t)$ will converge to zero. Therefore, there must be instants after the fault occurrence that the signs of $\varepsilon^{a(ij)}(t)$ and of $\varepsilon^{b(ij)}(t)$ are the same, that is to say the condition $v^{(ij)} = -1$ is not satisfied. If $k_\theta^{a(ij)} \neq k_\theta^{b(ij)}$, the difference $\varepsilon^{ab(ij)}(t)$ cannot converge to zero, and if Assumption 10.3 is not well satisfied, the amplitudes of $\varepsilon^{a(ij)}(t)$ and $\varepsilon^{b(ij)}(t)$ are small and one may possibly fail in the isolation procedure, in other words the robustness of the isolation is not good.

Remark 10.10   *To summarize the above, if $k_\theta^{a(ij)} = k_\theta^{b(ij)}$, after the fault occurrence, for the interval which does not contain the faulty parameter value of the post-fault system, there must be moments in which the condition $v^{(ij)}(t) = -1$ is not satisfied, that is to say, the faulty parameter value must be excluded from this interval, even if Assumption 10.3 is not well satisfied. On the other hand, if $k_\theta^{a(ij)} \neq k_\theta^{b(ij)}$ and if Assumption 10.3 is not well satisfied, one may possibly fail in the isolation procedure.*

As soon as condition $v^{(ij)}(t) = -1$ is not satisfied, the decision that the interval does not contain the faulty parameter value can be made. So we can carry out isolation in the same way as in the non-adaptive method. It is assumed that, after fault isolation, the ith interval of the jth parameter contains the value of the faulty parameter θ_j^f. Because the adjusted parameters $\theta_j^{oba(ij)}(t)$ and $\theta_j^{obb(ij)}(t)$ converge to θ_j^f without overshoot, we can calculate the estimated value of θ_j^f by Equation (10.153) and $\theta_j^{oba(ij)}(t), \theta_j^{obb(ij)}(t)$ can be used as the two bounds of parameter θ_j^f. Along with the identification operations, $\theta_j^{oba(ij)}(t), \theta_j^{obb(ij)}(t)$ and $\hat{\theta}_j^f(t)$ will converge to θ_j^f. That is to say the estimation of the faulty parameter value becomes more and more precise and the parameter bounds become tighter and tighter.

Remark 10.11   *It should be mentioned that although the classic parameter identification principle is used after t_f, our method of parameter identification is much faster than that of the classic method. In our method the initial parameter value of the parameter identification process is $(\theta_j^{a(ij)}(t) + \theta_j^{b(ij)}(t)(t_f))/2$, which is already very close to the faulty parameter value θ_j^f (because of the small size of the interval), while the initial parameter value of the classic identification method, θ_j^0, is far from θ_j^f.*

As discussed above, if we want to get good robustness of the fault isolation, in each interval, $k_\theta^{a(ij)} = k_\theta^{b(ij)}$ should be satisfied. Because each observer works for two neighboring intervals, $k_\theta^{(ij)} = k_\theta^{b(i+1,j)}$. So the parameter identification gains of all the intervals in series should be the same. On the other hand, for a nonlinear dynamic system, in order to ensure no overshoot and ideal converging speed for each of the adaptive observers in series, a single identification gain for all the adaptive observers is not usually suitable. For two neighboring intervals, if we want to choose different identification gains, we should build two observers with different gains rather than only one.

10.4.9   Simulation studies

The alcohol-fermentation process model obtained from mass balance considerations [62, 64] is as follows:

$$\frac{dC(t)}{dt} = \mu(t)C(t) - u(t)C(t),$$

$$\frac{dS(t)}{dt} = \frac{1}{Y_{c/s}}\mu(t)C(t) + u(t)S_a - u(t)S(t), \qquad (10.159)$$

where $C(t)$ and $S(t)$ represent, respectively, the biomass and the substrate concentrations in the bioreactor. $u(t)$ is the dilution rate which is used as the control variable and is selected as a rectangular wave varying between 0.1 and 0.27 with a period of 30 h in the simulation. S_a is the substrate concentration in the feeding. The yield coefficient $Y_{c/s}$ is known and constant. $S(t)$ is the only measurable state of the system.

The term $\mu(t)$ is the growth rate of the biomass concentration. In general, it is a nonlinear function of the variables $C(t)$, $S(t)$ and a parameter vector θ. In this section, we choose a model based on the "Monod law" [65] for this nonlinear function:

$$\mu(t) = \mu(\theta, S(t)) = \mu_m \frac{S(t)}{K_s + S(t)},$$

where μ_m and K_s are the maximum growth rate and the saturation constant, respectively. $\theta = [\mu_m, K_s]$ represents the system parameter vector and $\theta^0 = [\mu_m^0, K_s^0]$ is used to represent the nominal value of the parameter vector. The fault is defined as a single parameter change in the system parameter vector θ. It is assumed that the fault occurs at time $t_f = 70$ h. In the simulation, the following data obtained from real experiments [62] are used:

$$\theta^0 = [0.38, 5.0], \quad S_a = 100g/1, \quad Y_{c/s} = 0.07.$$

Table 10.2   The parameter filter values for μ_m

No.	1	2	3	4	5	6
μ_m^b	0.20	0.26	0.32	0.41	0.45	0.49
μ_m^a	0.26	0.32	0.36	0.45	0.49	0.53

The possible value domains of the parameters in practice are $\mu_m \in [0.2, 0.53]$ and $K_s \in [0.5, 5.1]$. Each of them is divided into six parameter intervals. (The two intervals which contain the nominal values of the parameters have been eliminated.) For each parameter interval, a parameter filter is built. The values of the parameter filters for μ_m are shown in Table 10.2 and for K_s are shown in Table 10.3.

Let $j = 1$ correspond to parameter μ_m and $j = 2$ correspond to parameter K_s. $\psi = a$ corresponds to the parameter interval bound $\theta_j^{a(ij)}$ and $\psi = b$ corresponds to the parameter interval bound $\theta_j^{b(ij)}$. The isolation observer for the ith interval of the jth parameter on bound $\theta_j^{a(ij)}$ or $\theta_j^{b(ij)}$ is given as

$$\frac{d\hat{C}^{\psi(ij)}(t)}{dt} = \mu^{\psi(ij)}(t)\hat{C}^{\psi(ij)}(t) - u(t)\hat{C}^{\psi(ij)}(t) + \alpha(\hat{S}^{\psi(ij)}(t) - S(t)),$$

$$\frac{d\hat{S}^{\psi(ij)}(t)}{dt} = -\frac{1}{Y_{c/s}}\mu^{\psi(ij)}(t)\hat{C}^{\psi(ij)}(t) - u(t)\hat{S}^{\psi(ij)} + u(t)S_a(t) + \beta(\hat{S}^{\psi(ij)}(t) - S(t)),$$

$$\mu^{\psi(i1)}(t) = \mu_m^{\psi(i)}(t)\frac{S(t)}{K_s^0 + S(t)},$$

$$\mu^{\psi(i2)}(t) = \mu_m^0(t)\frac{S(t)}{K_s^{\psi(i)} + S(t)},$$

$$\mu_m^{a(i)}(t) = \mu_m^{b(i)}(t) = \mu_m^0, \quad t < t_f,$$

$$\mu_m^{a(i)}(t) = \mu_m^{(i)}, \quad \mu_m^{b(i)}(t) = \mu_m^{(i-1)}, \quad t = t_f,$$

$$\frac{d\mu_m^{\psi(i)}(t)}{dt} = wk_\theta^{\psi(i1)}(\hat{S}^{\psi(i1)}(t) - S(t)), \quad t > t_f,$$

$$K_s^{a(i)}(t) = K_s^{b(i)}(t) = K_s^0, \quad t < t_f,$$

$$K_s^{a(i)}(t) = K_s^{(i)}, \quad K_s^{b(i)}(t) = K_s^{(i-1)}, \quad t = t_f,$$

$$\frac{dK_s^{\psi(i)}(t)}{dt} = -wk_\theta^{\psi(i2)}(\hat{S}^{\psi(i2)}(t) - S(t)), \quad t > t_f,$$

Table 10.3   The parameter filter values for K_s

No.	1	2	3	4	5	6
K_s^b	0.50	1.60	2.70	3.70	4.00	4.40
K_s^a	1.60	2.70	3.70	4.00	4.40	4.90

Table 10.4 Isolation time (h) when the fault is on μ_m

Value of μ_m^f	Non-adaptive	Adaptive	Zhang
0.20	0.10	0.10	56
0.25	0.10	0.10	51
0.30	0.10	0.10	49
0.35	0.10	0.10	30.5
0.42	5.75	5.65	49
0.45	1.75	1.70	54
0.50	1.10	1.10	64

where α and β are the constant gains, $\mu_m^{(i-1)}$ and $\mu_m^{(i)}$ are two bounds of the ith interval of the parameter μ_m, and $K_s^{(i-1)}$ and $K_s^{(i)}$ are two bounds of the ith interval of the parameter K_s. For the adaptive manner, $w = 1$, while for the non-adaptive manner $w = 0$.

If the value of the system parameter K_s is equal to its nominal value, the system faulty parameter is μ_m. For various values of the faulty parameter μ_m, the isolation times are shown in Table 10.4.

If the value of the system parameter μ_m is equal to its nominal value, the system faulty parameter is K_s. For various values of the faulty parameter K_s, the isolation times are shown in Table 10.5.

In Tables 10.4 and 10.5, the "non-adaptive" column represents the results using the non-adaptive manner, the "adaptive" column the results using the adaptive manner and the "Zhang" column the results using Zhang's method [35]. The tables show that the proposed method is faster than Zhang's method.

To show the fault isolation in detail, we choose an example where the system parameter μ_m is equal to its nominal value and the system faulty parameter value is $K_s = 2.0$. Since the fault is not on parameter μ_m, then the fault filter of parameter μ_m sends the fault exclusion signal at time $t = 71.85$ h (the isolation times of this example are the same for both the adaptive manner and the non-adaptive manner) and the fault filter of parameter K_s does not send the fault exclusion signal. Figure 10.16 shows the non-containing signals sent by the parameter filters of parameter μ_m and Figure 10.17 shows the non-containing signals sent by the parameter

Table 10.5 Isolation time (h) when the fault is on K_s

Value of K_s^f	Non-adaptive	Adaptive	Zhang
0.5	0.35	0.35	245
1.0	0.85	0.85	50
1.5	1.80	1.80	50
2.0	1.85	1.85	43
2.5	5.75	5.65	33
3.0	5.35	5.35	36
3.5	9.05	8.35	41
4.0	0.30	0.30	36
4.5	0.10	0.10	40
4.8	0.10	0.10	31

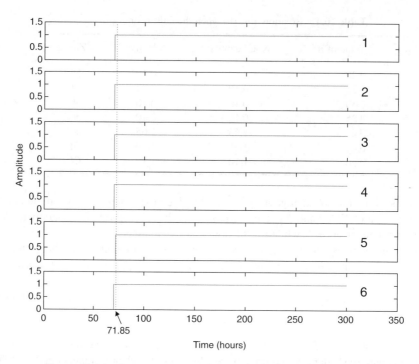

Figure 10.16   Non-containing signals sent by parameter filters of μ_m.

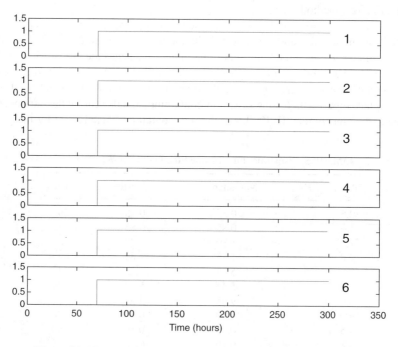

Figure 10.17   Non-containing signals sent by parameter filters of K_s.

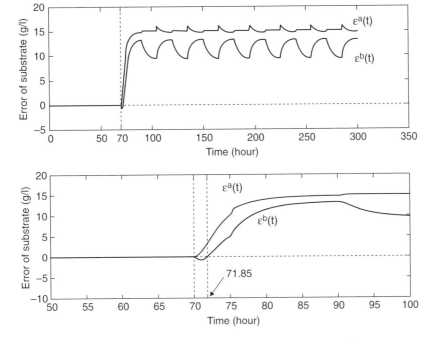

Figure 10.18   Results in the fifth interval of μ_m, non-adaptive

filters of parameter K_s. In the fault filter of μ_m, the fifth parameter filter is the last one which sends the non-containing signal, therefore, the isolation time is determined by this parameter filter.

Figure 10.18 shows the results of the non-adaptive manner of the fifth parameter filter of μ_m and Figure 10.19 shows the results of the adaptive manner of the fifth parameter filter of μ_m. Since the fault is not on parameter μ_m, the signs of the residuals $\varepsilon^a(t)$ and $\varepsilon^b(t)$ become the same after a period of time after the fault occurrence. It also shows that, because we chose $k_\theta^{a(5,1)} = k_\theta^{b(5,1)} = 0.0015$, for the adaptive manner over time, the difference $\varepsilon^a(t) - \varepsilon^b(t)$ converges to zero.

Figure 10.20 shows the results of the third parameter filter of parameter K_s in the non-adaptive manner, and Figure 10.21 shows the results of third parameter filter of the parameter K_s in the adaptive manner. Because the value 2.0 of the faulty parameter K_s is not in the 3rd parameter interval [$K_s^b = 2.7$, $K_s^a = 3.7$], so the signs of the residuals $\varepsilon^a(t)$ and $\varepsilon^b(t)$ are the same after the fault occurrence.

Figures 10.22 and 10.23 show the results (non-adaptive and adaptive, respectively) of the second parameter filter of parameter K_s. Because the value 2.0 of the faulty parameter K_s is in the second parameter interval [$K_s^b = 1.6$, $K_s^a = 2.7$], then the signs of the residuals $\varepsilon^a(t)$ and $\varepsilon^b(t)$ are different all the time after fault occurrence.

Figure 10.24 shows the results of the parameter identifications in the second parameter filter of parameter K_s (adaptive). $\hat{K}_s(t)$ represents the estimation of the faulty parameter calculated by Equation (10.153). $K_s^a(t)$ and $K_s^b(t)$ represent the two bounds of the faulty parameter K_s.

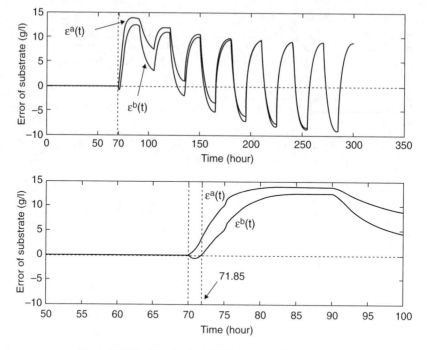

Figure 10.19   Results in the fifth interval of μ_m, adaptive

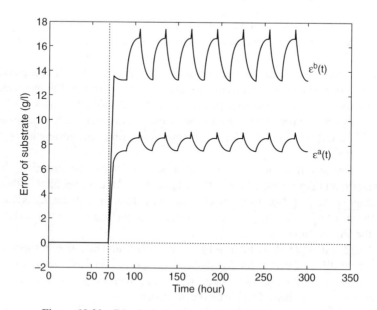

Figure 10.20   Results in the third interval of K_s, non-adaptive

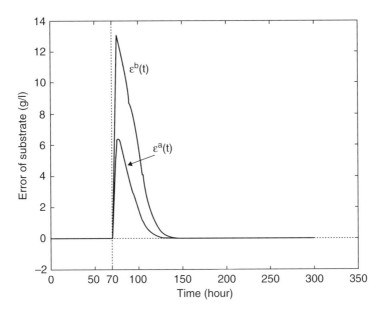

Figure 10.21 Results in the third interval of K_s, adaptive

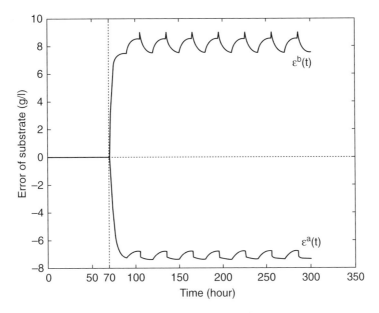

Figure 10.22 Results in the second interval of K_s, non-adaptive

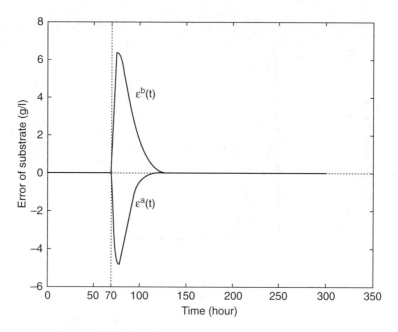

Figure 10.23   Results in the second interval of K_s, adaptive

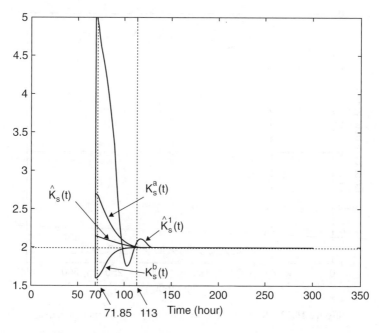

Figure 10.24   The results of parameter identification in the second interval of K_s

$\hat{K}_s^1(t)$ represents the estimation of the faulty parameter obtained by using Zhang's method. We can get the estimation of the faulty parameter only 1.85 h after the fault occurrence using the proposed method, but not until 43 h after the fault occurrence using Zhang's method.

To illustrate the effect of the difference of the parameter identification gains from the convergence of the parameter estimations difference in Equation (10.158), we use an example in which the value of K_s is nominal and the value of the faulty parameter is $\mu_m = 0.3$.

- *Case 1:* Between the first and second intervals of K_s, only one adaptive observer is used; its gain of identification is $k_\theta^{a(1,2)} = k_\theta^{b(2,2)} = 0.002$. The gain of another adaptive observer of the second interval is $k_\theta^{a(2,2)} = 0.005$. Therefore, in the second interval: $k_\theta^{b(2,2)} \neq k_\theta^{a(2,2)}$.

  Figure 10.25 shows the results of parameter identifications on the second interval. It shows that the parameter estimations $K_s^a(t)$ and $K_s^b(t)$ ($K_s^{oba}(t)$ and $K_s^{obb}(t)$) cannot converge to the accurate values $K_s = 5.0$ and the difference of the parameter estimations $K_s^a(t) - K_s^b(t)$ cannot converge to zero. Figure 10.26 shows the residuals $\varepsilon^a(t)$ and $\varepsilon^b(t)$ of the adaptive observers. Because $K_s^a(t) - K_s^b(t)$ cannot converge to zero, so the difference $\varepsilon^a(t) - \varepsilon^b(t)$ cannot converge to zero.
- *Case 2:* Between the first and the second intervals of K_s, two different adaptive observers are used. Their gains of identification are $k_\theta^{a(1,2)} = 0.002$ and $k_\theta^{b(2,2)} = k_\theta^{a(2,2)} = 0.005$; therefore, in the second interval $k_\theta^{b(2,2)} = k_\theta^{a(2,2)}$.

  Figure 10.27 shows the results of the parameter identifications on the second interval of K_s. It shows that though the parameter estimations $K_s^a(t)$ and $K_s^b(t)$ cannot converge to the

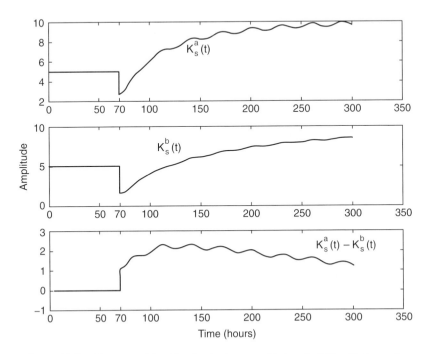

Figure 10.25 Parameter identification in the second interval of K_s with two gains

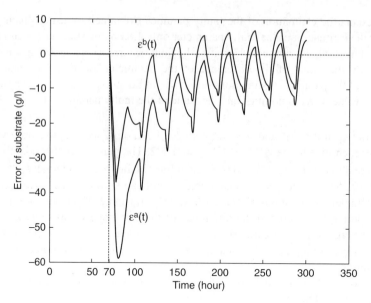

Figure 10.26   Residuals $\varepsilon^a(t)$ and $\varepsilon^b(t)$ in the second interval of K_s when the identification gains are different

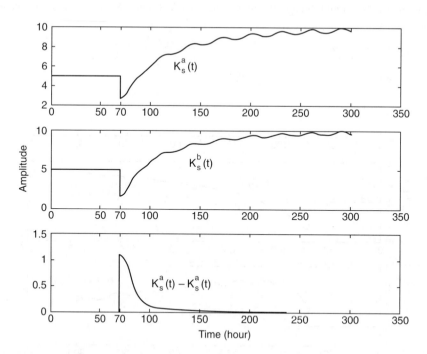

Figure 10.27   Parameter identification in the second interval of K_s when the identification gains are the same

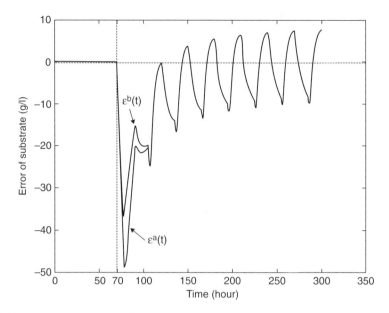

Figure 10.28   Residuals $\varepsilon^a(t)$ and $\varepsilon^b(t)$ in the second interval of K_s when the identification gains are the same

correct values $K_s = 5.0$, the difference of the parameter estimations, $K_s^a(t) - K_s^b(t)$, can converge to zero, therefore, the residual difference $\varepsilon^a(t) - \varepsilon^b(t)$ can converge to zero even if the residuals $\varepsilon^a(t)$ and $\varepsilon^b(t)$ cannot converge to zero, as shown in Figure 10.28.

10.5   Flexible-Joint Robots

Joint flexibility becomes important as the operating bandwidth of an industrial robot is increased. This flexibility of joints can cause instability and inaccuracy at high speeds and it is considered a predominant source of mechanical resonance and inaccurate tracking of robot trajectories. An important issue in robot control systems is to continuously monitor machines and detect faults that cause damaging effects to the performance of the systems. In this section, we present results pertaining to the application of the robust dynamic surface design approach for the purpose of fault detection followed by an adaptive fault accommodation control scheme.

10.5.1   Problem formulation

Control design methodologies for robotic manipulators require a nominal robot model. This model may describe the physical system correctly, but it is only an approximation to the physical system. There are several types of uncertainty which are prevalent and must be considered in the design of the control system. The uncertainty may be caused by unmodeled dynamics (e.g. flexibility), neglected nonlinearities, the compliance effect of the gear transmission mechanism in the joint, measurement noise, or uncertainty in the environment. Therefore, control design methods must be developed to account for robot model inaccuracies or uncertainties. Consider the dynamic model of n-link flexible-joint (FJ) robots described by Tomei [66]:

$$M(v)\ddot{v} + C(v, \dot{v})\dot{v} + G(v) + F\dot{v} + K_m(v - v_m) + \phi_1(v, \dot{;}v_m, t) = 0$$

$$J\ddot{v}_m + B\dot{v}_m + K_m(v_m - v) + \phi_2(v, \dot{v}, v_m, \dot{v}_m, t) = u$$

$$y = q \tag{10.160}$$

where $v, \dot{v}, \ddot{v} \in \mathfrak{R}^{3 \times 3}$ denote the joint position, velocity and acceleration vectors, respectively, $M(v) \in \mathfrak{R}^{3 \times 3}$ is a symmetric positive definite inertia matrix, $C(v, \dot{v}) \in \mathfrak{R}^{3 \times 3}$ denotes the Coriolis-centripetal matrix, $G(q)$ and F are the gravity vector and the coefficient of friction, respectively, at each joint. $v_m, \dot{v}_m, \ddot{v}_m \in \mathfrak{R}^n$ denote the actuator position, velocity, and acceleration vectors. The constant positive definite, diagonal matrices $K_m \in \mathfrak{R}^{(n \times n)}$, $J \in \mathfrak{R}^{(n \times n)}$, and $B \in \mathfrak{R}^{(n \times n)}$ represent the joint flexibility, the actuator inertia and the natural damping term, respectively. The control vector $u = ([u1, \ldots, un])^T \in \mathfrak{R}^{(n \times n)}$ is used as the torque input at each actuator, and $y = [y1, \ldots, yn]^T \in \mathfrak{R}^{(n \times n)}$ is an output vector. $\phi_1(v, \dot{v}, v_m, t) \in \mathfrak{R}^{(n \times n)}$ and $\phi_2(v, \dot{v}, v_m, \dot{v}_m, t) \in \mathfrak{R}^{(n \times n)}$ represent model uncertainty vectors including external disturbances. The ith actuator input, u_i, has the following fault type

$$u_i = \theta_{pi}u_{ni} + \theta_{bi}, \quad \forall t > t_{fi} \tag{10.161}$$

where $i = 1, \ldots, n; 0 < \theta_{pi} \leq 1$ is an unknown constant denoting partial loss of effectiveness; the constant θ_{bi} denotes an unknown actuator bias; and the ith actuator fails suddenly from an unknown time t_{fi}. It is assumed that a fault occurs only once on the ith actuator, so t_{fi} is unique. Hence, there exist time instants T_j for $j = 1, \ldots, l$ $(l \leq n)$, where T_1 and T_l are the unknown time instants for the first fault and the last fault, respectively. This assumption of the finite number of actuator faults can be found in several published works [67]–[70].

Remark 10.12 $\theta_{pi} = 1$ and $\theta_{bi} = 0 (i.e. u_i = u_{ni})$ *represent actuators working in a fault-free system.*

The following standard properties and assumptions are recalled.
Property 1 [71]: The inertia matrix $M(v)$ is a symmetric, uniformly bounded, positive definite matrix. In addition, $M'^1(v)$ is uniformly bounded i.e. $\|M^{-1}(v)\|_2 \leq \bar{M}_M$ where $\bar{M}_M$ is a positive constant.
Property 2: $\|M^{-1}(v)K_m\|_2 \leq \bar{M}_M$ where M_K is a known positive constant.

Assumption 10.4   *The functions* $\phi_1(v, \dot{v}, v_m, t)$ *and* $\phi_2(v, \dot{v}, v_m, \dot{v}_m, t)$ *representing the unstructured model uncertainties are bounded as* $\|\phi_1(v, \dot{v}, u, t)\| \leq \bar{\phi}_1(v, \dot{v}, v_m, t)$ *and* $\|\phi_2(v, \dot{v}, v_m, \dot{v}_m, t)\| \leq \bar{\phi}_2(v, \dot{v}, v_m, \dot{v}_m, t)$ *where* ϕ_1 *and* ϕ_2 *are known bounding functions.*

Assumption 10.5   *The desired trajectory vector,* y_d, *and its first and second derivatives* $y_d^{(1)}, y_d^{(1)}$ *are only available and bounded where* $y_d^{(i)}$ *denotes the ith derivative of* y_d.

Remark 10.13   *The fault detection problem for FJ robots and an adaptive fault accommodation control problem are considered with prescribed performance bounds; in previous fault* *detection and accommodation control schemes for robotic systems, bounds on performances are ignored.*

Let us consider the defining state variables as $x_1 = v$, $x_2 = \dot{v}$, $x_3 = v_m$ and $x_4 = \dot{v}_m$

$$
\begin{aligned}
\dot{x}_1 &= x_2 \\
\dot{x}_2 &= M^{-1}[-C(x_1, x_2)x_2 - G(x_1) - Fx_2 \\
&\quad - K_m(x_1 - x_3) - \phi_1(x_1, x_2, x_3, t)] \\
\dot{x}_3 &= x_4 \\
\dot{x}_4 &= J^{-1}\big[-Bx_4 - K_m(x_3 - x_1) \\
&\quad - \phi_2(x_1, x_2, x_3, x_4, t) + \theta_p U_n + \theta_b\big] \\
y &= x_1
\end{aligned}
\tag{10.162}
$$

where $\theta_p = diag[\theta_{p1}, \ldots, \theta_{pn}]$, $un = [u_{n1}, \ldots, u_{nn}]^T$ and $\theta_b = [\theta_{b1}, \ldots, \theta_{bn}]^T$. Here $diag[\cdot]$ denotes a diagonal matrix.

Remark 10.14 *There are two main objectives to be accomplished in this section. Firstly, we need to design the actuator fault detection scheme and the adaptive fault accommodation control law for FJ robots, so that the effects of actuator faults can be detected and then the adaptive tracking controller for the output vector y to track the desired trajectory vector y_d can be reconfigured to accommodate the faults.*

The second objective is to minimize the tracking control error $S_1 = y - y_d$ within certain given prescribed performance bounds regardless of actuator faults and improve its performance using tuning design parameters.

10.5.2   Fault detection scheme

This section introduces the robust fault detection observer to detect the occurrence of actuator faults in uncertain FJ robots. Fault detectability is analyzed in Theorem 10.4.

According to the system dynamics in (10.162), the fault detection observer is selected as

$$
\begin{aligned}
\dot{\hat{x}} &= A\hat{x} + f(x) + U + L(y - H\hat{x}) \\
\hat{y} &= H\hat{x}
\end{aligned}
\tag{10.163}
$$

where

$$
\begin{aligned}
x &= [x_1^T, \ldots, x_4^T]^T, \quad \hat{x} = [\hat{x}_1^T, \ldots, \hat{x}_4^T]^T, \\
f(x) &= [f_1^T(x), f_2^T(x), f_3^T(x), f_4^T]^T; \quad f_1 = f_2 = f_4 = 0_{n \times 1} \; 0_{n \times 1}, \\
f_3 &= M^{-1}(x_1)[-C(x_1, x_2)x_2 - G(x_1) - Fx_2 - K_m(x_1 - x_3)], \\
U &= [0_{n \times 1}^T, 0_{n \times 1}^T, 0_{n \times 1}^T, (J^{-1}U_n)^T]^T, \\
H &= [I_{n \times n} \; 0_{n \times n} \; 0_{n \times 1} \; 0_{n \times 1}]
\end{aligned}
$$

and $I_{n \times n}$ is an identity matrix with the dimensions $n \times n$, and the observer gain matrix $L \in \mathfrak{R}^{4n \times n}$ is chosen to make $A - LH$ stable.

Before the occurrence of the first fault, meaning for $t < T_1$, we have the fault detection error dynamics

$$\dot{\tilde{x}} = \bar{A}\tilde{x} + \phi(x, t) \tag{10.164}$$

$$\tilde{y} = H\tilde{x} \tag{10.165}$$

The threshold for monitoring fault occurrence is designed using the upper bound of the each component of the output error during the interval $[0, T_1)$. From Equation (10.163), we know that its solution can be expressed as

$$\tilde{x}(t) = e^{\bar{A}t}\tilde{x}(0) + \int_0^t e^{\bar{A}(t-\tau)}\phi(x, \tau)\,d\tau \tag{10.166}$$

where $\tilde{x}(0)$ is an initial value of $\tilde{x}(t)$. The analysis of each component of the output error $(\tilde{y}_i(t) = H_i\tilde{x}(t), i = 1, \ldots, n$, where H_i is the *ith* row vector of the matrix H) yields

$$|\tilde{y}_i(t)| \leq \alpha_i e^{-\Upsilon_i t}\|\tilde{x}(0)\| + \int_0^t \alpha_i e^{-\Upsilon_i(t-\tau)}\bar{\phi}\,d\tau \tag{10.167}$$

where α_i and Υ_i are positive constants chosen such that $\|H_i e^{\bar{A}t}\| \leq \alpha_i e^{-\Upsilon_i t}$ (since $\bar{A}$ is Hurwitz, constants α_i and v_i always exist). According to (10.168), the decision scheme for fault detection is as follows: the decision on the occurrence of a fault is made when at least one component of the output error (i.e. $|\tilde{y}_i(t)|$) exceeds its corresponding threshold $\gamma_i(t)$ defined as

$$\gamma_i(t) = \alpha_i e^{\Upsilon_i t}\|\tilde{x}(0)\| + \int_0^t \alpha_i e^{(-\Upsilon_i(t-\tau))}\bar{\phi}\,d\tau \tag{10.168}$$

Then, the fault detection time T_d is defined as the time instant such that $|\tilde{y}_i(T_d)| > \gamma_i(T_d)$ for some $T_d \geq T_1$ and some $i \in \{1, \ldots, n\}$.

Remark 10.15 *The main purpose of the fault detection observer is to design the threshold $\gamma_i(t)$ for the fault detection by separating the actuator faults and uncertainty terms ϕ of FJ robots. In the proposed scheme, the first occurrence of the fault is detected and then the adaptive accommodation controller with prescribed performance for actuator faults is applied to compensate for the actuator fault. Faults that occur after the occurrence of the first fault (i.e. $t > T_1$) can be overcome by the adaptive accommodation controller used after the detection of the first fault.*

In order to prove that the faults that are detectable by the proposed actuator fault detection scheme for uncertain FJ robots, the following result on fault detectability is derived.

Theorem 10.4 *Suppose that the faults of the FJ robots described by Equation (10.160) are detected using the fault detection observer scheme of Equation (10.163). Then, if there exist* *some time instant $T_d > T_1$ and some $i \in \{1, \ldots, n\}$, such that the actuator fault satisfies this inequality*

$$\int_{T_1}^{T_d} \|H_i e^{\bar{A}(T_d - \tau)}\| \| J^{-1}(\theta_p - I)u_n + \theta_b \| d\tau$$

$$> [\alpha_i \|\tilde{x}(T_1)\| + \gamma_i(T_1)]e^{-\Upsilon_i(T_d - T_1)}$$

$$+ 2 \int_{T_1}^{T_d} \alpha_i e^{-\Upsilon_i(T_d - \tau)} \bar{\phi} d\tau \tag{10.169}$$

then the actuator fault is detected at time $t = T_d$ (that is $|\tilde{y}_i(T_d)| > \gamma_i(T_d)$).

Proof. After the first occurrence of the fault (that is $t \geq T_1$), based on Equation (10.162) and Equation (10.163), the fault detection error dynamics becomes

$$\dot{\tilde{x}} = \bar{A}\tilde{x} + \phi(x, t) + (U_f - U)$$

$$\tilde{y} = H\tilde{x} \tag{10.170}$$

From Equation (10.170), each component of the output error is given by

$$\tilde{y}_i(t) = H_i e^{\bar{A}(t - T_1)}\tilde{x}(T_1) + \int_{T_1}^{t} H_i e^{-\bar{A}(t - T_1)} \phi(x, \tau) d\tau$$

$$+ \int_{T_1}^{t} H_i e^{\bar{A}(t - T_1)}(U_f - U) d\tau \tag{10.171}$$

where $i = 1, \ldots, n$. Then, applying the triangular inequality and standard assumptions, we get the following relationship

$$|\tilde{y}_i(t)| \geq \int_{T_1}^{t} \|H_i e^{\bar{A}(t - \tau)}\| \| J^{-1}(\theta_p - I)u_n + \theta_b \| d\tau$$

$$- \alpha_i e^{-\Upsilon_i(t - T_1)} \|\tilde{x}(T_1)\| - \int_{T_1}^{t} \alpha_i e^{-\Upsilon_i(t - \tau)} \bar{\phi} d\tau \tag{10.172}$$

Here, we know from Equation (10.168) that, for $t > T_1$,

$$\gamma_i(t) = e^{-\Upsilon_i(t - T_1)}(T_1) + \int_{T_1}^{t} \alpha_i e^{-\Upsilon_i(t - \tau)} \bar{\phi} d\tau \tag{10.173}$$

Therefore, using Equations (10.172) and (10.173), if there exists $T_d > T_1$ satisfying the inequality of Equation (10.169), we can conclude that $|\tilde{y}_i(T_d)| > \gamma_i(T_d)$, that is, the fault is detected at time $t = T_d$. ∎

10.5.3   Adaptive fault accommodation control

In this section we develop and design the adaptive fault accommodation control system and analyze its stability. For this, we first design the nominal control system for FJ robots operating without a fault condition and use this during the period $t < T_d$. After the detection of the first fault, the adaptive fault accommodation controller is used based on the combination of the dynamic surface design and the prescribed performance bounds.

A nominal controller design for FJ robots uses the dynamic surface design approach. Here we consider that the actuator operates in a fault-free condition. We first introduce error surfaces

$$S_1 = y - y_d$$
$$S_i = x_i - \Upsilon_{if} \quad i = 2, \dots, 4 \tag{10.174}$$

where $\Upsilon_{if} \in R_n$ is the filtered virtual controller of the virtual controller $\Upsilon_i \in R_n$.

Then, we propose the nominal control system as follows:

$$\Upsilon_2 = -k_1 S_1 y + \dot{y}_d \tag{10.175}$$
$$k_2 \dot{\Upsilon}_{2f} + \Upsilon_{2f} = \Upsilon_2 \tag{10.176}$$
$$\Upsilon_3 = K_m^{-1} \Big[K_m x_1 + F x_2 + G(x_1) + C(x_1, x_2) x_2$$
$$+ M(x_1) \Big\{ -k_2 S_2 + \frac{(\Upsilon_2 - \Upsilon_{2f})}{k_2} - \bar{M}_M \bar{\phi}_1 \frac{S_2}{(\|S_2\| + \epsilon_1)} \Big\} \Big] \tag{10.177}$$
$$k_3 \dot{\Upsilon}_{3f} + \Upsilon_{3f} = \Upsilon_3 \tag{10.178}$$
$$\Upsilon_4 = -k_3 S_3 + \frac{\Upsilon_3 - \Upsilon_{3f}}{k_3} \tag{10.179}$$
$$k_4 + \dot{\Upsilon}_{4f} + \Upsilon_{4f} = \Upsilon_4 \tag{10.180}$$
$$u = u_n = B x_4 + K_m (x_3 - x_1) + j \Big[-k_4 S_4$$
$$+ \frac{(\Upsilon_4 - \Upsilon_{4f})}{k_4} - J_M \bar{\phi}_2 \frac{S_4}{(\|S_4\| + \epsilon_2)} \Big] \tag{10.181}$$

where k_j, $j = 1, \dots, 4$, $l = 2, 3, 4$ and ϵ_1, ϵ_2 are positive design parameters.

Remark 10.16   *In the backstepping technique there is a need for repeated differentiation of the virtual controller. This situation is referred to as an "explosion of complexity". It can be overcome by using a first-order, low-pass filter at each step of design. For details, please refer to [72–75].*

For the stability analysis, we define the boundary layer error as

$$\bar{w}_i = \Upsilon_{if} - \Upsilon_i, \quad i = 2, 3, 4 \tag{10.182}$$

By applying Equations (10.175)–(10.181), the derivative of error surfaces can be represented by

$$\dot{S}_1 = S_2 + \bar{w}_2 - K_1 S_1 \tag{10.183}$$

$$\dot{S}_2 = M^{-1}(x_1) K_m (S_3 + \bar{w}_3) + M^{-1}(x_1)\phi_1 - k_2 S_2$$
$$- \bar{M}_M \bar{\phi}_1 \frac{S_2}{\|S_2\| + \epsilon_1} \tag{10.184}$$

$$\dot{S}_3 = S_4 + \bar{w}_4 - K_3 S_3 \tag{10.185}$$

$$\dot{S}_4 = J^- \phi_2 - k_4 S_4 - J_M \bar{\phi}_2 \frac{S_4}{\|S_4\| + \epsilon_2} \tag{10.186}$$

Then, differentiating Equation (10.182) yields

$$\dot{\bar{w}}_2 = -\frac{\bar{w}_2}{k_2} + \omega_2(S_1, S_2, \bar{w}_2, Y_d) \tag{10.187}$$

$$\dot{\bar{w}}_3 = -\frac{\bar{w}_3}{k_3} + \omega_3(S_1, S_2, S_3, \bar{w}_2, \bar{w}_3, Y_d) \tag{10.188}$$

$$\dot{\bar{w}}_4 = -\frac{\bar{w}_4}{k_4} + \omega_4(S_1, S_2, S_3, S_4, \bar{w}_2, \bar{w}_3, \bar{w}_4, Y_d) \tag{10.189}$$

where

$$Y_d = [y_d^T, \dot{y}_d^T, \ddot{y}_d^T]^T,$$

$$\omega_2 = k_1 \dot{S}_1 - \ddot{y}_d,$$

$$\omega_3 = -k_m^{-1}\Big[K_m \dot{x}_1 + F\dot{x}_2 + \frac{\partial G}{\partial x_1 \dot{x}_1}$$
$$+ (\frac{\partial C}{\partial x_1})\dot{x}_1 + (\frac{\partial C}{\partial x_2}\dot{x}_2)x_2 + C\dot{x}_2$$
$$+ \dot{x}_1^T \frac{\partial M}{\partial x_1}\big\{ -k_2 S_2 + (\Upsilon_2 - \Upsilon_{2f})/k_2 - \bar{M}_M \bar{\phi}_1 \vartheta \big\}$$
$$+ M\big\{ -k_2 \dot{S}_2 - \dot{\bar{w}}_2/k_2 - \bar{M}_M \bar{\phi}_1 \vartheta_1 \big\}\Big]$$

$$\vartheta_1 = S_2/\|S_2\| + \epsilon_1,$$

$$\omega_4 = k_3 \dot{S}_3 + \bar{\bar{w}}_3/k_3$$

and $\omega_2, \ldots, \omega_4 \omega_2$, ϑ_1 are continuous functions.

We now consider the following Lyapunov function candidate V_0 for $t \in [0, T_1)$

$$V_0 = \frac{1}{2}\Big[\sum_{l=1}^{4} S_l^T S_l + \sum_{l=1}^{3} \bar{w}_{l+1}^T \bar{w}_{l+1} \Big] \tag{10.190}$$

Theorem 10.5   *Consider the total closed-loop system consisting of the FJ robots of Equation (10.160) in the absence of actuator faults (i.e. $\theta_p = I$ and $\theta_b = 0$) and the dynamic surface controller of Equations (10.175)–(10.181). Under standard assumptions, for any initial conditions satisfying $V_0(0) \leq \mu_1$, where μ_1 is any positive constant, all signals of the closed-loop system are semiglobally uniformly ultimately bounded. Moreover, the tracking error satisfies the following property*

$$\|S_1(t)\| \geq \sqrt{2e^{-2\xi t}V_{(0)} + \frac{D}{\xi}(1 - e^{-2\xi t})}, \quad t \in [0, T_1) \tag{10.191}$$

where ξ and D are parameters of the set of attraction.

Remark 10.17   *The above equation makes it very clear that, before the fault occurrence $t < T_1$, the nominal control signal is sufficient and performance of tracking error S_1 can be improved by adjusting the design parameters. However, if the same control signal is continuously used even after the first fault occurrence $t \geq T_1$, then it is impossible to perform trajectory tracking at each fault time instant T_j, $j = 1, \ldots, l$ because the actuator fault time and magnitude are unknown. Therefore the transient performance of tracking error S_1 cannot be guaranteed during $t \geq T_1$. This motivates us to propose the adaptive fault accommodation control system with prescribed performance bounds.*

10.5.4   Control with prescribed performance bounds

After the detection of the first fault, (that is, $t \geq T_d$), the fault accommodation controller is required to maintain the control performance. The tracking error S_1 should be preserved within specified prescribed performance bounds all the time no matter when actuator faults occur. In brief, an adaptive fault accommodation controller for FJ robots has to be designed by defining the tracking error with performance function bounds as

$$-\delta_{L,i}\rho_i(t) < S_{1,i}(t) < \delta_{U,i}\rho_i(t), \quad \forall t \geq T_d \tag{10.192}$$

where $i = 1, \ldots, n$; $S_{1,i}$ is the ith component of S_1; $0 < \delta_{L,i}, \delta_{U,i} \leq 1$ are design constants; and $\rho_i(t) > 0$ is a decreasing smooth function with $\lim_{t \to +\infty} \rho_i(t) = \rho_{i,\infty} > 0$. Note that we choose the performance function as $\rho_i(t) = (\rho_{i,0} - \rho_{i,\infty})e^{\prime ai(t^\prime T_d)} + \rho_{i,\infty}$ with a_i, $\rho_{i,0}$ and $\rho_{i,\infty}$ as positive constants and $\rho_{i,0} > \rho_{i,\infty}$.

Remark 10.18   *The tracking errors $s_{1,i}(t)$ are bounded by the upper bounds $\delta_{U,i}, \rho_i(T_d)$ and lower bounds $\delta_{L,i}, \rho_i(T_d)$. This means that, even if there is a sudden actuator fault with unknown times and magnitudes, transient performance of the system can be guaranteed at the fault time instants T_j $j = 2, \ldots, l$. The performance function $\rho_i(t)$ and the design parameters $\delta_{L,i}$ and $\delta_{U,i}$ play a vital role in improving the transient performance of the system in terms of the convergence rate and overshoot. So, careful selection of these parameters is recommended.*

By considering the condition of tracking error Equation (10.191) instead of s_1 in the nominal controller, the original unconstrained control problem is transformed into a constrained tracking problem with bounds. For this error transformation, we can rewrite Equation (10.190) as

$$S_{1,i}(t) = \rho_i(t)\Psi_i(\Phi_i) \tag{10.193}$$

where $i = 1, \ldots, n$, Φ_i is a transformed error, and $\Psi_i(\Phi_i)$ is a smooth and strictly increasing function with $-\delta_{L,i} < \Psi_i(\Phi_i) < \delta_{U,i}$, $\lim_{\Phi_i \to +\infty} = \delta_{U,i}$, and $\lim_{\Phi_i \to +\infty} = \delta_{L,i}$. From the strictly increasing property of $\Psi_i(\Phi_i)$ and the fact $\rho_i(t) > 0$, the inverse function is well defined as

$$\Phi_i = \Psi_i^{-1}\left(\frac{S_{1,i}(t)}{\rho_i(t)}\right) \tag{10.194}$$

We now use the dynamic surface design approach to design the adaptive accommodation controller with prescribed performance bounds of uncertain FJ robots for $t \in [T_d, \infty]$. We apply the following procedure:

1. The first error surface is defined as Φ where $\Phi = [\Phi_1, \ldots, \Phi_n]^T$ and its derivative is

$$\dot{\Phi} = \Pi_\beta (x_2 - \dot{y}_d - \Pi_\rho S_1)$$

where $\Pi_\rho = diag[\dot{\rho}_1/\rho, \ldots, \dot{\rho}/\rho_n]$ and $\Pi_\beta = diag[\beta_1, \ldots, \beta_n]$ with

$$\beta_i = \frac{1}{2}\left[\frac{1}{S_{1,i}(t) + \delta_{L,i}(t)} - \frac{1}{S_{1,i}(t) - \delta_{U,i}(t)}\right] \tag{10.195}$$

Note that since the element of π_β is not zero and is well defined because of Equation (10.193), it is invertible.

Choose the virtual control Υ_2 as

$$\Upsilon_2 = -k_1 \Pi_\beta^{-1} \Phi + \dot{y}_d + \Pi_\rho S_1$$

where k_1 is a positive design parameter. Then, to obtain a filtered virtual controller Υ_{2f}, we pass Υ_2 through a low-pass first-order filter with a time constant k_2, i.e. we use Equation (10.175).

2. Design the virtual control laws Υ_3 for x_3. The second error surface is defined as $S_2 = x_2 - \Upsilon_{2f}$. Using its derivative, we can easily see that the virtual controller can be obtained as Equation (10.176). Then, pass Υ_3 through the low-pass first-order filter Equation (10.177) to obtain the filtered virtual control law Υ_{3f}.

3. In a similar way to Step 2, the third error surface, the virtual controller, and the first-order filter are defined as $S_3 = x_3 - \Upsilon_{3f}$, Equation (10.178), and Equation (10.179), respectively.

4. We consider the fourth equation of dynamics Equaton (10.160) with actuator faults. The error surface is defined as $S_4 = x_4 - \Upsilon_{4f}$. Differentiating it, we have

$$\dot{S}_4 = J^{-1}\left[-Bx_4 - K_m(x_3 - x_1)\right.$$
$$\left. - \phi_2(x_1, x_2, x_3, x_4, t) + \theta_p U_n + \theta_b\right] - \dot{\Upsilon}_{4f} \tag{10.196}$$

Then, the adaptive actual accommodation controller is proposed as

$$U_n = \hat{\Theta} W \tag{10.197}$$

where $\hat{\Theta} = [\hat{\Theta}_1 \ \ \hat{\Theta}_2]$; $\hat{\Theta}_1$ is an estimate matrix of $\hat{\Theta}_1 = \theta_p^{-1}$ and $\hat{\Theta}_2$ is an estimate vector of $\hat{\Theta}_2 = -\theta_p^{-1}\theta_b$ and $W = [U_{n0}, 1]^T$; U_{n0} is the nominal controller Equaton (10.181). Here, the adaptive parameter $\hat{\Theta}$ is tuned by the adaptive laws

$$\dot{\hat{\Theta}}_{1,i} = -\eta_{1,i} U_{n0,i} S_{4,i} - \sigma_1 \eta_{1,i}(\hat{\Theta}_{1,i} - \Theta_{1,i}^0) \tag{10.198}$$

$$\dot{\hat{\Theta}}_2 = \eta_2 S_4 - \sigma_2 \eta_2(\hat{\Theta}_2 - \hat{\Theta}_2^0) \tag{10.199}$$

where $i = 1, \dots, n$; $\hat{\Theta}_{1,i}$ the ith diagonal element of the matrix $\hat{\Theta}_1$; $U_{n0,i}$ and $S_{4,i}$ are the ith components of U_{n0} and S_4, respectively; $\eta_{1,i} > 0$ and $\eta_2 > 0$ are tuning parameters; $\Theta_{1,i}^0$ and Θ_2^0 are good guesses of $\Theta_{1,i}$ and Θ_2, respectively; and $\sigma_{1,i}$ and σ_2 are design constants for a σ modification [76].

Since the actuator faults occur at time instants T_j, we can define the time intervals $[T_j, T_{j+1})$ where $j = 1, \dots, l$, $T_1 \triangleq T_d$, and $T_{l+1} = \infty$. Note that since the adaptive accommodation controller is used after the detection of the first fault, we regard T_1 as T_d for the analysis and suppose that the fault patterns will not change until time T_{j+1}. For the time interval $[T_j, T_{j+1})$, we can define the Lyapunov function as

$$V_j = \frac{1}{2}\Big[\Phi^T \Phi + \sum_{h=2}^{4} S - h^T S_h + \sum_{h=1}^{3} \bar{w}_{h+1}^T w_{h+1}$$

$$+ tr(\tilde{\Theta}_1^T J^{-1}\theta_p \eta_1^{-1}\tilde{\Theta}_1) + \frac{1}{\eta_2}\tilde{\Theta}_2^T J^{-1}\theta_p \tilde{\Theta}_2\Big] \tag{10.200}$$

where $\tilde{\Theta}_1 - \hat{\Theta}_1$, $\tilde{\Theta}_2 = \Theta_2 - \hat{\Theta}_2$; Θ_1 and Θ_2 denote the matrix θ_p^{-1} and the vector $-\theta_p^{-1}\theta_b$ including the faults that occur at time instants T_j, $\bar{w}_{h+1}$ is the boundary layer error defined in Equation (10.181), and $\eta_1 = diag[\eta_{1,1}, \dots, \eta_{1,n}]$.

Theorem 10.6 *Consider the total closed-loop system consisting of FJ robots Equation (10.160) in the presence of actuator faults and the adaptive fault accommodation controller Equation (10.198) based on the prescribed performance bounds with adaptive laws of Equations (10.199) and (10.200). Under Assumptions 10.4 and 10.5, for any conditions satisfying $V_j(T_j) \le \mu_{2,j}$, where $\mu_{2,j}$ is any positive constant, all signals of the closed-loop system are semiglobally uniformly ultimately bounded and the tracking error S_1 converges to an adjustable neighborhood of the origin. Moreover, the transient performance of the system is guaranteed within specified prescribed performance bounds Equation (10.191) for $t \ge T_d$ regardless of the actuator faults that occur at time instants, T_j, $j = 2, \dots, l$.*

Remark 10.19 *It is very crucial to select the design parameters $(\rho_{i,0}, \rho_{i,\infty}, a_i)$ in order to guarantee the prescribed performance.*

- *Performance function ρ_i can be selected appropriately to adjust transient and steady-state performance bounds.*
- *A good transient performance largely depends on the initial estimate errors. Thus, we initially set small errors and choose appropriately big parameters k_i, μ_1 and μ_2 to obtain a large decay rate ζ_j where $i = 1, \dots, 4$ and $j = 1, \dots, \ell$.*
- *Parameters σ_1 and σ_2 are fixed to small values to get the small D_j and, accordingly, the tracking errors can be reduced within the prescribed bounds.*

10.5.5 Simulation results

We consider a three-link FJ manipulator with complex nonlinearities. The control objective is to design the fault detection observer detect actuator faults and the adaptive fault accommodation control law u_n for the output vector y to track the reference trajectory $y_d = [y_{d1}, y_{d2}, y_{d3}]^T$ as $y_{d1} = cos(1.5t + \pi/3)$, $y_{d2} = 1.2cos(1.5t)$ and $y_{d3} = 0.8cos(1.5t)$. The matrices $M(v)$, $C(v, \dot{v})$, $G(v)$, and system parameters of the three-link FJ manipulator are defined as

$$M(v) = \begin{bmatrix} 2d_1 + d_4C_2 + d_5C_{23} & 2d_2 + d_4C_2 + d_6C_3 & 2d_5 + d_5C_{23} + d_6C_3 \\ d_4C_2 + d_5C_{23} & 2d_2 + d_6C_3 & 2d_3 + d_6C_3 \\ d_5C_{23} & d_5C_3 & 2_d3 \end{bmatrix} \begin{bmatrix} 1 & 0 & 0 \\ 1 & 1 & 0 \\ 1 & 1 & 1 \end{bmatrix}$$

$$C(v, \dot{v}) = \begin{bmatrix} C_{m11} & C_{m12} & C_{m13} \\ C_{m21} & C_{m22} & C_{m23} \\ C_{m31} & C_{m32} & C_{m33} \end{bmatrix}$$

$$G(v) = \begin{bmatrix} \frac{1}{2}a_1C_1 & \frac{1}{2}a_2C_{12} + a_1C_1 & \frac{1}{2}a_3C_{123} + a_1C_1 + a_2C_{12} \\ 0 & \frac{1}{2}a_2C_{12} & \frac{1}{2}a_3C_{123} + a_2C_{12} \\ 0 & 0 & \frac{1}{2}a_3C_{123} \end{bmatrix}$$

$C_{m11} = -\dot{q}_2d_4S_2 - \dot{q}_2d_5S_{23} - \dot{q}_3d_5S_{23} - \dot{q}_3d_6S_3$

$C_{m12} = -\dot{q}_2d_4S_2 - \dot{q}_2d_5S_{23} - \dot{q}_3d_6S_3 - \dot{q}_3d_5S_{23} - \dot{q}_1d_4S_2 - \dot{q}_1d_5S_{23}$

$C_{m13} = -\dot{q}_2d_5S_{23} - \dot{q}_3d_5S_{23} - \dot{q}_3d_6S_3 - \dot{q}_1d_5S_{23}\dot{q}1d6S3\dot{q}2d6S3$

$C_{m21} = -\dot{q}_3d_6S_3 + \dot{q}_1d_4S_2 + \dot{q}_1d_5S_{23}$

$C_{m22} = -\dot{q}_3d_6S_3$

$C_{m23} = -d_6S_3(\dot{q}_1 + \dot{q}_2 + \dot{q}_3)$

$C_{m31} = \dot{q}_1d_5S_{23} + \dot{q}_1d_6S_3 + \dot{q}_2d_6S_3$

$C_{m32} = d_6S_3(\dot{q}_1 + \dot{q}_2)$

$C_{m33} = 0$

$$d_1 = \frac{1}{2}[(\frac{1}{4}m_1 + m_2 + m_3)a_1^2 + I_{01}]$$

$$d_2 = \frac{1}{2}[(\frac{1}{4}m_2 + m_3)a_2^2 + I_{02}]$$

$$d_3 = \frac{1}{2}[(\frac{1}{4}m_3)a_3^2 + I_{03}]$$

$$d_4 = (\frac{1}{2}m_2 + m_3)a_1a_2$$

$$d_5 = \frac{1}{2}m_3a_1a_3$$

$$d_6 = 0$$

Table 10.6 Simulation parameters for the robot dynamics

	Mass (kg)	Link (m)	Moment of inertia (kg/m^2)
Joint 1	1.0	0.5	43.33×10^{-3}
Joint 2	0.7	0.5	25.08×10^{-3}
Joint 3	1.4	0.5	32.67×10^{-3}

where m_{is} are the link masses (kg); $a_i S$ are the link lengths (m); $I_{oi} S$ are the moment of inertia about the centre of gravity (kg/m^2); g is the acceleration of gravity; q_i are the joint angular positions (rad); S_{ij} and C_{ij} denote $sin(q_i + q_j)$ and $cos(q_i + q_j)$, respectively. The parameters for the robot dynamics of the three-link FJ manipulator are shown in Table 10.6 and the FJ parameters are chosen as $J = diag[0.5, 0.5, 0.5]$, $B = diag[0.02, 0.02, 0.02]$, and $K_m = diag[15, 15, 15]$. It is assumed that the model uncertainty terms $\phi_1 = [0.3sin(x_{1,1}x_{2,2}), 0, 0.5cos(x_{2,2})x_{1,3}^2]^T$ and $\phi_2 = [0.2x_{1,1}, 0.7cos(x_{1,1}x_{1,2}), 0]^T$ influence the FJ robots.

In this simulation, the actuator faults are considered as

$$\theta_p = \begin{cases} diag[1, 1, 1] & \forall t \in [0, 5)\,s \\ diag[1, 0.7, 0.7] & \forall t \in [5, 8)\,s \\ diag[0.6, 1, 0.5] & \forall t \in [8, 20)\,s \end{cases}$$

$$\theta_b = \begin{cases} diag[0, 0, 10] & \forall t \in [0, 5)\,s \\ diag[0, 0, 0] & elsewhere \end{cases}$$

The initial positions of the three-link FJ manipulator and the fault detector are set to $y_1(0) = y_2(0) = y_3(0) = 0$ and $\hat{y}_1(0) = \hat{y}_2(0) = \hat{y}_3(0) = 0$. The design parameters for the fault detection scheme are selected as $L = 10[I_{3\times3}, I_{3\times3}, 0, I_{3\times3}]^T$, $\alpha_1 = \alpha_2 = \alpha_3 = 1$ and $v_1 = v_2 = v_3 = 5.6$. The design parameters of the nominal controller are chosen as $k_1 = k_2 = 10$, $k_3 = 5$, $k_4 = 30$, $\epsilon_1 = \epsilon_2 = 0.5$, and $\tau_2 = \tau_3 = \tau_4 = 0.005$. These parameters are also used for the adaptive accommodation controller. The other parameters for the adaptive accommodation controller are set to $\delta_{L,i} = \delta_{U,i} = 1$, $\rho_{i,0} = 2$, $\rho_{i,\infty} = 0.1$, $a_i = 1$, $\eta_{2,i} = 0.5$, $\eta_{1,1} =, \eta_{1,2} = 0.01\eta_{1,3} = 0.5$, $\sigma_1 = \sigma_2 = 0.001$, where $i = 1, 2, 3$. The initial values of $\Theta_{1,i}$ and $\Theta_{2,i}$ are set as $\Theta_{1,i}(0) = 1$ and $\Theta_{2,i}(0) = 0$.

The simulation considers three cases:

- *Case 1:* The nominal controller is used even in the presence of actuator faults. This is done to depict the effectiveness of the proposed adaptive accommodation scheme.
- *Case 2:* The nominal controller is used in fault-free conditions and adaptive accommodation controller in faulty situations.
- *Case 3:* The proposed adaptive fault accommodation controller without fault detection is used at all times.

The fault detection errors and their thresholds are shown in Figures 10.29–10.31. The proposed fault detection scheme detects the fault almost immediately, at approximately

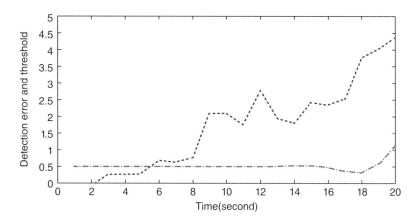

Figure 10.29 Actuator fault detection errors for link 1

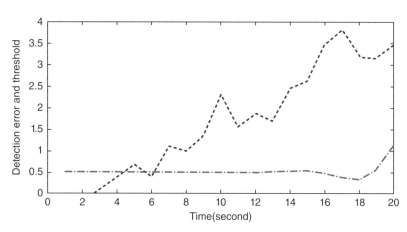

Figure 10.30 Actuator fault detection errors for link 2

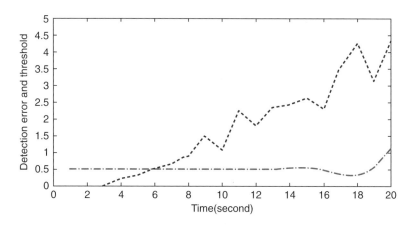

Figure 10.31 Actuator fault detection errors for link 3

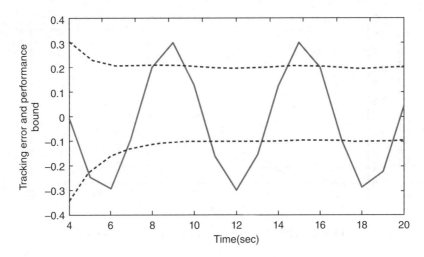

Figure 10.32   Control error with prescribed performance bound: nominal controller

$T_d = 5.37$. After the fault detection, the proposed adaptive accommodation controller with the prescribed performance bounds is used.

Figure 10.32 shows the simulation result obtained for Case 1 (using the nominal controller even after the fault is detected). From this result it can be concluded that the nominal controller alone is not sufficient in the event of actuator fault. The error tracking cannot be preserved within the prescribed bound after detection of the first fault at $t = 5.37$. This result motivates us to propose the adaptive fault accommodation control system with prescribed performance bounds.

Figure 10.33 reveals Case 2 (use of the nominal controller alongside the proposed adaptive accommodation controller). It can be seen that tracking error can be preserved within the

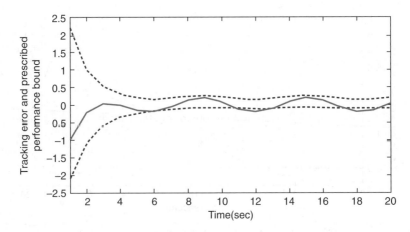

Figure 10.33   Control error with prescribed performance bound: nominal and adaptive controllers

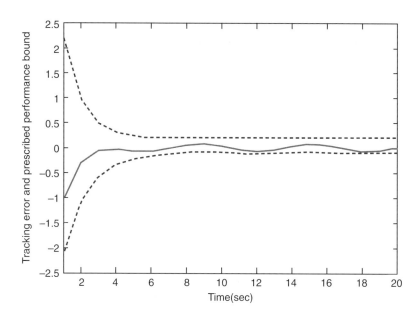

Figure 10.34   Control error with prescribed performance bound: adaptive controller

prescribed performance bound regardless of the actuator faults which occur after the initial fault detection. By this proposed scheme, we can overcome multiple actuator faults and guarantee transient performance throughout the process.

Figure 10.34 depicts the simulation result for Case 3 (use of the adaptive fault accommodation controller alone). We can see that the proposed controller can overcome multiple actuator faults and guarantee transient performance even at the moment that the faults occur; however, the adaptive scheme requires some computation effort even in the absence of faults.

This comparison clearly shows that the use of both controllers, that is both nominal and adaptive fault accommodation, is advantageous in terms of both stability and computational effort.

10.6   Notes

In this chapter, off-line and online GLR test methods were applied in biomedical signal processing and a windowed online GLR test scheme was proposed. In real data simulation, the RLS and Levinson methods were used to obtain the filter model parameters. Simulation shows the choice of threshold in a GLR test is important. It depends not only upon the system's signal-to-noise ratio but also on the actual noise levels. The proposed windowed online GLR test algorithm can be used easily in real applications, because of its fixed computing burden for each step.

A maximum likelihood estimation was presented in the frequency domain for abrupt state changes in MIMO linear systems. The state change signal in the frequency domain is modeled as a initial state disturbance by splitting the system into two parts at the hypothetical occurrence time instant, then the system in the frequency domain is free of the input signal of state changes.

The maximum likelihood estimation is applied to estimate the initial state disturbance, i.e. the magnitude of the abrupt state change. Finally, the occurrence time of the change is solved by maximizing the ratio of two likelihoods that are under two opposite hypotheses.

A fault isolation and identification scheme based on parameter intervals fits many kinds of nonlinear dynamic systems and has ideal fault isolation and identification speed. It is very useful in FTC systems. The objective of the work presented in this chapter was to find a fast fault isolation and identification method for nonlinear dynamic systems. When the fault is isolated, the primary estimation of the faulty parameter value is obtained bounded by the parameter interval. This primary estimation and the bounds are favorable for controller reconfiguration in FTC systems as it is provided very rapidly. Along with parameter identification, this estimation becomes more and more precise and the bounds become tighter and tighter. The adaptive manner of the method is interesting. For all the parameter filters of a parameter, no matter whether the fault is on this parameter or not, if the parameter identification procedures have the same gain, all the parameter differences between the adaptive isolation observers will converge to zero; in other words, the widths of all the parameter intervals will converge to zero. Consequently, the residuals of all the adaptive isolation observers will converge to a single curve. If the fault is not on the same parameter as the intervals, this curve cannot superpose with the abscissa axis, so there must be time when all the residuals have same sign. That is to say, the fault must be isolated even if Assumption 10.3 is not well satisfied.

The fault detection and adaptive accommodation control scheme for uncertain FJ robots with unknown multiple actuator faults was developed based on the dynamic surface design approach and prescribed performance bounds. The fault detection observer was designed to detect the actuator faults of FJ robots and fault detectability was analyzed. An adaptive fault accommodation scheme with prescribed performance bounds was presented to accommodate actuator faults. We proved the preservation within prescribed performance bounds of tracking errors and derived adaptive laws to compensate unknown actuator faults from a Lyapunov stability approach. Finally, we presented simulation results for the three-link FJ manipulator to verify the effectiveness of the proposed approach.

References

[1] Kral, C., Pirker, F., Pascoli, G., and Kapeller, H. (2008) "Robust rotor fault detection by means of the Vienna monitoring method and a parameter tracking technique", *IEEE Trans. Industrial Electronics*, **55**(12):4229–4237.

[2] Zhou, W., Habetler, T. G., and Harley, R. G. (2008) "Bearing fault detection via stator current noise cancellation and statistical control", *IEEE Trans. Industrial Electronics*, **55**(12):4260–4269.

[3] Boukas, E. K., and Xia, Y. (2008) "Descriptor discrete-time systems with random abrupt changes: stability and stabilisation", *Int. J. Control*, **81**:1311–1318.

[4] Amann, A., Tratnig, R., and Unterkofler, K. (2007) "Detecting ventricular fibrillation by time-delay methods", *IEEE Trans. Biomedical Engineering*, **54**(1):174–177.

[5] Rheinberger, K., Steinberger, T., Unterkofler, K., Baubin, M., *et al.* (2008) "Removal of CPR artifacts from the ventricular fibrillation ECG by adaptive regression on lagged reference signals", *IEEE Trans. Biomedical Engineering*, **55**(1):130–137.

[6] Aboukhalil, A., Nielsen, L., Saeed, M., Mark, R. G., Clifford, G. D. (2008) "Reducing false alarm rates for critical arrhythmias using the arterial blood pressure waveform", *J. Biomedical Informatics*, **41**:442–451.

[7] Matsumoto, T., and Yosui, K. (2007) "Adaptation and change detection with a sequential Monte Carlo scheme", *IEEE Trans. System, Man, and Cybernetics-Part B*, **37**(3):592–606.

[8] Willsky, A. S., and Jones, H. L. (1976) "A generalized likelihood ratio approach to the detection and estimation of jumps in linear systems", *IEEE Trans. Automatic Control*, **21**:108–112.

[9] Kerr, T. (1987) "Decentralized filtering and redundancy management for multisensor navigation", *IEEE Trans. Aerospace and Electronic Systems*, **23**:83–119.

[10] Xia, Y., and Han, J. (2005) "Robust Kalman filtering for systems under norm bounded uncertainties in all system matrices and error covariance constraints", *J. Systems Science and Complexity*, **18**(4):439–445.

[11] Shanableh, T., Assaleh, K., and Al-Rousan, M. (2007) "Spatio-temporal feature-extraction techniques for isolated gesture recognition in Arabic sign language", *IEEE Trans. System, Man, and Cybernetics-Part B*, **37**(3):641–650.

[12] Gustafsson, F. (2000) *Adaptive Filtering and Change Detection*, Chichester, UK: John Wiley & Sons.

[13] Harrison, P. J., and Stevens, C. F. (1976) "Bayesian forecasting (with discussion)", *J. Roy. Statist. Soc. B*, **38**(3):205–247.

[14] Yang, P., Dumont, G., and Ansermino, J. M. (2006) "Adaptive change detection in heart rate trend monitoring in anesthetized children", *IEEE Trans. Biomedical Engineering*, **53**(11):2211–2219.

[15] Simon, D. (2006) *Optimal State Estimation*, Chichester, UK: John Wiley & Sons.

[16] Ljung, L. (1987) *System Identification: Theory for the User*, New York: Prentice-Hall.

[17] Yin, Y., Shi, P., and Liu, F. (2011) "Gain-scheduled robust fault detection on time-delay stochastic nonlinear systems", *IEEE Trans. Industrial Electronics*, **58**:4908–4916.

[18] Wang, D., Shi, P., and Wang, W. (2010) "Robust fault detection for continuous-time switched delay systems: a linear matrix inequality approach", *IET Control Theory and Applications*, **4**(1):100–108.

[19] Mao, Z., Jiang, B., and Shi, P. (2009) "Protocol and fault detection design for nonlinear networked control systems", *IEEE Trans. Circuits and Systems–II: Express Briefs*, **56**:255–259.

[20] Zhong, M., Ding, Q., and Shi, P. (2009) "Parity space-based fault detection for Markovian jump systems", *Int. J. Systems Science*, **40**:421–428.

[21] Xia, Y., Amann, A., Liu, B. (2010) "Detection of abrupt changes in electrocardiogram with GLR algorithm", *IET Signal Processing*, **4**:650–657.

[22] Ukil, A., and Živanović, R. (2007) "Application of abrupt change detection in power systems disturbance analysis and relay performance monitoring", *IEEE Trans. Power Delivery*, **22**:59–66.

[23] Liu, Y., and Blostein, S. D. (1994) "Quickest detection of an abrupt change in a random sequence with finite change-time", *IEEE Trans. Information Theory*, **40**:1985–1993.

[24] Pintelon, R., and Schoukens, J. (2001) *System Identification: A Frequency Domain Approach*, New York: IEEE Press.

[25] McKelvey, T. (2002) "Frequency domain identification methods", *Circuits, Systems and Signal Processing*, **21**:39–55.

[26] Agüro, J. C., Yuz, J. I., Goodwin, G. C., Delgado, R. A. (2010) "On the equivalence of time and frequency domain maximum likelihood estimation", *Automatica*, **46**:260–270.

[27] Oppenheim, A. V., Schafer, R. W., and Buck, J. R. (1999) *Discrete-Time Signal Processing*, Englewood Cliffs, NJ, USA: Prentice-Hall.

[28] Brigham, E. O. (1988) *The Fast Fourier Transform and its Applications*, Englewood Cliffs, NJ, USA: Prentice-Hall.

[29] Basseville, M., and Nikiforov, I. V. (1993) *Detection of Abrupt Changes: Theory and Application*, Englewood Cliffs, NJ, USA: Prentice-Hall.

[30] Ersoy, O. (1985) "Real discrete Fourier transform", *IEEE Trans. Acoustics, Speech, and Signal Processing*, **33**:880–882.

[31] Blanke, M., Izadi-Zamanabadi, R., Bøgh, S. A., and Lunau, C. P. (1997) "Fault-tolerant control systems: a holistic view", *Control Engineering Practice*, **5**(5):693–702.

[32] Frank, P. M., and Ding, X. (1997) "Survey of robust residual generation and evaluation methods in observer-based fault detection systems", *J. Process Control*, **7**(6):403–424.

[33] Isermann, R. (2005) "Model-based fault-detection and diagnosis: status and applications", *Annual Reviews in Control*, **29**(1):71–85.

[34] Isermann, R., and Balle, P. (1997) "Trends in the application of model-based fault detection and diagnosis of technical processes", *Control Eng. Practice*, **5**:709–719.

[35] Zhang, Q. (2000) "A new residual generation and evaluation method for detection and isolation of faults in nonlinear systems", *Int. J. Adaptive Control Signal Processing*, **14**:759–773.

[36] Bonivento, C., Paoli, A., and Marconi, L. (2003) "Fault-tolerant control of the ship propulsion system benchmark", *Control Engineering Practice*, **11**:483–492.

[37] Chow, E. Y., and Willsky, A. S. (1984) "Analytical redundancy and the design of robust failure detection systems", *IEEE Trans. Automatic Control*, **29**:603–614.

[38] Gertler, J. J., and Singer, D. (1990) "A new structural framework for parity equation based failure detection and isolation", *Automatica*, **26**:381–388.

[39] Park, J., and Rizzoni, G. (1994) "An eigenstructure assignment algorithm for the design of fault detection filters", *IEEE Trans. Automatic Control*, **39**:1521–1524.

[40] Massoumnia, M. A. (1996) "A geometric approach to the synthesis of failure detection filters", *IEEE Trans. Automatic Control*, **31**:839–846.

[41] Hammouri, H., Kinnaert, M., and El Yaagoubi, E. H. (1999) "Observer-based approach to fault detection and isolation for nonlinear systems", *IEEE Trans. Automatic Control*, **44**(10):1879–1884.

[42] De Persis, C., and Isidori, A. (2001) "A geometric approach to nonlinear fault detection and isolation", *IEEE Trans. Automatic Control*, **45**(6):853–865.

[43] Hammouri, H., Kabore, P., Othman, S., and Biston, J. (2002) "Failure diagnosis and nonlinear observer application to a hydraulic process", *J. Franklin Institute*, **339**:455–478.

[44] Yu, D. N., Williams, D., Shields, D. N., and Gomm, J. B. (1995) "A parity space method of fault detection for bi-linear systems", in *Proc. American Control Conference*, Seattle, 1132–1133.

[45] Comtet-Varga, G., Cassar, J. P., and Staroswiecki, M. (1997) "Analytic redundancy relations for state affine systems", in *Proc. Fourth European Control Conference, ECC'97*, Brussels, Belgium, 375–380.

[46] Staroswiecki, M., and Comtet-Varga, G. (2001) "Analytical redundancy relations for fault detection and isolation in algebraic dynamic systems", *Automatica*, **37**:687–699.

[47] Armengol, J., Trave-Massuyes, L., Vehi, J., and Lluis de la Rosa, J. (2000) "A survey on interval model simulators and their properties related to fault detection", *Annual Review Control*, **24**:31–39.

[48] Fagarasan, I., Ploix, S., and Gentil, S. (2004) "Causal fault detection and isolation based on a set-membership approach", *Automatica*, **40**:2099–2110.

[49] Valdes-Gonzalez, H., Flaus, J. M., and Acuna, G. (2003) "Moving horizon state estimation with global convergence using interval technique: application to biotechnological processes", *J. Process Control*, **13**:325–336.

[50] Raissi, T., Ramdani, Y., and Candau, Y. (2005) "Bounded error moving horizon state estimator for nonlinear continuous-time systems: application to a bioprocess system", *J. Process Control*, **15**:37–545.

[51] Boukhris, A., Giuliani, S., and Mourot, G. (2001) "Rainfall-runoff multi-modelling for sensor fault diagnosis, *Control Engineering Practice*, **9**:659–671.

[52] Sainz, M. A., Armengol, J., and Vehi, J. (2002) "Fault detection and isolation of the three-tank system using the modal interval analysis", *J. Process Control*, **12**:325–338.

[53] Gouze, J. L., Rapaport, A., and Hadj-Sadok, M. Z. (2000) "Interval observers for uncertain biological systems", *Ecological Modeling*, **133**:45–56.

[54] Dochain, D. (2003) "State and parameter estimation in chemical and biochemical processes: a tutorial", *J. Process Control*, **13**:801–810.

[55] Alcorta Garcia, E., and Frank, P. M. (1997) "Deterministic nonlinear observer-based approaches to fault diagnosis: a survey", *Control Engineering Practice*, **5**(5):663–670.

[56] Adjallah, K., Maquin, D., and Ragot, J. (1994) "Nonlinear observer-based fault detection", in *Proc. 3rd IEEE Conference of Control Applications*, Glasgow, UK, 1115–1120.

[57] Gauthier, J. P., Hammouri, H., and Othman, S. (1992) "A simple observer for nonlinear systems applications to bioreactor", *IEEE Trans. Automatic Control*, **37**:875–880.

[58] Gauthier, J. P., and Kupka, I. A. K. (1994) "Observability and observers for nonlinear systems", *SIAM J. Control Optimization*, **32**:975–994.

[59] Kreisselmeier, G. (1977) "Adaptive observers with exponential rate of convergence", *IEEE Trans. Automatic Control*, **22**(1):2–8.

[60] Lynch, A. F., and Bortoff, S. A. (1997) "Nonlinear observer design by approximate error linearization", *Systems and Control Letters*, **32**(3):161–172.

[61] Zeng, F. Y., Dahhou, B., Nihtila, M. T., Goma, G. (1993) "Microbial specific growth rate control via MRAC method", *Int. J. Systems Science*, **24**:1973–1985.

[62] Kabbaj, N., Polit, M., Dahhou, B., and Roux, G. (2001) "Adaptive observers based fault detection and isolation for an alcoholic fermentation process", in *Proc. Eighth IEEE International Conference on Emerging Technologies and Factory Automation*, Antibes-Juan les Pins, France, 669–973.

[63] Stengel, R. F. (1991) "Intelligent failure-tolerant control", *IEEE Control Systems Magazine*, **11**(4):14–23.

[64] Zeng, F. Y., and Dahhou, B. (1993) "Adaptive control of a nonlinear fermentation process via MRAC technique", *Applied Mathematical Modeling*, **17**:58–69.

[65] Sauter, D., Mary, N., Sirou, F., and Thieltgen, A. (1994) "Fault diagnosis in systems using fuzzy logic", in *Proc. 3rd IEEE Conf. Control Application*, Glasgow, UK, 883–888.

[66] Tomei, P. (1991) "A simple PD controller for robots with elastic joints", *IEEE Trans. Automatic Control*, **36**(10):1208–1213.

[67] Tao, G., Chen, S. H., Joshi, S. M. (2002) "An adaptive failure compensation controller using output feedback", *IEEE Trans. Automatic Control*, **47**(3):506–511.

[68] Tang, X. D., Tao, G., and Joshi, S. M. (2005) "Adaptive output feedback actuator failure compensation for a class of nonlinear systems", *Int. J. Adaptive Control & Signal Processing*, **19**(6):419–444.

[69] Tang, X. D., Tao, G., and Joshi, S. M. (2007) "Adaptive actuator failure compensation for nonlinear MIMO systems with an aircraft control application", *Automatica*, **43**(11):1869–1883.

[70] Li, J. L. and Yang, G. H. (2012) "Adaptive actuator failure accommodation for linear systems with parameter uncertainties", *IET Control Theory & Application*, **6**(2):274–285.

[71] Lewis, F. L., Abdallah, C. T., and Dawson, D. M. (1993) *Control of Robot Manipulators*, New York: Macmillan.

[72] Swaroop, D., Hedrick, J. K., Yip, P. P., and Gerdes, J. C. (2000) "Dynamic surface control for a class of nonlinear systems", *IEEE Trans. Autom. Control*, **45**(10):1893–1899.

[73] Yip, P. P., and Hedrick, J. K. (1998) "Adaptive dynamic surface control: a simplified algorithm for adaptive backstepping control of nonlinear systems", *Int. J. Control*, **71**(5):959–979.

[74] Yoo, S. J., Park, J. B., and Choi, Y. H. (2006) "Adaptive dynamic surface control of flexible-joint robots using self-recurrent wavelet neural networks", *IEEE Trans. Syst., Man, Cybern. B*, **36**(6):1342–1355.

[75] Yoo, S. J., Park, J. B. and Choi, Y. H. (2008) "Adaptive output feedback control of flexible-joint robots using neural networks: dynamic surface design approach", *IEEE Trans. Neural Network*, **19**(10):1712–1726.

[76] Abouelsoud, A. A. (1998) "Robust regulator for flexible-joint robots using integrator backstepping", *J. Intell. Robot. Syst.*, **22**:23–38.

A

Supplementary Information

In this appendix, we provide some mathematical notations, collect useful algebraic inequalities and lemmas, and introduce some relevant topics which have been extensively used throughout the book.

A.1 Notation

Throughout the chapters, we denote by $\mathfrak{R}$ the field of real numbers and $\mathfrak{R}^{m \times n}$ the set of $m \times n$ real matrices. Let $\mathfrak{R}_+$ stands for the non-negative real numbers. Let $\mathbb{C}$ be the field of complex numbers and denote $\widehat{\mathbb{C}} = \{s \in \mathbb{C} : \mathfrak{R}(s) < 0\}$.

The transpose of a vector x and a matrix M is denoted by x^T and M^T, respectively. A matrix $M \in \mathfrak{R}^{n \times n}$ is symmetric if $M = M^T$.

1. If $M \in \mathfrak{R}^{m \times n}$, then $M[i : j, k : l]$ denotes a matrix of dimension $(j - i + 1) \times (l - k + 1)$ obtained by extracting rows i to j and columns k to l from the matrix M, with $m \geq j \geq i \geq 1, n \geq k \geq l \geq 1$.
2. I_m denotes the identity matrix of dimension m, $I_m \in \mathfrak{R}^{m \times m}$.
3. Let $\lambda_i(M)$ denote the ith eigenvalue of $M \in \mathfrak{R}^{n \times n}$, $i = 1, \ldots, n$. The spectrum of M is denoted by $\mathcal{S}(M) = \{\lambda_1(M), \ldots, \lambda_n(M)\}$.

Definition A.1 *Let matrices $A, B \in \mathfrak{R}^{n \times n}$ be symmetric.*

1. *A is positive definite if $x^T A x > 0$ for all nonzero $x \in \mathfrak{R}^n$, and A is positive semidefinite if $x^T A x \geq 0$ for all nonzero $x \in \mathfrak{R}^n$. We denote this by $A > 0$ and $A \geq 0$, respectively.*
2. *A is negative (semi)definite if $-A$ is positive (semi)definite.*
3. *$A < B$ and $A \leq B$ mean $A - B < 0$ and $A - B \leq 0$, respectively.*

Matrices, if their dimensions are not explicitly stated, are assumed to be compatible for algebraic operations. In symmetric block matrices or complex matrix expressions, we use the symbol • to represent a term that is induced by symmetry.

Analysis and Synthesis of Fault-Tolerant Control Systems, First Edition. Magdi S. Mahmoud and Yuanqing Xia.
© 2014 John Wiley & Sons, Ltd. Published 2014 by John Wiley & Sons, Ltd.

Definition A.2   *A matrix $M \in \mathfrak{R}^{n \times n}$ is called Hurwitz (or stable) if all its eigenvalues have negative real part, that is, $\lambda_i(M) \in \widehat{\mathbb{C}}, i = 1, \ldots, n$.*

A.1.1   Kronecker products

Let $A \in \mathfrak{R}^{m \times n}$, $B \in \mathfrak{R}^{p \times q}$. Then $A \otimes B$ denotes the Kronecker product of A and B, given by

$$A \otimes B = \begin{bmatrix} a_{11}B & \cdots & a_{1n}B \\ \vdots & \ddots & \vdots \\ a_{m1}B & \cdots & a_{mn}B \end{bmatrix} \in \mathfrak{R}^{mp \times nq}. \tag{A.1}$$

Proposition A.1   *Consider two matrices $A = \alpha I_n$ and $B \in \mathfrak{R}^{n \times n}$. Then $\lambda_i(A + B) = \alpha + \lambda_i(B), i = 1, \ldots, n$.*

Proof.   Take any eigenvalue $\lambda_i(B)$ and the corresponding eigenvector $v_i \in \mathbb{C}^n$. Then $(A + B)v_i = Av_i + Bv_i = \alpha v_i + \lambda_i v_i = (\alpha + \lambda_i)v_i$.   ■

Proposition A.2   *Given $A, C \in \mathfrak{R}^{m \times m}$ and $B \in \mathfrak{R}^{n \times n}$, consider two matrices $\bar{A} = I_n \otimes A$ and $\bar{C} = B \otimes C$, where $\bar{A}, \bar{C} \in \mathfrak{R}^{nm \times nm}$. Then $\mathcal{S}(A + \lambda_i(B)C)$, where $\lambda_i(B)$ is the ith eigenvalue of B.*

Proof.   Let $v \in \mathbb{C}^n$ be an eigenvector of B corresponding to $\lambda(B)$, and $u \in \mathbb{C}^m$ be an eigenvector of $M = (A + \lambda(B)C)$ with $\lambda(M)$ as the associated eigenvalue. Consider the vector $v \otimes u \in \mathbb{C}^{nm}$. Then

$$(\bar{A} + \bar{C})(v \otimes u) = v \otimes Au + Bv \otimes Cu$$

$$= v \otimes Au + \lambda(B)v \otimes Cu$$

$$= v \otimes (Au + \lambda(B)Cu)$$

Since $(A + \lambda(B)C)u = \lambda(M)u$, we get $(\bar{A} + \bar{C})(v \otimes u) = \lambda M(v \otimes u)$.   ■

In the following discussion, Ker(**M**) is used to denote the orthogonal complement of **M**. A block-diagonal matrix with sub-matrices $\mathbf{X}_1, \mathbf{X}_2, \ldots, \mathbf{X}_p$ on its diagonal is denoted by diag$\{\mathbf{X}_1, \mathbf{X}_2, \ldots, \mathbf{X}_p\}$. We use $\mathbf{S}^{n \times n}$ to denote the real, symmetric $n \times n$ matrices, and $\mathbf{S}_+^{n \times n}$ to denote positive definite matrices. If $\mathbf{M} \in \mathbf{S}^{n \times n}$, then $\mathbf{M} > 0$ ($M \geq 0$) indicates that $\mathbf{M}$ is positive definite (positive semi-definite), and ($\mathbf{M} < 0$ $M \leq 0$) denotes a negative definite (negative semi-definite) matrix. Given a symmetric matrix $\mathbf{A}$, its inertia is defined as In$(\mathbf{A}) = (\pi(\mathbf{A}), \nu(\mathbf{A}), \delta(\mathbf{A}))$ are the number of positive, negative and zero eigenvalues of $\mathbf{A}$. For a temporal and spatial variable $x(t, s)$, **T** and **S** are the temporal and spatial forward shift operators, defined as:

$$\mathbf{T}\mathbf{x}(t, s) = \mathbf{x}(t + 1, s), \quad \mathbf{S}\mathbf{x}(t, s) = \mathbf{x}(t, s + 1)$$

With a slight abuse of notation, the symbols **T** and **S** are also used to denote the time and spatial differential operators for continuous interconnected systems. The space of square summable functions is denoted by $\mathcal{L}_2$; that is, for any $u \in \mathcal{L}_2$, $\|u\|_2 := [\sum_{t=0}^{\infty} \sum_{s=-\infty}^{\infty} \mathbf{u}^T(t, s)\mathbf{u}(t, s)]^{\frac{1}{2}}$ is finite.

A.1.2 Some definitions

Let us now introduce some definitions from the work of Sepulchre, Janković, and Kokotović [1], that are frequently used in the treatment of nonlinear fault systems.

Definition A.3 *(see [1], page 45): The solution $x(t, x_0)$ of the system $\dot{x} = f(x)$, $x \in \mathfrak{R}^n$, f locally Lipschitz, is stable conditional on Z, if $x_0 \in Z$, and for each $\varepsilon > 0$ there exists $\delta(\varepsilon) > 0$ such that*

$$|\tilde{x}_0 - x_0| < \delta, \tilde{x}_0 \in Z \Rightarrow |x(t, \tilde{x}_0) - x(t, x_0)| < \epsilon, \quad \forall t \geq 0. \tag{A.2}$$

If, furthermore, there exist $r(x_0) > 0$, s.t. $|x(t, \tilde{x}_0) - x(t, x_0)| \Rightarrow 0$, for all $|\tilde{x}_0 - x_0| < r(x_0)$ and $\tilde{x}_0 \in Z$, the solution is asymptotically stable conditionally to Z. If $r(x_0) \to \infty$, the stability is global.

Definition A.4 *(see [1], page 48): Consider the system $H : \dot{x} = f(x, u)$, $y = h(x, u)$, $x \in \mathfrak{R}^n$, $u, y \in \mathfrak{R}^m$, with zero inputs, that is, $\dot{x} = f(x, 0)$, $y = h(x, 0)$, and let $Z \subset \mathfrak{R}^n$ be its largest positively invariant set contained in $\{x \in \mathfrak{R}^n | y = h(x, 0) = 0\}$. We say that H is globally zero-state detectable (GZSD) if $x = 0$ is globally asymptotically stable conditionally to Z. If $Z = \{0\}$, the system H is zero-state observable (ZSO).*

Definition A.5 *(see [1], page 27): We say that H is dissipative in $X \subset \mathfrak{R}^n$ containing $x = 0$, if there exists a function $S(x)$, $S(0) = 0$ such that for all $x \in X$*

$$S(x) \geq 0, S(x(T)) - S(x(0)) \leq \int_0^T \omega(u(t), y(t))dt, \tag{A.3}$$

for all $u \in U \subset \mathfrak{R}^m$ and all $T > 0$ such that $x(t) \in X$, for all $t \in [0, T)$, where the function $\omega : \mathfrak{R}^m \times \mathfrak{R}^m \to R$, called the supply rate, is locally integrable for every $u \in U$, that is, $\int_{t_0}^{t_1} |\omega(u(t), y(t))|dt < \infty, \forall t_0 \leq t_1$. S is called the storage function. If the storage function is differentiable, the previous conditions can be written as

$$\dot{S}(x(t)) \leq \omega(u(t), y(t)). \tag{A.4}$$

The system H is said to be passive if it is dissipative with the supply rate $w(u, y) = u^T y$.

Remark A.1 *The definitions of ZSD and ZSO are simply an extension to the nonlinear case of the classical notions of detectability and observability for linear systems [2].*

The following definitions are needed to study the case of time-varying faults in Section 5.3.12.

Definition A.6   *(see [3]): A function $\bar{x} : [0, \infty) \to \mathfrak{R}^n$ is called a limiting solution of the system $\dot{x} = f(t, x)$, and f a smooth vector function, with respect to an unbounded sequence t_n in $[0, \infty)$, if there exist a compact $\kappa \subset \mathfrak{R}^n$ and a sequence $\{x_n : [t_n, \infty) \to \kappa\}$ of solutions of the system such that the associated sequence $\{\hat{x}_n :\to x_n(t + t_n)\}$ converges uniformly to x on every compact subset of $[0, \infty)$.*

Definition A.7   *(see [4], page 144): A continuous function $\alpha[0, a) \to [0, \infty)$ is said to belong to class K if it is strictly increasing and $\alpha(0) = 0$. A continuous function $\beta : [0, a) \times [0, \infty) \to [0, \infty)$ is said to belong to class KL if for each fixed s the mapping $\beta(r, s)$ belongs to class K with respect to r and for each fixed r the mapping $\beta(r, s)$ is decreasing with respect to s and $\beta(r, s) \to 0$ as $s \to \infty$.*

Definition A.8   *A system is said to be of nonminimum phase, if it has internal dynamics and their associated zero dynamics are unstable in the Lyapunov sense.*

     Throughout this section it is said that a statement $P(t)$ holds a.e. if the Lebesgue measure of the set $\{t \in [0, \infty)| P(t)$ is false$\}$ is zero [3]. We also mean by semiglobal stability of the equilibrium point x^0 for the autonomous system $\dot{x} = f(x), x \in \mathfrak{R}^n$ with f a smooth function, that for each compact set $K \subset \mathfrak{R}^n$ containing x^0, there exists a locally Lipschitz state feedback, such that x^0 is asymptotically stable, with a basin of attraction containing K (see [5], Definition 3, page 1445):

A.1.3   Matrix lemmas

The first lemma relates the inertia of a non-singular symmetric matrix **N** on a subspace with the inertia of its inverse $\mathbf{N}^{-1}$ on the complementary subspace. It can be utilized to derive the dual formulation from its original test. For its proof, readers can refer to the work of Scherer [6].

Lemma A.1   *Suppose that N is symmetric and non-singular, and S is a negative subspace of $\mathbf{N}$ with maximal dimension. Then:*

$$N^{-1} > 0 \text{ on } S_\perp$$

where $S_\perp$ is the complimentary space of S.

     The second result is basically an extension of the well-known elimination lemma [7, 8] to quadratic matrix inequality. It is convenient for eliminating controller parameters from the synthesis conditions. Its proof follows closely the one in the work of D'Andrea et al. [8, 9] and is omitted here for lack of space.

Lemma A.2   *Assume a non-singular symmetric matrix*

$$\begin{bmatrix} \mathbf{Q} & \mathbf{S} \\ \bullet & \mathfrak{R} \end{bmatrix}$$

has its inverse

$$\begin{bmatrix} \tilde{\mathbf{Q}} & \tilde{\mathbf{S}} \\ \bullet & \tilde{\mathfrak{R}} \end{bmatrix}$$

with $In(\tilde{\mathbf{Q}}) = In(-\mathfrak{R})$ *and* $\delta(\mathfrak{R}) = 0$. *The quadratic inequality:*

$$\begin{bmatrix} \mathbf{I} & \mathbf{B}^T\mathbf{X}^T\mathbf{A} + \mathbf{C}^T \end{bmatrix} \begin{bmatrix} \mathbf{Q} & \mathbf{S} \\ \bullet & \mathfrak{R} \end{bmatrix} \begin{bmatrix} \mathbf{I} \\ \mathbf{A}^T\mathbf{X}\mathbf{B} + \mathbf{C} \end{bmatrix} < 0 \tag{A.5}$$

has a solution $\mathbf{X}$ *if and only if:*

$$\begin{bmatrix} \mathbf{B}_\perp^T & \mathbf{B}_\perp^T\mathbf{C}^T \end{bmatrix} \begin{bmatrix} \mathbf{Q} & \mathbf{S} \\ \bullet & \mathfrak{R} \end{bmatrix} \begin{bmatrix} \mathbf{B}_\perp \\ \mathbf{C}\mathbf{B}_\perp \end{bmatrix} < 0 \tag{A.6}$$

and

$$\begin{bmatrix} -\mathbf{A}_\perp^T\mathbf{C} & \mathbf{A}_\perp^T \end{bmatrix} \begin{bmatrix} \tilde{\mathbf{Q}} & \tilde{\mathbf{S}} \\ \tilde{\mathbf{S}}^T & \tilde{\mathfrak{R}} \end{bmatrix} \begin{bmatrix} -\mathbf{C}^T\mathbf{A}_\perp \\ \mathbf{A}_\perp \end{bmatrix} > 0 \tag{A.7}$$

Lemma A.3 *(Projection lemma [10]): Given a symmetric matrix* $Z \in S_m$ *and two matrices* U *and* V *of column dimension m, there exists an unstructured matrix* X *that satisfies*

$$U^T X V + V^T X^T U + Z < 0, \tag{A.8}$$

if and only if the following projection inequalities with respect to X *are satisfied*

$$N_U^T Z N_U < 0, \tag{A.9a}$$

$$N_V^T Z N_V < 0, \tag{A.9b}$$

where N_U *and* N_V *are arbitrary matrices whose columns form a basis of the null spaces of* U *and* V, *respectively.*

Lemma A.4 *Consider the switched system*

$$\begin{cases} \dot{x}(t) = A_\sigma x(t) + B_\sigma u(t) \\ y(t) = C_\sigma x(t) + D_\sigma u(t) \end{cases}, \tag{A.10}$$

and let $\alpha > 0$, $\mu \geq 1$ *and* $\gamma_i > 0, \forall i \in l$ *be given constants. Suppose that there exist positive definite* C^1 *functions* $V_{\sigma(t)} : \mathfrak{R}^n \to R$, $\sigma(t) \in l$ *with* $V_{\sigma(t_0)}(x_{t0}) \equiv 0$ *such that*

$$V_i(x(t_k)) \leq \mu V_j(x(t_k)), \forall (i, j) \in l \times l, i \neq j \tag{A.11}$$

$$\dot{V}_i(x(t)) \leq -\alpha V_i(x(t)) - y^T(t)y(t) + \gamma_i^2 u^T(t)u(t), \forall i \in l \tag{A.12}$$

then the switched system of Equation (A.10) is globally uniformly asymptotically stable and satisfies H_∞ performance with index γ no greater than $\max\{\gamma_i\}$, for any switching signal with average dwell time (ADT)

$$\tau_a > \tau_a^* = In\mu/\alpha. \tag{A.13}$$

Lemma A.5   *(Matrix inversion formula [11]): Suppose A_{11}, A_{22}, and $\begin{bmatrix} A_{11} & A_{12} \\ A_{21} & A_{22} \end{bmatrix}$ are nonsingular. Define $\Delta := A_{11} - A_{12A_{22}^{-1}}A_{21}$. Then,*

$$\begin{bmatrix} A_{11} & A_{12} \\ A_{21} & A_{22} \end{bmatrix}^{-1} = \begin{bmatrix} \Delta^{-1} & -\Delta^{-1}A_{12}A_{22}^{-1} \\ -A_{22}^{-1}A_{21}\Delta^{-1} & A_{22}^{-1} + A_{22}^{-1}A_{21}\Delta^{-1}A_{12}A_{22}^{-1} \end{bmatrix}. \tag{A.14}$$

A.2   Results from Probability Theory

A.2.1   Results-A

For complex signals in the frequency domain, it is very important to notice that there are some different issues in probability theory from real random variables. Suppose that X and Y are real random vectors in $\mathfrak{R}^k$, and $Z = X + jY$ is a complex vector in C^k, then the probability density function (PDF) of Z is defined as the joint PDF of X and Y:

$$p_Z(Z) = p_{X,Y}(X, Y) = \left. \frac{\partial^2 F(u, v)}{\partial u \partial v} \right|_{(u,v)=(X,Y)} \tag{A.15}$$

where $F(u, v) = p(U \le u, V \le v)$ is the probability distribution of the random vectors X and Y. Furthermore, if the random vectors X and Y are both with Gaussian distribution, then Z has complex Gaussian distribution. This distribution can be described with three parameters:

$$\mu \triangleq E\{Z\} \tag{A.16}$$

$$\Sigma \triangleq E\{(Z - \mu)(Z - \mu)^H\} \tag{A.17}$$

$$\Delta \triangleq E\{(Z - \mu)(Z - \mu)^T\} \tag{A.18}$$

where μ denotes the mathematical expectation of Z, covariance matrix Σ must be Hermitian and non-negative definite; the relation matrix Δ should be symmetric. Moreover, matrices Σ and Δ are such that the matrix

$$P_c = \bar{\Sigma} - \Delta^H \Sigma \Delta \tag{A.19}$$

is also non-negative definite, where $\bar{\Sigma}$ denotes the complex conjugate of Σ. The PDF for complex Gaussian distribution can be computed as

$$f(Z) = \frac{1}{\pi^k \sqrt{\det(\Sigma)\det(P_c)}} \exp\left\{ -\frac{1}{2}\xi^H R_c^{-1}\xi \right\} \tag{A.20}$$

where $\xi^H = [(Z - \mu)^H \ (Z - \mu)^T]$ and

$$R_c = \begin{bmatrix} \Sigma & \Delta \\ \Delta^H & \bar{\Sigma} \end{bmatrix} \tag{A.21}$$

Notice that any affine transformation of Z, which has complex Gaussian distribution denoted by $Z \sim \mathcal{CN}(\mu, \Sigma, \Delta)$, is with complex Gaussian distribution, i.e,

$$AZ + b \sim \mathcal{CN}(A\mu + b, A\Sigma A^H, A\Delta A^T) \tag{A.22}$$

Circular symmetric complex Gaussian distribution corresponds to the case in which the imaginary part and the real part of a complex Gaussian vector are uncorrelated, and the covariance of the real and imaginary parts are the same, i.e., $\Delta = 0$, then

$$\begin{bmatrix} X \\ Y \end{bmatrix} \sim \mathcal{N}\left(\begin{bmatrix} \Re(\mu) \\ \Im(\mu) \end{bmatrix}, \frac{1}{2} \begin{bmatrix} \Re(\Sigma) & -\Im(\Sigma) \\ \Im(\Sigma) & \Re(\Sigma) \end{bmatrix} \right). \tag{A.23}$$

This case is usually denoted $Z \sim \mathcal{CN}(\mu, \Sigma)$ and its distribution can also be simplified as

$$f(z) = \frac{1}{\pi^k \det(\Sigma)} \exp\{[z - \mu]^H \Sigma^{-1} [z - \mu]\}. \tag{A.24}$$

A.2.2 Results-B

Lemma A.6 *[12] If $A_S \in C^{m \times n}$, $b \in C^n$, and positive-definite matrix $W = W^H \in C^{m \times m}$, then the WLS problem*

$$\arg_x \min(A_S x - b)^H W (A_S x - b) \tag{A.25}$$

has a solution $x = Xb$, where X satisfies the following two conditions:

$$(i) \quad A_S X A_S = A_S \tag{A.26}$$

$$(ii) \quad (W A_S X)^H = W A_S X \tag{A.27}$$

A.2.3 Results-C

Lemma A.7 *If $P_i = P_i^H \in C^{n \times n}$, $i = 1, 2, \ldots, N$, $P_i \neq P_j$, $i \neq j$, $P = P^H \in C^{n \times n}$,*

$$\vec{P}_s = diag\{P_1, P_2, \ldots, P_N\}$$
$$\vec{P} = diag\{P, P, \ldots, P\}$$

then the finite DFT (defined in Equation (10.47)) and RDFT (defined in Equation (10.67)) have following properties:

(i) $(\mathcal{F}_N^n)\vec{P}(\mathcal{F}_N^n)^H = \vec{P}$

(ii) $(\mathcal{T}_N^n)\vec{P}(\mathcal{T}_N^n)^H = \vec{P}$

(iii) $(\mathcal{T}_N^n)\vec{P}_s(\mathcal{T}_N^n)^H$ *is not always with diagonal form when $N > 2$.*

Proof. A.2.2(i):

$$
\begin{aligned}
(\mathcal{F}_N^n)\vec{P}(\mathcal{F}_N^n)^H &= (\mathcal{F}_N \otimes I_n)(I_N \otimes P)(\mathcal{F}_N \otimes I_n)^H \\
&= [(\mathcal{F}_N I_N) \otimes (I_n P)](\mathcal{F}_N^H \otimes I_n) \\
&= (\mathcal{F}_N \otimes P)(\mathcal{F}_N^H \otimes I_n) \\
&= (\mathcal{F}_N \mathcal{F}_N^H) \otimes (P I_n) \\
&= \vec{P}
\end{aligned}
$$

A.2.2(ii): This is the same as for A.2.2(i).

A.2.2(iii): Define $N \times N$ matrices $E_i \triangleq diag\{0, \ldots, 0, 1, 0, \ldots, 0\}$, i.e. the ith diagonal element is 1 and all others are zero, then we have

$$
\begin{aligned}
(\mathcal{T}_N^n)\vec{P}_s(\mathcal{T}_N^n)^H &= (\mathcal{T}_N \otimes I_n)\left(\sum_{i=1}^N E_i \otimes P_i\right)(\mathcal{T}_N \otimes I_n)^H \\
&= \left[\sum_{i=1}^N (\mathcal{T}_N E_i) \otimes (I_n P_i)\right](\mathcal{T}_N^H \otimes I_n) \\
&= \sum_{i=1}^N (\mathcal{T}_N E_i \mathcal{T}_N^H) \otimes P_i \\
&= \sum_{i=1}^N (v_i v_i^H) \otimes P_i
\end{aligned}
$$

where v_i is the ith column of matrix $\mathcal{T}_N$. Let v_{ij} denote the jth element of v_i, then the elements of symmetric matrix $(v_i v_i^H)$ can be described as $v_{ij} v_{ik}$, if $N > 2$, there exist a column v_i with two nonzero elements v_{ij} and v_{ik}, $j \neq k$, i.e., there are two nonzero elements $v_{ij} v_{ik}$ located above and below the diagonal, respectively. Hence, matrix $(v_i v_i^H) \otimes P_i$ is not diagonal, and $(\mathcal{T}_N^n)\vec{P}_s(\mathcal{T}_N^n)^H$ is not always diagonal with diagonal $n \times n$ blocks.   ∎

A.2.4   Minimum mean square estimate

Assume a random variable Y depends on another random variable X. Of interest is the minimum mean square error estimate (MMSEE) which, simply stated, says $\hat{X}$ the estimate of X such that the mean square error given by $\mathbf{E}[X - \hat{X}]^2$ is minimized where the expectation is taken over the random variables X and Y. One of the standard results is given below.

Proposition A.3 *The minimum mean square error estimate is given by the conditional expectation $\mathbf{E}[X|Y = y]$.*

Proof. Consider the functional form of the estimator as $g(Y)$. Let $f_{X,Y}(x, y)$ denote the joint probability density function of X and Y. Then the cost function is given by

$$C := \mathbf{E}\left[X - \hat{X}\right]^2 = \int_x \int_y (x - g(y))^2 f_{X,Y}(x, y)\, dx\, dy$$

$$= \int_y dy\, f_Y(y) \int_x (x - g(y))^2 f_{X|Y}(x|y)\, dx$$

Taking the derivative of the cost function with respect to the function $g(y)$:

$$\frac{\partial C}{\partial g(y)} = \int_y dy\, f_Y(y) \int_x 2(x - g(y)) f_{X|Y}(x|y)\, dx$$

$$= 2 \int_y dy\, f_Y(y)\left(g(y) - \int_x x f_{X|Y}(x|y)\, dx\right)$$

$$= 2 \int_y dy\, f_Y(y)\left(g(y) - \mathbf{E}[X|Y = y]\right).$$

Therefore the only stationary point is $g(y) = \mathbf{E}[X|Y = y]$ and it can be easily verified that it is a minimum. ∎

Remark A.2 *It is noted that the result established in Proposition A.3 holds for vector random variables as well. Observe that MMSE estimates are important because, for Gaussian variables, they coincide with the maximum likelihood (ML) estimates. It is a standard result that, for Gaussian variables, the MMSE estimate is linear in the state value.*

In what follows, we assume zero mean values for all the random variables with R_X being the covariance of X and R_{XY} being the cross-covariance between X and Y.

Proposition A.4 *The best linear MMSE estimate of X given $Y = y$ is*

$$\hat{x} = R_{XY} \mathfrak{R}_Y^{-1} y$$

with the error covariance

$$P = R_X - R_{XY} \mathfrak{R}_Y^{-1} R_{YX}$$

Proof. Let the estimate be $\hat{x} = Ky$. Then the error covariance is

$$P := \mathbf{E}\big[(x - Ky)(x - Ky)^T$$
$$= R_X - K R_{YX} - R_{XY} K^T + K R_Y K^T$$

Differentiating P with respect to K and setting it equal to zero yields

$$-2R_{XY} + 2K\Re_Y^{-1}$$

The result follows immediately.   ∎

Extending Proposition A.3 to the case of linear measurements $y = Hx + v$, we have the following standard result.

Proposition A.5   *Let $y = Hx + v$, where H is a constant matrix and v is a zero mean Gaussian noise with covariance R_V independent of X. Then the MMSE estimate of X given $Y = y$ is*

$$\hat{x} = R_X H^T (H R_X H^T + R_V)^{-1} y$$

with the corresponding error covariance

$$P = R_X - R_X H^T (H R_X H^T + R_V)^{-1} H R_X$$

A.3   Stability Notions

In this section, we present some definitions and results pertaining to the stability of dynamical systems. A detailed account of this topic can be found in the work of [13].

Definition A.9   *A function of x and t is a* carathedory function *if, for all $t \in \Re$, it is continuous in x and for all $x \in \Re^n$, it is Lebesgue measurable in t.*

A.3.1   *Practical stabilizability*

Given the uncertain dynamical system

$$\dot{x}(t) = [A + \Delta A(r) + M]x(t) + [B + \Delta B(s)]u(t)$$
$$+ Cv(t) + H(t, x, r), \quad x(0) = x_o \tag{A.28}$$
$$y(t) = x(t) + w(t) \tag{A.29}$$

where $x \in \Re^n$, $u \in \Re^m$, $y =\in \Re^n$, $v \in \Re^s$, $w \in \Re^n$ are the state, control, measured state, disturbance, and measurement error of the system, respectively, and $r \in \Re^p$, $s \in \Re^q$ are the uncertainty vectors. The system of Equations (A.28)–(A.29) is said to be "practically stabilizable" if, given $\mathbf{d} > 0$, there is a control law $g(.,.) : \mathfrak{R}^m \times \mathfrak{R} \to \mathfrak{R}^m$, for which, given any admissible uncertainties r, s, disturbances $w \in \mathfrak{R}^n$, $v \in \mathfrak{R}^s$, any initial time $t_o \in \mathfrak{R}$ and any initial state $x_o \in \mathfrak{R}^n$, the following conditions hold

- The closed-loop system

$$\dot{x}(t) = [A + \Delta A(r) + M]x(t) + [B + \Delta B(s)]g(y, t)$$
$$+ Cv(t) + H(t, x, r) \tag{A.30}$$

possesses a solution $x(.) : [t_o, t_1] \to \mathfrak{R}^n$, $x(t_o) = x_o$.
- Given any $v > 0$ and any solution $x(.) : [t_o, t_1] \to \mathfrak{R}^n$, $x(t_o) = x_o$ of Equation (A.30) with $||x_o|| \le v$, there is a constant $d(v) > 0$ such that $||x(t)|| \le d(v)$, $\forall t \in [t_o, t_1]$.
- Every solution $x(.) : [t_o, t_1] \to \mathfrak{R}^n$ can be continued over $[t_o, \infty)$.
- Given any $\bar{d} \ge \mathbf{d}$, any $v > 0$ and solution $x(.) : [t_o, t_1] \to \mathfrak{R}^n$, $x(t_o) = x_o$ of Equation (A.30) with $||x_o|| \le v$, there exists a finite time $T(\bar{d}, v) < \infty$, possibly dependent on v but not on t_o, such that $||x(t)|| \le \bar{d}$, $\forall t \ge t_o + T(\bar{d}, v)$.
- Given any $d \ge \mathbf{d}$ and any solution $x(.) : [t_o, t_1] \to \mathfrak{R}^n$, $x(t_o) = x_o$ of Equation (A.30), there is a constant $\delta(d) > 0$ such that $||x(t_o)|| \le \delta d$ implies $||x(t)|| \le \bar{d}$, $\forall t \ge t_o$.

A.3.2   Razumikhin stability

We recall that a continuous function $\alpha : [0, a) \longmapsto [0, \infty)$ is said to belong to class $\mathcal{K}$ if it is strictly increasing and $\alpha(0) = 0$. Further, it is said to belong to class $\mathcal{K}_\infty$ if $a = \infty$ and $\lim_{r \to \infty} \alpha(r) = \infty$.

Consider a time-delay system

$$\dot{x}(t) = f(t, x(t - d(t))) \tag{A.31}$$

with an initial condition

$$x(t) = (t), \quad t \in [-\bar{d}, 0]$$

where the function vector $f : \mathfrak{R}^+ \times \mathcal{C}_{[-\bar{d},0]} \mapsto \mathfrak{R}^n$ takes $\mathcal{R} \times$ (bounded sets of $\mathcal{C}_{[-\bar{d},0]}$) into bounded sets in $\mathfrak{R}^n$; $d(t)$ is the time-varying delay and $d := \sup_{t \in \mathfrak{R}^+}\{d(t)\} < \infty$. The symbol $\mathcal{C}_{[a,b]}$ represents the set of $\mathfrak{R}^n$-valued continuous functions on $[a, b]$.

Lemma A.8   *If there exist class $\mathcal{K}_\infty$ functions $\zeta_1(.)$ and $\zeta_2(.)$, a class $\mathcal{K}$ function $\zeta_3(.)$, and a function $V_1(.) : [-\bar{d}, \infty] \times \mathfrak{R}^n \mapsto \mathfrak{R}^+$ satisfying*

$$\zeta_1(||x||) \le V_1(t, x) \le \zeta_2(||x||), \quad t \in \mathfrak{R}^+, \ x \in \mathfrak{R}^n$$

such that the time derivative of V_1 along the solution of the system of Equation (A.31) satisfies

$$\dot{V}_1(t, x) \le -\zeta_3(||x||) \quad if \quad V_1(t + d, x(t + d)) \le V_1(t, x(t)) \tag{A.32}$$

for any $d \in [-\bar{d}, 0]$, then the system of Equation (A.31) is uniformly stable. If in addition,

$$\zeta_3(\tau) > 0, \tau > 0$$

and there exists a continuous nondecreasing function $\xi(\tau) > 0, \tau > 0$ such that Equation (A.32) is strengthened to

$$\dot{V}_1(t, x) \leq -\zeta_3(\|x\|) \quad if \quad V_1(t + d, x(t + d)) \leq \xi(V_1(t, x(t))) \tag{A.33}$$

for any $d \in [-\bar{d}, 0]$, then the system of Equation (A.31) is uniformly asymptotically stable. If in addition, $\lim_{\tau \to \infty} \zeta_1(\tau) = \infty$, then the system of Equation (A.31) is globally uniformly asymptotically stable.

The proof of this lemma can be found in the work of Mahmoud [14].

Lemma A.9   *Consider the system of Equation (A.31). If there exists a function*

$$V_o(x) = x^T P x, \ P > 0$$

such that for $d \in [-\bar{d}, 0]$ the time derivative of V_o along the solution of Equation (A.31) satisfies

$$\dot{V}_o(t, x) \leq -q_1 \|x\|^2 \quad if \quad V_o(x(t + d)) \leq q_2 V_o(x(t)) \tag{A.34}$$

for some constants $q_1 > 0$ and $q_2 > 1$, then Equation (A.31) is globally uniformly asymptotically stable.

Proof. Since $P > 0$, it is clear that

$$\lambda_{\min}(P)\|x\|^2 \leq V_o(x) \leq \lambda_{\max}(P)\|x\|^2$$

Let $\zeta_1(\tau) = \lambda_{\min}(P)\tau^2$ and $\zeta_2(\tau) = \lambda_{\max}(P)\tau^2$. It is easy to see that both $\zeta_1(.)$ and $\zeta_2(.)$ are class $\mathcal{K}_\infty$ functions and

$$\zeta_1(\|x\|) \leq V_o(x) \leq \zeta_2(\|x\|), \quad x \in \mathfrak{R}^n$$

Further, let $\zeta_3(.) = -q_1 \tau^2$ and $\xi(\tau) = q_2 \tau$. It is evident from $q_1 > 0$ and $q_2 > 1$ that for $\tau > 0$.

$$\xi(\tau) > \quad and \quad \zeta_3(\tau) > 0$$

Hence, the conclusion follows from Equation (A.34).   ∎

A.4 Basic Inequalities

All mathematical inequalities are proved for completeness. They are termed "facts" because of their high frequency of usage in the analytical developments.

A.4.1 Schur complements

Given a matrix Ω composed of constant matrices Ω_1, Ω_2, Ω_3 where $\Omega_1 = \Omega_1^T$ and $0 < \Omega_2 = \Omega_2^T$ as follows

$$\Omega = \begin{bmatrix} \Omega_1 & \Omega_3 \\ \Omega_3^T & \Omega_2 \end{bmatrix}$$

We have the following results

    (A) $\Omega \geq 0$ if and only if either

$$\begin{cases} \Omega_2 \geq 0 \\ \Pi = \Upsilon \Omega_2 \\ \Omega_1 - \Upsilon \, \Omega_2 \, \Upsilon^T \geq 0 \end{cases} \tag{A.35}$$

or

$$\begin{cases} \Omega_1 \geq 0 \\ \Pi = \Omega_1 \Lambda \\ \Omega_2 - \Lambda^T \, \Omega_1 \, \Lambda \geq 0 \end{cases} \tag{A.36}$$

holds where Λ, Υ are some matrices of compatible dimensions.

    (B) $\Omega > 0$ if and only if either

$$\begin{cases} \Omega_2 > 0 \\ \Omega_1 - \Omega_3 \, \Omega_2^{-1} \, \Omega_3^T > 0 \end{cases}$$

or

$$\begin{cases} \Omega_1 \geq 0 \\ \Omega_2 - \Omega_3^T \, \Omega_1^{-1} \, \Omega_3 > 0 \end{cases}$$

holds where Λ, Υ are some matrices of compatible dimensions.

    Matrix $\Omega_3 \, \Omega_2^{-1} \, \Omega_3^T$ is often called the Schur complement $\Omega_1(\Omega_2)$ in Ω.

Proof. **(A)**: To prove Equation (A.35), we first note that $\Omega_2 \geq 0$ is necessary. Let $z^T = [z_1^T \quad z_2^T]$ be a vector partitioned in accordance with Ω. Thus we have

$$z^T \, \Omega \, z = z_1^T \Omega_1 z_1 + 2 z_1^T \Omega_3 z_2 + z_2^T \Omega_2 z_2 \tag{A.37}$$

Select z_2 such that $\Omega_2 z_2 = 0$. If $\Omega_3 z_2 \neq 0$, let $z_1 = -\pi \Omega_3 z_2$, $\pi > 0$. Then it follows that

$$z^T \Omega z = \pi^2 z_2^T \Omega_3^T \Omega_1 \Omega_3 z_2 - 2\pi z_2^T \Omega_3^T \Omega_3 z_2$$

which is negative for a sufficiently small $\pi > 0$. We thus conclude $\Omega_1 z_2 = 0$ which then leads to $\Omega_3 z_2 = 0$, $\forall z_2$ and consequently

$$\Omega_3 = \Upsilon \Omega_2 \tag{A.38}$$

for some Υ.

Since $\Omega \geq 0$, the quadratic term $z^T \Omega z$ possesses a minimum over z_2 for any z_1. By differentiating $z^T \Omega z$ from Equation (A.37) with respect to z_2^T, we get

$$\frac{\partial(z^T \Omega z)}{\partial z_2^T} = 2\Omega_3^T z_1 + 2\Omega_2 z_2 = 2\Omega_2 \Upsilon^T z_1 + 2\Omega_2 z_2$$

Setting the derivative to zero yields

$$\Omega_2 \Upsilon z_1 = -\Omega_2 z_2 \tag{A.39}$$

Using Equations (A.38) and (A.39) in Equation (A.37), it follows that the minimum of $z^T \Omega z$ over z_2 for any z_1 is given by

$$\min_{z_2} \; z^T \Omega z = z_1^T [\Omega_1 - \Upsilon \Omega_2 \Upsilon^T] z_1$$

which proves the necessity of $\Omega_1 - \Upsilon \Omega_2 \Upsilon^T \geq 0$.

On the other hand, we note that the conditions of Equation (A.35) are necessary for $\Omega \geq 0$ and since together they imply that the minimum of $z^T \Omega z$ over z_2 for any z_1 is nonnegative, they are also sufficient.

Using a similar argument, the conditions of Equation (A.36) can be derived as those of Equation (A.35) by starting with Ω_1.   ∎

The proof of **(B)** follows as direct corollary of **(A)**.

Let us introduce a lemma and a definition which are useful for the analysis of this multiplicative fault estimation problem.

Lemma A.10   *(Barbalat's Lemma [15]) If $\lim_{t\to\infty} \int_0^T f(\tau)d\tau$ exists and is finite, and $f(t)$ is a uniformly continuous function, then $\lim_{t\to\infty} f(\tau) = 0$.*

Definition A.10   *(Persistence of excitation [15]) A piecewise continuous signal vector ϕ : $\Re^+ \mapsto \Re^n$ is called "persistence of excitation" in $\Re^n$ with a level of excitation $\alpha_0 > 0$ if there exist constants $\alpha_1, T_0 > 0$ such that*

$$\alpha_1 I \geq \frac{1}{T_0} \int_t^{t+T_0} \phi(\tau)\phi^T(\tau) \geq \alpha_0 I, \quad \forall t \geq 0.$$

A.4.2   Bounding inequalities

Given matrices $0 < Q^T = Q, P = P^T$, then it follows that

$$- PQ^{-1}P \leq -2P + Q \tag{A.40}$$

This can be easily established by considering the algebraic inequality

$$(P - Q)^T Q^{-1}(P - Q) \geq 0$$

and expanding to get

$$PQ^{-1}P - 2P + Q \geq 0 \tag{A.41}$$

which when manipulating, yields Equation (A.40). An important special case is obtained when $P \equiv I$, that is

$$-Q^{-1} \leq -2I + Q \tag{A.42}$$

This inequality proves useful when using Schur complements to eliminate the quantity Q^{-1} from the diagonal of an LMI without alleviating additional math operations.

Lemma A.11   *The matrix inequality*

$$-\Lambda + S\,\Omega^{-1}\,S^T < 0 \tag{A.43}$$

holds for some $0 < \Omega = \Omega^T \in \Re^{n \times n}$, if and only if

$$\begin{bmatrix} -\Lambda & S\mathcal{X} \\ \bullet & -\mathcal{X} - \mathcal{X}^T + \mathcal{Z} \end{bmatrix} < 0 \tag{A.44}$$

holds for some matrices $\mathcal{X} \in \Re^{n \times n}$ and $\mathcal{Z} \in \Re^{n \times n}$.

Proof. ($\Longrightarrow$) By Schur complements, the inequality of Equation (A.43) is equivalent to

$$\begin{bmatrix} -\Lambda & S\Omega^{-1} \\ \bullet & -\Omega^{-1} \end{bmatrix} < 0 \tag{A.45}$$

Setting $\mathcal{X} = \mathcal{X}^T = \mathcal{Z} = \Omega^{-1}$, we readily obtain the inequality of Equation (A.44).
  ($\Longleftarrow$) Since the matrix $[I \quad S]$ is of full rank, we obtain

$$\begin{bmatrix} I \\ S^T \end{bmatrix}^T \begin{bmatrix} -\Lambda & S\mathcal{X} \\ \bullet & -\mathcal{X} - \mathcal{X}^T + \mathcal{Z} \end{bmatrix} \begin{bmatrix} I \\ S^T \end{bmatrix} < 0 \Longleftrightarrow$$

$$-\Lambda + S\,\mathcal{Z}\,S^T < 0 \Longleftrightarrow -\Lambda + S\,\Omega^{-1}\,S^T < 0\,, \ \mathcal{Z} = \Omega^{-1} \tag{A.46}$$

which completes the proof.   ∎

Lemma A.12   *The matrix inequality*

$$AP + PA^T + D^T \mathcal{R}^{-1} D + \mathcal{M} < 0 \tag{A.47}$$

holds for some $0 < P = P^T \in \Re^{n \times n}$, if and only if

$$\begin{bmatrix} A\mathcal{V} + \mathcal{V}^T A^T + \mathcal{M} & P + A\mathcal{W} - \mathcal{V} & D^T \mathcal{R} \\ \bullet & -\mathcal{W} - \mathcal{W}^T & 0 \\ \bullet & \bullet & -\mathcal{R} \end{bmatrix} < 0 \tag{A.48}$$

holds for some $\mathcal{V} \in \Re^{n \times n}$ and $\mathcal{W} \in \Re^{n \times n}$.

Proof. ($\Longrightarrow$) By Schur complements, the inequality of Equation (A.47) is equivalent to

$$\begin{bmatrix} AP + PA^T + \mathcal{M} & D^T \mathcal{R} \\ \bullet & -\mathcal{R} \end{bmatrix} < 0 \tag{A.49}$$

Setting $\mathcal{V} = \mathcal{V}^T = P, \mathcal{W} = \mathcal{W}^T = \mathcal{R}$, it follows from Lemma A.11 with Schur complements that there exists $P > 0, \mathcal{V}, \mathcal{W}$ such that the inequality of Equation (A.48) holds.
($\Longleftarrow$) In a similar way, Schur complements to the inequality of Equation (A.48) imply that:

$$\begin{bmatrix} A\mathcal{V} + \mathcal{V}^T A^T + \mathcal{M} & P + A\mathcal{W} - \mathcal{V} & D^T \mathcal{R} \\ \bullet & -\mathcal{W} - \mathcal{W}^T & 0 \\ \bullet & \bullet & -\mathcal{R} \end{bmatrix} < 0$$

$$\Longleftrightarrow \begin{bmatrix} I \\ A \end{bmatrix} \begin{bmatrix} A\mathcal{V} + \mathcal{V}^T A^T + \mathcal{M} + D^T P^{-1} D & P + A\mathcal{W} - \mathcal{V} \\ \bullet & -\mathcal{W} - \mathcal{W}^T \end{bmatrix} \begin{bmatrix} I \\ A \end{bmatrix}^T < 0$$

$$\Longleftrightarrow AP + PA^T + D^T P^{-1} D + \mathcal{M} < 0, \ \mathcal{V} = \mathcal{V}^T \tag{A.50}$$

which completes the proof.   ■

Lemma A.13   *Let $0 < L = L^T$ and X, Y be given matrices with appropriate dimensions. Then it follows that the inequality*

$$L(z) + X(z) \, P \, Y(z) + Y^T(z) \, P^T \, X^T(z) \ > \ 0 \tag{A.51}$$

holds for some P and $z = z_o$ if and only if the following inequalities

$$X_*^T(z) \, L(z) \, X_*(z) \ > \ 0, \ \ Y_*^T(z) \, L(z) \, Y_*(z) \ > \ 0 \tag{A.52}$$

hold with $z = z_o$.

It is significant to observe that the feasibility of the matrix inequality of Equation (A.51) with variables P and z is equivalent to the feasibility of Equation (A.52) with variable z and thus the matrix variable P has been eliminated from Equation (A.51) to form Equation (A.52). Using Finsler's lemma, we can express Equation (A.52) in the form

$$L(z) - \beta\, X(z)\, X^T(z) > 0, \quad L(z) - \beta\, Y(z)\, Y^T(z) > 0 \tag{A.53}$$

for some $\beta \in \mathfrak{R}$.

The following is a statement of the reciprocal projection lemma.

Lemma A.14   *Let $P > 0$ be a given matrix. The following statements are equivalent:*
(i) $\mathcal{M} + Z + Z^T < 0$
(ii) the LMI problem

$$\begin{bmatrix} \mathcal{M} + \mathcal{P} - (\mathcal{V} + \mathcal{V}') & V^T + Z^T \\ V + Z & -\mathcal{P} \end{bmatrix} < 0$$

is feasible with respect to the general matrix V.

The following lemmas show how to produce equivalent LMIs by an elimination procedure.

Lemma A.15   *There exists $\mathcal{X}$ such that*

$$\begin{bmatrix} \mathcal{P} & \mathcal{Q} & \mathcal{X} \\ \bullet & \mathcal{R} & \mathcal{Z} \\ \bullet & \bullet & \mathcal{S} \end{bmatrix} > 0 \tag{A.54}$$

if and only if

$$\begin{bmatrix} \mathcal{P} & \mathcal{Q} \\ \bullet & \mathcal{R} \end{bmatrix} > 0, \quad \begin{bmatrix} \mathcal{R} & \mathcal{Z} \\ \bullet & \mathcal{S} \end{bmatrix} > 0 \tag{A.55}$$

Proof. Since the LMIs of Equation (A.55) form sub-blocks on the principal diagonal of the LMI of Equation (A.54), necessity is established. To show sufficiency, apply the congruence transformation

$$\begin{bmatrix} I & 0 & 0 \\ \bullet & I & 0 \\ 0 & -V^T \mathfrak{R}^{-1} & I \end{bmatrix}$$

to the LMI of Equation (A.54); it is evident that Equation (A.54) is equivalent to

$$\begin{bmatrix} \mathcal{P} & \mathcal{Q} & \mathcal{X} - \mathcal{Q}\mathcal{R}^{-1}\mathcal{Z} \\ \bullet & \mathcal{R} & 0 \\ \bullet & \bullet & \mathcal{S} - \mathcal{Z}^T \mathcal{R}^{-1} \mathcal{Z} \end{bmatrix} > 0 \tag{A.56}$$

Clearly Equation (A.55) is satisfied for $\mathcal{X} = \mathcal{Q}\mathcal{R}^{-1}\mathcal{Z}$ if Equation (A.55) is satisfied in view of Schur complements.   ∎

Lemma A.16   *There exists $\mathcal{X}$ such that*

$$\begin{bmatrix} \mathcal{P} & \mathcal{Q}+\mathcal{X}\mathcal{G} & \mathcal{X} \\ \bullet & \mathcal{R} & \mathcal{Z} \\ \bullet & \bullet & \mathcal{S} \end{bmatrix} > 0 \qquad\qquad (\text{A.57})$$

if and only if

$$\begin{bmatrix} \mathcal{P} & \mathcal{Q} \\ \bullet & \mathcal{R}-\mathcal{V}\mathcal{G}-\mathcal{G}^T\mathcal{V}^T+\mathcal{G}^T\mathcal{Z}\mathcal{G} \end{bmatrix} > 0,$$

$$\begin{bmatrix} \mathcal{R}-\mathcal{V}\mathcal{G}-\mathcal{G}'\mathcal{V}'+\mathcal{G}'\mathcal{Z}\mathcal{G} & \mathcal{V}-\mathcal{G}^T\mathcal{Z} \\ \bullet & \mathcal{Z} \end{bmatrix} > 0 \qquad\qquad (\text{A.58})$$

Proof. Applying the congruence transformation

$$\begin{bmatrix} I & 0 & 0 \\ 0 & I & 0 \\ 0 & -\mathcal{G} & I \end{bmatrix}$$

to the LMI of Equation (A.57) and using Lemma A.15, we readily obtain the results.   ■

Lemma A.17   *There exists $0 < \mathcal{X}^T = \mathcal{X}$ such that*

$$\begin{bmatrix} \mathcal{P}_a + \mathcal{X} & \mathcal{Q}_a \\ \bullet & \mathcal{R}_a \end{bmatrix} > 0,$$

$$\begin{bmatrix} \mathcal{P}_c - \mathcal{X} & \mathcal{Q}_c \\ \bullet & \mathcal{R}_c \end{bmatrix} > 0 \qquad\qquad (\text{A.59})$$

if and only if

$$\begin{bmatrix} \mathcal{P}_a + \mathcal{P}_c & \mathcal{Q}_a & \mathcal{Q}_c \\ \bullet & \mathcal{R}_a & 0 \\ \bullet & \bullet & \mathcal{R}_c \end{bmatrix} > 0 \qquad\qquad (\text{A.60})$$

Proof. It is obvious from Schur complements that the LMI of Equation (A.60) is equivalent to

$$\mathcal{R}_a > 0, \quad \mathcal{R}_c > 0$$

$$\Xi = \mathcal{P}_a + \mathcal{P}_c - \mathcal{Q}_a \mathcal{R}_a^{-1} \mathcal{Q}_a^T - \mathcal{Q}_c \mathcal{R}_c^{-1} \mathcal{Q}_c^T > 0 \qquad\qquad (\text{A.61})$$

On the other hand, the LMI of Equation (A.59) is equivalent to

$$\mathcal{R}_a > 0, \quad \mathcal{R}_c > 0$$

$$\Xi_a = \mathcal{P}_a + \mathcal{X} - \mathcal{Q}_a \mathcal{R}_a^{-1} \mathcal{Q}_a^T > 0,$$

$$\Xi_c = \mathcal{P}_c - \mathcal{X} - \mathcal{Q}_c \mathcal{R}_c^{-1} \mathcal{Q}_c^T > 0 \qquad\qquad (\text{A.62})$$

It is readily evident from Equations (A.61) and (A.62) that $\Xi = \Xi_a + \Xi_c$ and hence the existence of $\mathcal{X}$ satisfying Equation (A.62) implies Equation (A.61). By the same token, if Equation (A.61) is satisfied, $\mathcal{X} = \mathcal{Q}_a R_a^{-1} \mathcal{Q}_a^T - \mathcal{P}_a - \frac{1}{2}\Xi$ yields $\Xi_a = \Xi_c = \Xi_a = \frac{1}{2}\Xi$ and Equation (A.62) is satisfied. ∎

A.5   Linear Matrix Inequalities

It has been shown that a wide variety of problems arising in system and control theory can be conveniently reduced to a few standard convex or quasi convex optimization problems involving linear matrix inequalities (LMIs). The resulting optimization problems can then be solved numerically very efficiently using commercially available interior-point methods.

A.5.1   Basics

One of the earliest LMIs arises in Lyapunov theory. It is well-known that the differential equation

$$\dot{x}(t) = Ax(t) \tag{A.63}$$

has all of its trajectories converge to zero (it is stable) if and only if there exists a matrix $P > 0$ such that

$$A^T P + AP \quad < \quad 0 \tag{A.64}$$

This leads to the LMI formulation of stability, that is, a linear time-invariant system is asymptotically stable if and only if there exists a matrix $0 < P = P^T$ satisfying the LMIs

$$A^T P + AP < 0, \quad P > 0$$

Given a vector variable $x \in \mathfrak{R}^n$ and a set of matrices $0 < G_j = G_j^T \in \mathfrak{R}^{n \times n}$, $j = 0, \ldots, p$, then a basic compact formulation of a linear matrix inequality is

$$G(x) := G_0 + \sum_{j=1}^{p} x_j G_j \quad > \quad 0 \tag{A.65}$$

Notice that Equation (A.65) implies that $v^T G(x) v > 0$, $\forall 0 \neq v \in \mathfrak{R}^n$. More importantly, the set $x \,|G(x) > 0$ is convex. Nonlinear (convex) inequalities are converted to LMI form using Schur complements in the sense that

$$\begin{bmatrix} Q(x) & S(x) \\ \bullet & R(x) \end{bmatrix} > 0 \tag{A.66}$$

where $Q(x) = Q^T(x)$, $R(x) = \mathfrak{R}^T(x)$, $S(x)$ dependent affinely on x, is equivalent to

$$R(x) > 0, \quad Q(x) - S(x)\mathfrak{R}^{-1}(x)S^T(x) > 0 \tag{A.67}$$

More generally, the constraint

$$Tr[S^T(x)\, P^{-1}(x)\, S(x)] < 1, \quad P(x) > 0$$

where $P(x) = P^T(x) \in \Re^{n \times n}$, $S(x) \in \Re^{n \times p}$ dependent affinely on x, is handled by introducing a new (slack) matrix variable $Y(x) = Y^T(x) \in \Re^{p \times p}$ and the LMI (in x and Y):

$$TrY < 1, \quad \begin{bmatrix} Y & S(x) \\ \bullet & P(x) \end{bmatrix} > 0 \tag{A.68}$$

Most of the time, our LMI variables are matrices. It should be clear from the preceding discussions that a quadratic matrix inequality (QMI) in the variable P can be readily expressed as a linear matrix inequality (LMI)in the same variable.

A.5.2   Some standard problems

Here we provide some common convex problems that we have encountered throughout the book. Given an LMI $G(x) > 0$, the corresponding LMI problem (LMIP) is: *to find a feasible* $x \equiv x^f$ *such that* $G(x^f) > 0$ *or to determine that the LMI is infeasible*. It is obvious that this is a convex feasibility problem.

The generalized eigenvalue problem (GEVP) is to minimize the maximum generalized eigenvalue of a pair of matrices that depend affinely on a variable, subject to an LMI constraint. GEVP has the general form

$$\begin{aligned} &minimize\ \ \lambda \\ &subject\ to\ \ \ \lambda B(x) - A(x) > 0, \quad B(x) > 0, \\ &\qquad\qquad\qquad\qquad\qquad\qquad C(x) > 0 \end{aligned} \tag{A.69}$$

where A, B, C are symmetric matrices that are affine functions of x. Equivalently stated

$$\begin{aligned} &minimize\ \ \lambda_M[A(x), B(x)] \\ &subject\ to\ \ B(x) > 0, \ C(x) > 0 \end{aligned} \tag{A.70}$$

where $\lambda_M[X, Y]$ denotes the largest generalized eigenvalue of the pencil $\lambda Y - X$ with $Y > 0$. This is problem is a quasi-convex optimization problem since the constraint is convex and the objective, $\lambda_M[A(x), B(x)]$, is quasi convex.

The eigenvalue problem (EVP) is to minimize the maximum eigenvalue of a matrix that depends affinely on a variable, subject to an LMI constraint. EVP has the general form

$$\begin{aligned} &minimize\ \ \lambda \\ &subject\ to\ \ \ \lambda I - A(x) > 0, \quad B(x) > 0 \end{aligned} \tag{A.71}$$

where A, B are symmetric matrices that are affine functions of the optimization variable x. This is a convex optimization problem.

EVPs can appear in the equivalent form of minimizing a linear function subject to an LMI, that is

$$minimize \;\; c^T x$$

$$subject \; to \;\; G(x) \; > \; 0 \tag{A.72}$$

where $G(x)$ is an affine function of x. Examples of $G(x)$ include

$$PA + A^T P + C^T C + \gamma^{-1} PBB^T P \; < \; 0, \quad P \; > \; 0$$

It should be stressed that the standard problems (LMIPs, GEVPs, EVPs) are tractable, from both theoretical and practical viewpoints:

- They can be solved in polynomial time.
- They can be solved in practice very efficiently using commercial software.

A.5.3   The S-procedure

In some design applications, we faced the constraint that some quadratic function be negative whenever some other quadratic function is negative. In such cases, this constraint can be expressed as an LMI in the data variables defining the quadratic functions.

Let $G_o, \ldots, G_p$ be quadratic functions of the variable $\xi \in \mathfrak{R}^n$:

$$G_j(\xi) := \xi^T R_j \xi + 2u_j^T \xi + v_j , \quad j = 0, \ldots, p, \quad R_j = \mathfrak{R}_j^T$$

We consider the following condition on $G_o, \ldots, G_p$:

$$G_o(\xi) \; \leq \; 0 \;\; \forall \xi \;\; such \; that \;\; G_j(\xi) \; \geq \; 0, \;\; j = 0, \ldots, p \tag{A.73}$$

It is readily evident that if there exist scalars $\omega_1 \geq 0, \ldots, \omega_p \geq 0$ such that

$$\forall \xi, \quad G_o(\xi) - \sum_{j=1}^{p} \omega_j \, G_j(\xi) \; \geq \; 0 \tag{A.74}$$

then the inequality of Equation (A.73) holds. Observe that if the functions $G_o, \ldots, G_p$ are affine, then Farkas' lemma states that Equation (A.73) and Equation (A.74) are equivalent. Interestingly enough, the inequality of Equation (A.74) can be written as

$$\begin{bmatrix} R_o & u_o \\ \bullet & v_o \end{bmatrix} - \sum_{j=1}^{p} \omega_j \begin{bmatrix} R_j & u_j \\ \bullet & v_j \end{bmatrix} \; \geq \; 0 \tag{A.75}$$

The preceding discussions were stated for non-strict inequalities. In case of strict inequality, we let $R_o, \ldots, R_p \in \Re^{n \times n}$ be symmetric matrices with the following qualifications

$$\xi^T R_o \xi > 0 \;\; \forall \xi \quad \text{such that} \quad \xi^T G_j \xi \geq 0, \;\; j = 0, \ldots, p \tag{A.76}$$

Once again, it is obvious that there exist scalars $\omega_1 \geq 0, \ldots, \omega_p \geq 0$ such that

$$\forall \xi, \quad G_o(\xi) - \sum_{j=1}^{p} \omega_j \, G_j(\xi) > 0 \tag{A.77}$$

and the inequality of Equation (A.76) holds. Observe that Equation (A.77) is an LMI in the variables $R_o, \omega_1, \ldots, \omega_p$.

It should be remarked that the S-procedure deals with non-strict inequalities that allow the inclusion of constant and linear terms. In the strict version, only quadratic functions can be used.

A.6 Some Formulas on Matrix Inverses

This section concerns some useful formulas for inverting matrix expressions in terms of the inverses of their constituents.

A.6.1 Inverse of block matrices

Let A be a square matrix of appropriate dimension and partitioned in the form

$$A = \begin{bmatrix} A_1 & A_2 \\ A_3 & A_4 \end{bmatrix} \tag{A.78}$$

where both A_1 and A_4 are square matrices. If A_1 is invertible, then the Schur complement of A_1 is

$$\Delta_1 = A_4 - A_3 \, A_1^{-1} \, A_2$$

If A_4 is invertible, then the Schur complement of A_4 is

$$\Delta_4 = A_1 - A_2 \, A_4^{-1} \, A_3$$

It is well-known that matrix A is invertible if and only if either

$$A_1 \quad and \quad \Delta_1 \quad are \;\; invertible$$

or

$$A_4 \quad and \quad \Delta_4 \quad are \;\; invertible.$$

Specifically, we have the following equivalent expressions

$$\begin{bmatrix} A_1 & A_2 \\ A_3 & A_4 \end{bmatrix}^{-1} = \begin{bmatrix} \Upsilon_1 & -A_1^{-1}A_2\Delta_1^{-1} \\ -\Delta_1^{-1}A_3A_1^{-1} & \Delta_1^{-1} \end{bmatrix} \qquad (A.79)$$

or

$$\begin{bmatrix} A_1 & A_2 \\ A_3 & A_4 \end{bmatrix}^{-1} = \begin{bmatrix} \Delta_4^{-1} & -\Delta_4^{-1}A_2A_4^{-1} \\ -A_4^{-1}A_3\Delta_4^{-1} & \Upsilon_4 \end{bmatrix} \qquad (A.80)$$

where

$$\Upsilon_1 = A_1^{-1} + A_1^{-1}A_2\Delta_1^{-1}A_3A_1^{-1}$$
$$\Upsilon_4 = A_4^{-1} + A_4^{-1}A_3\Delta_4^{-1}A_2A_4^{-1} \qquad (A.81)$$

Important special cases are

$$\begin{bmatrix} A_1 & 0 \\ A_3 & A_4 \end{bmatrix}^{-1} = \begin{bmatrix} A_1^{-1} & 0 \\ -A_4^{-1}A_3A_1^{-1} & A_4^{-1} \end{bmatrix} \qquad (A.82)$$

and

$$\begin{bmatrix} A_1 & A_2 \\ 0 & A_4 \end{bmatrix}^{-1} = \begin{bmatrix} A_1^{-1} & -A_1^{-1}A_2A_4^{-1} \\ 0 & A_4^{-1} \end{bmatrix} \qquad (A.83)$$

A.6.2   *Matrix inversion lemma*

Let $A \in \Re^{n \times n}$ and $C \in \Re^{m \times m}$ be nonsingular matrices. Then we have

$$[A + B\, C\, D]^{-1} = A^{-1} - A^{-1}\, B\, [DA^{-1}B + C^{-1}]^{-1}\, DA^{-1} \qquad (A.84)$$

Proof. Consider the block matrix

$$S = \begin{bmatrix} A & B \\ C & D \end{bmatrix}$$

By performing the LDU and UDL decompositions of S and equating them, we get

$$\begin{bmatrix} I & 0 \\ CA^{-1} & 0 \end{bmatrix}\begin{bmatrix} A & 0 \\ 0 & D - CA^{-1}B \end{bmatrix}\begin{bmatrix} I & A^{-1}B \\ 0 & I \end{bmatrix} = $$
$$\begin{bmatrix} I & BD^{-1} \\ 0 & I \end{bmatrix}\begin{bmatrix} A - BD^{-1}C & 0 \\ 0 & D \end{bmatrix}\begin{bmatrix} I & 0 \\ D^{-1}C & I \end{bmatrix}$$

Inverting both sides yields

$$\begin{bmatrix} I & -A^{-1}B \\ 0 & I \end{bmatrix} \begin{bmatrix} A^{-1} & 0 \\ 0 & (D - CA^{-1}B)^{-1} \end{bmatrix} \begin{bmatrix} I & 0 \\ -CA^{-1} & 0 \end{bmatrix} =$$
$$\begin{bmatrix} I & 0 \\ -D^{-1}C & I \end{bmatrix} \begin{bmatrix} (A - BD^{-1}C)^{-1} & 0 \\ 0 & D^{-1} \end{bmatrix} \begin{bmatrix} I & -BD^{-1} \\ 0 & I \end{bmatrix}$$

Equating the $(1, 1)$ block shows

$$[A - B \ C \ D]^{-1} = A^{-1} + A^{-1} \ B \ [D - CA^{-1} \ B]^{-1} \ CA^{-1}$$

Substituting $C \longrightarrow -D$ and $D \longrightarrow C^{-1}$, we readily obtain Equation (A.84). ∎

References

[1] Sepulchre, R., Janković, M., and Kokotović, P. V. (1997) *Constructive Nonlinear Control*, Communications and Control Engineering Series, Berlin, Germany: Springer.

[2] Byrnes, C. I., Isidori, A., and Willems, J. C. (1991) "Passivity, feedback equivalence, and the global stabilization of minimum phase nonlinear systems", *IEEE Trans. Automatic Control*, **36**(11):1228–1240.

[3] Lee, T. C., and Jiang, Z. P. (2005) "A generalization of Krasovskii–LaSalle theorem for nonlinear time-varying systems: converse results and applications", *IEEE Trans. Automatic Control*, **50**(8):1147–1163.

[4] Khalil, H. K. (2002) *Nonlinear Systems*, 3rd edition, Englewood Cliffs, NJ, USA: Prentice-Hall.

[5] Teel, A., and Praly, L. (1995) "Tools for semiglobal stabilization by partial state and output feedback", *SIAM Journal on Control and Optimization*, **33**(5):1443–1488.

[6] Scherer, C. W. (2001) "LPV control and full block multiplier", *Automatica*, **37**:361–375.

[7] Packard, A. K. (1994) "Gain scheduling via linear fractional transformations", *System and Control Letters*, **22**(2):79–92.

[8] D'Andrea, R., Dullerud, G. E., and Lall, S. (1998) "Convex $\mathcal{L}_2$ synthesis for multidimensional systems", in *Proc. 37th IEEE Conf. Decision and Control*, Tampa, FL, 1883–1888.

[9] D'Andrea, R. (2001) "Extension of Parrott's theorem to nondefinite scalings", *IEEE Trans. Autom. Control*, **45**(5):937–940.

[10] Gahinet, P., and Apkarian, P. (1994) "A linear matrix inequality pproach to H_∞ control", *Int. J. Robust and Nonlinear Control*, **4**(4):421–448.

[11] Zhou, K., and Doyle, J. C. (1998) *Essentials of Robust Control*, Upper Saddle River, NJ, USA: Prentice Hall.

[12] Bretscher, O. (2004) *Linear Algebra with Applications*, 3rd edition, Englewood Cliffs, NJ, USA: Prentice-Hall.

[13] Bahnasawi, A. A., and Mahmoud, M. S. (1989) *Control of Partially Known Dynamical Systems*, Berlin: Springer-Verlag.

[14] Mahmoud, M. S. (2011) *Decentralized Systems with Design Constraints*, UK: Springer-Verlag.

[15] Blanke, M., Izadi-Zamanabadi, R., Bøgh, S. A., and Lunau, C. P. (1997) "Fault-tolerant control systems: A holistic view", *Control Engineering Practice*, **5**(5):693–702.

Index

Analysis and Synthesis of Fault-Tolerant Control Systems, First Edition. Magdi S. Mahmoud and Yuanqing Xia.
© 2014 John Wiley & Sons, Ltd. Published 2014 by John Wiley & Sons, Ltd.